Cardiovascular Hormone Systems

Edited by
Michael Bader

Related Titles

G. Krauss

Biochemistry of Signal Transduction and Regulation

2008
ISBN: 978-3-527-31397-6

Novartis Foundation

Heart Failure

Molecules, Mechanisms and Therapeutic Targets. No. 274

2006
ISBN: 978-0-470-01597-1

Q. Xu (Ed.)

A Handbook of Mouse Models of Cardiovascular Disease

2006
ISBN: 978-0-470-01610-7

G.W.A. Milne

Ashgate Handbook of Cardiovascular Agents

2004
ISBN: 978-0-566-08386-0

P. Curtis-Prior (Ed.)

The Eicosanoids

2004
ISBN: 978-0-471-48984-9

W.C. De Mello (Ed.)

Renin Angiotensin System and the Heart

2004
ISBN: 978-0-470-86292-6

J.E. Van Eyk, M.J. Dunn (Eds.)

Proteomic and Genomic Analysis of Cardiovascular Disease

2003
ISBN: 978-3-527-30596-4

Cardiovascular Hormone Systems

Edited by
Michael Bader

WILEY-BLACKWELL

WILEY-VCH Verlag GmbH & Co. KGaA

The Editors

Prof. Dr. Michael Bader
Max Delbrück Center for
Molecular Medicine (MDC)
Robert-Rössle-Strasse 10
13092 Berlin
Germany

■ All books published by Wiley-VCH are carefully produced. Nevertheless, authors, editors, and publisher do not warrant the information contained in these books, including this book, to be free of errors. Readers are advised to keep in mind that statements, data, illustrations, procedural details or other items may inadvertently be inaccurate.

Library of Congress Card No.: applied for

British Library Cataloguing-in-Publication Data
A catalogue record for this book is available from the British Library.

Bibliographic information published by the Deutsche Nationalbibliothek
Die Deutsche Nationalbibliothek lists this publication in the Deutsche Nationalbibliografie; detailed bibliographic data are available in the Internet at http://dnb.d-nb.de

© 2008 WILEY-VCH Verlag GmbH & Co. KGaA, Weinheim

All rights reserved (including those of translation into other languages). No part of this book may be reproduced in any form – by photoprinting, microfilm, or any other means – nor transmitted or translated into a machine language without written permission from the publishers. Registered names, trademarks, etc. used in this book, even when not specifically marked as such, are not to be considered unprotected by law.

Printed in the Federal Republic of Germany
Printed on acid-free paper

Typesetting SNP Best-set Typesetter Ltd., Hong Kong
Printing betz-Druck GmbH, Darmstadt
Bookbinding Litges & Dopf GmbH, Heppenheim

ISBN: 978-3-527-31920-6

Contents

Preface *XVII*

List of Contributors *XIX*

Part One   Steroid Hormones

1   Glucocorticoids and Mineralocorticoids *3*
Eilidh Craigie, John J. Mullins, and Matthew A. Bailey
1.1   Synthesis of the Corticosteroids *3*
1.2   Regulation of Corticosteroid Synthesis *7*
1.2.1   Aldosterone *7*
1.2.2   Glucocorticoids *9*
1.3   Corticosteroid Receptors and Control of Ligand Access *11*
1.3.1   Steroid Receptors *11*
1.3.2   Control of Ligand Access *12*
1.3.2.1   11βHSD1 *13*
1.3.2.2   11βHSD2 *14*
1.4   Cardiovascular Effects of Aldosterone *16*
1.4.1   Aldosterone and the Heart *16*
1.4.2   Vasculature *19*
1.5   Cardiovascular Effects of Glucocorticoids *21*
1.5.1   Transgenic Models *23*
1.5.2   Lessons from Human Disease *24*
1.5.2.1   Cushing's Syndrome *24*
1.5.2.2   The Metabolic Syndrome and Tissue-Specific Regulation of Glucocorticoids *25*
1.5.2.3   Glucocorticoid Resistance Syndrome *25*
1.5.2.4   GR Polymorphisms *26*
1.5.3   Endothelial Dysfunction, Vascular Tone and Atherosclerosis *26*
References *27*

Cardiovascular Hormone Systems. Edited by Michael Bader
Copyright © 2008 WILEY-VCH Verlag GmbH & Co. KGaA, Weinheim
ISBN: 978-3-527-31920-6

2	**Sex Steroid Hormones** *39*
	Vera Regitz-Zagrosek, Eva Becher, Sebastian Brokat,
	Shokoufeh Mahmoodzadeh, and Carola Schubert
2.1	Sex Differences in Cardiovascular Physiology and Disease and Effect of Hormone Therapy *39*
2.2	Sex Steroids: Estradiol, Estrone, Testosterone and Progesterone *40*
2.2.1	Synthesis and Endocrine Physiology *40*
2.2.2	Age-Dependent Blood Levels of Sex Hormones in Females and Males *40*
2.2.3	Levels of Estradiol and Testosterone in Rodents *41*
2.3	Sex Hormone Receptors: Structure and Function *41*
2.3.1	Hormone Response Element *42*
2.3.2	Molecular Mechanisms of Nuclear Receptor Activation *42*
2.3.2.1	Ligand-Dependent Transcription *42*
2.3.2.2	Ligand-Independent Genomic Pathway: Cross-Talk Between SHRs and Other Signal Transduction Pathways *42*
2.3.2.3	Nongenomic Pathway *44*
2.3.3	Role of Coregulators in Transcriptional Regulation by SHRs *44*
2.3.4	Role of Proteasomal Degradation Pathway in SHR-Mediated Gene Transcription *44*
2.3.5	Receptor Activity and Availability in the Cardiovascular System *45*
2.3.6	Localization of SHRs in the Heart and Vessels of Rodent and Men *45*
2.4	Receptor-Independent Effects of Sex Steroids *47*
2.5	Sex Hormone Effects on Cardiovascular Cells and Organs *47*
2.5.1	Sex Hormone Effects on Different Cardiovascular Cell Types *47*
2.5.1.1	Cardiomyocytes *47*
2.5.1.2	Fibroblasts *48*
2.5.1.3	Endothelial Cells *48*
2.5.1.4	Vascular Smooth Muscle Cells *48*
2.5.1.5	Platelets *48*
2.5.2	Sex Hormone Effects on the Heart *49*
2.5.2.1	Effects of Exercise *49*
2.5.2.2	Pressure Overload *50*
2.5.2.3	Hypertension *50*
2.5.2.4	Volume Overload *50*
2.5.2.5	Myocardial Infarction *51*
2.5.3	Sex Hormone Effects on Atherosclerosis, Plaque Formation and Rupture *51*
2.5.4	Sex Hormone Effects on Systemic and Circulating Mediators of the Cardiovascular System *52*
2.5.4.1	Lipid Levels *52*
2.5.4.2	Glucose and Insulin *53*
2.5.4.3	Blood Pressure *53*
2.5.4.4	Coagulatory Activity *53*
	References *54*

Part Two Peptide Hormones

3 Angiotensins *67*
Robson Augusto Souza Dos Santos, Anderson José Ferreira, and Ana Cristina Simões e Silva

3.1 Introduction *67*
3.2 RAS: The Classical and Updated View *67*
3.3 New Aspects of Classical RAS Enzymes *68*
3.3.1 Renin/Prorenin Receptor *68*
3.3.2 Signaling Through ACE *68*
3.4 Ang II *69*
3.4.1 Ang II and the Cardiovascular System *72*
3.4.2 Ang II and the Endocrine System *73*
3.4.3 Ang II and the Renal System *74*
3.5 Ang-(1–7) *77*
3.5.1 Ang-(1–7) and the Heart *77*
3.5.2 Ang-(1–7) and Blood Vessels *81*
3.5.3 Ang-(1–7) and the Kidney *83*
3.6 Ang III [Ang-(2–8)] *85*
3.7 Ang IV/AT_4 Receptor Axis *85*
3.8 Des-Asp^1-Ang I [Ang-(2–10)] *86*
3.9 Other Angiotensin Peptides *87*
References *87*

4 Kinins *101*
Suzana Macedo de Oliveira, Kely de Picoli Souza, Michael Bader, and João Bosco Pesquero

4.1 Introduction *101*
4.2 Kininogens: Precursors of Kinins *103*
4.3 Kinin-Forming Systems *104*
4.4 Kininases *104*
4.5 Kinin Receptors *106*
4.6 Physiological and Pathological Cardiovascular Roles of Kinins *107*
4.7 Kinins in Blood Pressure Control and Hypertension *109*
4.8 Kinins in the Heart *110*
4.9 Renal Effects of Kinins *112*
References *115*

5 Natriuretic Peptides *125*
Paula M. Bryan and Lincoln R. Potter

5.1 History *125*
5.2 Natriuretic Peptides: Structure, Processing and Expression *126*
5.3 Natriuretic Peptide Receptors *126*
5.3.1 NPR-A *128*
5.3.2 NPR-B *130*

5.3.3 NPR-C *131*
5.4 Physiological Effects of the Natriuretic Peptide System *132*
5.4.1 Effects on Blood Pressure *132*
5.4.2 Effects of ANP/NPR-A on Intravascular Volume and Endothelium Permeability *133*
5.4.3 Effects of ANP and BNP on Cardiac Hypertrophy and Fibrosis *133*
5.4.4 Effects of ANP and CNP on Vascular Relaxation and Remodeling *133*
5.4.5 Effects of ANP on Natriuresis and Diuresis *134*
5.4.6 Natriuretic Peptides and Renal Function *134*
5.5 Natriuretic Peptides as Diagnostic Indicators of Heart Failure *135*
5.6 NPRs and Heart Failure *135*
5.7 Therapeutic Applications and Future Directions *136*
References *136*

6 Endothelins *143*
Gian Paolo Rossi and Teresa M. Seccia
6.1 Endothelin System *143*
6.2 ET and Cardiovascular Disease *145*
6.3 Assessment of the ET System in Disease States *146*
6.4 ET in PAH *148*
6.5 ET in Systemic Arterial Hypertension *149*
6.5.1 Animal Studies *149*
6.5.2 Human Studies *151*
6.5.3 ET-1 in Renal Disease *152*
6.5.3.1 Blood Pressure, Proteinuria and ERAs *154*
6.6 ERAs in Human Renal Diseases *154*
6.7 ET and Heart Failure *155*
6.8 Role of ET in Heart Failure *156*
6.8.1 ET-1 Plasma Levels in Heart Failure *156*
6.8.2 ERAs in Heart Failure *157*
6.8.3 Selective or Nonselective ERAs in Heart Failure? *158*
6.9 Long-Term Effects of ERAs in Heart Failure *159*
6.10 Conclusions *160*
References *161*

7 Adrenomedullin *169*
István Szokodi and Heikki Ruskoaho
7.1 Molecular Aspects of AM *169*
7.1.1 Structure and Synthesis of AM *169*
7.1.2 Distribution and Sites of AM Production *170*
7.1.3 Regulators of AM Gene Expression *170*
7.1.4 AM Receptors *171*
7.2 Functional Role of AM in the Heart *172*
7.2.1 AM and Myocardial Contractility *172*
7.2.1.1 Effect of AM on Cardiac Contractility *172*

7.2.1.2	Signaling Mechanisms	*172*
7.2.2	AM and Coronary Blood Flow	*173*
7.2.2.1	Effect of AM on Coronary Vascular Tone	*173*
7.2.2.2	Signaling Mechanisms	*175*
7.2.3	AM in Myocardial Ischemia	*176*
7.2.3.1	AM Production in Myocardial Ischemia	*176*
7.2.3.2	Myocardial Cytoprotection by AM	*177*
7.2.3.3	Signaling Mechanisms	*178*
7.2.4	AM and Angiogenesis	*179*
7.2.4.1	Angiogenic Effect of AM	*179*
7.2.4.2	Signaling Mechanisms	*180*
7.2.5	AM in Heart Failure	*180*
7.2.5.1	Cardiac AM Production in Heart Failure	*180*
7.2.5.2	Acute Hemodynamic Effects of AM in Heart Failure	*181*
7.2.5.3	Chronic Effects of AM on Heart Failure Progression	*182*
	References	*184*
8	**Apelin and Vasopressin** *193*	
	Xavier Iturrioz, Annabelle Reaux-Le Goazigo, Françoise Moos, and Catherine Llorens-Cortes	
8.1	Discovery of Apelin	*193*
8.2	Structure and Processing of the Apelin Precursor	*194*
8.3	Apelin Receptor Signaling and Internalization	*195*
8.4	Distribution of Apelin and Its Receptor within the Adult Rat Brain	*196*
8.4.1	Topographical Distribution of Apelin	*196*
8.4.2	Distribution of Apelin Receptor mRNA Expression	*197*
8.5	Apelin: Physiological Actions within the Brain and Anterior Pituitary Gland	*197*
8.5.1	Involvement of Vasopressin and Apelin in the Maintenance of Water Balance	*197*
8.5.1.1	Vasopressinergic System	*197*
8.5.1.2	Apelinergic System	*199*
8.5.2	Apelin, like Vasopressin, is involved in Regulating the Hypothalamic–Adrenal–Pituitary Axis	*202*
8.6	Peripheral Cardiovascular Actions	*203*
8.7	Conclusions and Pathophysiological Implications	*204*
	References	*205*

Part Three Amines

9 Serotonin *211*
Michael Bader
9.1 Introduction *211*
9.2 Components of the Serotonin System *213*
9.2.1 Enzymes *213*
9.2.1.1 Tryptophan Hydroxylases *213*
9.2.1.2 Aromatic Amino Acid Decarboxylase *215*
9.2.1.3 Monoamine Oxidases *216*
9.2.2 Transporters *216*
9.2.2.1 Serotonin Transporter *216*
9.2.2.2 Vesicular Monoamine Transporters *217*
9.2.3 5-HT Receptors *217*
9.2.3.1 5-HT$_1$ Family *217*
9.2.3.2 5-HT$_2$ Family *218*
9.2.3.3 5-HT$_3$ Family *218*
9.2.3.4 5-HT$_4$ Family *218*
9.2.3.5 5-HT$_5$ Family *219*
9.2.3.6 5-HT$_6$ Family *219*
9.2.3.7 5-HT$_7$ Family *219*
9.3 Cardiovascular Actions *219*
9.3.1 Platelets *219*
9.3.2 Vessels *221*
9.3.3 Heart *222*
9.3.4 Brain *223*
9.4 Conclusions *225*
References *225*

10 Adrenaline and Noradrenaline *233*
Nadine Beetz and Lutz Hein
10.1 Introduction *233*
10.2 Biosynthesis and Degradation of Noradrenaline and Adrenaline *234*
10.2.1 Biosynthesis *234*
10.2.1.1 Tyrosine Hydroxylase (TH) *235*
10.2.1.2 Aromatic L-Amino Acid Decarboxylase (AADC) *235*
10.2.1.3 Dopamine β-Hydroxylase (DBH) *236*
10.2.1.4 Phenylethanolamine N-Methyltransferase (PNMT) *237*
10.2.2 Metabolism *238*
10.2.2.1 Noradrenaline Transporter (NET) *239*
10.2.2.2 Organic Cation Transporter 3 (OCT3) *240*
10.2.2.3 Catechol O-Methyltransferase (COMT) *240*
10.2.2.4 Monoamine Oxidase A (MAO-A) *240*
10.2.2.5 Vesicular Monoamine Transporter 2 (VMAT2) *241*

10.3	Adrenergic Receptors *241*	
10.3.1	α$_1$-Adrenoceptors *242*	
10.3.1.1	Mouse Models *242*	
10.3.1.2	Human Genetics and Function *242*	
10.3.2	α$_2$-Adrenoceptors *243*	
10.3.2.1	Mouse Models *243*	
10.3.2.2	Human Genetics and Function *243*	
10.3.3	β-Adrenoceptors *244*	
10.3.3.1	Mouse Models *244*	
10.3.3.2	Human Genetics and Function *244*	
10.4	Conclusions *244*	
	References *245*	
11	**Dopamine** *251*	
	Pedro Gomes and Patríio Soares-da-Silva	
11.1	Introduction *251*	
11.2	Dopamine Synthesis *251*	
11.2.1	L-DOPA Uptake and Decarboxylation *251*	
11.2.2	Amino Acid Transporters *252*	
11.2.3	Mechanisms Regulating Dopamine Availability *254*	
11.3	Dopamine Receptors *255*	
11.3.1	Classification *255*	
11.3.2	Tissue Distribution *257*	
11.4	Signaling Machinery and Effectors Downstream Dopamine Receptor Activation *257*	
11.4.1	Ion Transporters and Channels *257*	
11.4.2	G-Proteins *258*	
11.4.3	Adenylyl Cyclase/Protein Kinase A *259*	
11.4.4	Phospholipase C/Protein Kinase C *259*	
11.4.5	Other Pathways *260*	
11.5	Peripheral Effects of Dopamine *261*	
11.5.1	Renal Function *261*	
11.5.1.1	Renal Blood Flow *261*	
11.5.1.2	Glomerular Filtration Rate *261*	
11.5.1.3	Tubular Effect *262*	
11.5.2	Gastrointestinal Effect *263*	
11.6	Dopamine and Pathophysiology *263*	
11.6.1	Hypertension *263*	
11.6.2	Renal Failure *265*	
11.6.3	Heart Failure *266*	
11.6.4	Diabetes Mellitus *266*	
11.6.5	Aging *267*	
11.7	Clinical Applications of Dopamine *268*	
11.7.1	Heart Failure *268*	
11.7.2	Renal Failure *269*	

11.7.3	Surgery and Transplantation	*271*
11.7.4	Sepsis and Inflammation	*274*
	References	*275*

12 Histamine *295*
Izabela Rozenberg, Felix C. Tanner, and Thomas F. Lüscher

12.1	Introduction	*295*
12.2	Biochemistry	*296*
12.2.1	Synthesis	*296*
12.2.2	Degradation	*296*
12.3	Receptors	*297*
12.4	Vasomotion	*298*
12.4.1	EDRF	*298*
12.4.1.1	Signaling Role of Ca^{2+}	*300*
12.4.1.2	Activation of eNOS	*300*
12.4.1.3	Transcriptional Regulation of eNOS	*300*
12.4.2	EDCF	*300*
12.5	Thrombosis	*301*
12.5.1	Tissue Factor Expression	*301*
12.5.2	Weibel–Palade Bodies	*303*
12.5.3	Platelet Aggregation	*304*
12.6	Inflammation	*304*
12.6.1	Vascular Permeability	*305*
12.6.2	Adhesion Molecule Expression	*305*
12.6.3	Leukocyte Accumulation	*306*
12.6.4	Regulation of T_h1/T_h2 Balance	*306*
12.6.5	Macrophage Activation	*307*
12.6.6	Obesity	*307*
12.7	Atherosclerosis	*307*
12.8	Autoimmune Diseases and Allergy	*308*
12.8.1	Autoimmune Diseases	*308*
12.8.2	Allergy	*308*
12.8.2.1	Immediate Response	*309*
12.8.2.2	Long-Term Response	*309*
12.9	Conclusions	*310*
	References	*310*

13 Prostaglandins and Leukotrienes *315*
Katharina Lötzer and Andreas J. R. Habenicht

13.1	AA Metabolism by the COX and 5-LO Pathways	*315*
13.2	PGs in Cardiovascular Physiology and Pathophysiology	*317*
13.2.1	Diversity of Prostanoid Effects in the Cardiovascular System	*317*
13.2.2	PGs and Atherosclerosis	*317*
13.2.3	COX-2 Inhibition and Cardiovascular Risk in Humans	*319*

13.3	LTs in Cardiovascular Physiology and Pathophysiology *320*	
13.3.1	Activities of LTs in the Cardiovascular System *320*	
13.3.2	5-LO Atherosclerosis Hypothesis *321*	
13.3.3	5-LO Pathway in Mouse Models of Atherosclerosis *322*	
13.3.4	Population Genetic Studies Indicate a Role of the 5-LO Pathway in Cardiovascular Disease *323*	
	References *324*	

14 Cytochrome P450-Dependent Eicosanoids *333*
Wolf-Hagen Schunck and Cosima Schmidt

14.1	Introduction *333*
14.2	Prospects of the Research Field *333*
14.3	How CYP Enzymes Became Established Members of the AA Cascade *335*
14.4	Structure and Function of CYP Enzymes and Their Role in AA Metabolism *336*
14.4.1	Unique Spectral and Catalytic Features of CYP Enzymes *336*
14.4.2	CYP Systems and Their Reaction Cycle in the ER *337*
14.4.2.1	Membrane Integration and Substrate Access *337*
14.4.2.2	Electron Transfer and Activation of Molecular Oxygen *337*
14.4.2.3	Product Formation and Specificity *337*
14.4.3	Reaction Types and Primary Products of CYP-Dependent AA Metabolism *339*
14.4.4	AA Metabolizing CYP Isoforms and Their Orthologs Among Rodents and Human *340*
14.4.4.1	CYP Superfamily *340*
14.4.4.2	Identity of AA-Metabolizing CYP Isoforms *340*
14.4.4.3	Problem of Overlapping Substrate Specificities *341*
14.4.4.4	Problem of Orthologous Genes *341*
14.5	Physiological and Pathophysiological Context of CYP-Dependent Eicosanoid Formation and Action *341*
14.5.1	Extracellular Signal-Induced AA Release *341*
14.5.2	Second Messenger Function *342*
14.5.3	Cellular Context and the Multiplicity of Signaling *342*
14.5.4	Physiological Context *342*
14.5.5	Role of I/R *342*
14.6	Systemic and Tissue-Specific Metabolic Factors Modulating CYP-Dependent Eicosanoid Formation *343*
14.6.1	Essential Fatty Acids Compete as Precursors for Oxygenated Metabolites *343*
14.6.1.1	ω-6 Fatty Acids *343*
14.6.1.2	ω-3 Fatty Acids *343*
14.6.1.3	Health Benefits from ω-3 Fatty Acids *345*
14.6.1.4	ω-3 Fatty Acids Are the Precursors of Novel CYP-Dependent Eicosanoids *345*

- 14.6.2 Role of Nitric Oxide 345
- 14.6.3 Carbon Monoxide and Heme Oxygenase 346
- 14.6.4 CYP Enzymes as Targets, Sources and Utilizers of Reactive Oxygen Species 346
- 14.6.4.1 Reactive Oxygen Species Affect CYP Activities 346
- 14.6.4.2 Reactive Oxygen Species Production by CYP Enzymes 346
- 14.6.4.3 CYP Enzymes Can Use Reactive Oxygen Species and Hydroperoxides for Substrate Oxygenation 346
- 14.7 Biological Activities of EETs and 20-HETE 347
- 14.7.1 Regulation of Vascular Tone 347
- 14.7.2 Regulation of Renal Tubular Function 347
- 14.7.3 Cardiac Function 350
- 14.7.4 General Cell- and Organ-Protective Properties of EETs 351
- 14.7.5 Biological Activities of Eicosanoids Originating from CYP-Dependent n-3 PUFA Oxygenation 351
- 14.8 Secondary Product Formation and the Metabolic Fate of CYP-Dependent Eicosanoids 352
- 14.9 CYP-Dependent AA Metabolism in Animal Models of Hypertension and End-Organ Damage 353
- 14.9.1 Prohypertensive and Proinflammatory Role of Vascular 20-HETE 355
- 14.9.1.1 Androgen-Induced Hypertension 355
- 14.9.1.2 Cyclosporin A-Induced Hypertension 355
- 14.9.2 Antihypertensive Role of EETs in Salt-Sensitive Hypertension 355
- 14.9.3 Antihypertensive Role of EETs in Pregnancy 356
- 14.9.3.1 Renal and Vascular Alterations during Pregnancy 356
- 14.9.3.2 Role of EETs in Placenta, Decidua and Trophoblasts 357
- 14.9.4 20-HETE Deficiency and Salt-Sensitive Hypertension 357
- 14.9.5 Role of EETs, HEETs and PPARα in Inflammatory Renal Damage 357
- 14.10 Structure and Cardiovascular Functions of the Soluble Epoxide Hydrolase 358
- 14.10.1 Enzymatic Activities 358
- 14.10.2 sEH – A Novel Target for the Treatment of Cardiovascular Disease? 359
- 14.11 General Conclusions on Cause–Effect Relationships Associating Alterations in CYP-Dependent Eicosanoid Production and Cardiovascular Disease 359
- 14.11.1 Any Alteration in CYP-Dependent AA Metabolism May Contribute to Cardiovascular Disease 359
- 14.11.2 EET and 20-HETE Availability – A Bottleneck of Various Signaling Pathways? 360
- 14.11.3 Is There a Primary Role of CYP-Dependent Eicosanoids and of CYP Gene Polymorphism in the Development of Cardiovascular Disease? 360

14.11.4 What Is the Cause for Alterations in the Production and Effects of CYP-Dependent Eicosanoids in Disease States? *361*
References *362*

15   Nucleotides and the Purinergic System *373*
Vera Jankowski and Joachim Jankowski

15.1 Introduction *373*
15.2 Mononucleoside Polyphosphates *373*
15.3 Dinucleoside Polyphosphates *375*
15.4 Purinoceptor System *380*
15.5 Metabolism of Nucleotides *383*
15.6 Therapeutic Aspects of the Purinergic System *383*
References *384*

16   Nitric Oxide *395*
Valérie B. Schini-Kerth and Paul M. Vanhoutte

16.1 Regulation of the Endothelial Formation of NO *396*
16.1.1 Hemodynamic Forces *396*
16.1.2 Blood- and Platelet-Derived Factors *398*
16.1.3 Local and Circulating Hormones, Growth Factors, and Neurotransmitters *399*
16.1.4 Polyphenols *400*
16.2 Vasoprotective Effects of NO *400*
16.2.1 Regulation of Vascular Tone and Structure *400*
16.2.2 Regulation of Coagulant and Thrombotic Responses *402*
16.2.3 Regulation of Atherogenic Responses *402*
16.3 Conclusions *403*
References *403*

17   Acetylcholine *407*
Maria Cláudia Irigoyen, Catarina S. Porto, Pedro Paulo Soares, Fernanda Consolin-Colombo, and Antônio Cláudio Nóbrega

17.1 Muscarinic Acetylcholine Receptor: Subtypes and Intracellular Signaling *407*
17.1.1 Muscarinic Receptors in the Heart *409*
17.2 Physiological Effects of ACh on the Heart *410*
17.2.1 Reflex Control of Heart Rate by the Autonomic Nervous System *411*
17.3 Parasympathetic Dysfunction: Clinical Impact on the Cardiovascular System *414*
17.4 ACh and Vascular Function *417*
References *418*

Index *425*

Preface

Cardiovascular diseases have the highest morbidity and mortality world-wide. They comprise disorders like hypertension, myocardial infarction, cardiac and renal failure as well as stroke. Since the cardiovascular system is regulated by hormones and autacoids, which either are circulating or locally released in vascular tissues from endothelial cells and neurons, these hormones are of major pathophysiological importance for cardiovascular diseases.

The chemical nature of these hormones is quite diverse. There are steroids, such as the estrogens, androgens, mineralocorticoids and glucocorticoids; peptides, such as angiotensins, kinins, endothelins, vasopressin, apelin, natriuretic peptides, calcitonin gene-related peptide and adrenomedullin; biogenic amines synthesized from amino acids, such as serotonin, dopamine, norepinephrine, epinephrine and histamine; arachidonic acid products, such as prostaglandins, leukotrienes and cytochrome P450 metabolites; esters, such as acetylcholine; as well as nucleotides and even gases such as nitric oxide. The receptors with which they interact are markedly different. While the majority of the factors act on members of the huge family of G-protein-coupled receptors with seven-transmembrane domains, steroids bind to nuclear receptors, natriuretic peptides activate membrane-bound guanylate cyclases and nitric oxide acts on soluble guanylate cyclases. In addition, the direct activation of ion channels has been described for serotonin (5-HT$_3$ receptors), some steroids (nongenomic actions), acetylcholine (nicotinic receptor) and nucleotides (P$_{2X}$ receptors).

The kinetics with which the hormone systems interfere with cardiovascular regulation also vary drastically. Steroid hormones induce their effects mainly on the transcriptional level. They start to be effective from hours up to days. Peptide hormones are released within minutes and are effective for hours. Autacoids such as the cytochrome P450 products, prostaglandins and nitric oxide are released in seconds, and their effects last for minutes. This allows a high flexibility of the organism to react to cardiovascular challenges at different time scales.

Furthermore, these hormone systems enable intensive networking between the cardiovascular tissues. The major player is the brain, which regulates all other organs in the cardiovascular system using the sympathetic and parasympathetic nerves with noradrenaline and acetylcholine as transmitters, respectively. Furthermore, it releases vasopressin targeting the kidney, and factors influencing steroid

Cardiovascular Hormone Systems. Edited by Michael Bader
Copyright © 2008 WILEY-VCH Verlag GmbH & Co. KGaA, Weinheim
ISBN: 978-3-527-31920-6

hormone generation in the adrenal gland and the gonads. The heart can signal to the rest of the cardiovascular system via the natriuretic peptides (atrial natriuretic peptide and brain natriuretic peptide). The kidney employs the renin–angiotensin system for the same purpose.

Due to the complexity of the cardiovascular system and its regulation, the study of cardiovascular hormones has been mainly limited to whole-organism models. As a consequence, transgenic technology employing the targeted alteration of the genome of a rat or mouse had a major impact on the study of these systems, and the description of such relevant animal models is a major focus of this book.

With the help of such models it was discovered that some of the hormones are already essential for the normal development of the cardiovascular system. For example, animals deficient for endothelins show drastic developmental defects in heart and vessels, and die shortly after birth. Also, most animals deficient for dopamine, norepinephrine and adrenal steroids are not viable for more than a few days. Mice lacking angiotensin exhibit abnormalities in kidney morphology; nevertheless, some of them reach adulthood.

The majority of common drugs for cardiovascular diseases interfere with the generation or signaling of the factors described in this book. For example, the renin–angiotensin system is the target of three classes of drugs: angiotensin-converting enzyme inhibitors, angiotensin AT_1 receptor antagonists and recently also newly developed renin inhibitors. Norepinephrine and its β-receptor are inhibited by β-blockers. However, these drugs not only affect a single hormone system but also interfere with other systems summarized in this book. For example, angiotensin-converting enzyme inhibitors stabilize bradykinin and β-blockers inhibit renin release. Thus, the sole view on one hormone system is not sufficient to understand the actions of these classical cardiovascular drugs. Therefore, this book was designed to give a comprehensive overview about cardiovascular hormones, their metabolism, physiological actions and therapeutic value. Each hormone system is described separately by one of the leaders in the field of research about this system.

I am very grateful to all of the coauthors of *Cardiovascular Hormone Systems* whose excellent contributions created a very valuable and comprehensive source of information for clinical and basic scientists interested in cardiovascular regulation, endocrinology and pharmacology.

Berlin, July 2008 *Michael Bader*

List of Contributors

Michael Bader
Max Delbrück Center for
Molecular Medicine (MDC)
Robert-Rössle-Strasse 10
13092 Berlin
Germany

Matthew A. Bailey
University of Edinburgh
The Queen's Medical Research
Institute
Center for Cardiovascular Science
Molecular Physiology
47 Little France Crescent
Edinburgh EH164TJ
United Kingdom

Eva Becher
Charité University Medicine
Berlin Institute of Gender in
Medicine
Hessische Strasse 3–4
10115 Berlin
Germany

Nadine Beetz
University of Freiburg
Institute of Clinical and
Experimental Pharmacology
and Toxicology
Albertstrasse 25
79104 Freiburg
Germany

Sebastian Brokat
Charité University Medicine
Berlin Institute of Gender in Medicine
Hessische Strasse 3–4
10115 Berlin
Germany

Paula M. Bryan
Research Associate – Biochemistry
Ventria Bioscience
4110 N. Freeway Blvd.
Sacramento, CA 95834
USA

Fernanda Consolin-Colombo
Universidade de São Paulo
Hospital das Clínicas da
Faculdade de Medicina
Instituto do Coração (InCor)
Hipertensão Experimental
Avenida Doutor Enéas de Carvalho
Aguiar, 44
05403-000 São Paulo – SP
Brazil

Eilidh Craigie
University of Edinburgh
Queen's Medical Research Institute
Centre for Cardiovascular Science
Molecular Physiology Unit
47 Little France Crescent
Edinburgh EH16 4TJ
United Kingdom

Cardiovascular Hormone Systems. Edited by Michael Bader
Copyright © 2008 WILEY-VCH Verlag GmbH & Co. KGaA, Weinheim
ISBN: 978-3-527-31920-6

Kely de Picoli Souza
Universidade Federal de São Paulo
Departamento de Biofísica
Rua Botucatu 862, 7°, Andar
Vila Clementino
04023-062 São Paulo
Brazil

Anderson José Ferreira
Federal University of Minas Gerais
Biological Sciences Institute
Department of Morphology
Av. Antonio Carlos, 6627
31270-901 Belo Horizonte, MG
Brazil

Pedro Gomes
University of Porto
Faculty of Medicine
Institute of Pharmacology and Therapeutics
Al. Prof. Hernani Monteiro
4200-316 Porto
Portugal

Andreas J. R. Habenicht
Friedrich Schiller University
Institute for Vascular Medicine
Bachstrasse 18
07743 Jena
Germany

Lutz Hein
University of Freiburg
Institute of Experimental and Clinical Pharmacology and Toxicology
79104 Freiburg
Albertstrasse 25
Germany

Maria Cláudia Irigoyen
Universidade de São Paulo
Hospital das Clínicas da
Faculdade de Medicina
Instituto do Coração (InCor)
Hipertensão Experimental
Avenida Doutor Enéas de Carvalho Aguiar, 44
05403-000 São Paulo – SP
Brazil

Xavier Iturrioz
INSERM U 691
Collège de France
11, place Marcelin Berthelot
75005 Paris
France

Joachim Jankowski
Charité-Universitätsmedizin Berlin
Campus Benjamin Franklin
Medizinische Klinik IV
Hindenburgdamm 30
12200 Berlin
Germany

Vera Jankowski
Charité-Universitätsmedizin Berlin
Campus Benjamin Franklin
Medizinische Klinik IV
Hindenburgdamm 30
12200 Berlin
Germany

Catherine Llorens-Cortes
INSERM U 691
Collège de France
11, place Marcelin Berthelot
75005 Paris
France

Katharina Lötzer
Friedrich Schiller University
Institute for Vascular Medicine
Bachstrasse 18
07743 Jena
Germany

Thomas F. Lüscher
University Hospital
Clinic of Cardiology
Rämistrasse 100
8091 Zurich
Switzerland

Suzana Macedo de Oliveira
Universidade Federal de São
Paulo
Departamento de Biofísica
Rua Botucatu 862, 7°, Andar
Vila Clementino
04023-062 São Paulo
Brazil

Shokoufeh Mahmoodzadeh
Charité University Medicine
Berlin Institute of Gender in
Medicine
Hessische Strasse 3–4
10115 Berlin
Germany

Françoise Moos
Université Victor Ségalen
Institut François Magendie
CNRS-INRA
146, rue Léo Saignat
33077 Bordeaux Cedex
France

John J. Mullins
University of Edinburgh
Queen's Medical Research Institute
Center for Cardiovascular Science
Molecular Physiology
47 Little France Crescent
Edinburgh EH16 4TJ
United Kingdom

Antônio Cláudio Nóbrega
Universidade Federal Fluminense
Departamento de Fisiologia e
Farmacologia
Instituto Biomédico
Rua Professor Hernani Pires de
Melo, 101
24210-130 Niterói – RJ
Brazil

João Bosco Pesquero
Federal University of São Paulo
Department of Biophysics
Rua Botucatu 862, 7°, andar
Vila Clementino
04023-062 São Paulo
Brazil

Catarina S. Porto
Universidade Federal de São Paulo
Departamento de Farmacologia
Setor Endocrinologia Experimental
Rua Botucatu, 740
04023-900 São Paulo – SP
Brazil

Lincoln R. Potter
University of Minnesota, Twin Cities
Department of Biochemistry,
Molecular Biology and Biophysics
7-174 MCB Building
420 Washington Ave. S.E.
Minneapolis, MN 55455
USA

Annabelle Reaux-Le Goazigo
INSERM U 691
Collège de France
11, place Marcelin Berthelot
75005 Paris
France

Vera Regitz-Zagrosek
Charité University Medicine
Berlin Institute of Gender in Medicine
Hessische Strasse 3–4
10115 Berlin
Germany

Gian Paolo Rossi
University of Padua
DMCS Internal Medicine
Via Giustiniani, 2
35128 Padova
Italy

Izabela Rozenberg
University of Zurich
Physiology Institute
Cardiovascular Research
Winterthurerstrasse 190
8057 Zurich
Switzerland

Heikki Ruskoaho
University of Oulu
Biocenter Oulu
Institute of Biomedicine
Department of Pharmacology and Toxicology
P.O. Box 5000
90014 Oulu
Finland

Valérie B. Schini-Kerth
Université Louis Pasteur de Strasbourg
Faculté de Pharmacie
UMR CNRS 7175
Département de Pharmacology et Physicochimie
74, Route du Rhin
67401 Illkirch
France

Cosima Schmidt
Max Delbrück Center for Molecular Medicine
Cardiovascular Research Program
Laboratory CYP-Eicosanoid Research
Robert-Rössle-Strasse 10
13125 Berlin
Germany

Carola Schubert
Charité University Medicine
Berlin Institute of Gender in Medicine
Hessische Strasse 3–4
10115 Berlin
Germany

Wolf-Hagen Schunck
Max Delbrück Center for Molecular Medicine
Cardiovascular Research Program
Laboratory CYP-Eicosanoid Research
Robert-Rössle-Strasse 10
13125 Berlin
Germany

Teresa M. Seccia
University of Padua
DMCS Internal Medicine
Via Giustiniani, 2
35128 Padova
Italy

Ana Cristina Simões e Silva
Federal University of Minas
Gerais
Biological Sciences Institute
Department of Pediatrics
Av. Antonio Carlos, 6627
31270-901 Belo Horizonte, MG
Brazil

Pedro Paulo Soares
Universidade Federal Fluminense
Departamento de Fisiologia e
Farmacologia
Instituto Biomédico
Rua Professor Hernani Pires de
Melo, 101
24210-130 Niterói – RJ
Brazil

Patrício Soares-da-Silva
University of Porto
Faculty of Medicine
Institute of Pharmacology and
Therapeutics
Al. Prof. Hernani Monteiro
4200-316 Porto
Portugal

Robson Augusto Souza Dos Santos
Federal University of Minas
Gerais
Biological Sciences Institute
Department of Physiology and
Biophysics
Av. Antonio Carlos, 6627
31270-901 Belo Horizonte, MG
Brazil

István Szokodi
University of Pécs
Faculty of Medicine
Heart Institute
Ifjúság útja 13
7624 Pécs
Hungary

Felix C. Tanner
University of Zurich
Physiology Institute
Cardiovascular Research
Winterthurerstrasse 190
8057 Zurich
Switzerland

Paul M. Vanhoutte
University of Hong Kong
Li Ka Shing Faculty of Medicine
Department of Pharmacology
21 Sassoon Road
Hong Kong
PR of China

Part One Steroid Hormones

1
Glucocorticoids and Mineralocorticoids
Eilidh Craigie, John J. Mullins, and Matthew A. Bailey

Glucocorticoids and mineralocorticoids are members of the corticosteroid hormone family, synthesized in the adrenal gland from the precursor sterol cholesterol via the intermediate pregnenolone (Figure 1.1). The principal glucocorticoid in humans is cortisol (in rodents corticosterone) and the principal mineralocorticoid is aldosterone. Sharing a common synthesis pathway, cortisol and aldosterone are structurally similar (Figure 1.1), and exhibit a degree of cross-receptor affinity and function. Nevertheless, small differences in structure permit important differences in physiological function. Aldosterone classically acts via the mineralocorticoid receptor (MR) to promote sodium transport in the kidney and gut, thereby regulating long-term electrolyte homeostasis and blood pressure control. Cortisol, by comparison, exhibits a wide range of metabolic and stress-related response effects.

1.1
Synthesis of the Corticosteroids

Steroid synthesis occurs principally in the adrenal gland but also occurs in the steroidogenic cells of the testes, ovary, placenta and brain. The intramitochondrial delivery of cholesterol is the rate-limiting step for steroid synthesis and is mediated by steroidogenic acute regulatory protein (StAR) [1]. Defects in cholesterol transport associated with mutations in StAR [2] cause the autosomal recessive disorder of lipoid congenital adrenal hyperplasia (CAH; Online Mendelian Inheritance in Man (OMIM) #201710). This rare condition presents with large adrenal glands containing high levels of cholesterol. Lipoid CAH is lethal within a few days without hormone replacement therapy. Over 30 mutations in StAR have been reported to cause lipoid CAH, all of which result in varying degrees of defective cholesterol transport (for review, see [3]). Mice null for StAR, generated by homologous recombination, emphasize the key role of this protein. Homozygous null pups fail to thrive and die within a week of birth: corticosterone and aldosterone levels are very low despite elevated ACTH (adrenal corticotropic hormone) and CRH (corticotropin-releasing hormone) [4]. Lipoid CAH can also arise from

1 Glucocorticoids and Mineralocorticoids

(a)

Cholesterol
CYP11A ↓
Pregnenolone — CYP17 → 17-OH-Pregnenolone
3β-HSD2 ↓   3β-HSD2 ↓
Progesterone — CYP17 → 17-OH-Progesterone
CYP21A ↓   CYP21A ↓
11-Deoxycorticosterone   11-Deoxycortisol
CYP11B1/CYP11B2 ↓   CYP11B1 ↓
Corticosterone   **Cortisol**
CYP11B2 ↓
18-OH-Corticosterone
CYP11B2 ↓
Aldosterone

Human

Figure 1.1 Corticosteroid biosynthesis. The biosynthesis pathways of cortisol and aldosterone in (a) humans and (b) rodents.

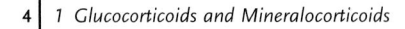

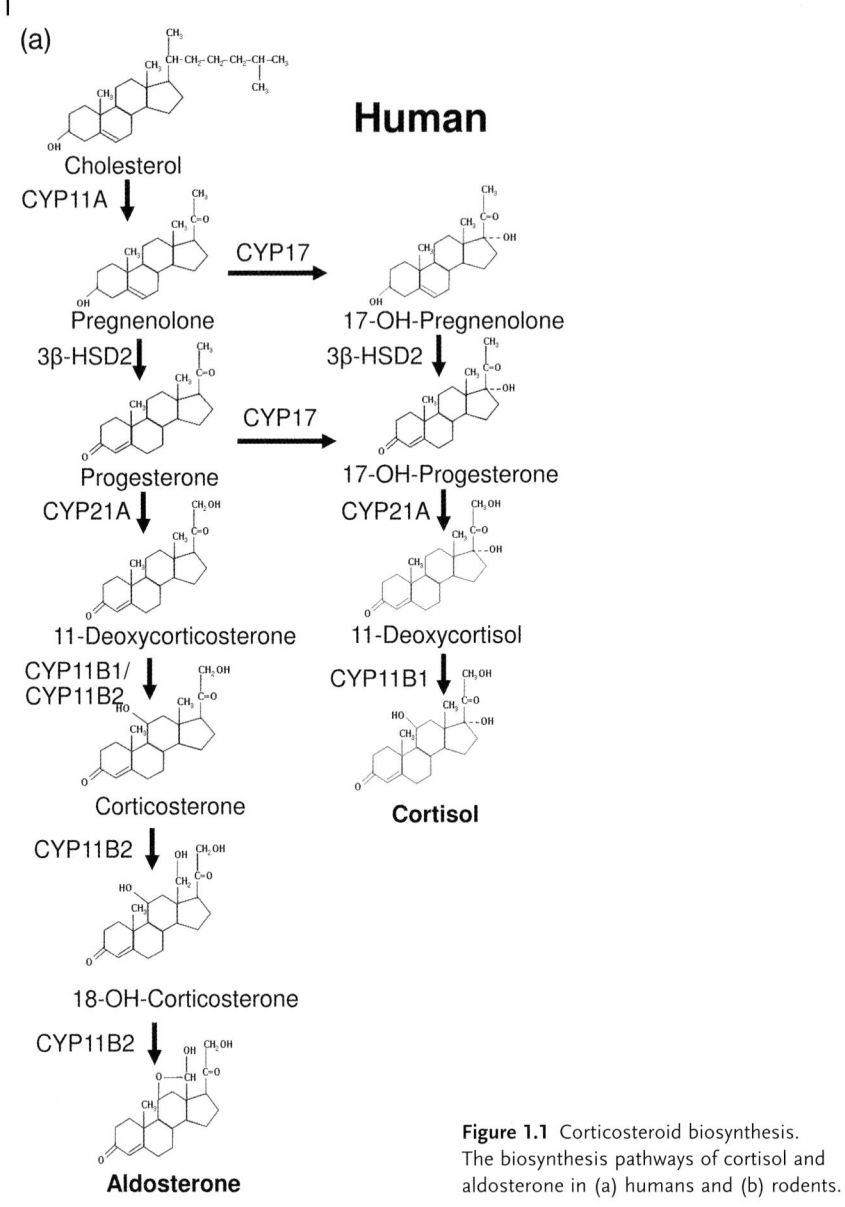

mutations in P450scc [5], an enzyme that cleaves cholesterol to produce pregnenolone – the common precursor for both cortisol and aldosterone synthesis (Figure 1.1). Indeed the biosysthetic pathways of both share a number of intermediates and enzymes (Figure 1.1), becoming fully exclusive only at 11-deoxycortisol (DOC; cortisol pathway) and 11-deoxycorticoisteroid (aldosterone pathway). In rodents, exclusivity occurs at 11-deoxycorticoisteroid (Figure 1.1b).

The final step in cortisol synthesis, the conversion of DOC to cortisol, is catalyzed by 11β-hydroxylase (CYP11B1 gene), while the final three stages of aldoste-

(b) **Rodent**

Cholesterol
CYP11A ↓
Pregnenolone
3β-HSD2 ↓
Progesterone
CYP21A ↓ — CYP11B1 → Corticosterone
11-Deoxycorticosterone
CYP11B2 ↓
Corticosterone
CYP11B2 ↓
18-OH-Corticosterone
CYP11B2 ↓
Aldosterone

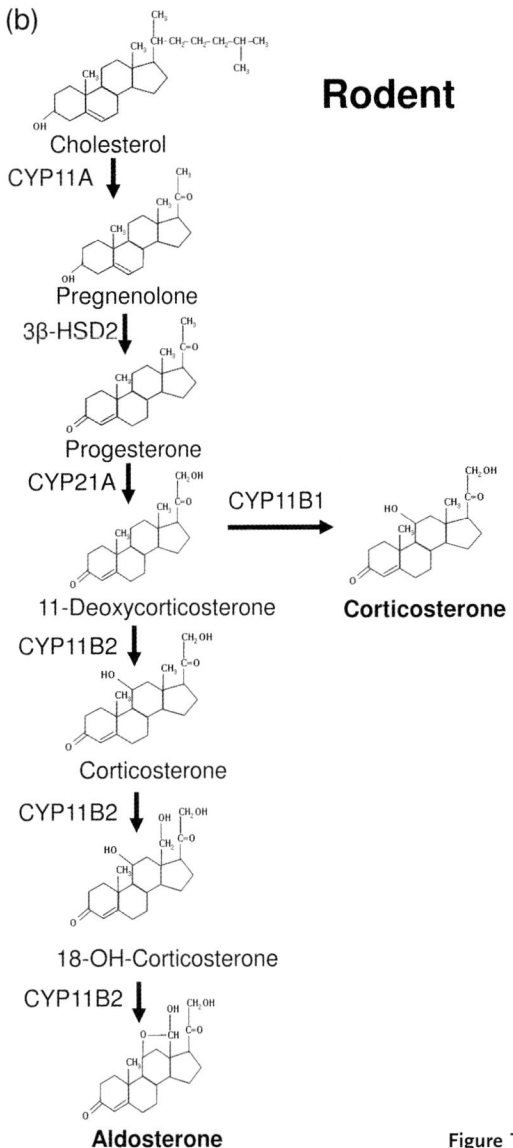

Figure 1.1 *Continued*

rone synthesis require aldosterone synthase (CYP11B2 gene). There is a differential spatial expression of these two enzymes in the cortex of the adrenal gland which is divided into three distinct zones: zona glomerulosa, zona fasciculata and zona reticularis. Cortisol is synthesized primarily in the zona fasciculata, with a small amount being produced by neighboring cells in the zona reticularis. 11β-Hydroxylase is present in both these zones. Aldosterone is produced in the zona glomerulosa, where aldosterone synthase expression is exclusively expressed.

Glucocorticoid remedial hypertension (GRA; OMIM #103900) is an autosomal dominant disorder that occurs when there is unequal crossing over between

CYP11B1 and CYP11B2, which are highly homologous and located in tandem at chromosome 8q24.3, approximately 45 kb apart [6]. In this situation, a chimeric gene is created in which the 5' regulatory regions of the 11β-hydroxylase gene are fused to the coding sequence of the aldosterone synthase gene. There is ectopic expression of the aldosterone synthase in the zona fasciculata, which is now strongly controlled by ACTH. GRA presents with constitutive release of aldosterone and hypertension associated with sodium retention and potassium wasting. Administration of exogenous glucocorticoids suppresses the HPA axis and alleviates the symptoms (see Section 1.2.2 and Figure 1.3).

CAH (OMIM +201910) is an autosomal recessive disorder of cortisol synthesis in which patients have low levels of cortisol and accumulation of DOC and 11-deoxycorticosterone. Approximately 11% of CAH arises from mutations in CYP11B1, the majority being caused by loss of 21-hydroxylase function. Regardless of genetic causality, CAH is associated with neonatal lethality, perhaps leading to an underestimation of the prevalence of the syndrome. Hypertension (DOC is a potent mineralocorticoid, see below) and symptoms of androgen excess, such as precocious puberty and the development of intersexual genitalia, are also features. The only mouse model of CAH available is the H-2(aw) strain which carries a variety of loss-of-function mutations in 21-hydroxylase [7, 8]. Homozygosity for the mutation causes neonatal death; mice heterozygous for mutations have compromised steroidogenesis and faithfully reproduce CAH.

Mutations in CYP11B2 cause the autosomal disorder of corticosterone methyoxidase deficiency (CMO), types 1 and 2 [9]. In CMO1 (OMIM #203400), there is no enzyme activity and aldosterone is undetectable. Patients have marked growth retardation and fail to thrive. Altered renal electrolyte balance leads to hyponatremia and hyperkalemia, and hypotension is evident. This is presumably secondary to volume depletion; however, since activation of MR in vascular smooth muscle potentiates the action of vasoconstrictors (see below), a contribution of vasodilation to the hypotension cannot be discounted. CMO2 (OMIM #610600) is a milder form of the disease: mutations impair but do not ablate aldosterone synthase activity. Lee *et al.* have modeled CMO1 in the mouse, replacing the first two of nine exons with enhanced green fluorescent protein (EGFP), thereby creating a gene expression reporter while concomitantly abolishing enzyme activity [10]. *cyp11b2* null mice were born in normal Mendelian ratios, but a third of the homozygous null animals died prior to weaning, with the rest showing marked retardation of growth, hyperkalemia and altered renal electrolyte handling [10, 11]. Plasma renin activity was elevated (45-fold) in *cyp11b2* null mice, and renin expression was induced in both the zona glomerulosa and fasciculata of the adrenal gland. That these changes failed to maintain blood pressure (null mice were mildly hypotensive), despite the high levels of angiotensin (Ang) II, underscores the essential role for aldosterone in blood pressure homeostasis. Salt supplementation rescued the electrolyte disturbances but did not correct blood pressure. In experimental animals, adrenalectomy or genetic ablation of MR will cause death unless salt therapy is administered. Aldosterone synthase null mice do not require salt supplements to survive with only modestly compromised blood pressure regulation, the implication being that a degree of MR activation persists. The *cyp11b2* null mouse has provided important data concerning the role and regulation of aldosterone synthesis. Induction of the EGFP construct was used to indicate gene activation. Surprisingly the strongest signals were present in the transition zone between cortex and medulla. It was demonstrated that this zone was rich in cells undergoing apoptotic cell death, suggesting that abnormal aldosterone synthesis has an extensive effect on adrenal gland structure and function. In addition to the expected expansion of the zona glomerulosa, cortical architecture becomes disorganized and there is significant accumulation of lipid in steroidogenic cells.

1.2 Regulation of Corticosteroid Synthesis

1.2.1 Aldosterone

The production of aldosterone from the zona glomerulosa is controlled by the renin–angiotensin system (RAS) (Figure 1.2) and plasma potassium. ACTH can also stimulate aldosterone secretion, particularly in rodents; however, since hypophysectomy or suppression of ACTH by dexamethasone administration does not alter basal aldosterone secretion (or, indeed, the response to salt deprivation), ACTH is not considered to be a key regulator. Other factors, such as plasma sodium, catecholamines, β-endorphins and serotonin, may also play a role, but compared to the RAS and potassium these are minor. Indeed, of all these agents, only Ang II and potassium exert a trophic effect on adrenal gland structure, promoting both hypertrophy of the zona glomerulosa and an increased sensitivity of secretion to their action [12]. Net secretion of aldosterone normally results from the integration of several signals.

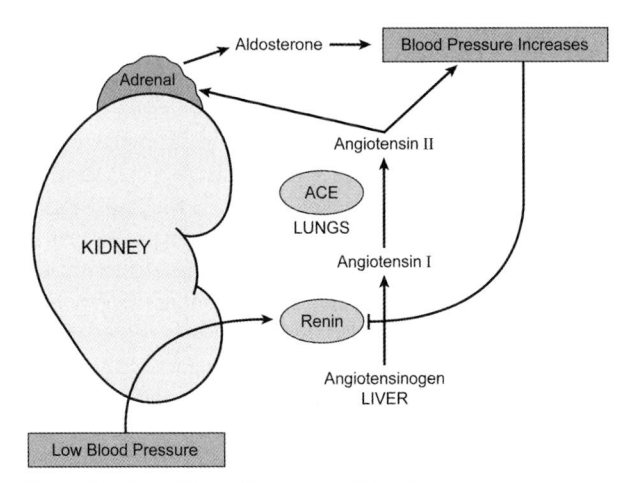

Figure 1.2 The RAS and the control of blood pressure.

Angiotensinogen, primarily synthesized in the liver, is cleaved by the aspartyl protease renin to produce Ang I. This is further cleaved by angiotensin-converting enzyme (ACE) to yield the octapeptide, Ang II. Ang II, acting via AT_1 and AT_2 receptors, will increase blood pressure due to effects on renal sodium reabsorption and vascular resistance. Furthermore, the RAS is a biologically significant regulator of angiogenesis [13]. Therefore, Ang II is an important cardiovascular hormone in its own right, quite separately from its effects on aldosterone synthesis, as covered in Section 3.1.

Ang II, and its metabolite Ang III, rapidly stimulate aldosterone production by activation of both early and late stages of steroid biosynthesis [14]. Both angiotensins are equally efficacious, but Ang II is present in the circulation at much higher concentrations and is, therefore, more important.

Classically, the RAS operates at a systemic level. Recent evidence, however, demonstrates that the RAS can operate independently at the level of the tissue and exert powerful cardiovascular effects quite independently of the systemic system [15]. Although there are no strong quantitative trait locus associations between the RAS and primary hypertension, the involvement of this system in the misregulation of blood pressure is undisputed: the beneficial effects of ACE inhibitors and AT_1 receptor blockers in patients with cardiovascular disease has been demonstrated many times in large-scale clinical trials [16]. More recently, a second form of ACE (ACEII) has been implicated in cardiovascular disease [17] and is a novel therapeutic target. Similarly, the new renin inhibitor, aliskiren, is effective in the treatment of moderate hypertension, although long-term outcome data are not yet available [18]. As for other systems, the use of transgenic animals has clearly demonstrated the major role of the RAS in cardiovascular homeostasis (for a detailed review of this subject, see [19]). Nevertheless, in the majority of these models, the primary abnormality in blood pressure relates to alterations in circulating Ang II, rather than aldosterone. That aldosterone synthesis occurs despite compromised RAS indicates the important regulatory role of plasma potassium levels. This is further supported by the observation that the circadian rhythm of aldosterone secretion does not coincide with that for renin, but for potassium.

An increase in plasma potassium concentration increases the synthesis of aldosterone and, conversely, potassium depletion reduces aldosterone synthesis. The regulation of aldosterone synthesis by potassium is very sensitive: changes of ±0.1 mM can alter the rate of production independently of either Ang II or plasma sodium [12]. There is, moreover, reciprocal regulation: if plasma potassium rises, the rapid increase in aldosterone synthesis promotes kaliuresis and a redistribution of potassium from the extracellular fluid into the cytosol, thereby returning plasma potassium levels to normal [20]. This feedback loop is so persuasive that it can, under conditions of sodium depletion for example, uncouple the secretion of aldosterone from control by Ang II [21]. Thus, the sodium-retaining (and pressor) effects of Ang II may be more important for blood pressure homeostasis than the effects of aldosterone on either the kidney or the vasculature, with the latter acting principally as a regulator of potassium [20].

1.2.2
Glucocorticoids

Glucocorticoid synthesis is regulated by the HPA axis via ACTH (Figure 1.3). ACTH is synthesized by the posterior pituitary mainly in response to two synergistic factors – CRH and antidiuertic hormone (ADH or vasopressin), both of which produced in the paraventricular nucleus of the hypothalamus. These peptides travel through the neurohypophyseal stalk to the median eminence from where they enter the portal circulation and stimulate release of ACTH via binding to the CRF type 1 receptor or V1b receptor, respectively. ACTH stimulates the synthesis of cortisol in the adrenal. Cortisol itself exerts negative feedback on the HPA axis by inhibiting both the release and actions of CRH. ACTH also exerts a short-loop negative feedback by inhibiting its own secretion.

Of the two peptides, CRH is the more important. Mice in which this is deleted have impaired HPA axis, ablated stress response and a loss of the normal circadian rhythm for glucocorticoid production [22]. CRH acts principally at the type 1 receptor with CRF-R1 null mice having a marked impairment of the HPA axis [23]. Unstressed ACTH levels in these animals are, however, normal and they are still able to mount a stress response. This is mediated in part through a second receptor for CRH, as shown by a double knockout strategy [24]. Nevertheless, injection of antisera to ADH was shown to reduce ACTH levels by 60%, indicating a key role for ADH [23] in the compensatory response. ADH acts synergistically to CRF but is not an absolute requirement for ACTH release: Brattleboro rats, which are congenitally devoid of ADH, have a normal HPA axis [25], and mice lacking the

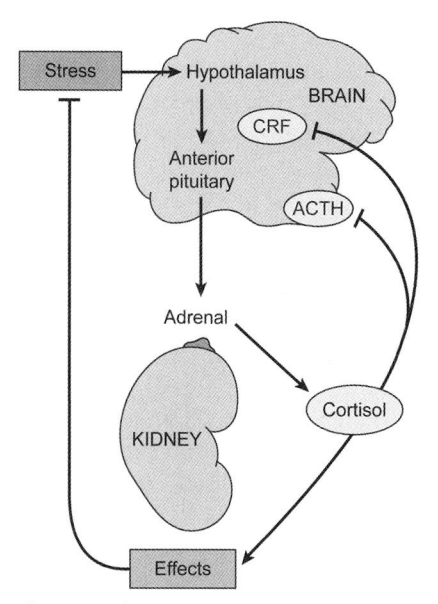

Figure 1.3 The HPA axis.

V1b receptor [26] have normal ACTH and corticosterone levels. Such studies do, however, demonstrate a key role for ADH in sustaining the ACTH response to stress.

That the HPA axis is not abolished by combined administration of antisera to CRF and AVP – or indeed by double knockout of the receptors – indicates other regulatory factors about which less is known. A number of neuroactive compounds, such as Ang II [27], catecholamines and glutamate [28], have been implicated.

ACTH circulates unbound to plasma with a half-life of approximately 15 minutes and exerts its effects via G-protein-coupled receptors belonging to the melanocortin receptor subfamily known as ACTH-R. ACTH-R is mainly expressed in the adrenal cell plasma membrane, with low expression levels being reported in skin and adipose tissue [29]. Although ACTH-R is specific for its ligand, ACTH itself is also recognized by the other four melanocortin receptors. The receptors are coupled to adenylyl cyclase: the cAMP–protein kinase A cascade causes the hydrolysis of cholesterol esters stored in the zona fasciculata and the synthesis of cortisol. The human inheritable condition of familial glucocorticoid deficiency [FGD; OMIM (202200)] has been attributed to mutations within the ACTH-R gene and several different FGD mutations within the gene have been so far identified [30]. FGD is characterized by glucocorticoid deficiency with high plasma ACTH levels and a normal RAS.

Administration of intravenous ACTH in humans is followed by a rapid (within minutes) increase in cortisol plasma levels [31], primarily due to *de novo* synthesis. Although the concentration of steroids is 2- to 3-fold higher in the adrenal gland than in the plasma, this does not act as a reservoir. A sustained increase in ACTH levels results in hypertrophy of the adrenal gland due to an increase in cell size, not number, thereby permitting increased storage of cholesterol. Conversely, adrenal atrophy occurs if ACTH levels remain chronically low.

Plasma cortisol levels fluctuate throughout the day as release occurs in an episodic, rather than constitutive, manner. Nevertheless, the episodes of release are more frequent in the late evening and early morning, and there is a true circadian rhythm (although light does have some effect on the cycle): most secretion occurs from the third hour of sleep to the early hours of wakefulness and plasma cortisol can be undetectable during the rest of the day. The rhythm synchronizes, to an extent, with plasma ACTH concentration and there is a peak in hypothalamic CRF preceding that of cortisol by 4–5 h. However, the circadian rhythm of glucocorticoid production persists even when CRF/ACTH levels are clamped [32] suggesting that the periodicity of release is entrained by other factors. Several agents have been suggested, although neither catecholamines nor serotonin appear to be involved. The rhythm is disrupted, however, by adrenal denervation, spinal chord transection or lesions in the ventromedial nucleus of the hypothalamus [33]. The circadian fluctuations are of major importance to the normal regulation of the HPA axis. Furthermore, glucocorticoids demonstrably influence the phase of peripheral oscillators in the kidney, heart and liver, although not in the central "clock" of the supra chiasmatic nuclei [34]. Studies in genetically modified mice suggest that disturbances in HPA signaling are connected with vulnerability to behavioral abnormalities [35]. Moreover, there is a well-established circadian variation in cardiovascular risk events, with an increase in events in the morning compared to other times of day [36], coinciding with elevated glucocorticoid and aldosterone levels. Indeed, there is growing literature showing cortisol secretion/metabolism to be directly associated with cardiovascular risk. In contrast, an association study found no link between a glucocorticoid receptor (GR) polymorphism (with a trend toward elevated cortisol) and adverse cardiovascular events [37].

1.3
Corticosteroid Receptors and Control of Ligand Access

1.3.1
Steroid Receptors

GR and MR are intracellular receptors responsible for binding and mediating the "classic" effects of cortisol and aldosterone, respectively. They belong to subfamily 3C of a large and diverse family of transcription factors known as the nuclear receptor family. Other members of subfamily 3C include the progesterone receptor (PR) and the androgen receptor (AR). GR and MR share a high degree of structural homology, reflecting the structural similarities between their corticosteroid ligands. The structural homology is highest at the DNA-binding domains (DBDs) (94%) and 56% between the ligand-binding domains (LBDs) [38]. This high degree of homology suggests that the two receptors are closely associated in evolutionary terms and are most likely descended from a common ancestral receptor. A polar surface within the ligand-binding pocket of MR, lacking in GR and other receptors of the family, permits preferential binding of aldosterone. Nevertheless, the cloning and expression of MR [39] revealed considerable ligand promiscuity with receptor specificity being governed by ligand access.

Unactivated, GR and MR are sequestered in the cytoplasm by complexing with heat-shock protein (HSP). The HSP acts to stop the receptors entering the nucleus in the absence of an appropriate activation signal. Cortisol and aldosterone circulate in plasma bound to plasma proteins, and can easily diffuse through the cell membrane into the cytoplasm. Here they will act to displace the HSP from their receptor to allow the formation of a hormone–receptor complex. This changes the conformation of the receptor, allowing it to form a homodimer, which can now readily enter the nucleus where it will recognize specific hormone response elements (HREs) associated with target genes, acting as a ligand-dependent transcription factor (Figure 1.4). Interestingly, a recent study has shown that GR and MR can form heterodimers [40], which can translocate to the nucleus: the downstream effects on DNA transcription of this complex are unknown. HREs are typically located within a gene enhancer, which can be several kilobases away from the gene promoter. GR and MR, along with PR and AR, recognize response elements whose

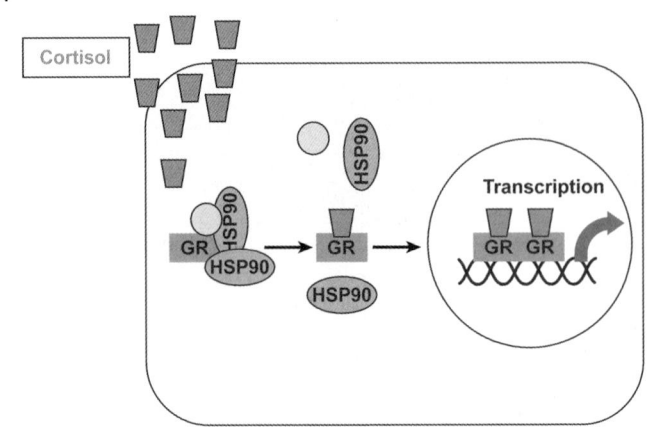

Figure 1.4 Nuclear translocation of GR on ligand binding. Schematic representation of the translocation of GR homodimer to the nucleus for transcriptional effects after ligand binding (cortisol) releases GR from its associated HSP inhibitory complex. MR activation behaves in a similar manner (see Figure 1.6).

HRE sequence consists of two hexameric half-sites (TGTTCT). Glucocorticoids can also affect transcription independent of direct DNA binding by interacting with protein transcription factors [41]. This allows the transcription of genes that mediate GR- and MR-induced responses to be tightly regulated by appropriate ligand binding and hormone–receptor complex conformation.

1.3.2
Control of Ligand Access

The two 11β-hydroxysteroid dehydrogenase (11βHSD) enzymes, types 1 and 2, are key determinants of ligand access to GR and MR, respectively. The enzymes interconvert cortisol (active) and cortisone (inactive), thereby controlling the local concentrations of glucocorticoids (Figure 1.5).

11βHSD1 is the product of HSD11B1 found on chromosome 1 in both mice and humans. The enzyme has a wide distribution but its major areas of action, in terms of both transcript expression and activity, are the liver, adipose and brain [42]. HSD11B2, found on chromosome 16 in humans and on chromosome 8 in mice, encodes the second isozyme. 11βHSD2 has a more limited distribution than 11βHSD1, being expressed predominantly in aldosterone target tissues such as the distal nephron and colon [43]. It is also found in the placenta [44, 45] and in the vascular endothelium [46]. 11βHSD1 was cloned from the liver [47] and 11BHSD2 from the kidney [48]. Although both enzymes belong to the same superfamily of short-chain alcohol dehydrogenase reductases [49], sequence comparison reveals little identity with the exception of the regions encompassing cofactor binding (NAD or NADP[H]) and the active site [50]. In cell culture systems, both

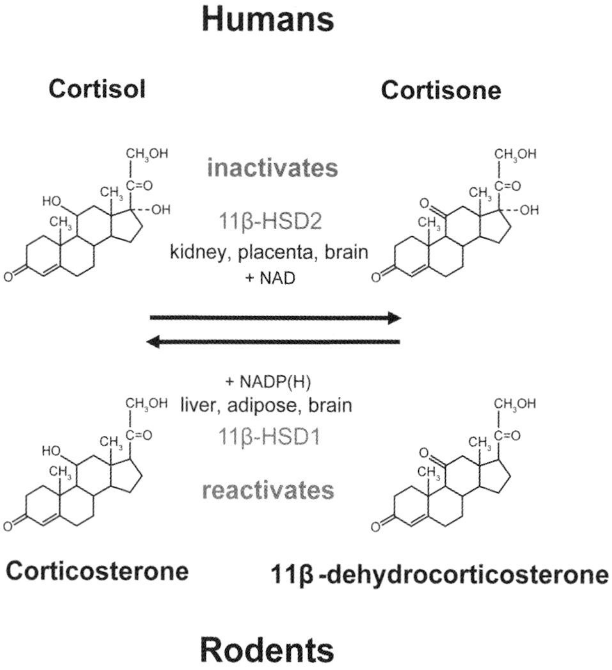

Humans

Cortisol   **Cortisone**

inactivates

11β-HSD2
kidney, placenta, brain
+ NAD

+ NADP(H)
liver, adipose, brain
11β-HSD1

reactivates

Corticosterone   **11β-dehydrocorticosterone**

Rodents

Figure 1.5 Glucocorticoid interconversion. Diagrammatic demonstration of the interconversion between active and inactive glucocorticoids by 11βHSD1 and 11βHSD2 in humans and rodents.

enzymes are single-chain polypeptides localized to the membrane of the endoplasmic reticulum, with opposing orientations of their catalytic sites [51, 52]. *In vivo*, however, homodimerization may provide additional regulation of enzyme activity. Dimerization supports full activity of 11βHSD1 [53], but inactivates 11βHSD2 [54].

1.3.2.1 11βHSD1

Although there appears to be no physical association of 11βHSD1 with GR, the enzyme governs *in vivo* the extent to which the receptor is activated by glucocorticoids. This occurs by converting inactive cortisone to active cortisol (see Figure 1.5) thereby maintaining glucocorticoid signaling at a local level [50]. Due to its widespread distribution and the lack of specific inhibitors, functional dissection of the role of the enzyme in specific tissues is difficult. However, significant advances in our understanding have come from the generation of genetically modified mice. For example, the 11βHSD1 null mouse has elucidated major functions for the enzyme in the response to stress and in regulation of the HPA [50]. For the latter, regeneration of glucocorticoids by the liver appears to be particularly important [55]. In addition, 11βHSD1 null mice are resistant to age-related cognitive impairment [56], indicating roles in the brain. This literature has recently been reviewed [50] and the role of 11βHSD1 in the regulation of metabolism and cardiovascular function is discussed later in the chapter. These studies have important implications for human disease, suggesting that 11βHSD1 is an exciting therapeutic target. This area will no doubt be advanced by the recent crystallization of the enzyme [57].

11βHSD1 null mice have improved glucose tolerance, a favorable lipoprotein profile, and increased sensitivity of the liver and fat to insulin [175]. Moreover, on the obese-prone C57BL/6J background, mice carrying the null mutation were resistant to the weight gain induced by high-fat feeding [176]. That loss of 11βHSD1 confers a cardioprotective metabolic profile is intriguing and most probably results from a lack of glucocorticoid regeneration in adipocytes. However, these mice also had an increase calorific intake, suggesting that energy expenditure was also stimulated (Morton *et al.*, 2004).

The generation of a mouse that overexpresses 11βHSD1 under the control of an adipocyte-specific promotor [58] further highlights the role of the enzyme in metabolic function. The amplification of glucocorticoids was relatively modest and confined to the adipocyte, circulating corticosterone being normal. Nevertheless, these transgenic mice developed central obesity, insulin resistance and glucose intolerance. In addition, these animals were hypertensive due to a chronic activation of the RAS [59]. This clustering of metabolic and cardiovascular phenotypes is characteristic of the metabolic syndrome, which is discussed later in this chapter. The level of enzyme function in the adipocyte appears therefore, to play a critical role in setting of metabolic profile.

1.3.2.2 11βHSD2

In vitro, MR can be activated with equal potency both by aldosterone and cortisol [39]. *In vivo*, ligand access to MR is determined by colocalization with 11βHSD2 (Figure 1.6). By catalyzing the rapid conversion of cortisol into cortisone (Figure 1.5), which does not activate MR, 11βHSD2 confers upon MR the specificity to aldosterone that it inherently lacks [60, 61]. MR and 11βHSD2 have overlapping distributions in those tissues classically held to be aldosterone selective [43]. In addition to protecting MR from activation by glucocorticoids, there is evidence of a direct association of the proteins [62] and 11βHSD2 may directly regulate MR activation by aldosterone.

Inactivating mutations in the gene *HSD11B2* cause the Syndrome of Apparent Mineralocorticoid Excess (SAME; OMIM #218030). In this setting, cortisol activates MR [63–67], resulting in severe hypertension thought to arise from volume expansion secondary to renal sodium retention [63, 64, 66]. Dexamethasone can be therapeutically effective [67] as it will suppress endogenous glucocorticoids, but does not activate MR. In addition, dexamethasone may act as a chaperone and stabilize mutant enzyme [68]. Nevertheless, neither synthetic glucocorticoid nor MR blockade has a consistent antihypertensive effect [69].

SAME has been modeled by targeted disruption of the 11βHSD2 locus, producing a mouse in which the cardinal features of the disorder were preserved [70].

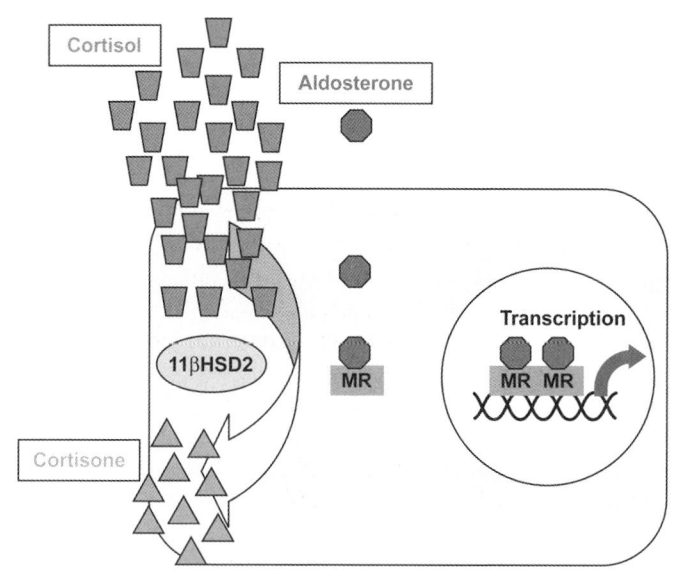

Figure 1.6 MR aldosterone specificity. Schematic representation of 11βHSD2-conferred MR aldosterone specificity.

Although animals were born in normal Mendelian ratios, there was high neonatal mortality in the homozygote null animals, and the remainder were hypertensive and severely hypokalemic. The RAS was suppressed and plasma aldosterone was also low [71, 70]. In one patient with SAME, the disorder was fully corrected by kidney transplant [72], indicating the disease is of renal origin. In support of this, 11βHSD2 null mice have excess renal sodium reabsorption due to activation of the epithelial sodium channel [71]. Nevertheless, sodium retention was found to be transient, and the hypertension moves from a renal to a central and ultimately vascular disorder through activation of the sympathetic nervous system [71]. It is possible that the sympathetic nervous system is activated by the hypernatremia that is sustained beyond the period of sodium retention. However, 11βHSD2 is also expressed in the nucleus of the solitary tract and amygdala of the mouse brain [73, 74], regions important to the central of blood pressure. Thus, SAME may reflect overactivation of MR in regions other than the kidney. This is supported by experiments showing that central administration of either aldosterone [75] or 11βHSD2 inhibitors [76] have a sustained hypertensive effect. In addition, inhibition (pharmacological or antisense) of 11βHSD2 sensitizes the vasculature to both Ang II [77] and catecholamines [78]. Vascular reactivity to noradrenalin is enhanced in a patient with SAME [79]. The 11βHSD2 null mice have endothelial dysfunction, with enhanced sensitivity to vasoactive agents being underpinned by a reduction in nitric oxide (NO) production [46, 80]. However, the extent of the endothelial dysfunction following targeted disruption of 11βHSD2 is dependent on the underlying background strain of the mouse [71] and may not, therefore, contribute in a major way to altered blood pressure homeostasis.

The enzyme is also expressed in the placenta where it serves to prevent maternal–fetal transfer of high levels of glucocorticoid. The deleterious effects on fetal development of *in utero* exposure to high levels of glucocorticoid are well documented, and such programming is associated with low birth weight and adverse cardiovascular risk. This subject has recently been reviewed [81].

Although SAME is an extreme phenotype and very rare, it illustrates the role that 11βHSD2 in the kidney, brain and vasculature has in the regulation of cardiovascular homeostasis. Indeed, it is possible that mild mutations are prevalent in the essential hypertensive population [82], particularly in those individuals with low-renin or salt-sensitive hypertension. Human molecular genetic studies in hypertensive populations have sought associations between blood pressure and loss-of-function polymorphisms in HSD11B2, with conflicting results (e.g. [83–85]). A more direct relationship between 11βHSD2 and blood pressure was obtained in nonhypertensive individuals subject to salt loading. For those individuals with salt-sensitive blood pressure, the extent of the rise in blood pressure following salt load was indirectly related to 11βHSD2 activity. That is, the lower the activity, the more exaggerated the response to salt. Our preliminary observations support this relationship finding that mice heterozygote null for *hsd11b2* have salt-sensitive blood pressure and an impaired ability to excrete sodium. These observations are of particular relevance to Western populations in which hypertension and excessive salt intake are common.

1.4
Cardiovascular Effects of Aldosterone

The well-documented effects of aldosterone on electrolyte transport in the epithelia of the distal nephron and colon [86, 87] can affect blood pressure and abnormal regulation of the RAS is implicated in hypertension [19, 88]. Mineralocorticoids also have actions in the heart [89], vasculature and brain [90] that can influence blood pressure homeostasis and cardiovascular control. Aldosterone is required for adaptation of blood pressure to postural changes and the clinical treatment of postural hypotension is fludrocortisone. These rapid changes in blood pressure occur long before any alteration in plasma volume and therefore lie outwith the control of renal MR. Furthermore, the antihypertensive effects of MR blockade do not correlate with any effects on renal salt balance [91]. The actions in nonepithelial tissue are informative in that they provide insights into "nonclassical" activation of MR and challenge the conventional view of receptor–ligand relationships.

1.4.1
Aldosterone and the Heart

Although aldosterone has both genomic and nongenomic effects on the biophysical properties of the cardiomyocyte (e.g. [92]), a physiological role has been discounted [93] on the basis that MR are antagonized by glucocorticoids under normal circumstances (see below). In contrast a pathological role for MR activation, particularly in the setting of mineralocorticoid excess or salt loading, has been demonstrated since the 1940s.

In the early 1990s a study by Brilla and Weber investigated the effects of mineralocorticoid excess with relation to cardiovascular function, observing that rats exposed to high levels of both aldosterone and salt developed hypertension and cardiac fibrosis [94]. This triggered resurgence in clinical interest with data suggesting that actual mineralocorticoid excess was associated with cardiac abnormalities [95]. Treatment of cardiac abnormalities through MR blockade was recommended following positive outcomes of two clinical trials: RALES and EPHESUS [96, 97]. In the RALES study, patients with severe heart failure were administered the MR antagonist spironolactone, alongside their continuing conventional medication. This produced a 30% reduction in mortality and a 35% lower frequency of hospitalization versus placebo-treated patients. Further verification of the beneficial effects of MR blockade and aldosterone antagonism was provided by the EPHESUS study in which eplerenone, an antagonist at MR more selective than spironolactone, was administered to patients who had suffered an acute myocardial infarction. Again, the results of MR blockade were particularly positive in terms of patient morbidity and mortality.

These trials show MR blockade to be beneficial in the treatment of heart disease, but the underlying mechanisms of action were not clear. The most straightforward explanation was that MR blockade inhibited the effects of aldosterone in the heart and was therefore cardioprotective. Indeed, it has been shown experimentally that increased aldosterone levels coupled with increased salt levels instigates deleterious cardiac and vascular pathologic responses [98] and, circumstantially, aldosterone levels are often raised in congestive heart failure [99]. However, in neither RALES nor EPHESUS were plasma aldosterone levels elevated [96, 97]. Similarly, in Dahl-salt-sensitive rats fed a high salt diet, MR blockade prevented the development of cardiac hypertrophy and the onset of chronic heart failure, despite the fact that aldosterone was lower in this group than controls [100]. In this case, the cardioprotective effects of eplerenone were independent of the antihypertensive effect of MR blockade, as has been reported elsewhere [91]. Together these data indicate that MR activation *per se*, rather than excess of agonist is critical to the developing pathology. Experiments using transgenic approaches are not so clear: mice overexpressing human MR display only mild cardiomyopathy [101] and mice in which MR has been knocked down via an inducible antisense transgene have severe heart failure (Figure 1.7) [103].

In contrast to classical aldosterone target tissues, occupancy of cardiac MR by glucocorticoids is the physiological norm [104]: 11βHSD2 is not normally expressed in cardiomyocytes at physiologically relevant levels. This would indicate that the benefits of MR blockade could be ascribed to relief from stimulation by glucocorticoids. However, the mode of glucocorticoid action at cardiac MR is not clear. Experiments designed to test this hypothesis followed the generation of a mouse

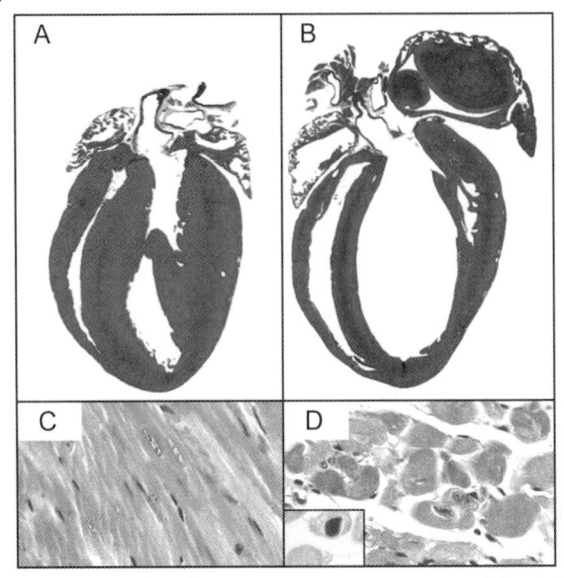

Figure 1.7 Cardiac damage in transgenic models of altered MR activation. Cardiac dilated hypertrophy in transgenic mice overexpressing 11βHSD2 in cardiomyocytes (B). (A) Heart of a nontransgenic mouse [102]. Histological analysis of cardiac remodeling induced by MR antisense mRNA expression in cardiomyocytes (D), showing large hyperchromatic nuclei, myocardial fiber disarray and myocyte hypertrophy, all of which are absent from the control mouse heart displayed in (C). (Adapted from [103]).

expressing 11βHSD2 selectively in cardiomyoctes [102]. Surprisingly, these mice developed severe cardiac hypertrophy and fibrosis, and died from accelerated heart failure (Figure 1.7). Moreover, an MR antagonist rescued the cardiopathology, whereas a GR antagonist did not. These data indicate that (i) glucocorticoids normally occupy cardiac MR, but act as an antagonist rather than agonist, and (ii) that aldosterone activation of MR – only observed when 11βHSD2 prevents glucocorticoid occupancy – is detrimental to heart function.

The data above are confusing and often conflicting, suggesting that MR blockade is both beneficial and damaging and that glucocorticoids can both activate and antagonize the MR. Recent data may reconcile these observations [105]. In isolated cardiomyocytes, aldosterone will activate the Na^+–K–$2Cl^-$ cotransporter, whereas cortisol will not [106]: coadministration of cortisol with aldosterone blocks the activation of the cotransporter. Moreover, if the redox state is altered to mimic production of reactive oxygen species, cortisol no longer antagonizes the actions of aldosterone and even acts as a mineralocorticoid. Thus, the question of what prompts the glucocorticoids to turn from tonic antagonists into pathological agonists may well rest with the generation of reactive oxygen species that can occur following cardiac trauma [89, 105].

1.4.2
Vasculature

MR have been located in freshly isolated vascular tissue and in both cultured vascular smooth muscle cells (VSMCs) and the vascular endothelium [107]. 11βHSD2 is also present in human VSMC, the adventitial fibroblasts and endothelial cells (ECs) [77, 108, 109]. In the mouse thoracic aorta, however, mRNA for 11βHSD2 is confined to the endothelium [46], as it is in cultured rat aortic cells [110]. Whether this is a species difference or reflects the sensitivity of enzyme expression to conditions of culture remains uncertain and resolution awaits the development of reliable antibodies.

Physiologically adrenal steroids–both aldosterone and glucocorticoids–potentiate the action of vasoconstrictors. This effect was first described in the 1950s for catecholamines but is also true for other vasoactive agents, notably Ang II [111]. There is some evidence to suggest that the potentiating effect of corticosteroids differs throughout the vasculature. For example, in the deoxycorticosterone acetate (DOCA)-salt-hypertensive rat model, the pressor effects of Ang II are exacerbated, indicating an increased sensitivity of the resistance vasculature to vasoconstrictors [112]. The conduit vasculature, in contrast, was not sensitized. However, the opposite has been reported for catecholamines: the conduit vasculature being sensitized and the resistance vessels desensitized to phenylephrine [113].

The mechanisms of potentiation have focused on increased receptor density, in part due to the actions of corticosteroids as transcription factors, but also because the effects are seen *ex vivo* and are therefore a property intrinsic to the vessel. For Ang II this appears to hold true, since both aldosterone and glucocorticoids greatly enhance receptor density [111]. Moreover, the increase in receptor number is transduced to a downstream effect, there being a more robust activation by Ang II of intracellular signaling cascades following exposure to mineralocorticoids [114]. These effects are exclusive to the AT_1 receptor [115], consistent with the fact that the promotor region for the gene contains several steroid response elements [116]. In addition to the effects on receptor density, mineralocorticoids also lead to activation of a localized RAS, with increased angiotensinogen formation [117] and ACE activity being found in both ECs and VSMCs [118].

Mineralocorticoid increases the expression of α-adrenergic receptors [119], whereas adrenalectomy reduces receptor density [120]. However, binding studies indicate that receptor affinity moves in the opposite direction to number, thereby offsetting greatly the theoretical "stimulatory" effects of mineralocorticoid. There are several conflicting reports in the literature, but overall convincing data to suggest a receptor number-based response to corticosteroids is lacking [111]. Neither does altered release or uptake of catecholamines at nerve terminals contribute to potentiation by aldosterone [121]. A clue to the underlying mechanism came from the observation that endothelium-dependent vasodilation was impaired in DOCA-salt rats [122]. This was initially attributed to damage secondary to chronic hypertension but other studies demonstrated that the heightened pressor responses to noradrenalin were, in fact, due to a reduced synthesis of the vasodilator prostaglandin E2 from the endothelium [123]. It is now clear that attenuation of endothelium-derived vasodilation contributes to the potentiation by corticosteroids of the response to catecholamines. This is NO-dependent in conduit vasculature, but not in resistance vessels. More recently, experiments have described direct effects of aldosterone on VSMCs. By improving the coupling of α_1-adrenoceptors to downstream signaling pathways, mineralocorticoids improve vascular tone [124].

The molecular mechanism of aldosterone's action in the vasculature involves genomic effects but these are observed more than 2 h after exposure. That vasoactive responses can be observed almost immediately after infusion of aldosterone (i.e. prior to the onset of *de novo* protein synthesis) suggests nongenomic action in the vasculature [104]. Although *in vivo* systemic infusions of aldosterone almost always promote an immediate pressor response [125], vasoconstriction is not a universal finding *ex vivo*. Indeed, local infusion of aldosterone into the forearm increases local blood flow [125]. This vasodilation results from stimulation of NO production by the endothelium [125]. There is, however, also an effect on the VSMCs to promote vasoconstriction and the net effect on vascular function depends on the balance between the two opposing forces. Thus, inhibition of endothelial NO synthase with N-monomethyl-L-arginine leads to a powerful and sustained vasoconstriction [126]. These observations explain why the effect of aldosterone on vascular tone may vary in different vascular beds. Moreover, local vasodilation may be offset by blunting of the baroreceptor reflex [127] and indirect activation of the sympathetic nervous system.

In addition to these well-documented responses, both mineralocorticoids and glucocorticoids will stimulate the local release of endothelin from the vasculature [128]. This is a potent vasoconstrictor (covered in Chapter 6) and could mediate aldosterone-stimulated increases in total peripheral resistance.

Aldosterone also exerts profound pathological effects in the vasculature. The accumulative and slowly developing disease atherosclerosis is a major cause of heart disease. Disruption of vascular homeostasis predisposes the endothelial vessel wall to vasoconstriction, inflammation and atherosclerosis, all of which can be contributors towards cardiac disease onset. The development and progression of atherosclerosis is largely associated with endothelial dysfunction (for review, see [129]), and it has been suggested that the positive clinical outcomes of the RALES and EPHESUS studies may be in part due alleviation of vascular endothelial aldosterone and/or MR antagonism. Using cultured human umbilical vein ECs, Oberleithner et al. demonstrated that aldosterone promotes remodeling of the endothelium *in vitro* [130]. They observed that aldosterone administration caused the cells to increase in both size and stiffness, which would *in vivo* lead to endothelial dysfunction and associated pathogenesis. Endothelial dysfunction can be rescued in the stroke-prone spontaneously hypertensive rat by administration of eplerenone [177]. Aldosterone also plays a major role in Ang II-induced vascular inflammation in the setting of high salt intake [98].

It is not easy to present a unifying view of the effect of aldosterone on vascular function since the literature is often divergent. It would appear, however, that the net effect of systemic aldosterone is to increase blood pressure by potentiating the action of vasoconstrictors, "activating" VSMCs and increasing sympathetic drive (either directly or indirectly). The action on the endothelium is less clear. Physiologically and in healthy individuals, it would appear that aldosterone promotes vasodilation, both acutely and chronically [131]. In the setting of hypertension or mineralocorticoid excess coupled with high salt, aldosterone might promote endothelial dysfunction [107]. Although controversial, some evidence suggests that the vasculature can synthesize aldosterone locally [132], adding a further level of complexity to the field. Indeed, locally activated RAS is proinflammatory and promotes detrimental remodeling of the vasculature in hypertension [133]. These findings, together with the positive outcomes of clinical trials, would advocate the use of MR antagonists in hypertension and cardiovascular disease. Indeed, the Framingham Heart Study reports a complex but positive correlation between serum aldosterone in the physiological range and cardiovascular risk [134].

The use of MR antagonists to treat cardiovascular disease is limited, as they tend to promote hyperkalemia due to actions in the kidney. Furthermore, targeted disruption in mice of the gene encoding MR has not been particularly informative in terms of the role of aldosterone in cardiovascular function. MR null mice die within 8 days of birth due to uncorrected salt wasting [135]. Despite significant activation of the RAS, MR null mice were unable to activate the epithelial sodium channel, modeling well the autosomal dominant form of pseudohypoaldosteronism type 1 (OMIM #177735), in which inactivating mutations in MR are reported. These experiments not only indicate the critical role for renal MR in the long-term regulation of blood pressure, but also demonstrate that activation of the GR does not compensate for loss of MR. In order to circumvent the problem of early postnatal death associated with global MR deficiency, the gene has been "floxed" allowing targeted deletion through use of the Cre–loxP system [136]. Surprisingly, when MR was deleted in the distal nephron, mice were able to thrive, albeit with much-elevated aldosterone [137]. This would suggest that MR in other systems could compensate for the loss of renal MR and this was indeed found in the colon. Nevertheless, the principal cell mutant mice were able to maintain near perfect salt balance, even on a low-sodium diet, via upregulation of the epithelial sodium channel and it was found that in a small percentage of principal cells in the early connecting tubule MR had not been deleted. Although this highlights a pitfall of the Cre–loxP system, it is expected that future experiments using the "floxed" MR will be informative.

1.5
Cardiovascular Effects of Glucocorticoids

Glucocorticoids are responsible for a wide range of physiological effects (Figure 1.8), the majority of which are united under the common subheading of stress

1 Glucocorticoids and Mineralocorticoids

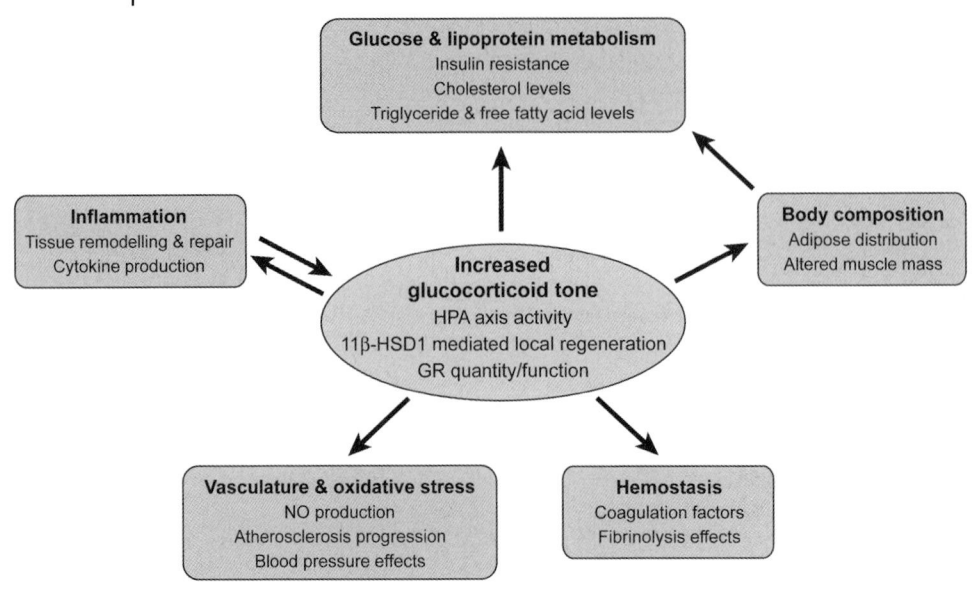

Glucose & lipoprotein metabolism
Insulin resistance
Cholesterol levels
Triglyceride & free fatty acid levels

Inflammation
Tissue remodelling & repair
Cytokine production

Increased glucocorticoid tone
HPA axis activity
11β-HSD1 mediated local regeneration
GR quantity/function

Body composition
Adipose distribution
Altered muscle mass

Vasculature & oxidative stress
NO production
Atherosclerosis progression
Blood pressure effects

Hemostasis
Coagulation factors
Fibrinolysis effects

Figure 1.8 Glucocorticoid cardiovascular effects. Glucocorticoid effects upon cardiovascular risk factors.

responses. The release of glucocorticoids following stress-induced stimulation of the HPA axis promotes coordination of endocrine, immune and nervous system responses to the initial stimuli. Examples of this include inducing the mobilization of energy resources in response to physical stresses such as starvation and the "fight or flight" response by stimulating gluconeogenesis and lipolysis, and inhibiting glucose uptake by peripheral tissues. Glucocorticoids also act to suppress inflammatory responses, cellular proliferation and tissue repair, suggesting a regulatory role to prevent these responses becoming undisciplined and destructive.

Several clinical disorders associated with cortisol deregulation – whether a consequence of synthesis, HPA axis or GR-mediated effects – have been associated with an increased rate of morbidity and mortality, which in turn is possibly corollary to an increased risk of cardiovascular events (for review, see [138]).

It is difficult to separate direct primary effects of glucocorticoids on the heart and vasculature from secondary changes arising from activation of GR in other systems (Figure 1.8). However, evidence from human patients and transgenic mice have helped to establish the nature of these primary responses. It appears that glucocorticoids, at physiological concentrations, may be beneficial to cardiac function, potentially by antagonizing the MR as described above. Furthermore, glucocorticoids potentiate the action of vasoactive substances and so can clearly influence vascular tone. That glucocorticoids are important cardiovascular hormones is illustrated through the extremes of altered glucocorticoid secretion:

Addison's disease presents with life-threatening hypotension and vascular collapse while high blood pressure is a common feature of Cushing's syndrome [139]

1.5.1
Transgenic Models

That no complete loss-of-function mutations in the GR are present in the human population indicates that glucocorticoid activity is essential for life. Transgenic mice with mutations in the GR gene, leading to partial or complete ablation of GR function, support this hypothesis (for review, see [140]) and have not been particularly informative due to high levels of mortality: GR null mice die a few hours after birth due to respiratory failure (glucocorticoids are critical in fetal lung maturation) [141].

Other models in which GR activity is only partially affected are more useful. For example, transgenic mice were generated with a point mutation in the GR gene that abolished DNA-binding capability without affecting the other actions of GR [142]. Approximately one-fifth of these mice survived until adulthood, suggesting that the DNA-binding capability of GR is not essential for survival. Surviving mice had impaired induction of gluconeogenic enzymes, increased ACTH levels as well as adrenal hypertrophy, hyperplasia and an overproduction of steroidogenic enzymes. Of interest, there were no obvious phenotypes usually associated in humans with glucocorticoid excess, such as altered fat disposition. This provides critical information regarding those pathways that require binding of GR to DNA. A second point mutation, introduced into exon 4 of mouse GR [143], impairs GR dimerization. The resulting mice are viable, again indicating that GR DNA binding is not essential for survival. Serum levels of corticosteroid and expression of key steroidogenic enzymes were upregulated, but no change was found in ACTH serum levels (although it was 2.2-fold increased in the anterior pituitary) or adrenal morphology. The data from this model indicate that certain downstream effects of GR activation do not require receptor dimerization in the classical mode.

GR tissue-specific null mice have also been developed using the Cre–loxP system. These mutations, so far used to target gene deletion to the either the liver or nervous system, circumvent the lethality of somatic GR ablation. Both have altered HPA axis regulation. In the liver-specific null, growth retardation was observed, even though serum glucocorticoids and growth hormone levels were normal [144]. Following deletion of GR in the nervous system, behavioral abnormalities were observed [145]. Utilizing this technology to develop a heart and/or vasculature GR-specific null model could reveal valuable insights into how glucocorticoids specifically affect the cardiovascular system.

Taking the opposite approach, a recent study has over-expressed GR exclusively in cardiomyocytes under a "tet-on, tet-off" system [146]. The mice displayed electrocardiogram abnormalities, which were completely reversible with GR overexpression shutoff. Isolated ventricular cardiomyocytes displayed major ion channel remodeling, as well as changes in cell calcium homeostasis. The electrophysical phenotyping of this model indicates that the cardiac GR overexpression produces defects in conduction with a high degree of atrio-ventricular block. This may reflect, in a magnified and acute setting, the physiological effects that excessive glucocorticoids levels have upon the heart.

1.5.2
Lessons from Human Disease

1.5.2.1 Cushing's Syndrome

Cushing's syndrome is the result of prolonged exposure to excessive cortisol, either as a consequence of hypersecretion of endogenous cortisol outwith the normal physiologic feedback of the HPA axis, or as a result of intensive exogenous exposure in the form of steroid treatment. The most common cause of endogenous Cushing's syndrome is pituitary adenoma, otherwise known as Cushing's disease (OMIM #219090), in which excess ACTH is secreted. Other causes involve ectopic ACTH secretion due to a carcinoid tumor and excess cortisol secretion from adrenal adenomas or carcinomas. Clinical symptoms of Cushing's syndrome include central obesity, hypertension, glucose intolerance, insulin resistance and dyslipidemia [147]. Cardiovascular disease is the main cause of death and disease in Cushing's syndrome patients, and an elevated risk remains even after successful treatment of other symptoms [148]. Epidemiological studies of Cushing's syndrome show patients have a mortality rate 4 times higher than average (age- and sex-matched controls) due to cardiovascular complications [149].

Hypertension is one of the most important cardiovascular risk factors associated with Cushing's syndrome, being present in around 80% of adult patients [150]. The hypertensive phenotype is an effect of interactions between several of the associated pathophysiological mechanisms underlying Cushing's syndrome including effects upon plasma volume, peripheral vascular resistance and cardiac output, all of which have a tendency to be increased [151]. Cushing's associated blood pressure abnormalities present initially with a deregulation of blood pressure circadian rhythm, characterized by a loss of the typical nocturnal fall. This transition from a "dipper" to a "nondipper" phenotype (nocturnal hypertension) is recognized as a cardiovascular risk in its own right [152].

The key underlying mechanism of hypertension is glucocorticoid excess. Part of this can be attributed to illicit MR occupation, but this is not the only cause as MR antagonists do not fully alleviate hypertension [148]. Other blood pressure effects are mediated by excessive activation of the RAS, potentiation of vasoactive substances (see above), and/or suppression of vasodilators (for further details, see [151]). Hypertension is generally resolved in patients after successful treatment, but in a few cases the hypertension persists, suggesting permanent damage to, and remodeling of, the renal and cardiovascular systems. The gain in body weight and partitioning of adipose tissue towards visceral obesity also plays a role in the cardiovascular risk associated with Cushing's syndrome.

1.5.2.2 The Metabolic Syndrome and Tissue-Specific Regulation of Glucocorticoids

The metabolic syndrome (or Syndrome X; OMIM &605552) is defined as the clustering of a plethora of metabolic and cardiovascular phenotypes, including hypertension, hyperglycemia (type 2 diabetes), dyslipidemia and abdominal obesity [153]. The metabolic syndrome is a major cardiovascular risk factor, the prevalence in society of which may be related to relatively recent lifestyle changes, such as dramatic increases in calorific/salt intake and sedentary life habits [154]. However, the metabolic interactions leading to the clustering if the metabolic syndrome phenotypes are not completely understood [155]. The phenotypic similarities between the metabolic syndrome and Cushing's disease cannot help but be noticed and glucocorticoid excess/resistance is certainly a viable as a candidate for an underlying cause of metabolic syndrome [156]. As GR are ubiquitously expressed throughout all tissues, the effects of a global increase in plasma glucocorticoids levels can have many varied consequences that may detract the needed effects away from the factor causing the initial stress. A mechanism to control this is to regulate the concentration of available glucocorticoids at a more local, tissue-specific level. As mentioned previously, 11βHSD1 is globally expressed and is responsible for the conversion of the inactive cortisone to the active cortisol. Studies have shown that 11βHSD1 acts as a regulator of locally available cortisol, as an upregulation in 11βHSD1 would result in increased generation of cortisone to cortisol at a local level. These phenotypes observed in 11βHSD1 transgenic mice are all characteristic of the metabolic syndrome as described above. The level of enzyme function in the adipocyte appears to play a critical role in setting of metabolic profile. Further support for 11βHSD1 involvement in local glucocorticoid regulation in relation to metabolic syndrome phenotypes can be found in other rodent models of obesity, such as the obese Zucker rat model. Corticosteroid metabolism was measured in these obese rats, and 11βHSD1 was enhanced in omental adipose tissue, an indicator that local GR activation may be underling the local promotion of obesity [157, 158]. Evidence of 11βHSD1 upregulation in human obesity has also been shown specifically in adipose tissue in both men [159] and women [160]. Observations such as these raise the possibility of 11βHSD1-specific inhibition as a potential novel treatment for the phenotypes of the metabolic syndrome.

1.5.2.3 Glucocorticoid Resistance Syndrome

Glucocorticoid resistance syndrome (GRS; OMIM +138040) is a rare syndrome characterized by diminished cortisol action mediated by altered GR [161]. This results in a compensatory increase in adrenal ACTH secretion through hyperactivation of the HPA axis, leading to increased levels of circulating glucocorticoids, mineralocorticoids and androgens. A number of mutations in the human GR (hGR) gene LBD or DBD have been identified as underlying molecular mechanisms of familial GRS [162]. Although the different locations of these GRS mutations within the hGR gene result in variable clinical phenotypes, some phenotypes are more persistently expressed than others. These vary from hypertension with or without hypokalemia to hirsutism and infertility [163]. Cardiovascular morbidity and mortality is increased if the disease is not successfully treated.

1.5.2.4 GR Polymorphisms

Glucocorticoid variation also exists in the general population, as has been experimentally shown by varied responses to dexamethasone (asynthetic glucocorticoid that has no affinity for the MR) by an elderly experimental cohort [164]. Several GR polymorphisms have been identified that seem to be associated with altered glucocorticoids sensitivity and metabolic parameters (for review, see [165]). The N363S polymorphism has been shown to increase glucose sensitivity and the insulin response to dexamethasone, as well as an increased body mass index (BMI). In certain populations there was found to be an increased cardiovascular risk due to high cholesterol and triglyceride concentrations [164]. The N363S variant is associated with obesity, angina and coronary artery disease [166], suggesting a role for altered GR variation in these conditions. The Bcl-1 restriction fragment length polymorphism increases glucocorticoid sensitivity and abdominal obesity in middle age, but then conversely towards old age BMI tends to be lower, perhaps due to accelerated muscle atrophy [167]. Both of these mutations could be described as glucocorticoid hypersensitivity.

Conversely, the ER22/23EK polymorphism [168] is associated with glucocorticoid-resistant phenotypes. Carriers display lower total cholesterol and improved insulin sensitivity, and elderly male carriers of the ER22/23EK polymorphism tend to be protected from cardiovascular damage and therefore have improved survival [169]. In young males and females, the ER22/23EK polymorphism is associated with "improved" body shape and lower body weight, compared to age- and sex-matched controls [170]. This evidence implies that the ER22/23EK polymorphism may predispose towards a healthier metabolic profile, which is regarded as an important aspect of reducing an individual's cardiovascular risk. A recent study identified another novel mutation in the LBD of the hGR which results in generalized glucocorticoid resistance [162], underlining the importance of the role that hGR polymorphisms may play in the general population for contributing towards an individuals cardiovascular risk profile.

1.5.3
Endothelial Dysfunction, Vascular Tone and Atherosclerosis

Atherosclerosis is a major cause of morbidity and mortality in cardiovascular disease, and occurs as a result of a chronic inflammatory response to the deposition of lipoproteins in arterial walls [171]. This leads to the formation of an atheromatous plaque in the arterial wall, which can protrude into the arterial lumen and alter blood flow through the artery. Narrowing of the arterial lumen causes an increase in blood pressure and the sheering stress caused by this can cause the atheromatous plaque to rupture, leading to an occlusion of blood flow and myocardial infarction. There is epidemiological evidence that long-term exposure to glucocorticoid excess (such as in Cushing's syndrome) is associated with accelerated atherosclerosis [172]. Administration of 11βHSD1-specific inhibitors in a mouse model of atherosclerosis resulted in a reduced progression of an atheromatous plaque, suggesting that intracellular inactivation of glucocorticoids can have a direct positive effect upon the progression atherosclerosis [173]. Moreover, chronic administration of glucocorticoids to patients with rheumatoid arthritis increases the incidence of atherosclerosis [174].

References

1 Clark, B.J., Wells, J., King, S.R. and Stocco, D.M. (1994) The purification, cloning, and expression of a novel luteinizing hormone-induced mitochondrial protein in MA-10 mouse Leydig tumor cells. Characterization of the steroidogenic acute regulatory protein (stAR). *Journal of Biological Chemistry*, **269**, 28314–22.

2 Lin, D., Sugawara, T., Strauss, J.F., 3rd, Clark, B.J., Stocco, D.M., Saenger, P., Rogol, A. and Miller, W.L. (1995) Role of steroidogenic acute regulatory protein in adrenal and gonadal steroidogenesis. *Science*, **267**, 1828–31.

3 Stocco, D.M. (2002) Clinical disorders associated with abnormal cholesterol transport: mutations in the steroidogenic acute regulatory protein. *Molecular and Cellular Endocrinology*, **191**, 19–25.

4 Caron, K.M., Soo, S.C., Wetsel, W.C., Stocco, D.M., Clark, B.J. and Parker, K.L. (1997) Targeted disruption of the mouse gene encoding steroidogenic acute regulatory protein provides insights into congenital lipoid adrenal hyperplasia. *Proceedings of the National Academy of Sciences of the United States of America*, **94**, 11540–5.

5 Bhangoo, A., Anhalt, H., Ten, S. and King, S.R. (2006) Phenotypic variations in lipoid congenital adrenal hyperplasia. *Pediatric Endocrinology Reviews*, **3**, 258–71.

6 Lifton, R.P., Dluhy, R.G., Powers, M., Rich, G.M., Cook, S., Ulick, S. and Lalouel, J.M. (1992) A chimaeric 11 beta-hydroxylase/aldosterone synthase gene causes glucocorticoid-remediable aldosteronism and human hypertension. *Nature*, **355**, 262–5.

7 Gotoh, H., Sagai, T., Hata, J., Shiroishi, T. and Moriwaki, K. (1988) Steroid 21-hydroxylase deficiency in mice. *Endocrinology*, **123**, 1923–7.

8 Riepe, F.G., Tatzel, S., Sippell, W.G., Pleiss, J. and Krone, N. (2005) Congenital adrenal hyperplasia: the molecular basis of 21-hydroxylase deficiency in H-2(aw18) mice. *Endocrinology*, **146**, 2563–74.

9 White, P.C., Curnow, K.M. and Pascoe, L. (1994) Disorders of steroid 11 beta-hydroxylase isozymes. *Endocrine Reviews*, **15**, 421–38.

10 Lee, G., Makhanova, N., Caron, K., Lopez, M.L., Gomez, R.A., Smithies, O. and Kim, H.S. (2005) Homeostatic responses in the adrenal cortex to the absence of aldosterone in mice. *Endocrinology*, **146**, 2650–6.

11 Makhanova, N., Lee, G., Takahashi, N., Sequeira Lopez, M.L., Gomez, R.A., Kim, H.S. and Smithies, O. (2006) Kidney function in mice lacking aldosterone. *American Journal of Physiology – Renal Physiology*, **290**, F61–69.

12 Quinn, S.J. and Williams, G.H. (1988) Regulation of aldosterone secretion. *Annual Review of Physiology*, **50**, 409–26.

13 Heffelfinger, S.C. (2007) The renin angiotensin system in the regulation of angiogenesis. *Current Pharmaceutical Design*, **13**, 1215–29.

14 Kramer, R.E., Gallant, S. and Brownie, A.C. (1980) Actions of angiotensin II on aldosterone biosynthesis in the rat adrenal cortex. Effects on cytochrome P-450 enzymes of the early and late pathway. *Journal of Biological Chemistry*, **255**, 3442–7.

15 Lazartigues, E., Feng, Y. and Lavoie, J.L. (2007) The two fACEs of the tissue renin–angiotensin systems: implication

in cardiovascular diseases. *Current Pharmaceutical Design*, **13**, 1231–45.
16 Stojiljkovic, L. and Behnia, R. (2007) Role of renin angiotensin system inhibitors in cardiovascular and renal protection: a lesson from clinical trials. *Current Pharmaceutical Design*, **13**, 1335–45.
17 Hamming, I., Cooper, M.E., Haagmans, B.L., Hooper, N.M., Korstanje, R., Osterhaus, A.D., Timens, W., Turner, A.J. and Navis G., and van Goor, H. (2007) The emerging role of ACE2 in physiology and disease. *Journal of Pathology*, **212**, 1–11.
18 Van Tassell, B.W. and Munger, M.A. (2007) Aliskiren for renin inhibition: a new class of antihypertensives. *Annals of Pharmacotherapy*, **41**, 456–64.
19 Mullins, L.J., Bailey, M.A. and Mullins, J.J. (2006) Hypertension, kidney, and transgenics: a fresh perspective. *Physiological Reviews*, **86**, 709–46.
20 Young, D.B. (1985) Analysis of long-term potassium regulation. *Endocrine Reviews*, **6**, 24–44.
21 Cowley, A.W. and McCaa, R.E. (1976) Acute and chronic dose–response relationships for angiotensin, aldosterone, and arterial pressure at varying levels of sodium intake. *Circulation Research*, **39**, 788–97.
22 Muglia, L.J., Bethin, K.E., Jacobson, L., Vogt, S.K. and Majzoub, J.A. (2000) Pituitary–adrenal axis regulation in CRH-deficient mice. *Endocrine Research*, **26**, 1057–66.
23 Turnbull, A.V., Smith, G.W., Lee, S., Vale, W.W., Lee, K.F. and Rivier, C. (1999) CRF type I receptor-deficient mice exhibit a pronounced pituitary–adrenal response to local inflammation. *Endocrinology*, **140**, 1013–7.
24 Bale, T.L., Picetti, R., Contarino, A., Koob, G.F., Vale, W.W. and Lee, K.F. (2002) Mice deficient for both corticotropin-releasing factor receptor 1 (CRFR1) and CRFR2 have an impaired stress response and display sexually dichotomous anxiety-like behavior. *Journal of Neuroscience*, **22**, 193–9.
25 Makara, G.B., Mergl, Z. and Zelena, D. (2004) The role of vasopressin in hypothalamo-pituitary–adrenal axis activation during stress: an assessment of the evidence. *Annals of the New York Academy of Sciences*, **1018**, 151–61.
26 Lolait, S.J., Stewart, L.Q., Jessop, D.S., Young, W.S., 3rd and O'Carroll, A.M. (2007) The hypothalamic–pituitary–adrenal axis response to stress in mice lacking functional vasopressin V1b receptors. *Endocrinology*, **148**, 849–56.
27 Phillips, M.I. (1987) Functions of angiotensin in the central nervous system. *Annual Review of Physiology*, **49**, 413–35.
28 Jezova, D., Makatsori, A., Smriga, M., Morinaga, Y. and Duncko, R. (2005) Subchronic treatment with amino acid mixture of L-lysine and L-arginine modifies neuroendocrine activation during psychosocial stress in subjects with high trait anxiety. *Nutritional Neuroscience*, **8**, 155–60.
29 Elias, L.L. and Clark, A.J. (2000) The expression of the ACTH receptor. *Brazilian Journal of Medical and Biological Research*, **33**, 1245–8.
30 Wu, S.M., Stratakis, C.A., Chan, C.H., Hallermeier, K.M., Bourdony, C.J., Rennert, O.M. and Chan, W.Y. (1998) Genetic heterogeneity of adrenocorticotropin (ACTH) resistance syndromes: identification of a novel mutation of the ACTH receptor gene in hereditary glucocorticoid deficiency. *Molecular Genetics and Metabolism*, **64**, 256–65.
31 Lachelin, G.C., Barnett, M., Hopper, B.R., Brink, G. and Yen, S.S. (1979) Adrenal function in normal women and women with the polycystic ovary syndrome. *Journal of Clinical Endocrinology and Metabolism*, **49**, 892–8.
32 Schulte, H.M., Chrousos, G.P., Gold, P.W., Booth, J.D., Oldfield, E.H., Cutler, G.B., Jr and Loriaux, D.L. (1985) Continuous administration of synthetic ovine corticotropin-releasing factor in man. Physiological and pathophysiological implications. *Journal of Clinical Investigation*, **75**, 1781–5.
33 Sherman, B., Wysham, C. and Pfohl, B. (1985) Age-related changes in the circadian rhythm of plasma cortisol in man. *Journal of Clinical Endocrinology and Metabolism*, **61**, 439–43.

34 Balsalobre, A., Brown, S.A., Marcacci, L., Tronche, F., Kellendonk, C., Reichardt, H.M., Schutz, G. and Schibler, U. (2000) Resetting of circadian time in peripheral tissues by glucocorticoid signaling. *Science*, **289**, 2344–7.

35 Howell, M.P. and Muglia, L.J. (2006) Effects of genetically altered brain glucocorticoid receptor action on behavior and adrenal axis regulation in mice. *Frontiers in Neuroendocrinology*, **27**, 275–84.

36 Giles, T.D. (2006) Circadian rhythm of blood pressure and the relation to cardiovascular events. *Journal of Hypertension. Supplement*, **24**, S11–16.

37 Rosmond, R., Chagnon, Y.C., Chagnon, M., Perusse, L., Bouchard, C. and Bjorntorp, P. (2000) A polymorphism of the 5′-flanking region of the glucocorticoid receptor gene locus is associated with basal cortisol secretion in men. *Metabolism*, **49**, 1197–9.

38 Lu, N.Z., Wardell, S.E., Burnstein, K.L., Defranco, D., Fuller, P.J., Giguere, V., Hochberg, R.B., McKay, L., Renoir, J.M., Weigel, N.L. et al. (2006) International Union of Pharmacology. LXV. The pharmacology and classification of the nuclear receptor superfamily: glucocorticoid, mineralocorticoid, progesterone, and androgen receptors. *Pharmacological Reviews*, **58**, 782–97.

39 Arriza, J.L., Weinberger, C., Cerelli, G., Glaser, T.M., Handelin, B.L., Housman, D.E. and Evans, R.M. (1987) Cloning of human mineralocorticoid receptor complementary DNA: structural and functional kinship with the glucocorticoid receptor. *Science*, **237**, 268–75.

40 Nishi, M. and Kawata, M. (2007) Dynamics of glucocorticoid receptor and mineralocorticoid receptor: implications from live cell imaging studies. *Neuroendocrinology*, **85**, 186–92.

41 Bamberger, C.M., Schulte, H.M. and Chrousos, G.P. (1996) Molecular determinants of glucocorticoid receptor function and tissue sensitivity to glucocorticoids. *Endocrine Reviews*, **17**, 245–61.

42 Seckl, J.R. and Walker, B.R. (2004) 11Beta-hydroxysteroid dehydrogenase type 1 as a modulator of glucocorticoid action: from metabolism to memory. *Trends in Endocrinology and Metabolism*, **15**, 418–24.

43 Brown, R.W., Diaz, R., Robson, A.C., Kotelevtsev, Y.V., Mullins, J.J., Kaufman, M.H. and Seckl, J.R. (1996) The ontogeny of 11 beta-hydroxysteroid dehydrogenase type 2 and mineralocorticoid receptor gene expression reveal intricate control of glucocorticoid action in development. *Endocrinology*, **137**, 794–7.

44 Thompson, A., Han, V.K. and Yang, K. (2002) Spatial and temporal patterns of expression of 11beta-hydroxysteroid dehydrogenase types 1 and 2 messenger RNA and glucocorticoid receptor protein in the murine placenta and uterus during late pregnancy. *Biology of Reproduction*, **67**, 1708–18.

45 Thompson, A., Han, V.K. and Yang, K. (2004) Differential expression of 11beta-hydroxysteroid dehydrogenase types 1 and 2 mRNA and glucocorticoid receptor protein during mouse embryonic development. *Journal of Steroid Biochemistry and Molecular Biology*, **88**, 367–75.

46 Christy, C., Hadoke, P.W., Paterson, J.M., Mullins, J.J., Seckl, J.R. and Walker, B.R. (2003) 11Beta-hydroxysteroid dehydrogenase type 2 in mouse aorta: localization and influence on response to glucocorticoids. *Hypertension*, **42**, 580–7.

47 Agarwal, A.K., Monder, C., Eckstein, B. and White, P.C. (1989) Cloning and expression of rat cDNA encoding corticosteroid 11 beta-dehydrogenase. *Journal of Biological Chemistry*, **264**, 18939–43.

48 Agarwal, A.K., Rogerson, F.M., Mune, T. and White, P.C. (1995) Analysis of the human gene encoding the kidney isozyme of 11 beta-hydroxysteroid dehydrogenase. *Journal of Steroid Biochemistry and Molecular Biology*, **55**, 473–9.

49 Grundy, W.N., Bailey, T.L., Elkan, C.P. and Baker, M.E. (1997) Hidden Markov model analysis of motifs in steroid dehydrogenases and their homologs.

Biochemical and Biophysical Research Communications, **231**, 760–6.
50 Paterson, J.M., Seckl, J.R. and Mullins, J.J. (2005) Genetic manipulation of 11beta-hydroxysteroid dehydrogenases in mice. *American Journal of Physiology– Regulatory Integrative and Comparative Physiology*, **289**, R642–52.
51 Naray-Fejes-Toth, A. and Fejes-Toth, G. (1996) Subcellular localization of the type 2 11beta-hydroxysteroid dehydrogenase. A green fluorescent protein study. *Journal of Biological Chemistry*, **271**, 15436–42.
52 Odermatt, A., Arnold, P., Stauffer, A., Frey, B.M. and Frey, F.J. (1999) The N-terminal anchor sequences of 11beta-hydroxysteroid dehydrogenases determine their orientation in the endoplasmic reticulum membrane. *Journal of Biological Chemistry*, **274**, 28762–70.
53 Maser, E., Volker, B. and Friebertshauser, J. (2002) 11Beta-hydroxysteroid dehydrogenase type 1 from human liver: dimerization and enzyme cooperativity support its postulated role as glucocorticoid reductase. *Biochemistry*, **41**, 2459–65.
54 Gomez-Sanchez, E.P., Ganjam, V., Chen, Y.J., Liu, Y., Clark, S.A. and Gomez-Sanchez, C.E. (2001) The 11beta hydroxysteroid dehydrogenase 2 exists as an inactive dimer. *Steroids*, **66**, 845–8.
55 Paterson, J.M., Holmes, M.C., Kenyon, C.J., Carter, R., Mullins, J.J. and Seckl, J.R. (2007) Liver-selective transgene rescue of hypothalamic–pituitary–adrenal axis dysfunction in 11beta-hydroxysteroid dehydrogenase type 1-deficient mice. *Endocrinology*, **148**, 961–6.
56 Yau, J.L., Noble, J., Kenyon, C.J., Hibberd, C., Kotelevtsev, Y., Mullins, J.J. and Seckl, J.R. (2001) Lack of tissue glucocorticoid reactivation in 11beta-hydroxysteroid dehydrogenase type 1 knockout mice ameliorates age-related learning impairments. *Proceedings of the National Academy of Sciences of the United States of America*, **98**, 4716–21.
57 Zhang, J., Osslund, T.D., Plant, M.H., Clogston, C.L., Nybo, R.E., Xiong, F., Delaney, J.M. and Jordan, S.R. (2005) Crystal structure of murine 11 beta-hydroxysteroid dehydrogenase 1: an important therapeutic target for diabetes. *Biochemistry*, **44**, 6948–57.
58 Masuzaki, H., Paterson, J., Shinyama, H., Morton, N.M., Mullins, J.J., Seckl, J.R. and Flier, J.S. (2001) A transgenic model of visceral obesity and the metabolic syndrome. *Science*, **294**, 2166–70.
59 Masuzaki, H., Yamamoto, H., Kenyon, C.J., Elmquist, J.K., Morton, N.M., Paterson, J.M., Shinyama, H., Sharp, M.G., Fleming, S., Mullins, J.J. et al. (2003) Transgenic amplification of glucocorticoid action in adipose tissue causes high blood pressure in mice. *Journal of Clinical Investigation*, **112**, 83–90.
60 Funder, J.W., Pearce, P.T., Smith, R. and Smith, A.I. (1988) Mineralocorticoid action: target tissue specificity is enzyme, not receptor, mediated. *Science*, **242**, 583–5.
61 Stewart, P.M., Wallace, A.M., Valentino, R., Burt, D., Shackleton, C.H. and Edwards, C.R. (1987) Mineralocorticoid activity of liquorice: 11-beta-hydroxy-steroid dehydrogenase deficiency comes of age. *Lancet*, **2**, 821–4.
62 Odermatt, A., Arnold, P. and Frey, F.J. (2001) The intracellular localization of the mineralocorticoid receptor is regulated by 11beta-hydroxysteroid dehydrogenase type 2. *Journal of Biological Chemistry*, **276**, 28484–92.
63 Dave-Sharma, S., Wilson, R.C., Harbison, M.D., Newfield, R., Azar, M.R., Krozowski, Z.S., Funder, J.W., Shackleton, C.H., Bradlow, H.L., Wei, J.Q. et al. (1998) Examination of genotype and phenotype relationships in 14 patients with apparent mineralocorticoid excess. *Journal of Clinical Endocrinology and Metabolism*, **83**, 2244–54.
64 Monder, C., Shackleton, C.H., Bradlow, H.L., New, M.I., Stoner, E., Iohan, F. and Lakshmi, V. (1986) The syndrome of apparent mineralocorticoid excess: its association with 11 beta-dehydrogenase and 5 beta-reductase deficiency and some consequences for corticosteroid metabolism. *Journal of Clinical Endocrinology and Metabolism*, **63**, 550–7.

65 Morineau, G., Sulmont, V., Salomon, R., Fiquet-Kempf, B., Jeunemaitre, X., Nicod, J. and Ferrari, P. (2006) Apparent mineralocorticoid excess: report of six new cases and extensive personal experience. *Journal of the American Society of Nephrology*, **17**, 3176–84.

66 New, M.I. and Wilson, R.C. (1999) Steroid disorders in children: congenital adrenal hyperplasia and apparent mineralocorticoid excess. *Proceedings of the National Academy of Sciences of the United States of America*, **96**, 12790–7.

67 Stewart, P.M., Corrie, J.E., Shackleton, C.H. and Edwards, C.R. (1988) Syndrome of apparent mineralocorticoid excess. A defect in the cortisol-cortisone shuttle. *Journal of Clinical Investigation*, **82**, 340–9.

68 Atanasov, A.G., Ignatova, I.D., Nashev, L.G., Dick, B., Ferrari, P., Frey, F.J. and Odermatt, A. (2007) Impaired protein stability of 11beta-hydroxysteroid dehydrogenase type 2: a novel mechanism of apparent mineralocorticoid excess. *Journal of the American Society of Nephrology*, **18**, 1262–70.

69 Milford, D.V. (1999) Investigation of hypertension and the recognition of monogenic hypertension. *Archives of Disease in Childhood*, **81**, 452–5.

70 Kotelevtsev, Y., Brown, R.W., Fleming, S., Kenyon, C., Edwards, C.R., Seckl, J.R. and Mullins, J.J. (1999) Hypertension in mice lacking 11beta-hydroxysteroid dehydrogenase type 2. *Journal of Clinical Investigation*, **103**, 683–9.

71 Bailey, M.A., Paterson, J.M., Hadoke, P.W., Wrobel, N., Bellamy, C.O., Brownstein, D.G., Seckl, J.R. and Mullins, J.J. (2008) Switch from sodium-dependent to sodium-independent hypertension in the syndrome of apparent mineralocorticoid excess. *Journal of the American Society of Nephrology*, **19**, 47–58.

72 Palermo, M., Delitala, G., Sorba, G., Cossu, M., Satta, R., Tedde, R., Pala, A. and Shackleton, C.H. (2000) Does kidney transplantation normalise cortisol metabolism in apparent mineralocorticoid excess syndrome? *Journal of Endocrinological Investigation*, **23**, 457–62.

73 Robson, A.C., Leckie, C.M., Seckl, J.R. and Holmes, M.C. (1998) 11 Beta-hydroxysteroid dehydrogenase type 2 in the postnatal and adult rat brain. *Brain Research. Molecular Brain Research*, **61**, 1–10.

74 Roland, B.L., Li, K.X. and Funder, J.W. (1995) Hybridization histochemical localization of 11 beta-hydroxysteroid dehydrogenase type 2 in rat brain. *Endocrinology*, **136**, 4697–700.

75 Gomez-Sanchez, E.P. (1986) Intracerebroventricular infusion of aldosterone induces hypertension in rats. *Endocrinology*, **118**, 819–23.

76 Gomez-Sanchez, E.P. and Gomez-Sanchez, C.E. (1992) Central hyper-tensinogenic effects of glycyrrhizic acid and carbenoxolone. *American Journal of Physiology*, **263**, E1125–1130.

77 Hatakeyama, H., Inaba, S., Takeda, R. and Miyamori, I. (2000) 11Beta-hydroxysteroid dehydrogenase in human vascular cells. *Kidney International*, **57**, 1352–7.

78 Souness, G.W., Brem, A.S. and Morris, D.J. (2002) 11Beta-hydroxysteroid dehydrogenase antisense affects vascular contractile response and glucocorticoid metabolism. *Steroids*, **67**, 195–201.

79 Walker, B.R., Connacher, A.A., Webb, D.J. and Edwards, C.R. (1992) Glucocorticoids and blood pressure: a role for the cortisol/cortisone shuttle in the control of vascular tone in man. *Clinical Science (London)*, **83**, 171–8.

80 Hadoke, P.W., Christy, C., Kotelevtsev, Y.V., Williams, B.C., Kenyon, C.J., Seckl, J.R., Mullins, J.J. and Walker, B.R. (2001) Endothelial cell dysfunction in mice after transgenic knockout of type 2, but not type 1, 11beta-hydroxysteroid dehydrogenase. *Circulation*, **104**, 2832–7.

81 Seckl, J.R. and Holmes, M.C. (2007) Mechanisms of disease: glucocorticoids, their placental metabolism and fetal "programming" of adult pathophysiology. *Nature Clinical Practice. Endocrinology & Metabolism*, **3**, 479–88.

82 Bocchi, B., Kenouch, S., Lamarre-Cliche, M., Muffat-Joly, M., Capron, M.H., Fiet, J., Morineau, G., Azizi, M., Bonvalet, J.P.

and Farman, N. (2004) Impaired 11-beta hydroxysteroid dehydrogenase type 2 activity in sweat gland ducts in human essential hypertension. *Hypertension*, **43**, 803–8.

83 Melander, O., Frandsen, E., Groop, L. and Hulthen, U.L. (2003) No evidence of a relation between 11beta-hydroxysteroid dehydrogenase type 2 activity and salt sensitivity. *American Journal of Hypertension*, **16**, 729–33.

84 Poch, E., Gonzalez, D., Giner, V., Bragulat, E. and Coca A., and de La Sierra, A. (2001) Molecular basis of salt sensitivity in human hypertension. Evaluation of renin–angiotensin–aldosterone system gene polymorphisms. *Hypertension*, **38**, 1204–9.

85 Williams, T.A., Mulatero, P., Filigheddu, F., Troffa, C., Milan, A., Argiolas, G., Parpaglia, P.P., Veglio, F. and Glorioso, N. (2005) Role of HSD11B2 polymorphisms in essential hypertension and the diuretic response to thiazides. *Kidney International*, **67**, 631–7.

86 Booth, R.E., Johnson, J.P. and Stockand, J.D. (2002) Aldosterone. *Advances in Physiology Education*, **26**, 8–20.

87 Good, D.W. (2007) Nongenomic actions of aldosterone on the renal tubule. *Hypertension*, **49**, 728–39.

88 Campese, V.M. and Park, J. (2006) The kidney and hypertension: over 70 years of research. *Journal of Nephrology*, **19**, 691–8.

89 Young, M.J., Lam, E.Y. and Rickard, A.J. (2007) Mineralocorticoid receptor activation and cardiac fibrosis. *Clinical Science*, **112**, 467–75.

90 Gomez-Sanchez, E.P. (1997) Central hypertensive effects of aldosterone. *Frontiers in Neuroendocrinology*, **18**, 440–62.

91 Levy, D.G., Rocha, R. and Funder, J.W. (2004) Distinguishing the antihypertensive and electrolyte effects of eplerenone. *Journal of Clinical Endocrinology and Metabolism*, **89**, 2736–40.

92 Grossmann, C., Benesic, A., Krug, A.W., Freudinger, R., Mildenberger, S., Gassner, B. and Gekle, M. (2005) Human mineralocorticoid receptor expression renders cells responsive for nongenotropic aldosterone actions. *Molecular Endocrinology*, **19**, 1697–710.

93 Funder, J.W. (2006) Minireview: aldosterone and the cardiovascular system: genomic and nongenomic effects. *Endocrinology*, **147**, 5564–7.

94 Brilla, C.G. and Weber, K.T. (1992) Mineralocorticoid excess, dietary sodium, and myocardial fibrosis. *Journal of Laboratory and Clinical Medicine*, **120**, 893–901.

95 Sato, A. and Saruta, T. (2004) Aldosterone-induced organ damage: plasma aldosterone level and inappropriate salt status. *Hypertension Research*, **27**, 303–10.

96 Pitt, B., Remme, W., Zannad, F., Neaton, J., Martinez, F., Roniker, B., Bittman, R., Hurley, S., Kleiman, J. and Gatlin, M. (2003) Eplerenone, a selective aldosterone blocker, in patients with left ventricular dysfunction after myocardial infarction. *New England Journal of Medicine*, **348**, 1309–21.

97 Pitt, B., Zannad, F., Remme, W.J., Cody, R., Castaigne, A., Perez, A., Palensky, J. and Wittes, J. (1999) The effect of spironolactone on morbidity and mortality in patients with severe heart failure. Randomized Aldactone Evaluation Study Investigators. *New England Journal of Medicine*, **341**, 709–17.

98 Rocha, R. and Funder, J.W. (2002) The pathophysiology of aldosterone in the cardiovascular system. *Annals of the New York Academy of Sciences*, **970**, 89–100.

99 Swedberg, K., Eneroth, P., Kjekshus, J. and Wilhelmsen, L. (1990) Hormones regulating cardiovascular function in patients with severe congestive heart failure and their relation to mortality. CONSENSUS Trial Study Group. *Circulation*, **82**, 1730–6.

100 Nagata, K., Obata, K., Xu, J., Ichihara, S., Noda, A., Kimata, H., Kato, T., Izawa, H., Murohara, T. and Yokota, M. (2006) Mineralocorticoid receptor antagonism attenuates cardiac hypertrophy and failure in low-aldosterone hypertensive rats. *Hypertension*, **47**, 656–64.

101 Le Menuet, D., Isnard, R., Bichara, M., Viengchareun, S., Muffat-Joly, M., Walker, F., Zennaro, M.C. and Lombes, M. (2001) Alteration of cardiac and renal functions in transgenic mice overexpressing human mineralocorticoid receptor. *Journal of Biological Chemistry*, **276**, 38911–20.

102 Qin, W., Rudolph, A.E., Bond, B.R., Rocha, R., Blomme, E.A., Goellner, J.J., Funder, J.W. and McMahon, E.G. (2003) Transgenic model of aldosterone-driven cardiac hypertrophy and heart failure. *Circulation Research*, **93**, 69–76.

103 Beggah, A.T., Escoubet, B., Puttini, S., Cailmail, S., Delage, V., Ouvrard-Pascaud, A., Bocchi, B., Peuchmaur, M., Delcayre, C., Farman, N. et al. (2002) Reversible cardiac fibrosis and heart failure induced by conditional expression of an antisense mRNA of the mineralocorticoid receptor in cardiomyocytes. *Proceedings of the National Academy of Sciences of the United States of America*, **99**, 7160–5.

104 Funder, J.W. (2004) Is aldosterone bad for the heart? *Trends in Endocrinology and Metabolism*, **15**, 139–42.

105 Funder, J.W. (2005) RALES, EPHESUS and redox. *Journal of Steroid Biochemistry and Molecular Biology*, **93**, 121–5.

106 Mihailidou, A.S. and Funder, J.W. (2005) Nongenomic effects of mineralocorticoid receptor activation in the cardiovascular system. *Steroids*, **70**, 347–51.

107 Duprez, D.A. (2007) Aldosterone and the vasculature: mechanisms mediating resistant hypertension. *Journal of Clinical Hypertension (Greenwich)*, **9**, 13–8.

108 Cai, T.Q., Wong, B., Mundt, S.S., Thieringer, R., Wright, S.D. and Hermanowski-Vosatka, A. (2001) Induction of 11beta-hydroxysteroid dehydrogenase type 1 but not -2 in human aortic smooth muscle cells by inflammatory stimuli. *Journal of Steroid Biochemistry and Molecular Biology*, **77**, 117–22.

109 Hatakeyama, H., Inaba, S. and Miyamori, I. (1999) 11Beta-hydroxysteroid dehydrogenase in cultured human vascular cells. Possible role in the development of hypertension. *Hypertension*, **33**, 1179–84.

110 Brem, A.S., Bina, R.B., King, T.C. and Morris, D.J. (1998) Localization of 2 11beta-OH steroid dehydrogenase isoforms in aortic endothelial cells. *Hypertension*, **31**, 459–62.

111 Ullian, M.E. (1999) The role of corticosteriods in the regulation of vascular tone. *Cardiovascular Research*, **41**, 55–64.

112 Couture, R. and Regoli, D. (1980) Vascular reactivity to angiotensin and noradrenaline in rats maintained on a sodium free diet or made hypertensive with desoxycorticosterone acetate and salt (DOCA/salt). *Clinical and Experimental Hypertension*, **2**, 25–43.

113 White, R.E. and Carrier, G.O. (1986) Alpha 1- and alpha 2-adrenoceptor agonist-induced contraction in rat mesenteric artery upon removal of endothelium. *European Journal of Pharmacology*, **122**, 349–52.

114 Ullian, M.E. and Fine, J.J. (1994) Mechanisms of enhanced angiotensin II-stimulated signal transduction in vascular smooth muscle by aldosterone. *Journal of Cellular Physiology*, **161**, 201–8.

115 Schelling, J.R., DeLuca, D.J., Konieczkowski, M., Marzec, R., Sedor, J.R., Dubyak, G.R. and Linas, S.L. (1994) Glucocorticoid uncoupling of antiogensin II-dependent phospholipase C activation in rat vascular smooth muscle cells. *Kidney International*, **46**, 675–82.

116 Uno, S., Guo, D.F., Nakajima, M., Ohi, H., Imada, T., Hiramatsu, R., Nakakubo, H., Nakamura, N. and Inagami, T. (1994) Glucocorticoid induction of rat angiotensin II type 1A receptor gene promoter. *Biochemical and Biophysical Research Communications*, **204**, 210–5.

117 Chan, J.S., Ming, M., Nie, Z.R., Sikstrom, R., Lachance, S. and Carriere, S. (1992) Hormonal regulation of expression of the angiotensinogen gene in cultured opossum kidney proximal tubular cells. *Journal of the American Society of Nephrology*, **2**, 1516–22.

118 Sim, M.K. and Chan, C.S. (1992) Effect of experimentally-induced hypertension on angiotensin converting enzyme activity in

the aortic endothelium and smooth muscle cum adventitia of the Sprague Dawley rat. *Life Sciences*, **50**, 1821–5.

119 Meggs, L.G., Stitzel, R., Ben-Ari, J., Chander, P., Gammon, D., Goodman, A.I. and Head, R. (1988) Upregulation of the vascular alpha-1 receptor in malignant DOCA-salt hypertension. *Clinical and Experimental Hypertension A.*, **10**, 229–47.

120 Haigh, R.M. and Jones, C.T. (1990) Effect of glucocorticoids on alpha 1-adrenergic receptor binding in rat vascular smooth muscle. *Journal of Molecular Endocrinology*, **5**, 41–8.

121 Longhurst, P.A., Rice, P.J., Taylor, D.A. and Fleming, W.W. (1988) Sensitivity of caudal arteries and the mesenteric vascular bed to norepinephrine in DOCA-salt hypertension. *Hypertension*, **12**, 133–42.

122 Lockette, W., Otsuka, Y. and Carretero, O. (1986) The loss of endothelium-dependent vascular relaxation in hypertension. *Hypertension*, **8**, II61–66.

123 Handa, M., Kondo, K., Suzuki, H. and Saruta, T. (1984) Dexamethasone hypertension in rats: role of prostaglandins and pressor sensitivity to norepinephrine. *Hypertension*, **6**, 236–41.

124 Losel, R., Schultz, A., Boldyreff, B. and Wehling, M. (2004) Rapid effects of aldosterone on vascular cells: clinical implications. *Steroids*, **69**, 575–8.

125 Schmidt, B.M., Oehmer, S., Delles, C., Bratke, R., Schneider, M.P., Klingbeil, A., Fleischmann, E.H. and Schmieder, R.E. (2003) Rapid nongenomic effects of aldosterone on human forearm vasculature. *Hypertension*, **42**, 156–60.

126 Schmidt, B.M., Sammer, U., Fleischmann, I., Schlaich, M., Delles, C. and Schmieder, R.E. (2006) Rapid nongenomic effects of aldosterone on the renal vasculature in humans. *Hypertension*, **47**, 650–5.

127 Yee, K.M. and Struthers, A.D. (1998) Aldosterone blunts the baroreflex response in man. *Clinical Science (London, England)*, **95**, 687–92.

128 Schiffrin, E.L., Lariviere, R., Li, J.S., Sventek, P. and Touyz, R.M. (1995) Deoxycorticosterone acetate plus salt induces overexpression of vascular endothelin-1 and severe vascular hypertrophy in spontaneously hypertensive rats. *Hypertension*, **25**, 769–73.

129 Landmesser, U., Hornig, B. and Drexler, H. (2004) Endothelial function: a critical determinant in atherosclerosis? *Circulation*, **109**, II27–33.

130 Oberleithner, H. (2005) Aldosterone makes human endothelium stiff and vulnerable. *Kidney International*, **67**, 1680–2.

131 Nietlispach, F., Julius, B., Schindler, R., Bernheim, A., Binkert, C., Kiowski, W. and Brunner-La Rocca, H.P. (2007) Influence of acute and chronic mineralocorticoid excess on endothelial function in healthy men. *Hypertension*, **50**, 82–8.

132 Williams, G.H. (2005) Aldosterone biosynthesis, regulation, and classical mechanism of action. *Heart Failure Reviews*, **10**, 7–13.

133 Savoia, C. and Schiffrin, E.L. (2006) Inflammation in hypertension. *Current Opinion in Nephrology and Hypertension*, **15**, 152–8.

134 Kathiresan, S., Larson, M.G., Benjamin, E.J., Corey, D., Murabito, J.M., Fox, C.S., Wilson, P.W., Rifai, N., Meigs, J.B., Ricken, G. et al. (2005) Clinical and genetic correlates of serum aldosterone in the community: the Framingham Heart Study. *American Journal of Hypertension*, **18**, 657–65.

135 Berger, S., Bleich, M., Schmid, W., Cole, T.J., Peters, J., Watanabe, H., Kriz, W., Warth, R., Greger, R. and Schutz, G. (1998) Mineralocorticoid receptor knockout mice: pathophysiology of Na$^+$ metabolism. *Proceedings of the National Academy of Sciences of the United States of America*, **95**, 9424–9.

136 Berger, S., Wolfer, D.P., Selbach, O., Alter, H., Erdmann, G., Reichardt, H.M., Chepkova, A.N., Welzl, H., Haas, H.L., Lipp, H.P. et al. (2006) Loss of the limbic mineralocorticoid receptor impairs behavioral plasticity. *Proceedings of the National Academy of Sciences of the United States of America*, **103**, 195–200.

137 Ronzaud, C., Loffing, J., Bleich, M., Gretz, N., Grone, H.J., Schutz, G. and

Berger, S. (2007) Impairment of sodium balance in mice deficient in renal principal cell mineralocorticoid receptor. *Journal of the American Society of Nephrology*, **18**, 1679–87.

138 Girod, J.P. and Brotman, D.J. (2004) Does altered glucocorticoid homeostasis increase cardiovascular risk? *Cardiovascular Research*, **64**, 217–26.

139 Hammer, F. and Stewart, P.M. (2006) Cortisol metabolism in hypertension. *Best Practice and Research Clinical Endocrinology and Metabolism*, **20**, 337–53.

140 Wintermantel, T.M., Berger, S., Greiner, E.F. and Schutz, G. (2005) Evaluation of steroid receptor function by gene targeting in mice. *Journal of Steroid Biochemistry and Molecular Biology*, **93**, 107–12.

141 Cole, T.J., Blendy, J.A., Monaghan, A.P., Krieglstein, K., Schmid, W., Aguzzi, A., Fantuzzi, G., Hummler, E., Unsicker, K. and Schutz, G. (1995) Targeted disruption of the glucocorticoid receptor gene blocks adrenergic chromaffin cell development and severely retards lung maturation. *Genes and Development*, **9**, 1608–21.

142 Kellendonk, C., Tronche, F., Reichardt, H.M. and Schutz, G. (1999) Mutagenesis of the glucocorticoid receptor in mice. *Journal of Steroid Biochemistry and Molecular Biology*, **69**, 253–9.

143 Reichardt, H.M., Kaestner, K.H., Tuckermann, J., Kretz, O., Wessely, O., Bock, R., Gass, P., Schmid, W., Herrlich, P., Angel, P. *et al.* (1998) DNA binding of the glucocorticoid receptor is not essential for survival. *Cell*, **93**, 531–41.

144 Tronche, F., Opherk, C., Moriggl, R., Kellendonk, C., Reimann, A., Schwake, L., Reichardt, H.M., Stangl, K., Gau, D., Hoeflich, A. *et al.* (2004) Glucocorticoid receptor function in hepatocytes is essential to promote postnatal body growth. *Genes and Development*, **18**, 492–7.

145 Tronche, F., Kellendonk, C., Kretz, O., Gass, P., Anlag, K., Orban, P.C., Bock, R., Klein, R. and Schutz, G. (1999) Disruption of the glucocorticoid receptor gene in the nervous system results in reduced anxiety. *Nature Genetics*, **23**, 99–103.

146 Sainte-Marie, Y., Cat, A.N., Perrier, R., Mangin, L., Soukaseum, C., Peuchmaur, M., Tronche, F., Farman, N., Escoubet, B., Benitah, J.P. *et al.* (2007) Conditional glucocorticoid receptor expression in the heart induces atrio-ventricular block. *FASEB Journal*, **21**, 3133–41.

147 Newell-Price, J., Bertagna, X., Grossman, A.B. and Nieman, L.K. (2006) Cushing's syndrome. *Lancet*, **367**, 1605–17.

148 Whitworth, J.A., Williamson, P.M., Mangos, G. and Kelly, J.J. (2005) Cardiovascular consequences of cortisol excess. *Vascular Health and Risk Management*, **1**, 291–9.

149 Etxabe, J. and Vazquez, J.A. (1994) Morbidity and mortality in Cushing's disease: an epidemiological approach. *Clinical Endocrinology*, **40**, 479–84.

150 Arnaldi, G., Mancini, T., Polenta, B. and Boscaro, M. (2004) Cardiovascular risk in Cushing's syndrome. *Pituitary*, **7**, 253–6.

151 Magiakou, M.A., Smyrnaki, P. and Chrousos, G.P. (2006) Hypertension in Cushing's syndrome. *Best Practice and Research Clinical Endocrinology and Metabolism*, **20**, 467–82.

152 Birkenhager, A.M. and van den Meiracker, A.H. (2007) Causes and consequences of a non-dipping blood pressure profile. *Netherlands Journal of Medicine*, **65**, 127–31.

153 Aizawa, Y., Kamimura, N., Watanabe, H., Aizawa, Y., Makiyama, Y., Usuda, Y., Watanabe, T. and Kurashina, Y. (2006) Cardiovascular risk factors are really linked in the metabolic syndrome: this phenomenon suggests clustering rather than coincidence. *International Journal of Cardiology*, **109**, 213–8.

154 Chew, G.T., Gan, S.K. and Watts, G.F. (2006) Revisiting the metabolic syndrome. *Medical Journal of Australia*, **185**, 445–9.

155 Grundy, S.M. (2007) Metabolic syndrome: a multiplex cardiovascular risk factor. *Journal of Clinical Endocrinology and Metabolism*, **92**, 399–404.

156 Walker, B.R. (2006) Cortisol–cause and cure for metabolic syndrome? *Diabetic Medicine*, **23**, 1281–8.

157 Livingstone, D.E., Jones, G.C., Smith, K., Jamieson, P.M., Andrew, R., Kenyon, C.J. and Walker, B.R. (2000) Understanding the role of glucocorticoids in obesity: tissue-specific alterations of corticosterone metabolism in obese Zucker rats. *Endocrinology*, **141**, 560–3.

158 Livingstone, D.E., Kenyon, C.J. and Walker, B.R. (2000) Mechanisms of dysregulation of 11 beta-hydroxysteroid dehydrogenase type 1 in obese Zucker rats. *Journal of Endocrinology*, **167**, 533–9.

159 Rask, E., Olsson, T., Soderberg, S., Andrew, R., Livingstone, D.E., Johnson, O. and Walker, B.R. (2001) Tissue-specific dysregulation of cortisol metabolism in human obesity. *Journal of Clinical Endocrinology and Metabolism*, **86**, 1418–21.

160 Rask, E., Walker, B.R., Soderberg, S., Livingstone, D.E., Eliasson, M., Johnson, O., Andrew, R. and Olsson, T. (2002) Tissue-specific changes in peripheral cortisol metabolism in obese women: increased adipose 11beta-hydroxysteroid dehydrogenase type 1 activity. *Journal of Clinical Endocrinology and Metabolism*, **87**, 3330–6.

161 van Rossum, E.F. and Lamberts, S.W. (2006) Glucocorticoid resistance syndrome: a diagnostic and therapeutic approach. *Best Practice and Research Clinical Endocrinology and Metabolism*, **20**, 611–26.

162 Charmandari, E., Kino, T., Souvatzoglou, E., Vottero, A., Bhattacharyya, N. and Chrousos, G.P. (2004) Natural glucocorticoid receptor mutants causing generalized glucocorticoid resistance: molecular genotype, genetic transmission, and clinical phenotype. *Journal of Clinical Endocrinology and Metabolism*, **89**, 1939–49.

163 Chrousos, G.P., Detera-Wadleigh, S.D. and Karl, M. (1993) Syndromes of glucocorticoid resistance. *Annals of Internal Medicine*, **119**, 1113–24.

164 Huizenga, N.A., Koper, J.W., De Lange, P., Pols, H.A., Stolk, R.P., Burger, H., Grobbee, D.E., Brinkmann, A.O., De Jong, F.H. and Lamberts, S.W. (1998) A polymorphism in the glucocorticoid receptor gene may be associated with and increased sensitivity to glucocorticoids *in vivo*. *Journal of Clinical Endocrinology and Metabolism*, **83**, 144–51.

165 van Rossum, E.F. and Lamberts, S.W. (2004) Polymorphisms in the glucocorticoid receptor gene and their associations with metabolic parameters and body composition. *Recent Progress in Hormone Research*, **59**, 333–57.

166 Lin, R.C., Wang, X.L. and Morris, B.J. (2003) Association of coronary artery disease with glucocorticoid receptor N363S variant. *Hypertension*, **41**, 404–7.

167 Murray, J.C., Smith, R.F., Ardinger, H.A. and Weinberger, C. (1987) RFLP for the glucocorticoid receptor (GRL) located at 5q11–5q13. *Nucleic Acids Research*, **15**, 6765.

168 Koper, J.W., Stolk, R.P., de Lange, P., Huizenga, N.A., Molijn, G.J., Pols, H.A., Grobbee, D.E., Karl, M., de Jong, F.H., Brinkmann, A.O. et al. (1997) Lack of association between five polymorphisms in the human glucocorticoid receptor gene and glucocorticoid resistance. *Human Genetics*, **99**, 663–8.

169 van Rossum, E.F., Feelders, R.A., van den Beld, A.W., Uitterlinden, A.G., Janssen, J.A., Ester, W., Brinkmann, A.O., Grobbee, D.E., de Jong, F.H., Pols, H.A. et al. (2004) Association of the ER22/23EK polymorphism in the glucocorticoid receptor gene with survival and C-reactive protein levels in elderly men. *American Journal of Medicine*, **117**, 158–62.

170 van Rossum, E.F., Voorhoeve, P.G., Velde, S.J., Koper, J.W., Delemarre-van de Waal, H.A., Kemper, H.C. and Lamberts, S.W. (2004) The ER22/23EK polymorphism in the glucocorticoid receptor gene is associated with a beneficial body composition and muscle strength in young adults. *Journal of Clinical Endocrinology and Metabolism*, **89**, 4004–9.

171 Ross, R. (1999) Atherosclerosis is an inflammatory disease. *American Heart Journal*, **138**, S419–420.

172 Colao, A., Pivonello, R., Spiezia, S., Faggiano, A., Ferone, D., Filippella, M., Marzullo, P., Cerbone, G., Siciliani, M. and Lombardi, G. (1999) Persistence of increased cardiovascular risk in patients with Cushing's disease after five years of successful cure. *Journal of Clinical Endocrinology and Metabolism*, **84**, 2664–72.

173 Hermanowski-Vosatka, A., Balkovec, J.M., Cheng, K., Chen, H.Y., Hernandez, M., Koo, G.C., Le Grand, C.B., Li, Z., Metzger, J.M., Mundt, S.S. *et al.* (2005) 11betaHSD1 inhibition ameliorates metabolic syndrome and prevents progression of atherosclerosis in mice. *Journal of Experimental Medicine*, **202**, 517–27.

174 del Rincon, I., O'Leary, D.H., Haas, R.W. and Escalante, A. (2004) Effect of glucocorticoids on the arteries in rheumatoid arthritis. *Arthritis and Rheumatism*, **50**, 3813–22.

175 Morton, N.M., Holmes, M.C., Fiévet, C., Staels, B., Tailleux, A., Mullins, J.J., Seckl, J.R. (2001) Improved lipid and lipoprotein profile, hepatic insulin sensitivity, and glucose tolerance in 11beta-hydroxysteroid dehydrogenase type 1 null mice. *Journal of Biological Chemistry*, **276**, 41293–300.

176 Morton, N.M., Paterson, J.M., Masuzaki, H., Holmes, M.C., Staels, B., Fiévet, C., Walker, B.R., Flier, J.S., Mullins, J.J., Seckl, J.R. (2004) Novel adipose tissue-mediated resistance to diet-induced visceral obesity in 11 beta-hydroxysteroid dehydrogenase type 1-deficient mice. *Diabetes*, **53**, 931–8.

177 Endemann, D.H., Touyz, R.M., Iglarz, M., Savoia, C., Schiffrin, E.L. (2004) Eplerenone prevents salt-induced vascular remodeling and cardiac fibrosis in stroke-prone spontaneously hypertensive rats. *Hypertension*, **43**, 1252–7.

2
Sex Steroid Hormones

Vera Regitz-Zagrosek, Eva Becher, Shokoufeh Mahmoodzadeh, and Carola Schubert

2.1
Sex Differences in Cardiovascular Physiology and Disease and Effect of Hormone Therapy

Major sex and gender differences have been found in cardiovascular growth, cardiac and vascular function, in the age-dependent manifestation and mortality in myocardial infarction, in the development and consequences of myocardial and vascular hypertrophy and hypertension as well as clinical manifestations and outcome of heart failure and arrhythmia [1]. It has been assumed that these sex differences are due to differences in sex steroid hormone levels (SH) and in the function of sex steroid hormone receptors (SHRs). Steroid hormone levels have been linked with cardiovascular outcomes in women and men. In women, clinical syndromes that are associated with alterations in sex hormones are associated with increased numbers of cardiovascular complications and events. The polycystic ovarian syndrome in young women is associated with an increased risk of atherosclerosis. The number of cardiovascular events in women increases strikingly with menopause [2]. In men, decreased estrogen receptor (ER) expression or lack of endogenous estrogen formation by aromatase has been linked with early myocardial infarction [3]. However, even though alterations in estrogen levels are linked to cardiovascular disease, the latest outcome trials with hormone replacement therapy (HRT) made it clear that replacement of hormones does not lead to clinical cardiovascular benefit [4, 5]. Outcome may be influenced by differences between exogenous and endogenous hormones, first-pass effects in the liver, effects of synthetic progesterones, which must, in most cases, be coadministered with estrogen, basal hormone status, phytoestrogens in the diet and timing of hormone intake, and these factors may have prevented therapeutic benefit from HRT so far [6]. Nevertheless, the manifold interactions of estrogens and androgens with cellular functions may severely influence the response of the cardiovascular system to pathophysiological stimuli. Therefore, we will discuss in this chapter the effects

---

Cardiovascular Hormone Systems. Edited by Michael Bader
Copyright © 2008 WILEY-VCH Verlag GmbH & Co. KGaA, Weinheim
ISBN: 978-3-527-31920-6

2.2
Sex Steroids: Estradiol, Estrone, Testosterone and Progesterone

2.2.1
Synthesis and Endocrine Physiology

The naturally occurring estrogens 17β-estradiol, estrone and estriol are C_{18} steroids derived from the cholesterol molecule. The side-chain of cholesterol is first cleaved by cytochrome P450, leading to pregnenolone, which is followed by an aromatization, catalyzed by the P450 aromatase monooxygenase enzyme complex, resulting in progesterone. In three consecutive hydroxylating reactions estrone and estradiol are formed from their obligatory precursors androstenedione and testosterone, respectively. The final hydroxylation step does not require enzymatic action and is not product-sensitive [7]. Several plant compounds have structural and functional similarities, and are therefore referred to as phytoestrogens (e.g. genestein and daidzein are isoflavoids found in soybeans and clover). Some studies suggest that diets rich in phytoestrogens protect against different types of cancer, cardiovascular disease and osteoporosis [8].

2.2.2
Age-Dependent Blood Levels of Sex Hormones in Females and Males

The primary source of endogenous estradiol in women are the theca and granulosa cells of the ovaries, and the luteinized derivates of these cells. Estrone and estriol are primarily formed in the liver from estradiol. However, aromatase activity has also been detected in heart, muscle [9], fat [10], nervous tissue [11] and the Leydig cells [12] of the testes. During puberty estradiol levels in serum of girls rise up to 15–35 pg/ml (55–128 pmol/l). During menstrual cycles estradiol production varies cyclically, with the highest rates and serum concentrations in the preovulatory phase at 250–500 pg/ml (200–400 pmol/l) [13]. In postmenopausal woman, serum estradiol concentrations are often lower than 20 pg/ml (73 pmol/l) and most of the estradiol is formed by extragonadal conversion of testosterone. These estrogen levels are comparable with estrogen serum concentrations in age-matched men [14].

Production rates of testosterone and androstenedione average about 6000 and 3000 µg/day. This results in plasma testosterone levels of 300–100 ng/dl and levels of dihydroxytestosterone around 50 ng/dl. All about estrone production (average about µg/day) can be accounted for by formation from circulating precursors. The mean estrogen production rate is about 45 µg/day; about 35% of this amount is derived from circulating testosterone, 50% is derived from the weak estrogen estrone and 15% is secreted directly into the circulation by the testis. This results in plasma estrogen levels of less than 50 pg/ml and plasma estrone levels of less than 80 pg/ml in men. Circulating dihydroxytestosterone is formed principally in the androgen target tissues and estrogen formation takes place in many tissues, the most significant being adipose tissue. The overall rate of extraglandular estrogen formation increases with increasing amounts of fat tissue and with age [15].

2.2.3
Levels of Estradiol and Testosterone in Rodents

Serum estradiol levels in normal female rats vary during the estrous cycle between 61 and 38 pg/ml [16]. Physiological concentrations in mice are between 40 and 122 pg/ml [17].

2.3
Sex Hormone Receptors: Structure and Function

The biological effects of steroid hormones are mediated by intracellular receptors that directly mediate the action of their cognate hormone [18, 19]. SHRs belong to the class I nuclear receptor superfamily that includes the steroid receptors for estradiol: ERs (α and β); cortisol: glucocorticoid receptors (GRs) (α and β); progesterone: progesterone receptors (PRs) (A and B); dihydroxytestosterone: androgen receptors (ARs) (α and β); and aldosterone: mineralocorticoid receptors (MRs) (a and b) (see reviews, see [20, 21]). The human GR was cloned and sequenced in the early 1980s [22], and subsequently MR [23], ER [24, 25], PR [26] and AR [27, 28]. Splice variants of steroid receptors and receptor isoforms generated by differential promoter usage have been described for nearly all steroid receptors; however, the distinctive role of each of those variants is not known in detail [29–35].

The SHRs share common structural features and are organized in different domains. (i) The variable N-terminal A/B domain of SHRs contains a ligand-independent transcriptional activating function [activating function (AF)-1]. This region is the target for phosphorylation mediated by different signaling pathways (for a review, see [36]. (ii) The highly conserved, cysteine-rich DNA-binding domain (DBD) or region C, is comprised of two zinc-finger motifs, and is responsible for binding the receptor to specific DNA sequences known as hormone response elements (HREs) that are located within the promoter of target genes [37, 38]. (iii) The "hinge" domain or D region is not well conserved among the different receptors and controls the translocation of receptors to the nucleus [39]. (iv) The ligand-binding domain located at the C-terminal half of the receptor (E region) is moderately conserved and contains a second transcriptional activating function domain termed AF-2, which is strictly ligand dependent and highly conserved among members of the nuclear receptor superfamily. This region mediates hormone binding, nuclear translocation, and homo- and heterodimerization of receptors [40] (for a review, see [41]). (v) The end of the C-terminal part of the receptors can harbor the F domain or not. This region is thought to regulate ligand selection (agonists versus antagonists) and receptor half-life (proteasomal degradation) [42, 43].

2.3.1
Hormone Response Elements

Steroid receptors regulate transcription by binding as homo- or heterodimers to specific DNA sequences in target genes referred to as HREs [37, 44]. In general, HREs are composed of variably spaced hexameric half sites (AGGTCA) arranged as direct, indirect or inverted repeats separated by three nucleotides [37, 44]. While AR, GR and PR subfamily recognize the core motif AGAACA, ER recognizes the consensus AGGTCA [45].

2.3.2
Molecular Mechanisms of Nuclear Receptor Activation

2.3.2.1 Ligand-Dependent Transcription
The binding of the ligand to the receptors induces the dissociation of heat-shock proteins, nuclear translocation, receptor dimerization and binding to DNA response elements in target gene promoters, which is called the classical ligand (hormone)-dependent pathway or HRE-dependent genomic action [46, 47] (Figure 2.1a).

Steroid receptors can also regulate gene expression without binding directly to HREs by mechanisms in which ligand-bound SHRs interact directly with other transcription factors (TF), such as AP-1, SP-1 and NF-κB, a mechanism generally referred to as "transcriptional cross-talk" (Figure 2.1b) (for reviews, see [21, 48, 49]).

2.3.2.2 Ligand-Independent Genomic Pathway: Cross-Talk Between SHRs and Other Signal Transduction Pathways
In addition to ligand-mediated activation, the transcriptional activity of a SHR can also be affected in the absence of its cognate ligands by phosphorylation of the receptors themselves or of their coregulatory proteins (Figure 2.1c; see also review [50]). Stimulation of cellular activity by growth factors (GF), such as epidermal growth factor (EGF) and insulin-like growth factor (IGF)-1 or the intracellular effector analog 8-bromo-cAMP (cAMP analogs) can activate SHR transactivation in a ligand-independent manner accompanied by an increased receptor phosphorylation [51–53].

ERs have been found to be phosphorylated at different sites by various kinase pathways [mitogen-activated protein kinase (MAPK)/extracellular signal-regulated kinase (ERK) 1/2, protein kinase A (PKA)] after stimulation with

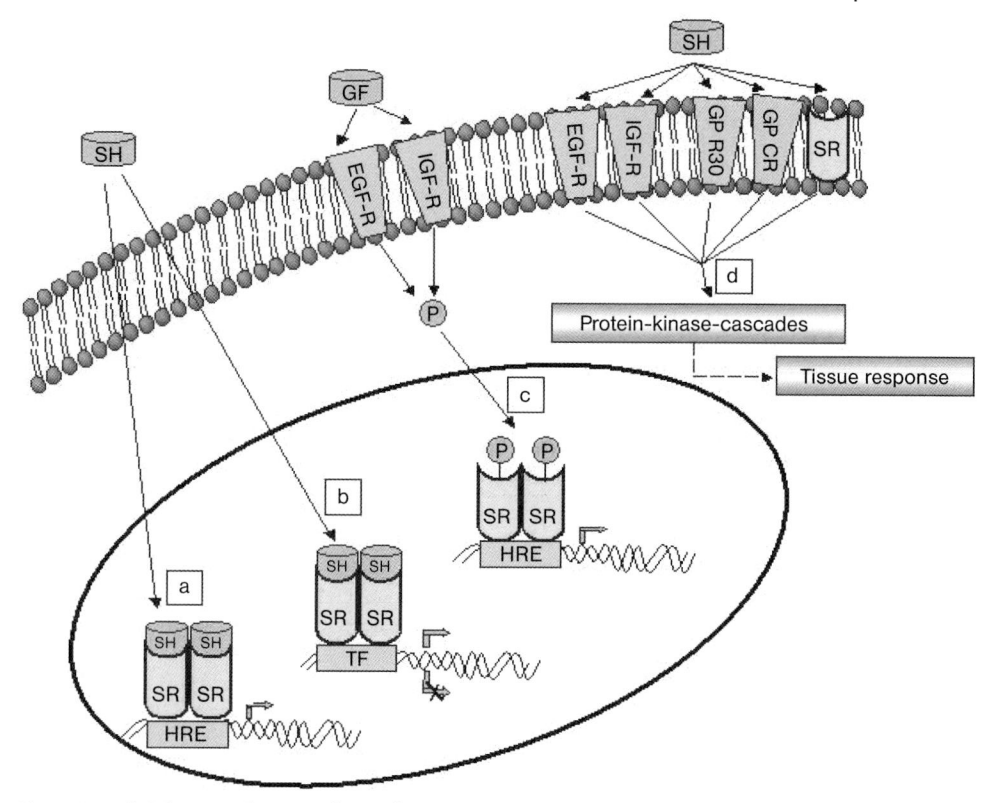

Figure 2.1a–d Schematic diagram of steroid receptor regulation of gene expression (for detailed descriptions see Section 2.3.2): (a) classical ligand-dependent genomic action, (b) nonclassical ligand-dependent genomic action, (c) ligand-independent genomic pathway and (d) nongenomic pathway.

EGF or IGF [54–59]. There is also some evidence that PR can be activated in the absence of progesterone by phosphorylation in response to EGF or cAMP [52]. It has been reported that AR can be phosphorylated by MAPK and PKA [53, 60]. Moreover, it has been reported that the phosphorylation of Ser236 by PKA, within the DBD, regulates the dimerization of ERα [61].

In addition to the modifications of the receptors themselves, nuclear coregulator proteins can also be phosphorylated by a variety of kinases leading to altered transcriptional activity of SHRs (for a review, see [21]). Phosphorylation may enhance interaction of coactivators with SHRs, modulate their efficiency to recruit histone acetyltransferase complexes, and enzymatic activity. Steroid receptor coactivator (SRC)-1 contains consensus sequences for ERK1/2, and EGF stimulation results in ERK1/2-mediated phosphorylation of SRC-1, which potentiates PRb transcriptional activity [62]. Phosphorylation of coactivators can also lead to their inactivation, which reduces its interaction with SHRs, resulting in repression of the SHR transactivation [63].

2.3.2.3 Nongenomic Pathway

It is widely accepted that almost all members of the steroid hormone family, namely progesterone, estrogen, androgen and glucocorticoids, can exhibit rapid membrane effects (within seconds or a few minutes) independent of the classical gene activation pathways of steroid action [64–66]. Nongenomic functions of steroids are mediated either by classical membrane-associated steroid receptors [67–71] or through membrane receptors that are not part of the nuclear receptor superfamily, such as GPR30, EGF receptor, IGF-1 receptors and G-protein-coupled receptors [72–75], (Figure 2.1d). These effects range from changes in activity of adenylyl cyclase, MAPKs, phosphatidylinositol 3-kinase and the increases in intracellular calcium concentrations [76–80].

2.3.3
Role of Coregulators in Transcriptional Regulation by SHRs

The transcriptional activity of SHRs is, in addition to the hormones, also regulated by several regulatory proteins called coactivators and corepressors [81, 82]. Biochemical and structural studies have shown that coactivators can physically interact with AF-2 region of receptors through highly conserved LxxLL motifs (also referred to as "nuclear receptor boxes"). More than 100 coactivator proteins have been identified for SHRs which act in a tissue-specific manner.

In general, in the absence of ligand, SHRs preferentially interact with corepressor proteins, such as NCoR and SMRT, to repress basal transcription of target genes [83]. Nevertheless, it has also been reported that ER upon ligand binding can suppress the basal transcription of target genes through interaction with corepressor proteins. It seems that these nuclear corepressors compete with coactivator proteins such as SRC by displacing them.

2.3.4
Role of Proteasomal Degradation Pathway in SHR-Mediated Gene Transcription

In the ubiquitin–proteasome pathway, proteins are targeted for degradation by different modification steps. Several recent studies have shown the involvement of the ubiquitin–proteasome pathway in ligand-dependent nuclear receptor degradation, which appears to be required to sustain the transcriptional activation of the SHR. The findings that E3-type ubiquitin ligases, components of the ubiquitin–proteasome degradation pathway, act as coactivators of ligand-bound SHRs, that the degradation of ERα can be blocked by specific inhibitors of proteasome [84] and that other coactivators involved in the ERα-mediated transcription such as SRC-1, SRC-3 and TIFII are also degraded (or co-degraded with ERα) through the proteasome pathway, support the hypothesis that proteasome-mediated degradation of steroid receptors plays an important role in contributing to (SHR-) ERα-mediated transcription.

2.3.5
Receptor Activity and Availability in the Cardiovascular System

ERα expression has been detected in several tissues including cardiovascular organs, with considerably different expression levels among these tissues [85]. Although different molecular mechanisms, such as post-transcriptional and translational, could be involved in the control of ER gene expression in the cell/tissue, several studies have indicated that the transcription of the ERα gene plays an important role in regulating the expression of ERα in different tissues. Previous reports have revealed that the human ERα mRNA is transcribed from at least seven different promoters with unique 5'-untranslated regions (UTRs) (A, B, C, D, E, F and T) (Figure 2.2) [86, 87]. These multiple promoters are utilized in a cell- and tissue-specific manner, strongly contributing to the regulation of ERα expression [88]. For example in endometrium, the predominant 5'-UTR variants are A and C, whereas C and F are the major forms in ovaries [85], and only F in osteoblasts [89]. A recent study indicated the existence of the 5'-UTR or promoter variants A, B, C and F in the human myocardium; however, variant F is the dominant form (Mahmoodzadeh et al., personal communication). These results suggest that the expression of the ERα gene is predominantly controlled by promoter F in the human heart. It appears that in each cell/tissue, there are a variety of specific factors that interact with the ERα promoter with trans-activating or trans-repressing functions, which also affect the regulation of transcription of the ERα gene [90, 91].

2.3.6
Localization of SHRs in the Heart and Vessels of Rodent and Men

ER, AR and GR have been identified in both vascular endothelial and smooth muscle cells, as well as in cardiac fibroblasts and myocytes, in human and rodents

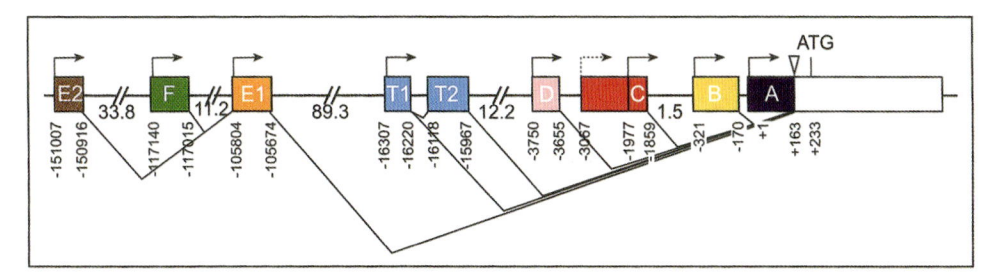

Figure 2.2 Illustration of the different promoter variants of the human ERα gene. (Adapted from [86]).

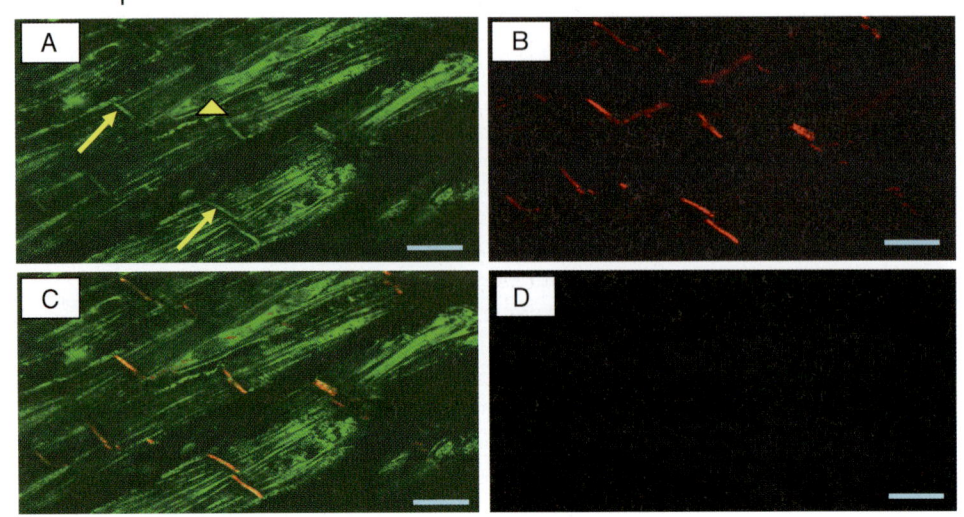

Figure 2.3 Detection of ERα in 5-μm paraffin sections of the left ventricle of a control human heart by immunofluorescent staining and confocal laser scanning microscopy. (a and b) The same section stained for (a) ERα (fluorescein isothiocyanate, green) and (b) β-catenin (Cy3, red). (c) Merged image of (a) and (b). Arrows indicate the intercalated disks and triangle indicates a nucleus. (d) Negative control – a myocardial section in which the primary antibodies were omitted. All scale bars represent 10 μm.

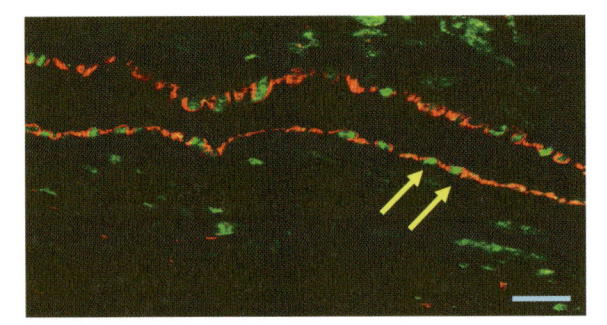

Figure 2.4 ERα localization in a vessel of a human hearts. The image shows the costaining of ERα (fluorescein isothiocyanate, green) and vimentin (Cy3, red) in a vessel of a normal heart. Arrows indicate the ERα-positive ECs.

[92–99]. In the heart of human and rats, ERα was visualized by confocal immunofluorescence microscopy, and localized to the cytoplasm, sarcolemma and intercalated disks of cardiomyocytes (Figure 2.3) [94, 96, 99], and in the nuclei of fibroblasts, endothelial cells (ECs) (Figure 2.4) and myocytes (Figure 2.3) [95, 99].

2.4
Receptor-Independent Effects of Sex Steroids

ERs are considered to mediate the protective effects of 17β-estradiol on the cardiovascular system, including rapid vasodilatation, reduction of vessel wall responses to injury and decreasing the development of atherosclerosis [77, 100]. However, the fact, that 17β-estradiol exhibits these protective effects against vascular injury in mice that lack either ERα [101] or ERβ [102] as well as in double-knockout mice (ERα$^{-/-}$ERβ$^{-/-}$) [103] suggests that the inhibitory effects of 17β-estradiol may be, at least in part, ER-independent. Several studies provided evidence that the inhibitory effects of estradiol on vascular smooth muscle cell (VSMC), cardiac fibroblast and mesangial cell migration, proliferation, and extracellular matrix production are mediated by a ER-independent mechanisms involving estradiol metabolisms [104–106]. These studies showed that the protective effects of estradiol are mediated by the hydroxylation of estradiol to catecholestradiols, catalyzed by CYP450 isoforms, followed by methylation of catecholestradiols to methoxyestradiols, catalyzed by the enzyme catechol-O-methyltransferase. The catecholestradiols and methoxyestradiols have little or no affinity for ERs [104].

2.5
Sex Hormone Effects on Cardiovascular Cells and Organs

2.5.1
Sex Hormone Effects on Different Cardiovascular Cell Types

2.5.1.1 **Cardiomyocytes**
Cardiac myocytes express functional ERs. Immunofluorescence demonstrated ER protein expression in both female and male rat cardiac myocytes [94]. Furthermore, estradiol induces a significant increase in expression of ERα/β and PR in cardiac myocytes. These data suggest that gender-based differences in cardiac diseases may in part be due to direct effects of estrogen on the heart [95]. For example, estradiol has been shown to prevent apoptosis in cardiac myocytes and could thereby attenuate the loss of cardiac myocytes in heart failure [107]. The protective effect of estrogen is mediated by inhibition of reactive oxygen species and differential regulation of p38 kinase isoforms [108, 109]. In addition, physiological concentrations of estradiol activate the protein kinase B/Akt via ERα and reduce thereby apoptosis [110]. The antiapoptotic effect of estrogen has been shown in an coronary occlusion model in mice and in cultured cardiomyocytes [110]. Furthermore, estrogen opposes mechanisms that lead to cardiac hypertrophy. 17β-Estradiol prevents angiotensin (Ang) II- or endothelin-1-induced new protein synthesis, skeletal muscle actin expression, and increased surface area in cultured rat cardiomyocytes, via inhibition of calcineurin phosphatase activity and induction of MCIP1, which is an inhibitor of calcineurin activity [109].

2.5.1.2 Fibroblasts

Cardiac fibroblasts have been identified as estrogen target cells in rat [95], mice and human hearts [111]. Controversial effects of estrogens on fibroblast functions are described: on the one hand, estrogen enhances proliferative capacity of fibroblasts via ER- and MAPK-dependent mechanism [112]. A stimulating effect on cardiac fibroblast proliferation has also been shown by estrone, whereas 17β-estradiol showed only a weak stimulatory capacity [113]. Progesterone has been shown to inhibit DNA synthesis in neonatal cardiac fibroblasts [114]. In contrast to the above-mentioned results, it has been shown in human male and female cardiac fibroblasts that 17β-estradiol, its metabolites and progesterone inhibit cardiac fibroblasts growth in a gender-independent fashion. Estradiol and progesterone inhibit extracellular matrix (collagen) synthesis by cardiac fibroblasts [105]. Estradiol inhibits tumor necrosis factor (TNF)-α production and limits the deleterious intercellular adhesion molecule (ICAM)-1-mediated binding of leukocytes to injured myocardium [115]. Thus, it has been suggested, that HRT using 17β-estradiol and progesterone may protect postmenopausal women against cardiovascular disease by inhibiting cardiac fibroblasts growth and cardiac remodeling. Finally, there is evidence that phytoestrogens inhibit cardiac fibroblasts growth, and may be clinically useful as a substitute for feminizing estrogens in preventing cardiovascular disease in both women and men [105].

2.5.1.3 Endothelial Cells

The above-mentioned upregulation of Akt pathway by estrogens leads to activation of endothelial nitric oxide (NO) synthase and vasodilation [116]. Estrogens cause short-term vasodilatation by increasing the formation and release of NO and prostacyclin in ECs [117]. A protective role of estrogen might be mediated by inhibition of TNF-α-induced apoptosis, which has been shown in ECs [118], and promotion of angiogenic activity, which has also been shown *in vitro* [119]. Furthermore 17β-estradiol increases EC water content and elasticity, which might also contribute to vasoprotective mechanisms postulated for estradiol [120]. Estradiol blocks the induction of CD40 and CD40 ligand expression on ECs and prevents neutrophil adhesion by an ERα-mediated pathway [121].

2.5.1.4 Vascular Smooth Muscle Cells

Estrogens reduce vascular smooth muscle tone by opening specific calcium channels by a mechanism which is dependent on guanosine monophosphate [122] and inhibits VSMC proliferation [123].

2.5.1.5 Platelets

Platelets contribute to the development of thrombotic events, and possibly to the development of atherosclerosis, by providing an active surface for procoagulant reactions and by secreting vasoactive and mitogenic cytokines. Both megakaryocytes, precursors of platelets and circulating platelets contain ERα/β and AR [124–126]. It has been shown that platelet aggregation and secretion changes with sexual maturation in pigs [127].

2.5.2
Sex Hormone Effects on the Heart

Cardiac hypertrophy is an important adaptive response to situations of increased workload, which enables the heart to increase its contractile power [128]. At increased load, the heart undergoes a profound and sex-specific physical remodeling that, based on the type of stimuli and duration, results in two classes of cardiac geometries and functional phenotypes [128]. If a pathologic stimulus persists, depending on the individual, maladaptive hypertrophy becomes irreversible, and frequently leads to cardiac systolic and diastolic dysfunctions and manifestation of heart failure. In contrast, in nondiseased or physiologic states such as exercise and pregnancy, heart hypertrophy fulfills its adaptive function, being reversible and beneficial without significant long-term detrimental effects on cardiac function [128]. Genetic and epigenetic mechanisms, including secreted substances (e.g. sexual hormones) released in each of these situations, determine the outcome of malignant or physiologic hypertrophy [128].

Animal models use different interventions to access cardiac remodeling. Most of this models employed rodents, especially mice and rats, but certain studies require larger animals like sheeps, dogs or pigs. In rodent models different forms of voluntary (cage wheel running) or forced (treadmill, swimming) exercise are used to induce a physiological/beneficial kind of cardiac hypertrophy. Pathological hypertrophy can be provoked by pressure overload induced by transverse aortic constriction, volume overload induced by aorto-caval shunt operation or subtotal nephrectomy in combination with salt- and water-retaining drugs. For analysis of the cardiac adaptation to ischemia and myocardial infarction different strategies are used. *In vivo* ligation of a coronary artery leads to a permanent infarction of the affected part of the heart resulting in the formation of scar tissue and hypertrophic adaptation of the remaining myocardium. Another approach is the ischemia/reperfusion model, which investigates very early reactions of the explanted heart to hypoxemia. These different animal models reveal numerous genes/proteins that are regulated by sexual hormones [129].

2.5.2.1 Effects of Exercise
Physical exercise leads to a mild reversible cardiac hypertrophy with increased capillary density which was strongly inhibited by testosterone [130]. During physical exercise this microvascular impairment may trigger an imbalance between the myocardial oxygen supply and demand [130]. Furthermore, androgenic steroid administration to exercised rats produces a further rightward shift in the left ventricular end diastolic pressure–volume relationship associated with an increased diameter intercept [131]. Loss of ovarian hormones causes adverse cardiac and aortic wall remodeling, including cardiomyocyte hypertrophy, myocardial interstitial reparative fibrosis and vascularization impairment with loss of cardiomyocytes, and aortic tunica media hypertrophy [132]. Exercise training attenuates these effects by reducing blood pressure and cardiac hypertrophy, and increasing myocardial vascularization [132]. Voluntary cage running is associated with higher

performance and a greater degree of hypertrophy in female than in male mice [133].

2.5.2.2 Pressure Overload

The early genomic response to pressure overload induced by transverse aortic constriction reveals that large numbers of genes are regulated in a sex-specific manner [134]. These genes are associated with biological functions including signal transduction, transcription, stress response, and extracellular matrix and cytoskeletal remodeling [134]. It has been claimed that sex hormones mediate the disparity between sexes in cardiovascular morphology, function, morbidity and mortality [135]. Experimental data from clinical studies and animal models suggest that estrogen limits hypertrophy and fibrosis in animal and human hearts [1, 129], whereas testosterone facilitates hypertrophy [136]. Estrogen deficiency potentiates, while estrogen replacement attenuates the development of both right and left ventricular hypertrophy in different rodent models [129]. Estradiol replacement in ovariectomized mice with pressure overload leads to a significant lower degree of left ventricular hypertrophy compared to placebo-treated animals [137, 138]. Pressure overload experiments with ER knockout mice indicate that estrogen plays an important role in attenuating the hypertrophic response in females [17, 138].

2.5.2.3 Hypertension

Left ventricular hypertrophy in spontaneously hypertensive rats is reduced by treatment with estrogen or specific ERα agonist [139]. Chronic administration of estrogen in rats with sustained heart failure reduced total peripheral resistance and left ventricular end diastolic pressure. Endogenous estrogens inhibit the activity of the renin–angiotensin system (RAS) [135]. Thus, estrogens can inhibit cardiac hypertrophy by counteracting hypertension, by direct effects on the heart and by triggering the release of cardioprotective factors [135]. Estrogens promote vasodilatation and cause favorable changes in the lipid profile [135].

2.5.2.4 Volume Overload

Volume load induced by aorto-caval shunt operation leads to concentric hypertrophy in female rats, but eccentric hypertrophy with dilatation in males. Chronic volume overload in ovariectomized females leads to a more extensive cardiac remodeling compared to intact females, characterized by significantly greater left ventricular hypertrophy and a significant increase in left ventricular dilatation [140]. This effect can be reversed by dietary phytoestrogens [141]. The extent of myocardial remodeling and decrease in left ventricular function in volume-loaded ovariectomized females was comparable to those changes reported for males with symptomatic heart failure, while intact females maintained chronic compensated ventricular function [140]. These findings suggest an influence of circulating ovarian hormones on the pattern of myocardial remodeling resulting from chronic volume overload [140].

2.5.2.5 Myocardial Infarction

After myocardial infarction high testosterone levels enhance acute myocardial inflammation, adversely affecting myocardial healing and early remodeling, as indicated by increased cardiac rupture, and possibly causing deterioration of cardiac function [142]. In contrast, estrogen prevents worsening of cardiac function and remodeling after myocardial infarction [129, 142, 143]. After ovariectomy physiologic replacement with 17β-estradiol reduces infarct size and cardiomyocyte apoptosis during the early phase after myocardial infarction [110, 144]. One probable mechanism of this protective effect could be the estradiol-mediated activation of the Akt signaling pathway within the myocardium [110]. Furthermore, estradiol preserves the integrity of ischemic tissue by augmenting the mobilization and incorporation of bone marrow-derived endothelial progenitor cells into sites of neovascularization [145]. ERβ seems to be the receptor involved in the protective effects of estradiol in the infarcted heart [146]. ERβ deficiency results in prolonged ventricular repolarization and decreased ventricular automaticity in female mice with chronic myocardial infarction [147]. Furthermore, systemic deletion of ERβ in female mice increases mortality, aggravates clinical and biochemical markers of heart failure, and contributes to impaired expression of Ca^{2+}-handling proteins in chronic heart failure after myocardial infarction [148].

2.5.3
Sex Hormone Effects on Atherosclerosis, Plaque Formation and Rupture

A protective role of estrogen against atherosclerosis is suggested by the finding that estrogen treatment reduced the progression of coronary artery atherosclerosis in oophorectomized monkeys [149]. In an animal model with apolipoprotein E deficiency, which leads spontaneously to atherosclerosis, pretreatment with 17β-estradiol leads to a significant inhibition of vascular inflammation and reduction of atherosclerotic lesions after Ang II treatment [150] and reduces formation of aortic aneurysms [151]. Similar protective effects of estrogen could also be observed in animal models with atherosclerosis after streptozotocin induced hypergylcemia [152] or by N^G-nitro-L-arginine methyl ester-induced inhibition of NO [153]. Accumulating evidence suggests that the antiatherosclerotic effects of estrogen can be attributed, at least in part, to anti-inflammatory actions. It has been shown that estradiol inhibits leukocyte adhesion and transendothelial migration [154] and reduces expression of monocytic chemoattractant protein (MCP)-1 in the descending thoracic aorta in rabbits [155]. Furthermore it has been demonstrated, that estrogen inhibits NF-κB-mediated induction of proinflammatory genes, which are involved in vascular inflammation and atherosclerosis, including ICAM-1, vascular cell adhesion molecule-1, E-selectin, MCP-1 and macrophage colony stimulating factor [151, 156, 157]. In contrast, the expression of both peroxisome proliferator-activated receptor (PPAR) α and γ are significantly increased by estrogen treatment [150]. PPARs play an important anti-inflammatory and defensive role [158]. Thus, upregulation of PPARs may contribute, at least in part, to the vasculoprotective effects of estrogen.

Despite theses findings, the question whether estrogen treatment in the post-menopausal women prevents atherosclerosis remains controversial [159]. It is hypothesized that the time point of hormone replacement is crucial for the outcome of atherosclerotic disease: in early-stage atherosclerosis hormone replacement might act beneficial on vasodilation and anti-inflammatory processes. In established atherosclerosis with already proliferating VSMCs, existing foam cells and established atherosclerotic plaque the deleterious effects might come to the fore: hormones inhibit vasodilatation and act proinflammatory, and increase the plaque instability by upregulation of matrix metalloproteinases and neovascularization [6, 160, 161].

2.5.4
Sex Hormone Effects on Systemic and Circulating Mediators of the Cardiovascular System

The direct effects of estradiol on the heart may be amplified by estradiol-induced changes in circulating and local factors such as components of the RAS, bradykinin, homocysteine and endothelin. In addition, estradiol downregulates the expression of angiotensin-converting enzyme in serum, as well as in the vasculature [162], and decreases renin release and Ang II formation [163]. As Ang II is a potent mitogen for VSMCs and inductor of vascular remodeling during hypertension, estradiol might lead to a reduced vascular growth by inhibition of the RAS. Furthermore, expression of Ang II type 1 receptors is downregulated by 17β-estradiol in VSMCs and thereby the angiotensin effects might be abrogated at least in part [164]. Homocysteine contributes to vascular disease by inducing EC damage, inhibition of EC growth and induction of smooth muscle cell growth [165]. Clinical studies provide evidence that estradiol reduces circulating levels of homocysteine in postmenopausal women and thereby improves the cardiovascular risk profile [166]. Endothelin is a vasoconstrictor and mitogenic peptide that is thought to play a role in various forms of vascular disease. Estradiol inhibits serum and Ang II-stimulated synthesis of endothelin-1 in ECs via an ER-independent mechanism [167, 168]. Compared with premenopausal women, plasma endothelin-1 levels are increased in postmenopausal women not taking estradiol and are reduced following estradiol replacement therapy [166]. These findings suggest, that estradiol may have antimitogenic effects on the vasculature in part by reducing endothelin-1 levels.

2.5.4.1 Lipid Levels
After menopause, low-density lipoprotein (LDL) and triglyceride levels rise, and high-density lipoprotein (HDL) levels fall. Hormone replacement has antiatherogenic effects on lipids, lowering LDL and raising HDL, but paradoxically also elevates triglycerides [161, 169, 170]. Androgens have variable effects on lipoproteins and other risk factors, depending on the hormone formulation and the population studied. Exogenous androgens generally lower HLD-cholesterol and lipoprotein a,

with only modest effects on LDL-cholesterol, and facilitate both macrophage lipoprotein uptake and efflux of cellular cholesterol to HLD [171].

2.5.4.2 Glucose and Insulin

The relationship between oral contraceptive use and glucose intolerance/insulin resistance is controversial. A review of large epidemiologic studies found no difference in the frequency of diabetes between users and nonusers of high-dose oral contraceptives, while data from smaller studies revealed a 2-fold increased risk of impaired glucose tolerance in current high-dose oral contraceptives users compared to nonusers [172].

2.5.4.3 Blood Pressure

Gender has an important influence on blood pressure, with premenopausal women having a lower arterial blood pressure than age matched men. Compared with premenopausal women postmenopausal women have higher blood pressure, suggesting that ovarian hormones may modulate blood pressure. Similar to humans, sex-associated differences in blood pressure also exist in animals. For example compared with females, male spontaneously hypertensive rats [173, 174], male Dahl salt-sensitive rats [175, 176], male deoxycorticosterone acetate-salt hypertensive rats [177] and male New Zealand genetically hypertensive rats [178] have higher blood pressure. In some animal models of hypertension, blood pressure is reduced in males after castration [179], but is not increased in females by ovariectomy [180]. Thus, sex-associated differences in blood pressure also may be due to changes in testicular hormones [181].

Support for the conclusion that estradiol lowers blood pressure in humans is provided by the observation, that during the menstrual cycle blood pressure is lower during the luteal phase (when estradiol levels peak) than during the follicular phase [182]. Observations made during pregnancy provide additional circumstantial evidence for a blood pressure lowering effect of estradiol. Estradiol levels increase 50- to 180-fold during pregnancy and these increases are associated with substantial reductions in blood pressure [183]. If endogenous estradiol does lower blood pressure, administration of estrogenic preparations to women might also be expected to reduce blood pressure. However, data on the effects of estrogenic preparations on blood pressure are inconsistent, and include reports of blood pressure-lowering [184–186], blood pressure-elevating [187] and blood pressure-neutral effects [188]. Whether blood pressure is decreased, increased or unchanged in response to estrogen treatment depends on the type of estrogenic preparation and the dose of estrogen. Contraceptive estrogenic preparations tend to increase blood pressure, conjugated equine estrogens appear to have little effect on blood pressure and estradiol tends to lower blood pressure [181].

2.5.4.4 Coagulatory Activity

Estradiol influences the synthesis of a number of factors associated with coagulation and fibrinolysis. For example, estradiol decreases plasma concentrations of clottable fibrinogen, soluble thrombomodulin, plasminogen activator inhibior-1, antithrombin III and protein S [189–191]. Moreover, most studies report that estradiol increases levels of von Willebrand factor in ECs [192] and during pregnancy in women [193]. Importantly, the effects of estradiol on coagulation and fibrinolytic factors are not the same as those of oral contraceptive ethinyl estradiol [194]. Taken together, oral contraceptives induce disorders of coagulation and increase blood pressure, whereas estradiol, when used for HRT tends to lower blood pressure and have neutral effects on the coagulation system [181].

References

1 Regitz-Zagrosek, V., Brokat, S. and Tschope, C. (2007) Role of gender in heart failure with normal left ventricular ejection fraction. *Progress in Cardiovascular Diseases*, **49**, 241–51.

2 Regitz-Zagrosek, V. (2006) Therapeutic implications of the gender-specific aspects of cardiovascular disease. *Nature Reviews Drug Discovery*, **5**, 425–239.

3 Sudhir, K. and Komesaroff, P.A. (1999) Clinical review 110: cardiovascular actions of estrogens in men. *Journal of Clinical Endocrinology and Metabolism*, **84**, 3411–15.

4 Rossouw, J.E., Anderson, G.L., Prentice, R.L. et al. (2002) Risks and benefits of estrogen plus progestin in healthy postmenopausal women: principal results From the Women's Health Initiative randomized controlled trial. *Journal of the American Medical Association*, **288**, 321–33.

5 Anderson, G.L., Limacher, M., Assaf, A.R. et al. (2004) Effects of conjugated equine estrogen in postmenopausal women with hysterectomy: the Women's Health Initiative randomized controlled trial. *Journal of the American Medical Association*, **291**, 1701–12.

6 Rossouw, J.E., Prentice, R.L., Manson, J.E. et al. (2007) Postmenopausal hormone therapy and risk of cardiovascular disease by age and years since menopause. *Journal of the American Medical Association*, **297**, 1465–77.

7 Gruber, C.J., Tschugguel, W., Schneeberger, C. et al. (2002) Production and actions of estrogens. *New England Journal of Medicine*, **346**, 340–52.

8 Mulligan, A.A., Welch, A.A., McTaggart, A.A. et al. (2007) Intakes and sources of soya foods and isoflavones in a UK population cohort study (EPIC-Norfolk). *European Journal of Clinical Nutrition*, **61**, 248–54.

9 Matsumine, H., Hirato, K., Yanaihara, T. et al. (1986) Aromatization by skeletal muscle. *Journal of Clinical Endocrinology and Metabolism*, **63**, 717–20.

10 Miller, W.R. (1991) Aromatase activity in breast tissue. *Journal of Steroid Biochemistry and Molecular Biology*, **39** (5B) 783–90.

11 Naftolin, F., Ryan, K.J., Davies, I.J. et al. (1975) The formation of estrogens by central neuroendocrine tissues. *Recent Progress in Hormone Research*, **31**, 295–319.

12 Brodie, A. and Inkster, S. (1993) Aromatase in the human testis. *Journal of Steroid Biochemistry and Molecular Biology*, **44**, 549–55.

13 Baird, D.T. and Fraser, I.S. (1974) Blood production and ovarian secretion rates of estradiol-17 beta and estrone in women throughout the menstrual cycle. *Journal of Clinical Endocrinology and Metabolism*, **38**, 1009–17.

14 Yentis, S.M., Steer, P.J. and Plaat, F. (1998) Eisenmenger's syndrome in pregnancy: maternal and fetal mortality in the 1990s. *British Journal of Obstetrics and Gynaecology*, **105**, 921–2.

15 Braunwald, E., Isselbacher, K.J., Petersdorf, R.G., Wilson, J.D., Martin, J.B. and Fauci, A.F. (1987) *Harrison's*

Principles of Internal Medicine, 11th edn, McGraw-Hill, New York.
16 Moreira, R.M., Lisboa, P.C., Curty, F.H. et al. (1997) Dose-dependent effects of 17-beta-estradiol on pituitary thyrotropin content and secretion in vitro. *Brazilian Journal of Medical and Biological Research*, **30**, 1129–34.
17 Babiker, F.A., Lips, D., Meyer, R. et al. (2006) Estrogen receptor β protects the murine heart against left ventricular hypertrophy. *Arteriosclerosis, Thrombosis, and Vascular Biology*, **26**, 1524–30.
18 Beato, M. (1989) Gene regulation by steroid hormones. *Cell*, **56**, 335–44.
19 Mangelsdorf, D.J., Thummel, C., Beato, M. et al. (1995) The nuclear receptor superfamily: the second decade. *Cell*, **83**, 835–9.
20 Evans, R.M. (1988) The steroid and thyroid hormone receptor superfamily. *Science*, **240**, 889–95.
21 Germain, P., Staels, B., Dacquet, C. et al. (2006) Overview of nomenclature of nuclear receptors. *Pharmacological Reviews*, **58**, 685–704.
22 Hollenberg, S.M., Weinberger, C., Ong, E.S. et al. (1985) Primary structure and expression of a functional human glucocorticoid receptor cDNA. *Nature*, **318**, 635–41.
23 Arriza, J.L., Weinberger, C., Cerelli, G. et al. (1987) Cloning of human mineralocorticoid receptor complementary DNA: structural and functional kinship with the glucocorticoid receptor. *Science*, **237**, 268–75.
24 Walter, P., Green, S., Greene, G. et al. (1985) Cloning of the human estrogen receptor cDNA. *Proceedings of the National Academy of Sciences of the United States of America*, **82**, 7889–93.
25 Mosselman, S., Polman, J. and Dijkema, R. (1996) ER beta: identification and characterization of a novel human estrogen receptor. *FEBS Letters*, **392**, 49–53.
26 Misrahi, M., Atger, M., Auriol, L. et al. (1987) Complete amino acid sequence of the human progesterone receptor deduced from cloned cDNA. *Biochemical and Biophysical Research Communications*, **143**, 740–8.
27 Bruning, P.F., Bonfrer, J.M., Hart, A.A. et al. (1988) Tamoxifen, serum lipoproteins and cardiovascular risk. *British Journal of Cancer*, **58**, 497–9.
28 Lubahn, D.B., Joseph, D.R., Sar, M. et al. (1988) The human androgen receptor: complementary deoxyribonucleic acid cloning, sequence analysis and gene expression in prostate. *Molecular Endocrinology*. **2**, 1265–75.
29 Kastner, P., Krust, A., Turcotte, B. et al. (1990) Two distinct estrogen-regulated promoters generate transcripts encoding the two functionally different human progesterone receptor forms A and B. *EMBO Journal*, **9**, 1603–14.
30 Kuiper, G.G., Enmark, E., Pelto-Huikko, M. et al. (1996) Cloning of a novel receptor expressed in rat prostate and ovary. *Proceedings of the National Academy of Sciences of the United States of America*, **93**, 5925–30.
31 Flouriot, G., Brand, H., Denger, S. et al. (2000) Identification of a new isoform of the human estrogen receptor-alpha (hER-alpha) that is encoded by distinct transcripts and that is able to repress hER-alpha activation function 1. *EMBO Journal*, **19**, 4688–700.
32 Wilson, C.M. and McPhaul, M.J. (1996) A and B forms of the androgen receptor are expressed in a variety of human tissues. *Molecular and Cellular Endocrinology*, **120**, 51–7.
33 Faber, P.W., van Rooij, H.C., Van der Korput, H.A. et al. (1991) Characterization of the human androgen receptor transcription unit. *Journal of Biological Chemistry*, **266**, 10743–9.
34 Zennaro, M.C., Farman, N., Bonvalet, J.P. et al. (1997) Tissue-specific expression of alpha and beta messenger ribonucleic acid isoforms of the human mineralocorticoid receptor in normal and pathological states. *Journal of Clinical Endocrinology and Metabolism*, **82**, 1345–52.
35 Leygue, E., Dotzlaw, H., Watson, P.H. et al. (1996) Identification of novel exon-deleted progesterone receptor variant mRNAs in human breast tissue.

Biochemical and Biophysical Research Communications, **228**, 63–8.

36 Shao, D. and Lazar, M.A. (1999) Modulating nuclear receptor function: may the phos be with you. *Journal of Clinical Investigation*, **103**, 1617–8.

37 Schwabe, J.W., Fairall, L., Chapman, L. et al. (1993) The cocrystal structures of two zinc-stabilized DNA-binding domains illustrate different ways of achieving sequence-specific DNA recognition. *Cold Spring Harbor Symposium on Quantitative Biology*, **58**, 141–7.

38 Klug, A. and Schwabe, J.W. (1995) Protein motifs 5. Zinc fingers. *FASEB Journal*, **9**, 597–604.

39 Hisamoto, K., Ohmichi, M., Kanda, Y. et al. (2001) Induction of endothelial nitric-oxide synthase phosphorylation by the raloxifene analog LY117018 is differentially mediated by Akt and extracellular signal-regulated protein kinase in vascular endothelial cells. *Journal of Biological Chemistry*, **276**, 47642–9.

40 Wurtz, J.M., Bourguet, W., Renaud, J.P. et al. (1996) A canonical structure for the ligand-binding domain of nuclear receptors. *Nature Structural and Molecular Biology*, **3**, 87–94.

41 Leclercq, G., Lacroix, M., Laios, I. et al. (2006) Estrogen receptor alpha: impact of ligands on intracellular shuttling and turnover rate in breast cancer cells. *Current Cancer Drug Targets*, **6**, 39–64.

42 Schwartz, J.A., Zhong, L., Deighton-Collins, S. et al. (2002) Mutations targeted to a predicted helix in the extreme carboxyl-terminal region of the human estrogen receptor-alpha alter its response to estradiol and 4-hydroxy-tamoxifen. *Journal of Biological Chemistry*, **277**, 13202–9.

43 Lonard, D.M., Nawaz, Z., Smith, C.L. et al. (2000) The 26S proteasome is required for estrogen receptor-alpha and coactivator turnover and for efficient estrogen receptor-alpha transactivation. *Mol. Cell*, **5**, 939–48.

44 Glass, C.K. (1994) Differential recognition of target genes by nuclear receptor monomers, dimers, and heterodimers. *Endocrine Reviews*, **15**, 391–407.

45 Nelson, C.C., Hendy, S.C., Shukin, R.J. et al. (1999) Determinants of DNA sequence specificity of the androgen, progesterone, and glucocorticoid receptors: evidence for differential steroid receptor response elements. *Molecular Endocrinology*, **13**, 2090–107.

46 Weigel, N.L. (1996) Steroid hormone receptors and their regulation by phosphorylation. *Biochemical Journal*, **319**, 657–67.

47 Glass, C.K. and Rosenfeld, M.G. (2000) The coregulator exchange in transcriptional functions of nuclear receptors. *Genes and Development*, **14**, 121–41.

48 Aranda, A. and Pascual, A. (2001) Nuclear hormone receptors and gene expression. *Physiological Reviews*, **81**, 1269–304.

49 Olefsky, J.M. (2001) Nuclear receptor minireview series. *Journal of Biological Chemistry*, **276**, 36863–4.

50 Weigel, N.L. and Zhang, Y. (1998) Ligand-independent activation of steroid hormone receptors. *Journal of Molecular Medicine*, **76**, 469–79.

51 Tremblay, A., Tremblay, G.B., Labrie, F. et al. (1999) Ligand-independent recruitment of SRC-1 to estrogen receptor beta through phosphorylation of activation function AF-1. *Molecular Cell*, **3**, 513–19.

52 Zhang, Y., Beck, C.A., Poletti, A. et al. (1994) Identification of phosphorylation sites unique to the B form of human progesterone receptor. In vitro phosphorylation by casein kinase II. *Journal of Biological Chemistry*, **269**, 31034–40.

53 Nazareth, L.V. and Weigel, N.L. (1996) Activation of the human androgen receptor through a protein kinase A signaling pathway. *Journal of Biological Chemistry*, **271**, 19900–7.

54 Gilliam, M.L. (2001) Local and systemic options for hormone replacement therapy. *International Journal of Fertility and Women's Medicine*, **46**, 222–7.

55 Ignar-Trowbridge, D.M., Nelson, K.G., Bidwell, M.C. et al. (1992) Coupling of dual signaling pathways: epidermal

growth factor action involves the estrogen receptor. *Proceedings of the National Academy of Sciences of the United States of America*, **89**, 4658–62.
56 Ignar-Trowbridge, D.M., Teng, C.T., Ross, K.A. *et al.* (1993) Peptide growth factors elicit estrogen receptor-dependent transcriptional activation of an estrogen-responsive element. *Molecular Endocrinology*, **7**, 992–8.
57 Chang, R.K., Chen, A.Y. and Klitzner, T.S. (2002) Female sex as a risk factor for in-hospital mortality among children undergoing cardiac surgery. *Circulation*, **106**, 1514–22.
58 Bunone, G., Briand, P.A., Miksicek, R.J. *et al.* (1996) Activation of the unliganded estrogen receptor by EGF involves the MAP kinase pathway and direct phosphorylation. *EMBO Journal*, **15**, 2174–83.
59 Hishikawa, K., Nakaki, T., Marumo, T. *et al.* (1995) Up-regulation of nitric oxide synthase by estradiol in human aortic endothelial cells. *FEBS Letters*, **360**, 291–3.
60 Yeh, S., Lin, H.K., Kang, H.Y. *et al.* (1999) From HER2/Neu signal cascade to androgen receptor and its coactivators: a novel pathway by induction of androgen target genes through MAP kinase in prostate cancer cells. *Proceedings of the National Academy of Sciences of the United States of America*, **96**, 5458–63.
61 Lannigan, D.A. (2003) Estrogen receptor phosphorylation. *Steroids*, **68**, 1–9.
62 Brown, M.A., Hague, W.M., Higgins, J. *et al.* (2000) The detection, investigation and management of hypertension in pregnancy: full consensus statement. *Australian and New Zealand Journal of Obstetrics and Gynaecology*, **40**, 139–55.
63 Legro, R.S., Bentley-Lewis, R., Driscoll, D. *et al.* (2002) Insulin resistance in the sisters of women with polycystic ovary syndrome: association with hyperandrogenemia rather than menstrual irregularity. *Journal of Clinical Endocrinology and Metabolism*, **87**, 2128–33.
64 Cato, A.C., Nestl, A. and Mink, S. (2002) Rapid actions of steroid receptors in cellular signaling pathways. *Science's STKE*, **2002**, RE9.
65 Losel, R. and Wehling, M. (2003) Nongenomic actions of steroid hormones. *Nature Reviews Molecular Cell Biology*, **4**, 46–56.
66 Mendiberri, J., Rauschemberger, M.B., Selles, J. *et al.* (2006) Involvement of phosphoinositide-3-kinase and phospholipase C transduction systems in the non-genomic action of progesterone in vascular tissue. *International Journal of Biochemistry and Cell Biology*, **38**, 288–96.
67 Migliaccio, A., Piccolo, D., Castoria, G. *et al.* (1998) Activation of the Src/p21^{ras}/ Erk pathway by progesterone receptor via cross-talk with estrogen receptor. *EMBO Journal*, **17**, 2008–18.
68 Chen, D., Pace, P.E., Coombes, R.C. *et al.* (1999) Phosphorylation of human estrogen receptor alpha by protein kinase A regulates dimerization. *Molecular and Cellular Biology*, **19**, 1002–15.
69 Simoncini, T. and Genazzani, A.R. (2000) Raloxifene acutely stimulates nitric oxide release from human endothelial cells via an activation of endothelial nitric oxide synthase. *Journal of Clinical Endocrinology and Metabolism*, **85**, 2966–9.
70 Chambliss, K.L., Yuhanna, I.S., Anderson, R.G. *et al.* (2002) ERbeta has nongenomic action in caveolae. *Molecular Endocrinology*, **16**, 938–46.
71 Li, L., Haynes, M.P. and Bender, J.R. (2003) Plasma membrane localization and function of the estrogen receptor alpha variant (ER46) in human endothelial cells. *Proceedings of the National Academy of Sciences of the United States of America*, **100**, 4807–12.
72 Filardo, E.J., Quinn, J.A., Bland, K.I. *et al.* (2000) Estrogen-induced activation of Erk-1 and Erk-2 requires the G protein-coupled receptor homolog, GPR30, and occurs via trans-activation of the epidermal growth factor receptor through release of HB-EGF. *Molecular Endocrinology*, **14**, 1649–60.
73 Kahlert, H., Grage-Griebenow, E., Stuwe, H.T. *et al.* (2000) T cell reactivity with allergoids: influence of the type of APC. *Journal of Immunology*, **165**, 1807–15.
74 Grazzini, E., Guillon, G., Mouillac, B. *et al.* (1998) Inhibition of oxytocin

receptor function by direct binding of progesterone. *Nature*, **392**, 509–12.
75 Thomas, P., Pang, Y., Filardo, E.J. *et al.* (2005) Identity of an estrogen membrane receptor coupled to a G protein in human breast cancer cells. *Endocrinology*, **146**, 624–32.
76 Razandi, M., Pedram, A., Greene, G.L. *et al.* (1999) Cell membrane and nuclear estrogen receptors (ERs) originate from a single transcript: studies of ERalpha and ERbeta expressed in Chinese hamster ovary cells. *Molecular Endocrinology*, **13**, 307–19.
77 Simoncini, T., Hafezi-Moghadam, A., Brazil, D.P. *et al.* (2000) Interaction of oestrogen receptor with the regulatory subunit of phosphatidylinositol-3-OH kinase. *Nature*, **407**, 538–41.
78 Filardo, E.J., Quinn, J.A., Frackelton A.R., Jr *et al.* (2002) Estrogen action via the G protein-coupled receptor, GPR30: stimulation of adenylyl cyclase and cAMP-mediated attenuation of the epidermal growth factor receptor-to-MAPK signaling axis. *Molecular Endocrinology*, **16**, 70–84.
79 Benten, W.P., Stephan, C., Lieberherr, M. *et al.* (2001) Estradiol signaling via sequestrable surface receptors. *Endocrinology*, **142**, 1669–77.
80 Stirone, C., Boroujerdi, A., Duckles, S.P. *et al.* (2005) Estrogen receptor activation of phosphoinositide-3 kinase, Akt, and nitric oxide signaling in cerebral blood vessels: rapid and long-term effects. *Molecular Pharmacology*, **67**, 105–13.
81 McKenna, N.J., Xu, J., Nawaz, Z. *et al.* (1999) Nuclear receptor coactivators: multiple enzymes, multiple complexes, multiple functions. *Journal of Steroid Biochemistry and Molecular Biology*, **69**, 3–12.
82 Barnes, C.J., Vadlamudi, R.K. and Kumar, R. (2004) Novel estrogen receptor coregulators and signaling molecules in human diseases. *Cellular and Molecular Life Sciences*, **61**, 281–91.
83 Matthews, J. and Gustafsson, J.A. (2003) Estrogen signaling: a subtle balance between ER alpha and ER beta. *Molecular Interventions*, **3**, 281–92.
84 El Khissiin, A. and Leclercq, G. (1999) Implication of proteasome in estrogen receptor degradation. *FEBS Letters*, **448**, 160–6.
85 Flouriot, G., Griffin, C., Kenealy, M. *et al.* (1998) Differentially expressed messenger RNA isoforms of the human estrogen receptor-alpha gene are generated by alternative splicing and promoter usage. *Molecular Endocrinology*, **12**, 1939–54.
86 Kos, M., Denger, S., Reid, G. *et al.* (2002) Upstream open reading frames regulate the translation of the multiple mRNA variants of the estrogen receptor alpha. *Journal of Biological Chemistry*, **277**, 37131–8.
87 Okuda, Y., Hirata, S., Watanabe, N. *et al.* (2003) Novel splicing events of untranslated first exons in human estrogen receptor alpha (ER alpha) gene. *Endocrine Journal*, **50**, 97–104.
88 Grandien, K. (1996) Determination of transcription start sites in the human estrogen receptor gene and identification of a novel, tissue-specific, estrogen receptor-mRNA isoform. *Molecular and Cellular Endocrinology*, **116**, 207–12.
89 Lambertini, E., Penolazzi, L., Giordano, S. *et al.* (2003) Expression of the human oestrogen receptor-alpha gene is regulated by promoter F in MG-63 osteoblastic cells. *Biochemical Journal*, **372**, 831–9.
90 Tang, Z., Treilleux, I. and Brown, M. (1997) A transcriptional enhancer required for the differential expression of the human estrogen receptor in breast cancers. *Molecular and Cellular Biology*, **17**, 1274–80.
91 Tanimoto, K., Eguchi, H., Yoshida, T. *et al.* (1999) Regulation of estrogen receptor alpha gene mediated by promoter B responsible for its enhanced expressionin human breast cancer. *Nucleic Acids Research*, **27**, 903–9.
92 Karas, R.H., Patterson, B.L. and Mendelsohn, M.E. (1994) Human vascular smooth muscle cells contain functional estrogen receptor. *Circulation*, **89**, 1943–50.
93 Venkov, C.D., Rankin, A.B. and Vaughan, D.E. (1996) Identification of authentic estrogen receptor in cultured endothelial cells. A potential mechanism for steroid

hormone regulation of endothelial function. *Circulation*, **94**, 727–33.
94 Nordmeyer, J., Eder, S., Mahmoodzadeh, S. *et al.* (2004) Upregulation of myocardial estrogen receptors in human aortic stenosis. *Circulation*, **110**, 3270–5.
95 Grohe, C., Kahlert, S., Lobbert, K. *et al.* (1997) Cardiac myocytes and fibroblasts contain functional estrogen receptors. *FEBS Letters*, **416**, 107–12.
96 Ropero, A.B., Eghbali, M., Minosyan, T.Y. *et al.* (2006) Heart estrogen receptor alpha: Distinct membrane and nuclear distribution patterns and regulation by estrogen. *Journal of Molecular and Cellular Cardiology*, **41**, 496–510.
97 Death, A.K., McGrath, K.C., Sader, M.A. *et al.* (2004) Dihydrotestosterone promotes vascular cell adhesion molecule-1 expression in male human endothelial cells via a nuclear factor-kappaB-dependent pathway. *Endocrinology*, **145**, 1889–97.
98 Hodges, Y.K., Richer, J.K., Horwitz, K.B. *et al.* (1999) Variant estrogen and progesterone receptor messages in human vascular smooth muscle. *Circulation*, **99**, 2688–93.
99 Mahmoodzadeh, S., Eder, S., Nordmeyer, J. *et al.* (2006) Estrogen receptor alpha up-regulation and redistribution in human heart failure. *FASEB Journal*, **20**, 926–34.
100 Mendelsohn, M.E. and Karas, R.H. (1999) The protective effects of estrogen on the cardiovascular system. *New England Journal of Medicine*, **340**, 1801–11.
101 Iafrati, M.D., Karas, R.H., Aronovitz, M. *et al.* (1997) Estrogen inhibits the vascular injury response in estrogen receptor alpha-deficient mice. *Nature Medicine*, **3**, 545–8.
102 Karas, R.H., Hodgin, J.B., Kwoun, M. *et al.* (1999) Estrogen inhibits the vascular injury response in estrogen receptor beta-deficient female mice. *Proceedings of the National Academy of Sciences of the United States of America*, **96**, 15133–6.
103 Karas, R.H., Schulten, H., Pare, G. *et al.* (2001) Effects of estrogen on the vascular injury response in estrogen receptor alpha, beta (double) knockout mice. *Circulation Research*, **89**, 534–9.
104 Dubey, R.K., Gillespie, D.G., Zacharia, L.C. *et al.* (2000) Methoxyestradiols mediate the antimitogenic effects of estradiol on vascular smooth muscle cells via estrogen receptor-independent mechanisms. *Biochemical and Biophysical Research Communications*, **278**, 27–33.
105 Dubey, R.K., Gillespie, D.G., Jackson, E.K. *et al.* (1998) 17Beta-estradiol, its metabolites, and progesterone inhibit cardiac fibroblast growth. *Hypertension*, **31** (*1 Pt 2*), 522–8.
106 Barchiesi, F., Jackson, E.K., Gillespie, D.G. *et al.* (2002) Methoxyestradiols mediate estradiol-induced anti-mitogenesis in human aortic SMCs. *Hypertension*, **39**, 874–9.
107 Pelzer, T., Schumann, M., Neumann, M. *et al.* (2000) 17Beta-estradiol prevents programmed cell death in cardiac myocytes. *Biochemical and Biophysical Research Communications*, **268**, 192–200.
108 Kim, J.K., Pedram, A., Razandi, M. *et al.* (2006) Estrogen prevents cardiomyocyte apoptosis through inhibition of reactive oxygen species and differential regulation of p38 kinase isoforms. *Journal of Biological Chemistry*, **281**, 6760–7.
109 Pedram, A., Razandi, M., Aitkenhead, M. *et al.* (2005) Estrogen inhibits cardiomyocyte hypertrophy *in vitro*. Antagonism of calcineurin-related hypertrophy through induction of MCIP1. *Journal of Biological Chemistry*, **280**, 26339–48.
110 Patten, R.D. and Karas, R.H. (2006) Estrogen replacement and cardiomyocyte protection. *Trends in Cardiovascular Medicine*, **16**, 69–75.
111 Dubey, R.K., Jackson, E.K., Gillespie, D.G. *et al.* (2005) Cytochromes 1A1/1B1- and catechol-O-methyltransferase-derived metabolites mediate estradiol-induced antimitogenesis in human cardiac fibroblast. *Journal of Clinical Endocrinology and Metabolism*, **90**, 247–55.
112 Lee, H.W. and Eghbali-Webb, M. (1998) Estrogen enhances proliferative capacity of cardiac fibroblasts by estrogen receptor- and mitogen-activated protein kinase-dependent pathways. *Journal of*

113 Grohe, C., Kahlert, S., Lobbert, K. et al. (1996) Modulation of hypertensive heart disease by estrogen. *Steroids*, **61**, 201–4.

114 Mercier, I., Colombo, F., Mader, S. et al. (2002) Ovarian hormones induce TGF-beta(3) and fibronectin mRNAs but exhibit a disparate action on cardiac fibroblast proliferation. *Cardiovascular Research*, **53**, 728–39.

115 Squadrito, F., Altavilla, D., Squadrito, G. et al. (1997) 17Beta-oestradiol reduces cardiac leukocyte accumulation in myocardial ischaemia reperfusion injury in rat. *European Journal of Pharmacology*, **335**, 185–92.

116 Florian, M., Lu, Y., Angle, M. et al. (2004) Estrogen induced changes in Akt-dependent activation of endothelial nitric oxide synthase and vasodilation. *Steroids*, **69**, 637–45.

117 Kim, H.P., Lee, J.Y., Jeong, J.K. et al. (1999) Nongenomic stimulation of nitric oxide release by estrogen is mediated by estrogen receptor alpha localized in caveolae. *Biochemical and Biophysical Research Communications*, **263**, 257–62.

118 Spyridopoulos, I., Sullivan, A.B., Kearney, M. et al. (1997) Estrogen-receptor-mediated inhibition of human endothelial cell apoptosis. Estradiol as a survival factor. *Circulation*, **95**, 1505–14.

119 Morales, D.E., McGowan, K.A., Grant, D.S. et al. (1995) Estrogen promotes angiogenic activity in human umbilical vein endothelial cells *in vitro* and in a murine model. *Circulation*, **91**, 755–63.

120 Hillebrand, U., Hausberg, M., Stock, C. et al. (2006) 17Beta-estradiol increases volume, apical surface and elasticity of human endothelium mediated by Na^+/H^+ exchange. *Cardiovascular Research*, **69**, 916–24.

121 Geraldes, P., Gagnon, S., Hadjadj, S. et al. (2006) Estradiol blocks the induction of CD40 and CD40L expression on endothelial cells and prevents neutrophil adhesion: an ERalpha-mediated pathway. *Cardiovascular Research*, **71**, 566–73.

122 White, R.E., Darkow, D.J. and Lang, J.L. (1995) Estrogen relaxes coronary arteries by opening BKCa channels through a cGMP-dependent mechanism. *Circulation Research*, **77**, 936–42.

123 Sullivan, T.R. Jr, Karas, R.H., Aronovitz, M. et al. (1995) Estrogen inhibits the response-to-injury in a mouse carotid artery model. *Journal of Clinical Investigation*, **96**, 2482–8.

124 Jayachandran, M. and Miller, V.M. (2002) Ovariectomy upregulates expression of estrogen receptors, NOS, and HSPs in porcine platelets. *American Journal of Physiology, Heart and Circulatory Physiology*, **283**, H220–226.

125 Khetawat, G., Faraday, N., Nealen, M.L. et al. (2000) Human megakaryocytes and platelets contain the estrogen receptor beta and androgen receptor (AR): testosterone regulates AR expression. *Blood*, **95**, 2289–96.

126 Tarantino, M.D., Kunicki, T.J. and Nugent, D.J. (1994) The estrogen receptor is present in human megakaryocytes. *Annals of the New York Academy of Sciences*, **714**, 293–6.

127 Jayachandran, M., Okano, H., Chatrath, R. et al. (2004) Sex-specific changes in platelet aggregation and secretion with sexual maturity in pigs. *Journal of Applied Physiology*, **97**, 1445–52.

128 Eghbali, M., Wang, Y., Toro, L. et al. (2006) Heart hypertrophy during pregnancy: a better functioning heart? *Trends in Cardiovascular Medicine*, **16**, 285–91.

129 Babiker, F.A., De Windt, L.J., van Eickels, M. et al. (2002) Estrogenic hormone action in the heart: regulatory network and function. *Cardiovascular Research*, **53**, 709–19.

130 Tagarakis, C.V., Bloch, W., Hartmann, G. et al. (2000) Testosterone-propionate impairs the response of the cardiac capillary bed to exercise. *Medicine and Science in Sports and Exercise*, **32**, 946–53.

131 Woodiwiss, A.J., Trifunovic, B., Philippides, M. et al. (2000) Effects of an androgenic steroid on exercise-induced cardiac remodeling in rats. *Journal of Applied Physiology*, **88**, 409–15.

132 Marques, C.M., Nascimento, F.A., Mandarim-de-Lacerda, C.A. et al. (2006) Exercise training attenuates cardio-vascular adverse remodeling in adult

ovariectomized spontaneously hypertensive rats. *Menopause*, **13**, 87–95.
133 Konhilas, J.P., Maass, A.H., Luckey, S.W. *et al.* (2004) Sex modifies exercise and cardiac adaptation in mice. *American Journal of Physiology, Heart and Circulatory Physiology*, **287**, H2768–76.
134 Weinberg, E.O., Mirotsou, M., Gannon, J. *et al.* (2003) Sex dependence and temporal dependence of the left ventricular genomic response to pressure overload. *Physiological Genomics*, **12**, 113–27.
135 Vuolteenaho, O. and Ruskoaho, H. (2003) Gender matters: estrogen protects from cardiac hypertrophy. *Trends in Endocrinology and Metabolism*, **14**, 52–4.
136 Li, Y., Kishimoto, I., Saito, Y. *et al.* (2004) Androgen contributes to gender-related cardiac hypertrophy and fibrosis in mice lacking the gene encoding guanylyl cyclase-A. *Endocrinology*, **145**, 951–8.
137 Babiker, F.A., De Windt, L.J., van Eickels, M. *et al.* (2004) 17Beta-estradiol antagonizes cardiomyocyte hypertrophy by autocrine/paracrine stimulation of a guanylyl cyclase A receptor-cyclic guanosine monophosphate-dependent protein kinase pathway. *Circulation*, **109**, 269–76.
138 van Eickels, M., Grohe, C., Cleutjens, J.P. *et al.* (2001) 17Beta-estradiol attenuates the development of pressure-overload hypertrophy. *Circulation*, **104**, 1419–23.
139 Pelzer, T., Jazbutyte, V., Hu, K. *et al.* (2005) The estrogen receptor-alpha agonist 16alpha-LE2 inhibits cardiac hypertrophy and improves hemodynamic function in estrogen-deficient spontaneously hypertensive rats. *Cardiovascular Research*, **67**, 604–12.
140 Brower, G.L., Gardner, J.D. and Janicki, J.S. (2003) Gender mediated cardiac protection from adverse ventricular remodeling is abolished by ovariectomy. *Molecular and Cellular Biochemistry*, **251**, 89–95.
141 Gardner, J.D., Brower, G.L. and Janicki, J.S. (2005) Effects of dietary phytoestrogens on cardiac remodeling secondary to chronic volume overload in female rats. *Journal of Applied Physiology*, **99**, 1378–83.
142 Cavasin, M.A., Tao, Z.Y., Yu, A.L. *et al.* (2006) Testosterone enhances early cardiac remodeling after myocardial infarction, causing rupture and degrading cardiac function. *American Journal of Physiology, Heart and Circulatory Physiology*, **290**, H2043–2050.
143 Cavasin, M.A., Sankey, S.S., Yu, A.L. *et al.* (2003) Estrogen and testosterone have opposing effects on chronic cardiac remodeling and function in mice with myocardial infarction. *American Journal of Physiology, Heart and Circulatory Physiology*, **284**, H1560–1569.
144 van Eickels, M., Patten, R.D., Aronovitz, M.J. *et al.* (2003) 17-Beta-estradiol increases cardiac remodeling and mortality in mice with myocardial infarction. *Journal of the American College of Cardiology*, **41**, 2084–92.
145 Iwakura, A., Shastry, S., Luedemann, C. *et al.* (2006) Estradiol enhances recovery after myocardial infarction by augmenting incorporation of bone marrow-derived endothelial progenitor cells into sites of ischemia-induced neovascularization via endothelial nitric oxide synthase-mediated activation of matrix metalloproteinase-9. *Circulation*, **113**, 1605–14.
146 Babiker, F.A., Lips, D.J., Delvaux, E. *et al.* (2007) Oestrogen modulates cardiac ischaemic remodelling through oestrogen receptor-specific mechanisms. *Acta Physiologica (Oxford, England)*, **189**, 23–31.
147 Korte, T., Fuchs, M., Arkudas, A. *et al.* (2005) Female mice lacking estrogen receptor beta display prolonged ventricular repolarization and reduced ventricular automaticity after myocardial infarction. *Circulation*, **111**, 2282–90.
148 Pelzer, T., Loza, P.A., Hu, K. *et al.* (2005) Increased mortality and aggravation of heart failure in estrogen receptor-beta knockout mice after myocardial infarction. *Circulation*, **111**, 1492–8.
149 Clarkson, T.B., Anthony, M.S. and Klein, K.P. (1996) Hormone replacement therapy and coronary artery

150 Wang, Y.X. (2005) Cardiovascular functional phenotypes and pharmacological responses in apolipoprotein E deficient mice. *Neurobiology of Aging*, **26**, 309–16.

151 Martin-McNulty, B., Tham, D.M., Da Cunha, V. *et al.* (2003) 17Beta-estradiol attenuates development of angiotensin II-induced aortic abdominal aneurysm in apolipoprotein E-deficient mice. *Arteriosclerosis, Thrombosis, and Vascular Biology*, **23**, 1627–32.

152 Tse, J., Martin-McNulty, B., Halks-Miller, M. *et al.* (1999) Accelerated atherosclerosis and premature calcified cartilaginous metaplasia in the aorta of diabetic male Apo E knockout mice can be prevented by chronic treatment with 17 beta-estradiol. *Atherosclerosis*, **144**, 303–13.

153 Kauser, K., Da Cunha, V., Fitch, R. *et al.* (2000) Role of endogenous nitric oxide in progression of atherosclerosis in apolipoprotein E-deficient mice. *American Journal of Physiology, Heart and Circulatory Physiology*. **278**, H1679–85.

154 Nathan, L., Pervin, S., Singh, R. *et al.* (1999) Estradiol inhibits leukocyte adhesion and transendothelial migration in rabbits *in vivo*: possible mechanisms for gender differences in atherosclerosis. *Circulation Research*, **85**, 377–85.

155 Pervin, S., Singh, R., Rosenfeld, M.E. *et al.* (1998) Estradiol suppresses MCP-1 expression In vivo: implications for atherosclerosis. *Arteriosclerosis, Thrombosis, and Vascular Biology*, **18**, 1575–82.

156 Marx, N., Sukhova, G.K., Collins, T. *et al.* (1999) PPARalpha activators inhibit cytokine-induced vascular cell adhesion molecule-1 expression in human endothelial cells. *Circulation*, **99**, 3125–31.

157 Evans, M.J., Eckert, A., Lai, K. *et al.* (2001) Reciprocal antagonism between estrogen receptor and NF-kappaB activity *in vivo*. *Circulation Research*, **89**, 823–30.

158 Tham, D.M., Wang, Y.X. and Rutledge, J.C. (2003) Modulation of vascular inflammation by PPARs. *Drug News and Perspectives*, **16**, 109–16.

159 Grodstein, F., Chen, J., Pollen, D.A. *et al.* (2000) Postmenopausal hormone therapy and cognitive function in healthy older women. *Journal of the American Geriatrics Society*, **48**, 746–52.

160 Rosenfeld, M.E., Kauser, K., Martin-McNulty, B. *et al.* (2002) Estrogen inhibits the initiation of fatty streaks throughout the vasculature but does not inhibit intra-plaque hemorrhage and the progression of established lesions in apolipoprotein E deficient mice. *Atherosclerosis*, **164**, 251–9.

161 Mendelsohn, M.E. and Karas, R.H. (2005) Molecular and cellular basis of cardiovascular gender differences. *Science*, **308**, 1583–7.

162 Proudler, A.J., Ahmed, A.I., Crook, D. *et al.* (1995) Hormone replacement therapy and serum angiotensin-converting-enzyme activity in postmenopausal women. *Lancet*, **346**, 89–90.

163 Schunkert, H., Danser, A.H., Hense, H.W. *et al.* (1997) Effects of estrogen replacement therapy on the renin–angiotensin system in postmenopausal women. *Circulation*, **95**, 39–45.

164 Nickenig, G., Baumer, A.T., Grohe, C. *et al.* (1998) Estrogen modulates AT1 receptor gene expression *in vitro* and *in vivo*. *Circulation*, **97**, 2197–201.

165 Tsai, J.C., Perrella, M.A., Yoshizumi, M. *et al.* (1994) Promotion of vascular smooth muscle cell growth by homocysteine: a link to atherosclerosis. *Proceedings of the National Academy of Sciences of the United States of America*, **91**, 6369–73.

166 van Baal, W.M., Smolders, R.G., van der Mooren, M.J. *et al.* (1999) Hormone replacement therapy and plasma homocysteine levels. *Obstetrics and Gynecology*, **94**, 485–91.

167 Morey, A.K., Razandi, M., Pedram, A. *et al.* (1998) Oestrogen and progesterone inhibit the stimulated production of endothelin-1. *Biochemical Journal*, **330**, 1097–105.

168 Dubey, R.K., Jackson, E.K., Keller, P.J. et al. (2001) Estradiol metabolites inhibit endothelin synthesis by an estrogen receptor-independent mechanism. *Hypertension*, **37** (2 Pt 2), 640–4.

169 Gorodeski, G.I. (2002) Update on cardiovascular disease in post-menopausal women. *Best Practice and Research Clinical Obstetrics and Gynaecology*, **16**, 329–55.

170 Seed, M. and Knopp, R.H. (2004) Estrogens, lipoprotcins, and cardiovascular risk factors: an update following the randomized placebo-controlled trials of hormone-replacement therapy. *Current Opinion in Lipidology*, **15**, 459–67.

171 Wu, F.C. and von Eckardstein, A. (2003) Androgens and coronary artery disease. *Endocrine Reviews*, **24**, 183–217.

172 Gaspard, U.J. and Lefebvre, P.J. (1990) Clinical aspects of the relationship between oral contraceptives, abnormalities in carbohydrate metabolism, and the development of cardiovascular disease. *American Journal of Obstetrics and Gynecology*, **163** (1 Pt 2), 334–43.

173 Reckelhoff, J.F., Zhang, H. and Srivastava, K. (2000) Gender differences in development of hypertension in spontaneously hypertensive rats: role of the renin–angiotensin system. *Hypertension*, **35** (1 Pt 2), 480–3.

174 Reckelhoff, J.F., Zhang, H., Srivastava, K. et al. (1999) Gender differences in hypertension in spontaneously hypertensive rats: role of androgens and androgen receptor. *Hypertension*, **34** (4 Pt 2), 920–3.

175 Crofton, J.T., Ota, M. and Share, L. (1993) Role of vasopressin, the renin–angiotensin system and sex in Dahl salt-sensitive hypertension. *Journal of Hypertension*, **11**, 1031–8.

176 Rowland, N.E. and Fregly, M.J. (1992) Role of gonadal hormones in hypertension in the Dahl salt-sensitive rat. *Clinical and Experimental Hypertension Part A–Theory and Practice*, **14**, 367–75.

177 Ouchi, Y., Share, L., Crofton, J.T. et al. (1987) Sex difference in the development of deoxycorticosterone-salt hypertension in the rat. *Hypertension*, **9**, 172–7.

178 Ashton, N. and Balment, R.J. (1991) Sexual dimorphism in renal function and hormonal status of New Zealand genetically hypertensive rats. *Acta Endocrinologica (Copenhagen)*, **124**, 91–7.

179 Ganten, U., Schroder, G., Witt, M. et al. (1989) Sexual dimorphism of blood pressure in spontaneously hypertensive rats: effects of anti-androgen treatment. *Journal of Hypertension*, **7**, 721–6.

180 Baylis, C. (1994) Age-dependent glomerular damage in the rat. Dissociation between glomerular injury and both glomerular hypertension and hypertrophy. Male gender as a primary risk factor. *Journal of Clinical Investigation*, **94**, 1823–9.

181 Dubey, R.K., Oparil, S., Imthurn, B. et al. (2002) Sex hormones and hypertension. *Cardiovascular Research*, **53**, 688–708.

182 Chapman, A.B., Zamudio, S., Woodmansee, W. et al. (1997) Systemic and renal hemodynamic changes in the luteal phase of the menstrual cycle mimic early pregnancy. *American Journal of Physiology*, **273** (5 Pt 2), F777–782.

183 Siamopoulos, K.C., Papanikolaou, S., Elisaf, M. et al. (1996) Ambulatory blood pressure monitoring in normotensive pregnant women. *Journal of Human Hypertension*, **10** (Suppl. 3), S51–54.

184 Seely, E.W., Walsh, B.W., Gerhard, M.D. et al. (1999) Estradiol with or without progesterone and ambulatory blood pressure in postmenopausal women. *Hypertension*, **33**, 1190–4.

185 Szekacs, B., Vajo, Z., Acs, N. et al. (2000) Hormone replacement therapy reduces mean 24-hour blood pressure and its variability in postmenopausal women with treated hypertension. *Menopause*, **7**, 31–5.

186 Butkevich, A., Abraham, C. and Phillips, R.A. (2000) Hormone replacement therapy and 24-hour blood pressure profile of postmenopausal women. *American Journal of Hypertension*, **13**, 1039–41.

187 Crane, M.G. and Harris, J.J. (1978) Estrogens and hypertension: effect of

discontinuing estrogens on blood pressure, exchangeable sodium, and the renin-aldosterone system. *American Journal of the Medical Sciences*, **276**, 33–55.

188 The Writing Group for the PEPI Trial. (1995) Effects of estrogen or estrogen/progestin regimens on heart disease risk factors in postmenopausal women. The Postmenopausal Estrogen/Progestin Interventions (PEPI) Trial. *Journal of the American Medical Association*, **273**, 199–208.

189 Gebara, O.C., Mittleman, M.A., Sutherland, P. *et al.* (1995) Association between increased estrogen status and increased fibrinolytic potential in the Framingham Offspring Study. *Circulation*, **91**, 1952–8.

190 Koh, K.K., Mincemoyer, R., Bui, M.N. *et al.* (1997) Effects of hormone-replacement therapy on fibrinolysis in postmenopausal women. *New England Journal of Medicine*, **336**, 683–90.

191 Nabulsi, A.A., Folsom, A.R., White, A. *et al.* (1993) Association of hormone-replacement therapy with various cardiovascular risk factors in postmenopausal women. The Atherosclerosis Risk in Communities Study Investigators. *New England Journal of Medicine*, **328**, 1069–75.

192 Harrison, R.L. and McKee, P.A. (1984) Estrogen stimulates von Willebrand factor production by cultured endothelial cells. *Blood*, **63**, 657–64.

193 Massouh, M., Jatoi, A., Gordon, E.M. *et al.* (1989) Heparin cofactor II activity in plasma during pregnancy and oral contraceptive use. *Journal of Laboratory and Clinical Medicine*, **114**, 697–9.

194 Helmerhorst, F.M., Rosendaal, F.R. and Vandenbroucke, J.P. (1998) Venous thromboembolism and the pill. The WHO technical report on cardiovascular disease and steroid hormone contraception: state-of-the-art. *Human Reproduction*, **13**, 2981–3.

Part Two　Peptide Hormones

3
Angiotensins

*Robson Augusto Souza Dos Santos, Anderson José Ferreira,
and Ana Cristina Simões e Silva*

3.1
Introduction

There have been significant advances in our knowledge regarding the renin–angiotensin system (RAS) over recent decades. Originally, the RAS was viewed as consisting of a single, biologically active hormone, angiotensin (Ang) II, generated by the sequential action of renin, an aspartyl protease, on angiotensinogen (AGT) forming the decapeptide Ang I and angiotensin-converting enzyme (ACE) on Ang I. Further degradation by aminopeptidase A and N produces Ang III [Ang-(2–8)], and Ang IV [Ang-(3–8)], respectively [1]. Ang II interacts with two plasma membrane receptors, AT_1 and AT_2. Recent advances in the RAS field have included the discovery of novel biologically active peptides, additional specific receptors, alternative pathways of Ang II generation, and additional roles for enzymes and precursor components, other than Ang II synthesis. In this chapter we will briefly review relevant aspects of the angiotensin peptides with a focus on their cardiovascular and renal effects.

3.2
RAS: The Classical and Updated View

The classical RAS is considered systemic in nature. This view considers that Ang II is synthesized in the circulation, and contributes to maintaining electrolyte balance, body fluid volume and arterial pressure primarily through vasoconstriction, a direct effect on sodium absorption in proximal tubules and stimulation of aldosterone release.

One of the most significant improvements in our understanding of the RAS has been the discovery of local or tissue RASs (for review, see [2]). A local system is characterized by the presence of RAS components, AGT, processing enzymes, angiotensins and specific receptors. Local systems appear to be regulated independently of the circulatory RAS, but can also interact with the latter.

The identification of a functional local RAS has been established through a variety of pharmacological and genetic approaches. In addition to the brain [3], other sites in which local systems have been demonstrated include the heart, blood vessels, kidney, pancreas, and reproductive, lymphatic and adipose tissues [1, 4–6]. Putative actions of the local RAS include cell growth and remodeling in the heart and blood vessels, control of blood pressure and kidney function, central stimulation of food and water intake, and hormone secretion.

Another important advance in the understanding of the RAS was the recognition that Ang-(1–7) is a biologically active product of the renin–angiotensin cascade [7]. Most evidence supports a counter-regulatory role for Ang-(1–7) by opposing many AT_1 receptor-mediated actions, especially vasoconstriction and proliferation [8].

Other important conceptual changes were introduced by the identification of other receptors [insulin-regulated aminopeptidase [9], Mas [10], Pro (renin) [11]] and the enzyme ACE2 [12, 13]. The possibility of signaling through the RAS enzymes, renin/prorenin [11] and ACE [14] as well as the Ang II-independent signaling through AT_1 receptor [15] are other landmark changes in our understanding of this fascinating system. Figure 3.1 shows the current view of the RAS.

3.3
New Aspects of Classical RAS Enzymes

3.3.1
Renin/Prorenin Receptor

Renin is considered a key enzyme of the RAS due to the rate-limiting nature of its hydrolytic activity on the precursor AGT. In recent years this view has acquired new and important features after the discovery of the renin/prorenin receptor by Nguyen *et al.* [11]. Thus, according to our new understanding of the role of renin within the RAS, it serves as an Ang II-generating enzyme through the limited proteolysis process started by its action on AGT and it can act as an agonist of the RAS-inducing signaling through its receptor. In addition, binding of renin to its receptor appears to increase its catalytic efficiency on its substrate. We are still searching for the biological and physiopathological implications of these actions, but there are suggestions that it may be implicated in target-organ lesion, especially in the kidney [16].

3.3.2
Signaling Through ACE

ACE plays a central role in the RAS since it is the main Ang II-generating enzyme. However, the role of ACE is not restricted to is hydrolytic activity. Fleming

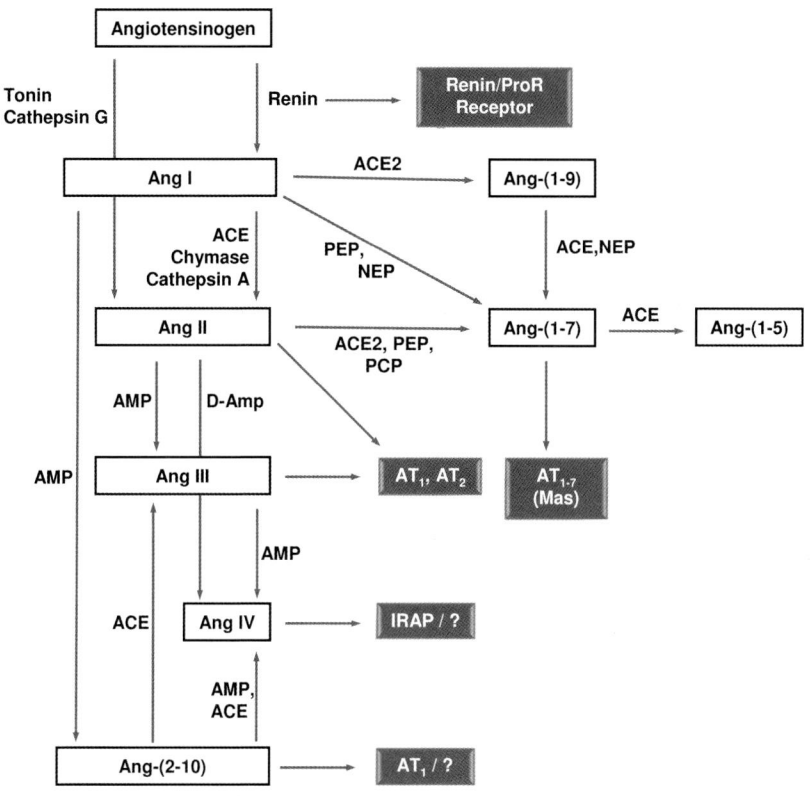

Figure 3.1 Updated view of the RAS cascade.
ACE, angiotensin-converting enzyme; Ang, angiotensin;
AMP, aminopeptidase; AT_1, Ang II type 1 receptor;
AT_2, Ang II type 2 receptor; AT_{1-7}, Ang-(1–7) receptor Mas;
D-Amp, dipeptidyl-aminopeptidase; IRAP, insulin-regulated aminopeptidase; NEP, neutral endopeptidase 24.11; PCP, prolyl-carboxypeptidase; PEP, prolyl-endopeptidase; ProR, prorenin.

et al. [14] have identified a new action of ACE – the out–in signaling. According to this novel view, upon binding by bradykinin (BK) and ACE inhibitors (ACEIs), ACE is phosphorylated at the amino acid Ser1270 and triggers a signal cascade which can lead to increased ACE synthesis and cycloxygenase-2 augmentation. More recently signaling through ACE was described in adipocytes and preadipocytes [14].

3.4
Ang II

Ang II was isolated in 1940 by Braun-Menendez et al. [17] and Page and Helmer [18], and first characterized as a potent vasoconstrictor that increases peripheral vascular resistance and consequently elevates arterial pressure. In situations of extracellular fluid volume contraction, Ang II reduces renal sodium excretion via alterations in renal hemodynamics, direct enhancement of proximal tubule sodium bicarbonate reabsorption, and aldosterone-mediated increases in late distal tubule and cortical collecting duct reabsorption [19–21]. Ang II also increases thirst, salt appetite and intestinal sodium absorption, all of which increase the extracellular fluid volume. In this context, the RAS was initially defined as an endocrine system in which circulating Ang II regulates blood pressure and electrolyte balance via its actions on vascular tone, aldosterone secretion, renal sodium handling, thirst, water intake, sympathetic activity and vasopressin release. Figure 3.2 illustrates some of the actions of Ang II.

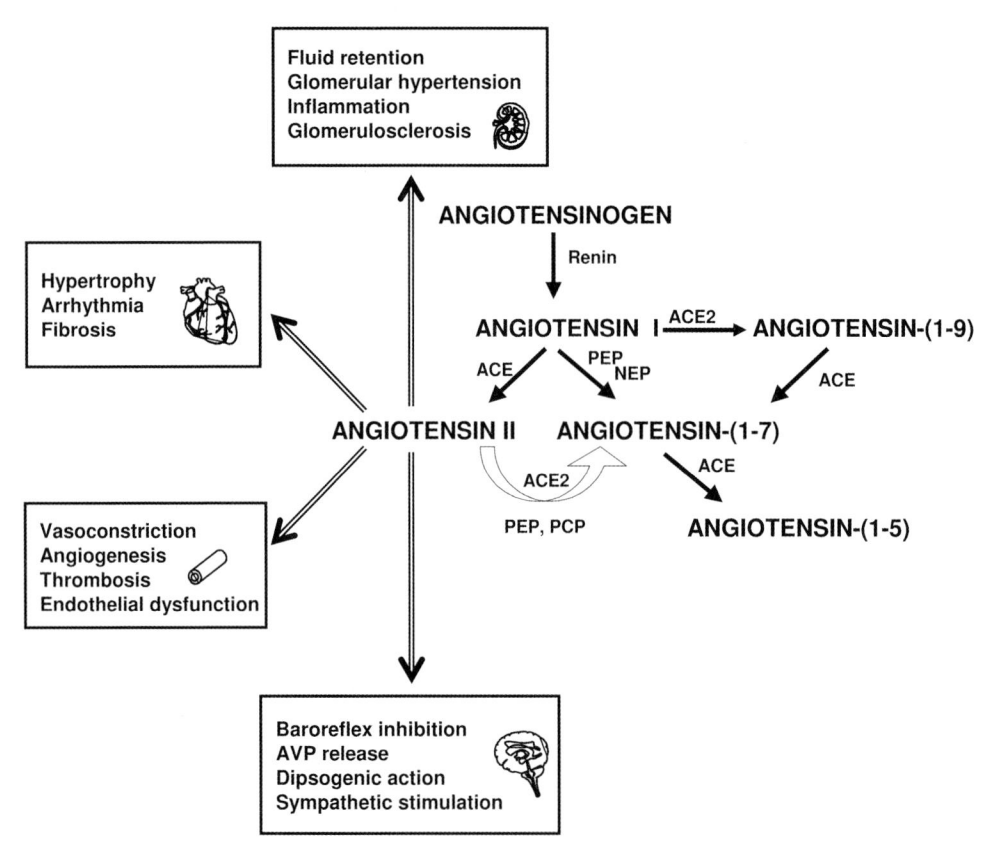

Figure 3.2 Schematic representation of the main actions of the Ang II in the heart, vessels, kidneys and brain. ACE, angiotensin-converting enzyme; AVP, arginine vasopressin; NEP, neutral endopeptidase 24.11; PCP, prolyl-carboxypeptidase; PEP, prolyl-endopeptidase.

Additional experimental studies as well as clinical trials using ACEIs and Ang II type 1 receptor blockers (ARBs) have shown that the actions of this system extend far beyond blood pressure control and electrolyte balance [22]. It has become clear that the deleterious actions of the RAS on cardiovascular remodeling and renal function account for the beneficial effects of these agents in patients with hypertension, left ventricular hypertrophy, heart failure and diabetic nephropathy [22]. Thus, apart from the circulating RAS, Ang II as well as other RAS mediators can also be locally produced, and exert paracrine, intracrine and autocrine effects [2]. Although the existence of the local RAS was proposed by Dzau [23] a long time ago, only more recently, with advances in measurement technology, the various components of the RAS have been clearly detected in different tissues [2]. One of the first pieces of evidence for the tissue formation of Ang II came from studies on angiotensin metabolism in sheep [24]. The octapeptide was subsequently found to be produced in numerous tissues, including the adrenal, brain, heart, kidney, vasculature, adipose tissue, gonads, pancreas, prostate, eye and placenta [1, 4–6]. Further investigations proposed that in the circulating RAS, renin originating from the kidney is a rate-limiting factor for Ang II generation in plasma, but in the vascular tissue, ACE and other peptidases such as chymase regulate Ang II generation [2]. Although not all components of the classical RAS are synthesized locally in some tissues, alternative enzymatic pathways or the presence of the renin receptor, which binds to and activates circulating renin and prorenin [11], may also permit local Ang II formation. In addition, the RAS is also able to exert intracrine effects due to the intracellular formation of Ang II [25, 26]. However, kinetic studies suggest that intracellular Ang II is largely derived from receptor-mediated uptake of the extracellular peptide [27].

The cellular Ang II receptor was identified in 1974 as a high-affinity plasma membrane-binding site that is sensitive to guanyl nucleotides [28]. Later, distinct AT_1 and AT_2 receptor subtypes were identified by selective ligands [29], and were subsequently characterized as seven-transmembrane receptors by molecular cloning [30–33]. Thus, Ang II mediates its effects via complex intracellular signaling pathways that are stimulated following binding of the peptide to its cell-surface receptors, AT_1 receptor and AT_2 receptor [30, 34]. In humans, AT_1 receptor is widely expressed in blood vessels, heart, kidney, adrenal glands and liver, whereas AT_2 receptor is present mainly in fetal tissue, decreasing rapidly after birth, with relatively low amounts normally expressed in adult tissue. The G-protein-coupled AT_1 receptor mediates the known physiological and pathological actions of Ang II, and undergoes rapid desensitization and internalization after agonist stimulation. In contrast, AT_2 receptor does not exhibit the latter features and acts mainly through G_i and tyrosine phosphatases to exert predominantly inhibitory actions on cellular responses mediated by AT_1 receptor and growth factor receptors [35]. For example, both receptors play a role in regulating vascular smooth muscle cell (VSMC) function, although they differ in their actions. While AT_1 receptor is associated with growth, inflammation and vasoconstriction, AT_2 receptor is generally associated with opposite actions stimulating apoptosis and vasodilatation [34, 35].

Animal models and clinical data have also helped to establish that inhibition of Ang II actions in nonclassical target sites, such as immune cells, explains some of the unanticipated therapeutic effects of ACEIs and ARBs [22]. Parallel studies on the molecular mechanism of action of Ang II have revealed that its main target, the AT_1 receptor, is one of the most versatile members of the G-protein-coupled receptor (GPCR) family. Ang II exerts several cytokine-like actions via the AT_1 receptor and can stimulate multiple signaling pathways, activate several growth factor receptors, and promote the formation of reactive oxygen species (ROS) and other proinflammatory responses.

3.4.1
Ang II and the Cardiovascular System

The potential regulatory role of the RAS in endothelial cell (EC) function is indicated by the ability of ACEIs and ARBs to improve EC parameters in patients with cardiovascular disease and diabetes. EC dysfunction is manifested by impaired control of vascular tone, increased adhesiveness of leukocytes, and increased formation of growth factors and cytokines. These processes have been implicated in the development of hypertension, heart failure, atherosclerosis, diabetes and renal failure. Ang II-induced endothelial dysfunction promotes the cytokine-mediated recruitment of monocyte/macrophages into the vascular wall by monocyte chemotactic protein (MCP)-1 and vascular cell adhesion molecule (VCAM)-1. Ang II also increases the expression of vascular endothelial growth factor (VEGF) and its receptors in aortic wall cells, and causes inflammatory changes with monocyte infiltration and vascular remodeling [36]. The octapeptide also potentiates VEGF-induced proliferation of endothelial progenitor cells and promotes their VEGF-dependent formation into networks, by increasing the expression of VEGF receptor kinase domain-containing receptors [37] as well as angiopoietin-2 and multiple growth factors such as fibroblast growth factor, endothelial growth factor, platelet-derived growth factor and transforming growth factor (TGF)-β [36, 37].

Apart from endothelial function, VSMCs are dynamic and multifunctional cells critically involved in maintaining vascular integrity, contributing to arterial remodeling through numerous processes, such as hyperplasia, hypertrophy, apoptosis, cell elongation, reorganization, altered production of extracellular matrix proteins and inflammation [38]. It has been recognized that inflammation participates in vascular remodeling and accelerates vascular damage in cardiovascular diseases in part due to the activation of RAS [39]. In this regard, Ang II has significant proinflammatory actions in the vascular wall, inducing the production of ROS, cytokines, adhesion molecules and activation of redox-sensitive inflammatory genes [40]. Ang II also modulates expression of proinflammatory molecules in the vessel wall, such as VCAM-1, intercellular adhesion molecule (ICAM) and E-selectin through redox-dependent pathways [41]. In VSMCs, Ang II stimulates VCAM-1 production, chemokine MCP-1 and the cytokine interleukin-6, which stimulates recruitment of mononuclear leukocytes into the vessel media. Many of these factors are increased in plasma from hypertensive patients, and may reflect vascular inflammation and target organ damage [42].

Animal studies have shown that the deleterious actions of Ang II in the cardiovascular system can be also caused by increases in aldosterone and other corticosteroids that activate the mineralocorticoid receptor (MR), particularly at inappropriate levels for the prevailing sodium status. These hormones have been implicated in the development of endothelial dysfunction, cardiac and vascular inflammation and remodeling, and cardiac arrhythmias. The harmful effects of aldosterone in the heart were discovered in studies on the perivascular and interstitial and ventricular fibrosis associated with arterial hypertension and cardiac failure [43]. The finding that Ang II and aldosterone regulate the accumulation of collagen within the rat heart led to the concept that ventricular remodeling in cardiac failure could be treated by aldosterone receptor antagonists [44]. This prediction was confirmed by the RALES and EPHESUS trials, in which blockade of the MR by spironolactone and epleronone, respectively, significantly reduced mortality in patients with advanced cardiac failure [45, 46]. Aldosterone has also been found to potentiate Ang II-stimulated proliferation of rat VSMCs, possibly by increasing expression of the AT_1 receptor [47, 48]. Recent studies on human coronary artery VSMCs have shown that Ang II and aldosterone regulate the transcription of genes involved in vascular fibrosis, inflammation and calcification [49].

3.4.2
Ang II and the Endocrine System

Ang II has pleiotropic effects on the endocrine system. Very low concentrations of Ang II are able to release vasopressin, adrenocorticotropic hormone, luteinizing hormone-releasing hormone and aldosterone. Regarding the female reproductive system, Ang II has been proposed as a mediator of ovarian function and follicle maturation, and also participates in the control of uterine blood flow during pregnancy. In the male reproductive system, local RAS function is thought to be relevant for fertility. This system is regulated independently from the circulating RAS, and the blood–testicular barrier prevents ACE inhibitors and AT_1 blockers affecting fertility. In addition, Ang II can modulate the actions of many hormones including adrenaline, aldosterone and insulin (for review, see [50]).

The association of the RAS with the endocrine system is particularly illustrated by the prominent role of Ang II in diabetes and metabolic syndrome. The frequent association of diabetes mellitus (DM) with hypertension, retinopathy, nephropathy and cardiovascular disease has implicated the RAS in the initiation and progression of these disorders. This has been demonstrated by clinical trials in which RAS inhibitors significantly reduced the incidence of vascular complications in DM patients [51]. These improvements appear to result from increased perfusion of the microcirculation in skeletal muscle and pancreatic islets, and also from enhanced insulin sensitivity associated with increased leptin and adiponectin production, as well as increased hexokinase activity and glucose transport. Ang II can also cause insulin resistance by interfering with the insulin-stimulated increase in insulin receptor substrate 1-associated phosphatidylinositol 3-kinase (PI3K) activity [52]. In addition, the renal RAS is clearly activated in DM, with increased tissue Ang II [53] that leads to the development of diabetic nephropathy – a major cause of end-stage renal disease. Blockade of the RAS could thus reduce tissue Ang II levels, with beneficial effects on cardiovascular and renal function.

The so-called metabolic syndrome, now regarded as a global epidemic, includes obesity, insulin resistance, dyslipidemia and hypertension, often with hypertriglyceridemia and reduced high-density lipoprotein, as well as hyperuricemia and increased C-reactive protein [54]. These features are often associated with EC dysfunction with impaired control of vascular tone, increased adhesiveness of leukocytes, and increased formation of growth factors and cytokines. These processes have been implicated in the development of hypertension, heart failure, atherosclerosis, diabetes and renal failure. In this context, RAS components, mostly Ang II, have a potential role in EC dysfunction, insulin resistance, inflammation and proliferative effects.

3.4.3
Ang II and the Renal System

Ang II is one of the most powerful regulators of sodium excretion, operating through extrarenal mechanisms, such as the stimulation of aldosterone secretion, as well as intrarenal mechanisms [19–21]. In the physiological concentration range, Ang II acts mainly as an antinatriuretic hormone by directly or indirectly producing changes in renal hemodynamics and in sodium reabsorption [55]. Considerable evidence suggests that the intrarenal actions of Ang II are quantitatively more important than the extrarenal mechanisms in the normal day-to-day regulation of sodium and water excretion [19–21].

The extrarenal actions of Ang II include stimulation of the sympathetic nervous system and aldosterone secretion [19]. The antinatriuretic effect of Ang II could be mediated in part by increased renal sympathetic nerve activity that in turn could increase renal tubular sodium reabsorption directly or indirectly by causing renal vasoconstriction [56]. The other well-known extrarenal mechanism by which Ang II controls sodium and water excretion is through its indirect renal effect mediated by aldosterone. Aldosterone biosynthesis and secretion are strongly influenced by Ang II, which acts directly on adrenal glomerulosa cells to stimulate both early and late steps in the steroid biosynthetic cascade [57].

Two intrarenal effects of Ang II that may contribute to fluid retention and that occur at very low concentrations include the constriction of efferent arterioles and increased sodium transport by the renal tubules. Ang II markedly raises efferent glomerular arteriolar resistance and does not change afferent arteriolar resistance unless the renal perfusion pressure rises [58–60]. The consequence of the disproportionate increase in efferent (over afferent) resistance is a marked increase in the mean transcapillary hydraulic pressure, which results in an increase in mean transcapillary ultrafiltration pressure. Thus, the Ang II-induced decrease in renal plasma flow is offset by the increase in mean transcapillary ultrafiltration pressure and this maintains the glomerular filtration rate (GFR) by increasing the filtration fraction. Apart from its effect on efferent arteriolar resistance, Ang II has a marked influence on glomerular mesangial tone. Ang II promotes mesangial cell contraction and thus a decrease in the glomerular capillary ultrafiltration coefficient (K_f) [61, 62]. The K_f-lowering effect of Ang II is attenuated by the stimulatory effect of Ang II on the production of vasodilatory prostaglandins [62]. Thus, the final effect of Ang II is a stable GFR despite the changes in renal perfusion pressure.

The vasoconstrictor effect of Ang II on efferent arterioles causes changes in peritubular capillary dynamics that could increase renal tubular fluid reabsorption [58]. The efferent arteriolar vasoconstriction would reduce the peritubular capillary hydrostatic pressure as well as increase peritubular capillary colloid osmotic pressure due to an increase in the filtration fraction. Both of these changes would tend to decrease renal interstitial fluid hydrostatic pressure and raise interstitial fluid colloid osmotic pressure, thereby increasing the driving force for fluid reabsorption [21, 58].

Another mechanism by which Ang II-mediated changes in renal hemodynamics could increase tubular reabsorption is by decreasing renal medullary blood flow [59, 63]. The constriction of efferent arterioles of juxtaglomerular nephrons or a direct action on the vasa recta may lower renal medullary blood flow and increase medullary interstitial fluid osmolality. The increased osmolality could raise the urine-concentrating ability and tend to enhance passive sodium chloride reabsorption in the thin ascending limb of Henle's loop [58].

Micropuncture and microperfusion studies have shown that the effect of Ang II on proximal tubule sodium transport is bimodal [21, 60, 64]. Ang II at concentrations of 1–100 pM significantly stimulates proximal sodium reabsorption, whereas higher concentrations of 0.1–1 μM inhibit transport [64]. Various studies suggest that at physiological concentrations (1–100 pM) Ang II promotes sodium bicarbonate reabsorption by activation of Na^+–H^+ exchange [65]. Ang II also participates in the regulation of distal nephron acidification by stimulating the Na^+–H^+ exchange in early and late distal segments via activation of AT_1 receptors, as well as vacuolar H^+-ATPase in the late distal segment [66].

In contrast to its antinatriuretic actions, Ang II produces natriuresis when infused at supraphysiological concentrations that increase mean arterial pressure [21]. This natriuretic effect of Ang II is mainly due to an increase in renal perfusion pressure and, additionally, to a direct inhibitory action on proximal tubular reabsorption. This effect, called pressure natriuresis, is thought to involve changes in peritubular Starling forces or interstitial pressures increasing backleakage of fluid into the tubule lumen [21]. Ang II-associated pressure natriuresis appears to serve as a negative feedback system on sodium retention produced by Ang II. Thus, the net effect of different levels of Ang II on sodium and water excretion depends critically on the balance between the antinatriuretic actions of Ang II and the effect of increased renal perfusion pressure that tends to cause natriuresis.

In human nondiabetic renal disease, combined treatment with ACEIs and ARBs seems to be more effective than either agent alone [67]. Interestingly, RAS blockade and receptor antagonists are also effective in reducing glomerular damage and renal chemokine expression in MRL/*1pr* mice with lupus-like autoimmune renal disease [68]. A slight regression of sclerosis was also observed after long-term treatment with AT_1 receptor antagonists in the aging rat kidney [69]. The proposed mechanisms involved include increased tubular epithelial cell turnover and inhibition of plasminogen activator inhibitor-1 expression. Administration of an ACEI accelerated the obstruction release-induced regression in the model of unilateral ureteral obstruction [70].

In an experimental model of nitric oxide (NO) deficiency that leads to hypertension and acute renal failure due to vascular and glomerular fibrosis, the administration of AT_1 receptor antagonist for 1 week decreased collagen I, collagen IV and TGF-β gene and protein expression without affecting the increased level of metalloproteinase-2 and -9 activities in glomeruli. These cellular alterations were accompanied by a gradual regression of glomerulosclerosis, a partial restoration of renal function and an arrest of mortality. When the treatment was extended to 1 month, all cellular, structural and functional parameters of the kidney were normalized, indicating that the progression of renal vascular fibrosis is a reversible process [71]. Based on these data, the authors proposed that the mechanism of the regression was dual: inhibition of the exaggerated synthesis of extracellular matrix (due to blockade of the Ang II-TGF-β pathway) and increased rate of matrix degradation (due to metalloproteinase activity, probably associated with the degree of fibrosis and independent of angiotensin blockade). Subsequently, other investigators confirmed the reversibility of fibrotic process after Ang II blockade in the model of acute nephronic reduction [72, 73]. However, the decreased number of podocytes, following renal ablation, was not restored by the pharmacological treatment, suggesting that glomerular regeneration largely depends on the degree of damage of glomerular podocytes [74]. In agreement with this notion, mesangial proliferation was reduced and interstitial changes were reversed after favorable treatment, whereas the number of sclerotic glomeruli remained unchanged in biopsies of IgA nephropathy patients [75].

In summary, the remarkable array of physiological and pathological actions elicited by Ang II reflects its extraordinarily diverse repertoire of signaling mechanisms and pathways, sometimes leading to progressive inflammatory and degenerative conditions that underlie major disease entities. On the other hand, the availability of highly effective agents for inhibition of Ang II formation, and selective blockade of the AT_1 receptor, has led to the identification of many of the hitherto unrecognized pathological actions of Ang II. In addition to hypertension, these include an impressive spectrum of disorders ranging from endothelial dysfunction and atheroma to cardiac hypertrophy, atrial fibrillation, cardiac failure, renal disease, obesity and the metabolic syndrome, DM, and insulin resistance. Over the recent past our view of Ang II has changed from being a simple vasoconstrictor to that of a complex growth factor mediating effects through diverse signaling pathways. It has also become clear that Ang II is a key player in vascular inflammation. Through increased generation of ROS and activation of redox-sensitive transcription factors, Ang II promotes expression of cell adhesion molecules and induces synthesis of proinflammatory mediators and growth factors. These processes facilitate increased vascular permeability, leukocyte recruitment and fibrosis leading to tissue injury and structural remodeling. Targeting some of these signaling events with novel therapeutic strategies may provide important tissue protection in many forms of cardiovascular, renal, metabolic and even neoplasic diseases.

3.5 Ang-(1–7)

Ang-(1–7) was regarded as an inactive component of the RAS for many years [8]. However, after the demonstration that Ang-(1–7) was the main product formed from Ang I through an ACE-independent pathway in dog brain micropunches homogenates [76] and the demonstration that this heptapeptide was equipotent to Ang II to elicit vasopressin release from neurohyphophyseal explants [7] and after decreasing blood pressure upon micro-injection into the dorsal medulla of chloralose-urethan anesthetized rats [77] the importance of Ang-(1–7) within the RAS was becoming increasingly evident. Ang-(1–7) is present in the circulation and in cardiovascular tissues including heart, vessels and kidney [8]. Figure 3.3 illustrates some of the actions of Ang-(1–7).

The recent identification of the ACE homolog enzyme ACE2 which forms Ang-(1–7) from Ang II [12, 13] and the G-protein-coupled receptor Mas as an Ang-(1–7) receptor [10] have provided biochemical and molecular evidence for the biological significance of the Ang-(1–7). In addition, the inhibitory effects of Ang-(1–7) on Ang II-induced vasoconstriction, and its growth-inhibitory, antiarrhythmogenic and antithrombogenic effects imply a counter-regulatory role for this angiotensin within the RAS and suggest that this may be a potential target for development of new cardiovascular drugs (Figure 3.4).

3.5.1 Ang-(1–7) and the Heart

Evidence for a functional RAS in the heart has been obtained in several species [78]. Ang-(1–7) is one of the RAS components present in the heart [79–84]. The presence and local generation of Ang-(1–7) in the canine heart was demonstrated by measurements performed in blood samples collected from the aortic root, coronary sinus and right atrium at basal conditions [79]. In isolated rat hearts, Ang I

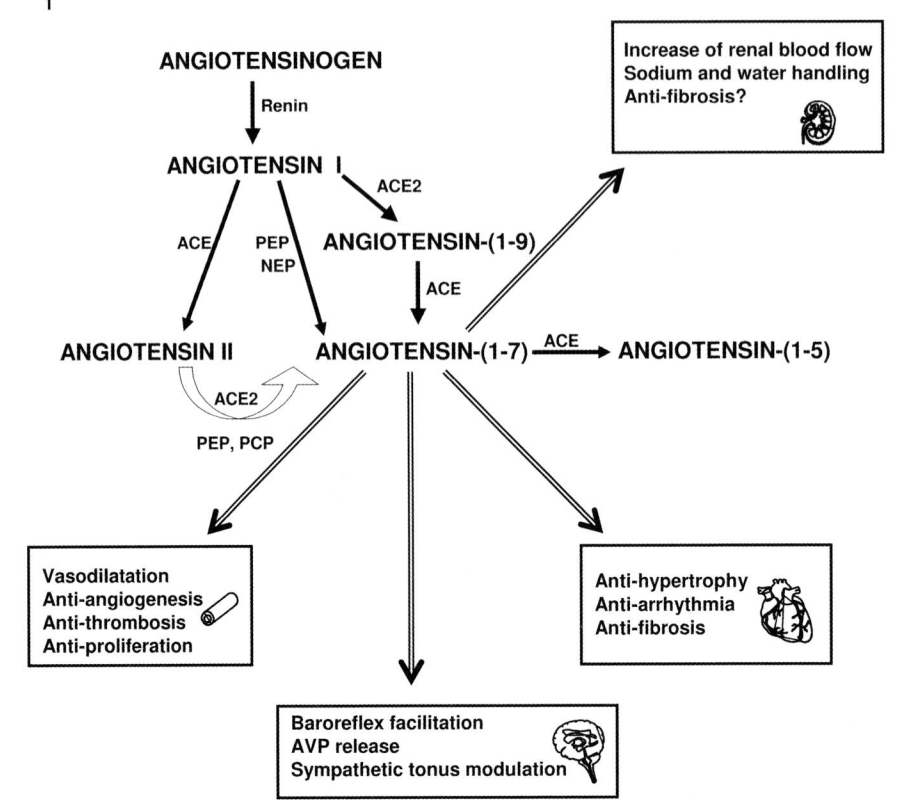

Figure 3.3 Schematic representation of the main actions of the Ang-(1–7) in the heart, vessels, kidneys and brain. ACE, angiotensin-converting enzyme; AVP, arginine vasopressin; NEP, neutral endopeptidase 24.11; PCP, prolyl-carboxy-peptidase; PEP, prolyl-endopeptidase.

is extensively metabolized during a single pass through the coronary bed leading to the formation of Ang II, Ang III, Ang IV and Ang-(1–7) [80, 81]. Averill et al. [84] found that Ang-(1–7) immunoreactivity was limited to cardiac myocytes and absent in interstitial cells and vessels. Myocardial infarction induced by left coronary artery occlusion significantly augmented the Ang-(1–7) immunoreactivity in the ventricular tissue surrounding the infarcted area. In addition, Ang-(1–7) is also formed in intact human myocardial circulation and its generation is markedly decreased by ACEI [82]. Furthermore, the novel ACE homolog (ACE2) is highly expressed in the heart [13]. Recent studies have indicated that ACE2 is an important regulator of the cardiac function [85, 86]. However, the genetic background appears to influence the role of ACE2 in heart function [87]. This enzyme can form Ang-(1–7) directly from Ang II or indirectly from Ang I. Interestingly, heart ACE2 is increased by treatment with AT_1 antagonists and in failing human hearts ventricles [86, 88].

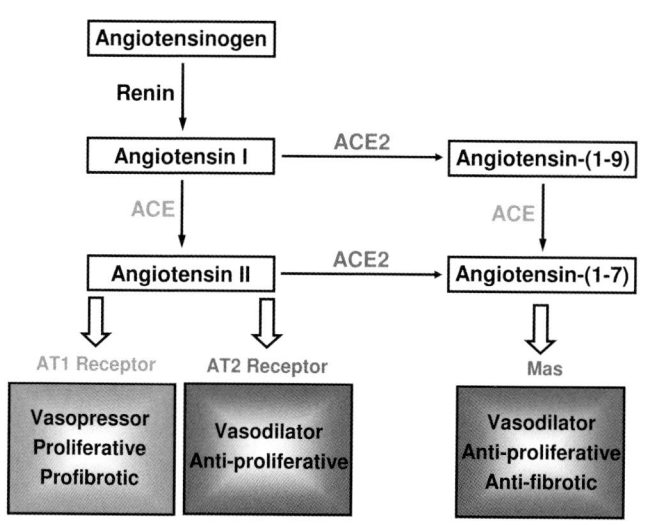

Figure 3.4 Opposing cardiovascular actions of the angiotensin receptors. ACE2 represents the intersection between the counter-regulatory arms of the RAS [ACE–Ang II–AT₁ receptor and ACE2–Ang-(1–7)–Mas], since this enzyme can cleave the vasoconstrictor/proliferative peptide Ang II to form the vasodilator/antiproliferative fragment Ang-(1–7). ACE, angiotensin-converting enzyme; AT$_1$, Ang II type 1 receptor; AT$_2$, Ang II type 2 receptor; Mas, Ang-(1–7) receptor.

It has been suggested that the heart is a main target for the Ang-(1–7) actions, and that this angiotensin induces vasodilation in the coronary bed [89–91] and improves cardiac function [92–96]. Ang-(1–7) reduced the growth of cardiomyocytes [97] and has an inhibitory effect on collagen synthesis [98, 99]. In isolated canine coronary artery rings precontracted with the thromboxane A$_2$ analog U46,619, Ang-(1–7) elicited a dose-dependent vasorelaxation (10 nM to 100 µM). This effect was completely blocked by the nonselective Ang II antagonist [Sar¹, Thr⁸]-Ang II, but not by the selective AT$_1$ or AT$_2$ Ang II antagonists CV11974 and PD 123319, respectively, suggesting the existence of a specific ligand site for Ang-(1–7) in dog coronary vessels. Moreover, the vasorelaxant activity of Ang-(1–7) was markedly attenuated by the NO synthase inhibitor N^G-nitro-L-arginine (L-NAME) and by the specific BK-B$_2$ receptor blocker HOE 140, but not by the cyclooxygenase inhibitor indometachin [90, 100, 101]. Ang-(1–7) also elicited an endothelium-dependent dilator response in porcine coronary artery rings which was markedly attenuated by a NO synthase inhibitor [91]. In keeping with the report concerning isolated canine coronary artery rings [90], the Ang-(1–7)-induced relaxation in porcine coronary artery was not affected by AT$_1$ or AT$_2$ receptor blockade or by cyclooxygenase inhibition, but was attenuated by the B$_2$ receptor antagonist. Moreover, the ACEI, quinaprilat, augmented this vascular effect. Other reports have also demonstrated that local kinin production, BK-B$_2$ receptors and NO release are involved in Ang-(1–7)-induced dilatation in canine coronary vessels [102, 103] and in porcine coronary artery [104]. Another possible mechanism involved in the coronary effects of Ang-(1–7) is the amplification of the vasodilator effects of BK. Thus, the vasodilator effect produced by BK was potentiated by Ang-(1–7) in isolated perfused rat hearts [89], in isolated canine coronary arteries [90], and in porcine coronary arteries [105]. On the other hand, Neves *et al.* [106] found that Ang-(1–7) induced a concentration-dependent (27, 70 and 210 nM) decrease in coronary flow in isolated perfused rat hearts. A similar finding was observed in isolated hamster hearts [107].

Ang-(1–7) has been described as a beneficial effector of the RAS in the heart. We have shown that Ang-(1–7) at low concentration (220 pM) decreases the incidence and duration of ischemia/reperfusion arrhythmias [95] and improves the postischemic contractile function [96] in isolated perfused rat hearts by a mechanism apparently involving its receptor Mas [10], and release of BK and prostaglandins. It is also well known that Ang II exerts important deleterious effects in the heart. Therefore, one possible additional mechanism that may contribute to the beneficial effects of Ang-(1–7) [108] in the heart is the modulation of cardiac Ang II levels, as recently described by Mendes *et al.* [109]. Of note is the fact that, at higher concentration (27 nM), Ang-(1–7) facilitates the development of reperfusion arrhythmias in isolated perfused rat hearts [106]. Accordingly, transgenic mice overexpressing ACE2 in the heart presented sudden death due to cardiac arrhythmias [110]. These observations suggest that high local concentrations of Ang-(1–7) exert deleterious effects in the heart.

In keeping with the beneficial effects of Ang-(1–7) in the heart, chronic infusion (8 weeks) of Ang-(1–7) improved endothelial aortic function and coronary perfusion, and preserved cardiac function in an experimental rat model of heart failure induced by ligation of the left coronary artery [93]. Accordingly, transgenic rats [TGR(A1-7)3292] which present a 2.5-fold increase in plasma Ang-(1–7) concentration, showed a slight but significant increase in daily and nocturnal dP/dt, a less pronounced cardiac hypertrophy induced by isoproterenol, a reduced duration of reperfusion arrhythmias and an improved postischemic function in isolated perfused hearts [92]. In addition, this peptide produced a significant increase in cardiac output and stroke volume in anesthetized Wistar rats, as assessed by a fluorescent microspheres methodology. These effects were partially attenuated by the Ang-(1–7) antagonist, A-779 [94].

Many of the Ang-(1–7) effects in the heart are completely blocked by the Ang-(1–7) antagonist A-779 [89, 94, 95, 111–113], suggesting the presence of a specific binding site for Ang-(1–7) in the heart. These indirect pharmacological evidence were recently reinforced by the identification of an Ang-(1–7) receptor, the GPCR Mas. Recent studies using Mas knockout mice have unraveled a pivotal role of the Mas receptor in the cardiac function. Isolated hearts of Mas-deficient mice presented a marked decrease in systolic tension, positive and negative dT/dt, heart rate, and an increase in coronary vascular resistance [114]. These findings are in keeping with the antifibrotic effect of Ang-(1–7) in many models, including Ang II infusion [115], DOCA-salt [116] and treatment with L-NAME [117]. Furthermore, the left ventricular pressure, positive dP/dt and cardiac output were significantly lower in Mas knockout mice, as assessed by means a micro conductance catheter placed into the left ventricle [118]. A reduced heart rate variability associated with an increased sympathetic tone was also found in Mas-deficient animals [119].

Recently, Wiemer et al. [120] described a nonpeptide compound AVE 0991 which mimics the Ang-(1–7) effects on the endothelium. It was observed that AVE 0991 evoked a similar NO/O_2^- releasing profile to Ang-(1–7) most probably through stimulation of a specific Ang-(1–7) receptor sensible to the selective Ang-(1–7) antagonist A-779. Afterwards, AVE 0991 has been demonstrated to be an Ang-(1–7) analog in several tissues such as kidney [121], vessels [122, 123] and heart [117, 124, 125]. In the heart, the protective effects of the AVE 0991 were qualitatively comparable to those of Ang-(1–7), that is, improvement of the postischemic contractile function in isolated perfused hearts from spontaneously hypertensive rats treated with L-NAME [117], decrease of the infarcted area induced by left coronary artery ligation [124] and attenuation of the fibrosis induced by isoproterenol treatment [125]. It is worth mentioning the fact that, in addition to preventing Ang II-induced morphological changes in the heart, Ang-(1–7) has been reported to decrease cardiac Ang II levels [109] – an observation that was recently confirmed using transgenic rats with chronic increases in Ang-(1–7) [TGR(A1-7)3292] [126].

An important role related to the cardioprotection produced by Ang-(1–7)-Mas is that it is blood pressure-independent as evidenced by the *in vitro* studies [97, 99] and many experimental models [114–116, 124, 125].

3.5.2
Ang-(1–7) and Blood Vessels

Vascular ECs are an important site for the formation [127] and metabolism [128] of Ang-(1–7), where it induces release of vasodilator active products, including vasodilatory prostanoids, NO and endothelium-derived hyperpolarizing factor [90, 91, 102, 111, 129–132]. Accordingly, Ang-(1–7) elicits relaxation in several vascular beds, including canine [90] and porcine coronary arteries [91], canine middle cerebral artery [133], porcine piglet pial arterioles [134], feline systemic vasculature [135], rabbit renal afferent arterioles [112], rat aortic rings [136], and mesenteric microvessels of normotensive [137] and hypertensive rats [111]. A vasodilator effect of this peptide was also reported in the neovasculature as well as in the skin [138]. However, contradictory effects have been reported in human vessels. Sasaki et al. [139] reported vasodilation in the human forearm, while Davie and McMurray [140] did not observe any effect of Ang-(1–7) in the same territory in ACEI-treated patients. In keeping with previous reports by Iyer et al. [141], a role for Ang-(1–7) in mediating the chronic hypotensive effect of losartan in normal rats has been proposed by Collister and Hendel [142].

Several mechanisms are involved in the Ang-(1–7) vasodilatory effect, depending on the vessel diameter, the vascular regional bed, and the species. A complex interaction between the AT_{1-7} receptor and B_2 and AT_2 receptors appears to be involved in mediating some of the Ang-(1–7) effects in blood vessels. It has been demonstrated that the vasodilator action of Ang-(1–7) can be negated by the B_2

receptor antagonist HOE 140 [90, 104]. Moreover, a recent study showed that the link between Ang II and BK in endothelium-dependent vasorelaxation could involve receptors that bind the AT_2 and Ang-(1–7) receptor antagonists, PD 123319 and A-779, respectively [143]. More recently, Sampaio et al. [132] demonstrated that Ang-(1–7) is present in human ECs and elicited NO release from human aortic ECs through a mechanism involving phosphorylation of endothelial NO synthase through the PI3/Akt pathway.

It has been described that Ang-(1–7) potentiates the vasodilatory effect induced by BK in several beds and species. Ang-(1–7) potentiates the vasodilator and hypotensive effects of BK in normotensive [137, 144] and hypertensive [111, 145] rats, in porcine coronary arteries [105], in canine coronary artery rings [102], in rat coronary vessels [89], and in endothelium-denuded segments of the rabbit jugular vein [146]. As mentioned above for the vasodilation, the BK-potentiating activity of Ang-(1–7) is controversial in humans. Ueda et al. [147] reported BK potentiation, whereas Wilsdorf et al. [148] were unable to show an effect at supraphysiological doses.

Several mechanisms have been proposed to explain the BK-potentiating activity of Ang-(1–7), including ACE inhibition [102, 105], nonhydrolytic interaction with ACE favoring a cross-talk between ACE and the BK-B_2 receptor [149] and AT_{1-7} receptor-mediated changes in the coupling and/or signaling of BK. The recent observation that the Ang-(1–7) [95, 111, 137, 144] antagonist, D-Ala^7-Ang-(1–7), reverts the potentiation of BK in mesenteric microvessels by enalapril or enalaprilat [111] or attenuates the potentiated hypotensive response to BK in captopril-treated rats [150] indicates that an Ang-(1–7)-related mechanism is importantly involved in the BK-related effects of ACEIs. In addition, vascular Ang-(1–7) actions could involve modulation of Ang II-induced changes in vascular resistance [151, 152].

The Ang-(1–7) receptor Mas is also involved in the vascular effects of Ang-(1–7). The endothelium-dependent relaxation induced by Ang-(1–7) in mouse aortic rings is absent in vessels derived from Mas-deficient mice. A link between Mas and NO release was recently described by Pinheiro et al. [121]. Either Ang-(1–7) and the Ang-(1–7) receptor agonist, AVE 0991 [120], elicited an increase in NO release in Mas-transfected CHO cells. This effect was blocked by A-779 but not by AT_1 or AT_2 antagonists [121]. Furthermore, Ang-(1–7) treatment elicited arachidonic acid release from CHO and COS cells transfected with the Mas receptor [10].

It has been described that Ang II and Ang-(1–7) exhibit opposite effects on the regulation of cell growth. In a recent report, Gallagher and Tallant [153] demonstrated that Ang-(1–7) inhibited the growth of lung cancer cells by a mechanism presumably involving its receptor Mas and through inhibition of the extracellular signal-regulated kinase (ERK) 1/2 signal transduction pathway. Ang-(1–7) also inhibited [^{3}H]thymidine incorporation in VSMCs in response to stimulation by fetal bovine serum, platelet-derived growth factor and Ang II [154]. In addition, using a murine sponge model of angiogenesis, it has been demonstrated an antiproliferative effect of Ang-(1–7) on fibrovascular tissue [155]. The receptor Mas appears to be also involved in the antiproliferative effect of Ang-(1–7) in VSMCs [156] and stent-induced neointima proliferation [157]. Furthermore, it has been demonstrated that Ang-(1–7) inhibits vascular growth through the production of prostaglandin-mediated intracellular events, which includes cAMP production and reduction of Ang II-stimulated ERK1/2 activities [158].

A still unsolved question is whether other receptors besides Mas could contribute to the vasorelaxant effect of Ang-(1–7). The intriguing observations that in some circumstances the effects of Ang-(1–7) on blood vessels cannot be blocked by A-779 raised the possibility of involvement of AT_2 or an yet unknown receptor in some of the Ang-(1–7) actions. Walters et al. [159] recently reported that in candesartan-treated spontaneously hypertensive rats, Ang-(1–7) produced a marked hypotensive effect which could not be blocked by A-779 but was completely prevent by the specific AT_2 antagonist PD 123319. Whether this is only a candesartan-related effect remains to be clarified. Silva et al. [160] have recently observed that in Sprague-Dawley isolated aorta the Ang-(1–7)-induced vasorelaxation could not be blocked by A-779. However, PD 123319 was also ineffective. In this particular case the only effective blockade was obtained with another Ang-(1–7) antagonist D-Pro^7-Ang-(1–7). Another intriguing observation was made by Castro et al. [161] who reported that in isolated mice hearts treated with losartan, Ang-(1–7), at subpicomolar concentration, produced Mas-mediated vasodilation (blunted in Mas knockout mice and by A-779). In contrast, in PD 123319-treated hearts Ang-(1–7) produced a slight but significant increase in the perfusion pressure which was reduced but not blocked by losartan or a combination of losartan with A-779. Taken together, these observations suggest that in particular situations the vascular effects of Ang-(1–7) may be determined by a complex and still elusive mechanism involving a cross-talk or physical interaction of Mas with AT_1 or AT_2 receptors. This possibility appears to be true, at least for AT_1 receptors [162–164].

3.5.3
Ang-(1–7) and the Kidney

Accumulating evidence suggest that besides Ang II, Ang-(1–7) plays a key role in renal function. Ang-(1–7) increased renal blood flow in anesthetized rats [94]. More important, it has been also reported that Ang-(1–7) produces renal vasodilation in hypertensive patients by a mechanism independent of the prevailing level of RAS activity [165]. These observations are in line with the increased renal blood flow observed in rats harboring an Ang-(1–7)-producing fusion protein [166]. Ang-(1–7) produces afferent arteriolar relaxation through specific receptor-mediated NO release in isolated kidneys of rabbits [112]. Of note is the fact that the vascular effects of Ang-(1–7) in the kidney appear to be importantly influenced by the experimental condition [167].

In addition to the effects on renal blood flow, studies in vitro and in anesthetized animals have suggested Ang-(1–7) as a natriuretic/diuretic hormone, since it has a putative role in the regulation of sodium transport across the proximal tubules. In this segment, a modulatory role of Ang-(1–7) on Na^+-ATPase [168] and a reduction in energy-dependent transcellular sodium transport have been described [169]. In contrast, Ang-(1–7) has an antidiuretic effect in water-loaded animals [121, 170, 171], probably due to a receptor mediated-increase in water transport across the intramedullary collecting ducts [170, 172]. The natriuresis and diuresis observed with acute [170, 171] or chronic [173] infusion of the Ang-(1–7) antagonist, A-779, and the natriuresis observed with acute A-779 administration in anesthetized rats [174] are in agreement with these findings.

In a spite of the controversial and complex findings, agonists and antagonists of the Ang-(1–7)–Mas axis probably possess a therapeutic potential for the modulation of sodium and water excretion in many physiological and pathological conditions, such as arterial hypertension, nephrogenic insipidus diabetic, nephrotic syndrome, chronic renal failure, chronic heart failure, ascites and hepatorenal syndrome. To date, the levels of Ang-(1–7) in plasma, renal tissue and urine are altered during physiological and pathophysiological conditions, including those associated with changes in blood pressure, blood volume and sodium intake [175–181]. For instance, Ferrario et al. [178] reported that Ang-(1–7) is excreted into the urine of normal healthy adult volunteers in amounts 2.5-fold higher than those measured in plasma. In the same study it was also observed that untreated essential hypertensive subjects exhibited a lower urinary excretion rate of Ang-(1–7) than normotensive controls [178]. Furthermore, urinary concentrations of Ang-(1–7) were inversely correlated with blood pressure [178]. In this regard, we have recently detected a significant increase in plasma Ang-(1–7) levels among hypertensive children with chronic renal failure and an even more pronounced elevation among patients with end-stage renal disease [181]. On the other hand, Ang-(1–7) levels were similar in normotensive chronic renal failure subjects and in healthy subjects, despite the significant difference in glomerular filtration rate [181]. Accordingly, our previous study showed significant differences in Ang II and Ang-(1–7) plasma levels among pediatric subjects with renovascular disease and essential hypertension [180]. In renovascular disease, plasma levels of Ang II were higher than Ang-(1–7) and the surgical treatment produced a return of circulating angiotensins to healthy subject values. In contrast, essential hypertensive subjects exhibited a selective and blood-independent elevation of plasma Ang-(1–7) [180]. Taken together, these results show changes in the RAS profile according to differences in hypertensive states and the degree of renal dysfunction. These findings do not allow any mechanistic inferences, but suggest that an alteration in the RAS metabolism or Ang-(1–7) activity may contribute toward the evolution of hypertension and human renal diseases.

In view of the aforementioned results, the relevance of the actions of Ang-(1–7) as a mechanism counteracting the effects of enhanced Ang II levels on renal function remains unknown. Future studies exploring the role of the GPCR Mas in its renal effects and using Ang-(1–7) receptor agonists such as the nonpeptide Ang-(1–7) agonist, AVE 0991 [120, 121], could be useful to elucidate the physiological and pathological role of this heptapeptide in the kidney.

3.6
Ang III [Ang-(2–8)]

Ang III is generated from the metabolism of Ang II by aminopeptidase A, which cleaves the Asp1–Arg2 bond of Ang II. Ang III is a biologically active peptide of the RAS and, currently, some actions initially attributed to Ang II have been proven to be mediated by Ang III [182]. Essentially, the Ang III effects are similar to those observed for Ang II. In fact, Ang III enhances blood pressure, vasopressin release and thirst when it is centrally administrated [183]. Ang III infusion increases blood pressure in healthy volunteers and hypertensive patients [184] as well as augments aldosterone release [185, 186]. Although Ang III does not change renal function in humans [186], it induces natriuresis in AT_1 receptor-blocked rats likely by binding to AT_2 receptors [187]. In addition, in cultured renal cells this peptide stimulates the expression of many growth factors, proinflammatory mediators and extracellular matrix proteins [188]. The similarity between both peptides is also revealed in terms of their cell surface ligand binding, since apparently Ang II and Ang III share the same receptors, that is, AT_1 and AT_2 receptors [33]. However, some studies suggest that Ang III possesses more affinity to AT_2 receptors [189].

In summary, although Ang III effects are similar to those described for Ang II, they are much less potent. However, whether the supposed Ang III effects are due to direct action of this peptide or mediated by smaller peptides formed as a consequence of its degradation remains to be established.

3.7
Ang IV/AT_4 Receptor Axis

Recent findings have suggested important functional roles for Ang IV [Ang-(3–8)] in several tissues. The aminopeptidase N removes the amino acid arginine from the N-terminus of the Ang III sequence to generate Ang IV. Alternatively, this angiotensin fragment can be also formed directly from Ang II by action of the D-aminopeptidase.

Ang IV may be involved in the learning and memory acquisition and retrieval in animal models of amnesia when it is centrally administered [190–192]. Interestingly, microinjection of Ang II into the same brain areas elicits opposite effects [193]. Renal infusions of Ang IV produce a dose-dependent increase in cortical blood flow without altering systemic blood pressure. This effect was completely abolished by divalinal – a specific antagonist of Ang IV binding sites [194]. However, Ang IV effects in renal function are not clear, since contrasting results have shown that this peptide can decrease total and regional renal blood flow in rats [195]. It has been reported that, contrasting to the Ang II-induced vasopressor effect, Ang IV induces vasodilation in precontracted endothelium-intact but not endothelium-denuded pulmonary artery through a specific binding site that is blocked by divalinal [196]. In addition, this peptide increases endothelial NO synthase activity and cellular cGMP content in porcine pulmonary arterial ECs [197]. However, the vasodilatory effect of Ang IV was not confirmed by other reports, at least in the renal vasculature [198, 199].

Ang IV has mixed effects in the heart. It reduced left ventricular pressure development and ejection capabilities, but increased the sensitivity of pressure development during the systole and speeded relaxation [200]. All together, it seems that Ang IV is a biologically active peptide of the RAS with many effects in the brain, kidney, vessels and heart. Nevertheless, its actions are poorly understood and, frequently, they are contrasting. Additionally, these findings also suggest that some Ang II effects could be mediated through its metabolism to Ang IV, as recently proposed by Braszko et al. [192].

Most of Ang IV's actions are mediated by a specific binding site commonly called the AT_4 receptor. It is important to note that Ang IV is not the only ligand for this receptor, since other peptides such as LVV-hemorphin 7 can bind to this binding site [201]. It has a broad distribution and is found in many organs, including brain, adrenal gland, kidney, lung and heart. This receptor is specifically blocked by the AT_4 receptor antagonist, divalinal, but not by the AT_1 receptor antagonist DuP753 or the AT_2 receptor antagonist PD 123177 [194]. Albiston et al. [9] have identified the AT_4 receptor as an insulin-regulated aminopeptidase (IC_{50} 32 nM). However, in some circumstances, Ang IV can induce its effects by interacting with Ang II type 1 receptors [195].

3.8
Des-Asp^1-Ang I [Ang-(2–10)]

The nonapeptide Ang-(2–10) is generated by N-terminal degradation of Ang I through specific aminopeptidase hydrolysis [202]. The first evidence for a biological role of this peptide was provided by Sim and Radhakrishnan [203]. These authors demonstrated that Ang-(2–10) attenuates the central pressor actions of Ang II and Ang III in spontaneously hypertensive rats and Wistar Kyoto rats. Recent reports suggest that Ang-(2–10) has a cardioprotective effect. Thus, Ang-(2–10) prevents the development of cardiac hypertrophy induced by coarctation of the abdominal aorta in rats [204] and reduces the infarct size in ischemic reperfused rat hearts [205]. In keeping with these data, this nonapeptide attenuates the Ang II-induced incorporation of [^{3}H]phenylalanine in cultured rat cardiomyocytes, but not in rat aortic smooth muscle cells. In addition, this peptide significantly attenuates the Ang II-induced [^{3}H]thymidine incorporation in the smooth muscle cells [206]. It has been suggested that Ang-(2–10) exerts its actions on an indomethacin-sensitive angiotensin AT_1 receptor subtype [207–209]. However, whether des-Asp^1-Ang I cardiovascular effects are due to a direct action or depend on its processing to smaller peptides remain to be elucidated.

3.9
Other Angiotensin Peptides

The issue whether Ang-(1–9) itself has direct biological effects or whether its supposed actions are elicited by generation of different metabolites, especially Ang-(1–7), remains unsolved. In some pathological conditions, its concentration changes significantly indicating that it could be involved in those diseases [210]. Ang-(1-9) can be formed direct from Ang I through catalytic action of ACE2 and it is metabolized by ACE and neutral endopeptidase 24.11 to generate Ang-(1–7) [211]. It has been reported that this peptide enhances BK actions, increases NO and arachidonic acid release [212], and is involved in the regulation of platelet function [213].

Recent reports have attributed some biological actions to the small fragment of the RAS, Ang-(3–7). Resulting from the degradation of Ang-(1–7), Ang II and Ang IV by aminopeptidases and carboxypeptidases it is believed that this angiotensin plays important role in the brain [214, 215] and kidney [216]. Indeed Ang-(3–7) has an affinity for AT_4 receptors that differs from Ang IV.

Differently from others angiotensin fragments, des-Asp^1-Ala^1-Ang II is generated by a decarboxylation reaction instead of cleavage process. Thus, the first amino acid of the Ang II sequence (aspartic acid) is converted into alanine. It has similar affinity to the AT_1 receptor as Ang II, but a higher affinity to the AT_2 receptor. However, its activity in *in vitro* preparations is lower when compared with Ang II. Interestingly, due to its high affinity to the AT_2 receptor, it could be an endogenous Ang II regulatory peptide [217].

The demonstration of endogenous Ang-(1–12) in the rat may foretell renin-independent pathways that lead to the formation of biologically active peptides [218]. Indeed, the peptide bond Tyr12–Tyr13 hydrolyzed to produce Ang-(1–12) from rat AGT is structurally distinct from the Leu10–Leu11 bond recognized by rat renin to form Ang I. Moreover, these bonds are also distinct for human angiotensinogen [218]. Despite the absence of any functional effect attributed to Ang-(1–12), the processing to this intermediate peptide may comprise another level of RAS regulation.

The expansion of RAS understanding in the next few years will be remarkable. There is little doubt that further studies will add new evidence to the relevance of these newly discovered endogenous pathways and reveal whether these pathways are of therapeutic value.

References

1 Lavoie, J.L. and Sigmund, C.D. (2003) Minireview: overview of the renin–angiotensin system-an endocrine and paracrine system. *Endocrinology*, **144**, 2179–83.

2 Miyazaki, M. and Takai, S. (2006) Tissue angiotensin II generating system by angiotensin-converting enzyme and chymase. *Journal of Pharmacy and Pharmaceutical Sciences*, **100**, 391–7.

3 Ganten, D., Marquez-Julio, A., Granger, P., Hayduk, K., Karsunky, K.P., Boucher, R. and Genest, J. (1971) Renin in dog brain. *American Journal of Physiology*, **221**, 1733–7.

4 Spät, A. and Hunyady, L. (2004) Control of aldosterone secretion: a model for convergence in cellular signaling pathways. *Physiological Reviews*, **84**, 489–539.

5 Sernia, C. (2001) A critical appraisal of the intrinsic pancreatic angiotensin-generating system. *Journal of the Pancreas*, **2**, 50–5.

6 Nielsen, A.H., Schauser, K.H. and Poulsen, K. (2000) Current topic: the uteroplacental renin–angiotensin system. *Placenta*, **21**, 468–77.

7 Schiavone, M.T., Santos, R.A.S., Brosnihan, K.B., Khosla, M.C. and Ferrario, C.M. (1988) Release of vasopressin from the rat hypothalamo-neurohypophysial system by angiotensin-(1–7) heptapeptide. *Proceedings of the National Academy of Sciences of the United States of America*, **85**, 4095–8.

8 Santos, R.A.S., Ferreira, A.J., Pinheiro, S.V., Sampaio, W.O., Touy, Z.R. and Campagnole-Santos, M.J. (2005) Angiotensin-(1–7) and its receptor as a potential targets for new cardiovascular drugs. *Expert Opinion on Investigational Drugs*, **14**, 1019–31.

9 Albiston, A.L., McDowall, S.G., Matsacos, D., Sim, P., Clune, E., Mustafa, T., Lee, J., Mendelsohn, F.A., Simpson, R.J., Connolly, L.M., Chai, S.Y. (2004) Evidence that the angiotensin IV (AT_4) receptor is the enzyme insulin-regulated aminopeptidase. *Journal of Biological Chemistry*, **276**, 48623–6.

10 Santos, R.A.S., Simoes e Silva, A.C., Maric, C., Silva, D.M., Machado, R.P., de Buhr, I., Heringer-Walther, S., Pinheiro, S.V., Lopes, M.T., Bader, M., Mendes, E.P., Lemos, V.S., Campagnole-Santos, M.J., Schultheiss, H.P., Speth, R. and Walther, T. (2003) Angiotensin-(1–7) is an endogenous ligand for the G protein-coupled receptor Mas. *Proceedings of the National Academy of Sciences of the United States of America*, **100**, 8258–63.

11 Nguyen, G., Delarue, F., Burckle, C., Bouzhir, L., Giller, T. and Sraer, J.D. (2002) Pivotal role of the renin/prorenin receptor in angiotensin II production and cellular responses to renin. *Journal of Clinical Investigation*, **109**, 1417–27.

12 Donoghue, M., Hsieh, F., Baronas, E., Godbout, K., Gosselin, M., Stagliano, N., Donovan, M., Woolf, B., Robison, K., Jeyaseelan, R., Breitbart, R.E. and Acton, S. (2000) A novel angiotensin-converting enzyme-related carboxypeptidase (ACE2) converts angiotensin I to angiotensin 1–9. *Circulation Research*, **87**, E1–9.

13 Tipnis, S.R., Hooper, N.M., Hyde, R., Karran, E., Christie, G. and Turner, A.J. (2000) A human homolog of angiotensin-converting enzyme. Cloning and functional expression as a captopril-insensitive carboxypeptidase. *Journal of Biological Chemistry*, **275**, 33238–43.

14 Kohlstedt, K., Brandes, R.P., Muller-Esterl, W., Busse, R. and Fleming, I. (2004) Angiotensin-converting enzyme is involved in outside-in signaling in endothelial cells. *Circulation Research*, **94**, 60–7.

15 Abdalla, S., Lother, H. and Quitterer, U. (2000) AT_1-receptor heterodimers show enhanced G-protein activation and altered receptor sequestration. *Nature*, **407**, 94–8.

16 Staessen, J.A., Li, Y. and Richart, T. (2006) Oral renin inhibitors. *Lancet*, **368**, 1449–56.

17 Braun-Menendez, E., Fasciolo, J.C., Leloir, L.F. and Munoz, J.M. (1940) The substance causing renal hypertension.

Journal of Physiology – London, **98**, 283–98.

18. Page, I.H. and Helmer, O.M. (1940) A crystalline pressor substance (angiotonin) resulting from the reaction between renin and renin activator. *Journal of Experimental Medicine*, **71**, 29–42.

19. Hall, J.E. (1991) The renin–angiotensin system: renal actions and blood pressure regulation. *Comprehensive Therapy*, **17**, 8–17.

20. Ichikawa, I. and Harris, R.C. (1991) Angiotensin actions in the kidney. Renewed insight into the old hormone. *Kidney International*, **40**, 583–96.

21. Hall, J.E., Brands, M.W. and Henegar, J.R. (1999) Angiotensin II and log-term arterial pressure regulation: the overriding dominance of the kidney. *Journal of the American Society of Nephrology*, **10**, S258–65.

22. Ferrario, C., Abdelhamed, A.I. and Moore, M. (2004) AII antagonists in hypertension, heart failure, and diabetic nephropathy: focus on losartan. *Current Medical Research and Opinion*, **20**, 279–93.

23. Dzau, V.J. (1986) Significance of the vascular renin–angiotensin pathway. *Hypertension*, **8**, 553–9.

24. Fei, D.T., Coghlan, J.P., Fernley, R.T., Scoggins, B.A. and Tregear, G.W. (1980) Peripheral production of angiotensin II and III in sheep. *Circulation Research*, **46**, I135–7.

25. Re, R.N. (2003) Intracellular renin and the nature of intracrine enzymes. *Hypertension*, **42**, 117–22.

26. Baker, K.M., Chernin, M.I., Schreiber, T., Sanghi, S., Haiderzaidi, S., Booz, G.W., Dostal, D.E. and Kumar, R. (2004) Evidence of a novel intracrine mechanism in angiotensin II-induced cardiac hypertrophy. *Regulatory Peptides*, **120**, 5–13.

27. Danser, A.H. (2003) Local renin–angiotensin systems: the unanswered questions. *International Journal of Biochemistry and Cell Biology*, **35**, 759–68.

28. Glossmann, H., Baukal, A. and Catt, K.J. (1974) Angiotensin II receptors in bovine adrenal cortex. Modification of angiotensin II binding by guanyl nucleotides. *Journal of Biological Chemistry*, **249**, 664–6.

29. De Gasparo, M., Catt, K.J., Inagami, T., Wright, J.W. and Unger, T. (2000) International union of pharmacology. XXIII. The angiotensin II receptors. *Pharmacological Reviews*, **52**, 415–72.

30. Murphy, T.J., Alexander, R.W., Griendling, K.K., Runge, M.S. and Bernstein, K.E. (1991) Isolation of a cDNA encoding the vascular type-1 angiotensin II receptor. *Nature*, **351**, 233–6.

31. Sasaki, K., Yamano, Y., Bardhan, S., Iwai, N., Murray, J.J., Hasegawa, M., Matsuda, Y. and Inagami, T. (1991) Cloning and expression of a complementary. DNA encoding a bovine adrenal angiotensin II type-1 receptor. *Nature*, **351**, 230–3.

32. Kambayashi, Y., Bardhan, S., Takahashi, K., Tsuzuki, S., Inui, H., Hamakubo, T. and Inagami, T. (1993) Molecular cloning of a novel angiotensin II receptor isoform involved in phosphotyrosine phosphatase inhibition. *Journal of Biological Chemistry*, **268**, 24543–6.

33. Mukoyama, M., Nakajima, M., Horiuchi, M., Sasamura, H., Pratt, R.E. and Dzau, V.J. (1993) Expression cloning of type 2 angiotensin II receptor reveals a unique class of seven-transmembrane receptors. *Journal of Biological Chemistry*, **268**, 24539–42.

34. Touy, Z.R. and Schiffrin, E.L. (2000) Signal transduction mechanisms mediating the physiological and pathophysiological actions of angiotensin II in vascular smooth muscle cells. *Pharmacological Reviews*, **52**, 639–72.

35. Nouet, S. and Nahmias, C. (2000) Signal transduction from the angiotensin II AT_2 receptor. *Trends In Endocrinology and Metabolism*, **11**, 1–6.

36. Zhao, Q., Ishibashi, M., Hiasa, K., Tan, C., Takeshita, A. and Egashira, K. (2004) Essential role of vascular endothelial growth factor in angiotensin II-induced vascular inflammation and remodeling. *Hypertension*, **44**, 264–70.

37. Imanishi, T., Hano, T. and Nishio, I. (2004) Angiotensin II potentiates vascular endothelial growth factor-induced proliferation and network formation of

endothelial progenitor cells. *Hypertension Research*, **27**, 101–8.
38 Berk, B.C. (2001) Vascular smooth muscle growth: autocrine growth mechanisms. *Physiological Reviews*, **81**, 999–1030.
39 Virdis, A. and Schiffrin, E.L. (2003) Vascular inflammation: a role in vascular disease in hypertension? *Current Opinion in Nephrology and Hypertension*, **12**, 181–7.
40 Suematsu, M., Suzuki, H., DeLano, F.A. and Schmid-Schonbein, G.W. (2002) The inflammatory aspect of the microcirculation in hypertension: oxidative stress, leukocytes/endothelial interaction, apoptosis. *Microcirculation*, **9**, 259–76.
41 Pueyo, M.E., Gonzalez, W., Nicoletti, A., Savoie, F., Arnal, J.F. and Michel, J.B. (2002) Angiotensin II stimulates endothelial vascular cell adhesion molecule-1 via nuclear factor-kappaB activation induced by intracellular oxidative stress. *Arteriosclerosis Thrombosis and Vascular Biology*, **20**, 645–51.
42 Hlubocka, Z., Umnerova, V., Heller, S., Peleska, J., Jindra, A., Jachymova, M., Kvasnicka, J., Horky, K. and Aschermann, M. (2002) Circulating intercellular cell adhesion molecule-1, endothelin-1 and von Willebrand factor-markers of endothelial dysfunction in uncomplicated essential hypertension: the effect of treatment with ACE inhibitors. *Journal of Human Hypertension*, **16**, 557–62.
43 Brilla, C.G., Pick, R., Tan, L.B., Janicki, J.S. and Weber, K.T. (1990) Remodeling of the rat right and left ventricles in experimental hypertension. *Circulation Research*, **67**, 1355–64.
44 Weber, K.T., Brilla, C.G. and Campbell, S.E. (1992) Regulatory mechanisms of myocardial hypertrophy and fibrosis: results of *in vivo* studies. *Cardiology*, **81**, 266–73.
45 Pitt, B., Zannad, F., Remme, W.J., Cody, R., Castaigne, A., Perez, A., Palensky, J. and Wittes, J. (1999) The effect of spironolactone on morbidity and mortality in patients with severe heart failure. Randomized Aldactone Evaluation Study Investigators. *New England Journal of Medicine*, **341**, 709–17.
46 Pitt, B., Remme, W., Zannad, F., Neaton, J., Martinez, F., Roniker, B., Bittman, R., Hurley, S., Kleiman, J. and Gatlin, M. (2003) Eplerenone, a selective aldosterone blocker, in patients with left ventricular dysfunction after myocardial infarction. *New England Journal of Medicine*, **348**, 1309–21.
47 Ullian, M.E., Hutchison, F.N., Hazen-Martin, D.J. and Morinelli, T.A. (1993) Angiotensin II–aldosterone interactions on protein synthesis in vascular smooth muscle cells. *American Journal of Physiology*, **264**, C1525–31.
48 Xiao, F., Puddefoot, J.R., Barker, S. and Vinson, G.P. (2004) Mechanism for aldosterone potentiation of angiotensin II-stimulated rat arterial smooth muscle cell proliferation. *Hypertension*, **44**, 340–5.
49 Jaffe, I.Z. and Mendelsohn, M.E. (2005) Angiotensin II and aldosterone regulate gene transcription via functional mineralocortocoid receptors in human coronary artery smooth muscle cells. *Circulation Research*, **96**, 643–50.
50 Paul, M., Mehr, A.P. and Kreut, Z.R. (2006) Physiology of local renin–angiotensin systems. *Physiological Reviews*, **86**, 747–803.
51 Jandeleit-Dahm, K.A., Tikellis, C., Reid, C.M., Johnston, C.I. and Cooper, M.E. (2005) Why blockade of the renin–angiotensin system reduces the incidence of new-onset diabetes. *Journal of Hypertension*, **23**, 463–73.
52 Folli, F., Saad, M.J., Velloso, L., Hansen, H., Carandente, O., Feener, E.P. and Kahn, C.R. (1999) Crosstalk between insulin and angiotensin II signalling systems. *Experimental and Clinical Endocrinology and Diabetes*, **107**, 133–9.
53 Giacchetti, G., Sechi, L.A., Rilli, S. and Carey, R.M. (2005) The renin–angiotensin–aldosterone system, glucose metabolism and diabetes. *Trends in Endocrinology and Metabolism*, **16**, 120–6.
54 Meigs, J.B. (2003) The metabolic syndrome. *British Medical Journal*, **327**, 61–2.
55 Olsen, M.E., Hall, J.E., Montani, J.P., Guyton, A.C., Langford, H.G. and

Cornell, J.E. (1985) Mechanisms of angiotensin II natriuresis and antinatriuresis. *American Journal of Physiology*, **249**, F299–307.
56. DiBona, G.F. and Koop, U.C. (1997) Neural control of renal function. *Physiological Reviews*, **77**, 75–197.
57. Fredlund, P., Saltman, S. and Catt, P. (1975) Aldosterone production by isolated adrenal glomerulosa cells. Stimulation by physiological concentrations of angiotensin II. *Endocrinology*, **97**, 1577–82.
58. Arendshorst, W.J., Brännström, K. and Ruan, X. (1999) Actions of angiotensin II on the renal microvasculature. *Journal of the American Society of Nephrology*, **10**, S149–61.
59. Navar, L.G. and Nishiyama, A. (2004) Why are angiotensin concentrations so high in the kidney? *Current Opinion in Nephrology and Hypertension*, **13**, 107–15.
60. Brewster, U.C. and Perazella, M.A. (2004) The renin–angiotensin–aldosterone system and the kidney: effects on kidney disease. *American Journal of Medicine*, **116**, 263–72.
61. Blant, Z.R.C., Konnen, K.S. and Tucker, B.J. (1976) Angiontensin II effects upon the glomerular microcirculation and ultrafiltration coefficient of the rat. *Journal of Clinical Investigation*, **57**, 419–24.
62. Schor, N., Ichikawa, I. and Brenner, B.M. (1981) Mechanisms of action of various hormones and vasoactive substances on glomerular ultrafiltration in the rat. *Kidney International*, **20**, 442–50.
63. Brezis, M., Greenfeld, Z., Shina, A. and Rosen, S. (1990) Angiotensin II augments medullary hypoxia and predisposes to acute renal failure. *European Journal of Clinical Investigation*, **20**, 199–207.
64. Harris, P.J. and Navar, L.G. (1985) Tubular transport responses to angiotensin. *American Journal of Physiology*, **248**, F621–30.
65. Liu, F.Y. and Cogan, M.G. (1988) Angiotensin II stimulation of hydrogen ion secretion in rat early proximal tubule. *Journal of Clinical Investigation*, **82**, 601–6.
66. Barreto-Chaves, M.L.M. and Mello-Aires, M. (1996) Effect of luminal angiotensin II and ANP on early and late cortical distal tubule HCO3- reabsorption. *American Journal of Physiology*, **271**, F977–84.
67. Doggrell, S.A. (2003) ACE inhibitors and AT-1-receptor antagonists cooperate in non-diabetic renal disease. *Expert Opinion on Pharmacotherapy*, **4**, 1185–8.
68. Pere, Z.D., De Wit, C., Cohen, C.D., Nieto, E., Molina, A., Banas, B., Luckow, B., Vicente, A.B., Mampaso, F. and Schlondorff, D. (2003) Angiotensin inhibition reduces glomerular damage and renal chemokine expression in MRL/lpr mice. *Journal of Pharmacology and Experimental Therapeutics*, **307**, 275–81.
69. Ma, L.J., Nakamura, S., Whitsitt, J.S., Marcantoni, C., Davidson, J.M. and Fogo, A.B. (2000) Regression of sclerosis in aging by an angiotensin inhibition-induced decrease in PAI-1. *Kidney International*, **58**, 2425–36.
70. Koo, J.W., Kim, Y., Rozen, S. and Mauer, M. (2003) Enalapril accelerates remodeling of the renal interstitium after release of unilateral ureteral obstruction in rats. *Journal of Nephrology*, **16**, 203–9.
71. Boffa, J.J., Ying, L., Placier, S., Stefanski, A., Dussaule, J.C. and Chatziantoniou, C. (2003) Regression of renal vascular and glomerular fibrosis. Role of angiotensin II receptor antagonism and metalloproteinases. *Journal of the American Society of Nephrology*, **14**, 1132–44.
72. Adamczak, M., Gross, M.L., Krtil, J., Koch, A., Tyralla, K., Amann, K. and Ritz, E. (2003) Reversal of glomerulosclerosis after high-dose enalapril treatment in subtotally nephrectomized rats. *Journal of the American Society of Nephrology*, **14**, 2833–42.
73. Adamczak, M., Gross, M.L., Amann, K. and Rit, Z.E. (2004) Reversal of glomerular lesions involves coordinated restructuring of glomerular microvasculature. *Journal of the American Society of Nephrology*, **15**, 3063–72.
74. Chatziantonion, C. and Dussaule, J.C. (2005) Insights into the mechanisms of renal fibrosis. is it possible to achieve

75 Hotta, O., Furuta, T., Chiba, S., Tomioka, S. and Taguma, Y. (2002) Regression of IgA nephropathy: a repeat biopsy study. *American Journal of Kidney Diseases*, **39**, 493–502.

76 Santos, R.A.S., Brosnihan, K.B., Chappell, M.C., Pesquero, J., Chernicky, C.L., Greene, L.J., Ferrario, C.M. (1988) Converting enzyme activity and angiotensin metabolism in the dog brainstem. *Hypertension*, **11**, I153–7.

77 Campagnole-Santos, M.J., Diz, D.I., Santos, R.A.S., Khosla, M.C., Brosnihan, K.B. and Ferrario, C.M. (1989) Cardiovascular effects of angiotensin-(1–7) injected into the dorsal medulla of rats. *American Journal of Physiology*, **257**, H324–9.

78 Grinstead, W.C. and Young, J.B. (1992) The myocardial renin–angiotensin system: existence, importance, and clinical implications. *American Heart Journal*, **123**, 1039–45.

79 Santos, R.A.S., Brum, J.M., Brosnihan, K.B. and Ferrario, C.M. (1990) The renin–angiotensin system during acute myocardial ischemia in dogs. *Hypertension*, **15**, I121–7.

80 Neves, L.A.A., Almeida, A.P., Khosla, M.C. and Santos, R.A.S. (1995) Metabolism of angiotensin I in isolated rat hearts. Effect of angiotensin converting enzyme inhibitors. *Biochemical Pharmacology*, **50**, 1451–9.

81 Mahmood, A., Jackman, H.L., Teplitz, L. and Igic, R. (2002) Metabolism of angiotensin I in the coronary circulation of normal and diabetic rats. *Peptides*, **23**, 1171–5.

82 Zisman, L.S., Meixell, G.E., Bristow, M.R. and Canver, C.C. (2003) Angiotensin-(1–7) formation in the intact human heart: *in vivo* dependence on angiotensin II as substrate. *Circulation*, **108**, 1679–81.

83 Zisman, L.S., Keller, R.S., Weaver, B., Lin, Q., Speth, R., Bristow, M.R. and Canver, C.C. (2003) Increased angiotensin-(1–7)-forming activity in failing human heart ventricles: evidence for up regulation of the angiotensin-converting enzyme homologue ACE2. *Circulation*, **108**, 1707–12.

84 Averill, D.B., Ishiyama, Y., Chappell, M.C. and Ferrario, C.M. (2003) Cardiac angiotensin-(1–7) in ischemic cardiomyopathy. *Circulation*, **106**, 2141–6.

85 Crackower, M.A., Sarao, R., Oudit, G.Y., Yagil, C., Kozieradzki, I., Scanga, S.E., Oliveira-dos-Santos, A.J., Costa, J., Zhang, L., Pei, Y., Scholey, J., Ferrario, C.M., Manoukian, A.S., Chappell, M.C., Backx, P.H., Yagil, Y. and Penninger, J.M. (2002) Angiotensin-converting enzyme 2 is an essential regulator of heart function. *Nature*, **417**, 822–8.

86 Raizada, M.K. and Ferreira, A.J. (2007) ACE2: a new target for cardiovascular disease therapeutics. *Journal of Cardiovascular Pharmacology*, **50**, 112–19.

87 Gurley, S.B., Allred, A., Le, T.H., Griffiths, R., Mao, L., Philip, N., Haystead, T.A., Donoghue, M., Breitbart, R.E., Acton, S.L., Rockman, H.A. and Coffman, T.M. (2006) Altered blood pressure responses and normal cardiac phenotype in ACE2-null mice. *Journal of Clinical Investigation*, **116**, 2218–25.

88 Vickers, C., Hales, P., Kaushik, V., Dick, L., Gavin, J., Tang, J., Godbout, K., Parsons, T., Baronas, E., Hsieh, F., Acton, S., Patane, M., Nichols, A. and Tummino, P. (2002) Hydrolysis of biological peptides by human angiotensin-converting enzyme-related carboxypeptidase. *Journal of Biological Chemistry*, **277**, 14838–43.

89 Almeida, A.P., Frábregas, B.C., Madureira, M.M., Santos, R.J.S., Campagnole-Santos, M.J. and Santos, R.A.S. (2000) Angiotensin-(1–7) potentiates the coronary vasodilatatory effect of bradykinin in the isolated rat heart. *Brazilian Journal of Medical and Biological Research*, **33**, 709–13.

90 Brosnihan, K.B., Li, P. and Ferrario, C.M. (1996) Angiotensin-(1–7) dilates canine coronary arteries through kinins and nitric oxide. *Hypertension*, **27**, 523–8.

91 Porsti, I., Bara, A.T., Busse, R. and Hecker, M. (1994) Release of nitric oxide by angiotensin-(1–7) from porcine coronary endothelium: implications for a novel angiotensin receptor. *British Journal of Pharmacology*, **111**, 652–4.

92 Santos, R.A.S., Ferreira, A.J., Nadu, A.P., Braga, A.N.G., Almeida, A.P., Campagnole-Santos, M.J., Baltatu, O., Iliescu, R., Reudelhuber, T.L. and Bader, M. (2004) Expression of an angiotensin-(1–7)-producing fusion protein produces cardioprotective effects in rats. *Physiological Genomics*, 17, 292–9.

93 Loot, A.E., Roks, A.J.M., Henning, R.H., Tio, R.A., Suurmeijer, A.J.H., Boomsma, F. and van Gilst, W.H. (2002) Angiotensin-(1–7) attenuates the development of heart failure after myocardial infarction in rats. *Circulation*, 105, 1548–50.

94 Sampaio, W.O., Nascimento, A.A. and Santos, R.A.S. (2003) Systemic and regional hemodynamics effects of angiotensin-(1–7) in rats. *American Journal of Physiology*, 284, H1985–94.

95 Ferreira, A.J., Santos, R.A.S. and Almeida, A.P. (2001) Angiotensin-(1–7): cardioprotective effect in myocardial ischemia/reperfusion. *Hypertension*, 38, 665–8.

96 Ferreira, A.J., Santos, R.A.S. and Almeida, A.P. (2002) Angiotensin-(1–7) improves the post-ischemic function in isolated perfused rat hearts. *Brazilian Journal of Medical and Biological Research*, 35, 1083–90.

97 Tallant, E.A., Ferrario, C.M. and Gallagher, P.E. (2005) Angiotensin-(1–7) inhibits growth of cardiac myocytes through activation of the Mas receptor. *American Journal of Physiology*, 289, H1560–6.

98 Diaz-Araya, G.A., Petrov, V.V., Fagard, R.H. and Lijnen, P.J. (2005) Effect of angiotensin (1–7) on collagen production in cardiac fibroblasts. *American Journal of Hypertension*, 18, A230.

99 Iwata, M., Cowling, R.T., Gurantz, D., Moore, C., Zhang, S., Yuan, J.X. and Greenberg, B.H. (2005) Angiotensin-(1–7) binds to specific receptors on cardiac fibroblasts to initiate anti-fibrotic and anti-trophic effects. *American Journal of Physiology*, 289, H2356–63.

100 Ferrario, C.M., Chappell, M.C., Tallant, E.A., Brosnihan, K.B. and Diz, D.I. (1997) Counterregulatory actions of angiotensin-(1–7). *Hypertension*, 30, 535–41.

101 Brosnihan, K.B., Li, P., Tallant, E.A. and Ferrario, C.M. (1998) Angiotensin-(1–7): a novel vasodilator of the coronary circulation. *Biological Research*, 31, 227–34.

102 Lima, C.V., Paula, R.D., Resende, F.L., Khosla, M.C. and Santos, R.A.S. (1997) Potentiation of the hypotensive effect of bradykinin by short-term infusion of angiotensin-(1–7) in normotensive and hypertensive rats. *Hypertension*, 30, 542–8.

103 Seyedi, N., Xu, X., Nasjletti, A. and Hintze, T.H. (1995) Coronary kinin generation mediates nitric oxide release after angiotensin receptor stimulation. *Hypertension*, 26, 164–70.

104 Gorelik, G., Carbini, L.A. and Scicli, A.G. (1998) Angiotensin 1–7 induces bradykinin-mediated relaxation in porcine coronary artery. *Journal of Pharmacology and Experimental Therapeutics*, 286, 403–10.

105 Tom, B., de Vries, R., Saxena, P.R. and Danser, A.H. (2001) Bradykinin potentiation by angiotensin-(1–7) and ACE inhibitors correlates with ACE C- and N-domain blockade. *Hypertension*, 38, 95–9.

106 Neves, L.A.A., Almeida, A.P., Khosla, M.C., Campagnole-Santos, M.J. and Santos, R.A.S. (1997) Effect of angio-tensin-(1–7) on reperfusion arrhythmias in isolated rat hearts. *Brazilian Journal o f Medical and Biological Research*, 30, 801–9.

107 Kumagai, H., Khosla, M., Ferrario, C.M. and Fouad-Tarazi, F.M. (1990) Biological activity of angiotensin-(1–7) heptapeptide in the hamster heart. *Hypertension*, 15, I29–33.

108 Senbonmatsu, T., Saito, T., Landon, E.J., Watanabe, O., Price, E. Jr, Roberts, R.L., Imboden, H., Fitzgerald, T.G., Gaffney, F.A. and Inagami, T. (2003) A novel angiotensin II type 2 receptor signaling pathway: possible role in cardiac hypertrophy. *EMBO Journal*, 22, 6471–82.

109 Mendes, A.C., Ferreira, A.J., Pinheiro, S.V. and Santos, R.A.S. (2005) Chronic infusion of angiotensin-(1–7) reduces

heart angiotensin II levels in rats. *Regulatory Peptides*, **125**, 29–34.

110 Donoghue, M., Wakimoto, H., Maguire, C.T., Acton, S., Hales, P., Stagliano, N., Fairchild-Huntress, V., Xu, J., Lorenz, J.N., Kadambi, V., Berul, C.I. and Breitbart, R.E. (2003) Heart block, ventricular tachycardia, and sudden death in ACE2 transgenic mice with downregulated connexins. *Journal of Molecular and Cellular Cardiology*, **35**, 1043–53.

111 Fernandes, L., Fortes, Z.B., Nigro, D., Tostes, R.C.A., Santos, R.A.S. and Carvalho, M.H.C. (2001) Potentiation of bradykinin by angiotensin-(1–7) on arterioles of spontaneously hypertensive rats studied *in vivo*. *Hypertension*, **37**, 703–9.

112 Ren, Y., Garvin, J.L. and Carretero, O.A. (2002) Vasodilator action of angiotensin-(1–7) on isolated rabbit afferent arterioles. *Hypertension*, **39**, 799–802.

113 Gironacci, M.M., Yujnovsky, I., Gorzalczany, S., Taira, C. and Pena, C. (2004) Angiotensin-(1–7) inhibits the angiotensin II-enhanced norepinephrine release in coarcted hypertensive rats. *Regulatory Peptides*, **118**, 45–9.

114 Santos, R.A.S., Castro, C.H., Gava, E., Pinheiro, S.V., Almeida, A.P., Paula, R.D., Cruz, J.S., Ramos, A.S., Rosa, K.T., Irigoyen, M.C., Bader, M., Alenina, N., Kitten, G.T. and Ferreira, A.J. (2006) Impairment of *in vitro* and *in vivo* heart function in angiotensin-(1–7) receptor Mas knockout mice. *Hypertension*, **47**, 996–1002.

115 Grobe, J.L., Mecca, A.P., Lingis, M., Shenoy, V., Bolton, T.A., Machado, J.M., Speth, R.C., Raizada, M.K., Katovich, M. (2007) Prevention of angiotensin II-induced cardiac remodeling by angiotensin-(1–7). *American Journal of Physiology*, **292**, H736–42.

116 Grobe, J.L., Mecca, A.P., Mao, H. and Katovich, M.J. (2006) Chronic angiotensin-(1–7) prevents cardiac fibrosis in DOCA-salt model of hypertension. *American Journal of Physiology*, **290**, H2417–23.

117 Benter, I.F., Yousif, M.H., Anim, J.T., Cojocel, C. and Diz, D.I. (2006) Angiotensin-(1–7) prevents development of severe hypertension and end-organ damage in spontaneously hypertensive rats treated with L-NAME. *American Journal of Physiology*, **290**, H684–91.

118 Gembardt, F., Westermann, D., Heringer-Walther, S., Schultheiss, H.P., Tschöpe, C. and Walter, T. (2003) Deficiency in the G protein-coupled receptor Mas leads to cardiomyopathy and demonstrates the importance of the endogenous ligand angiotensin-(1–7) for the cardiac function. *Naunyn-Schmiedebergs Archives of Pharmacology*, **369** (Suppl. 1), R68.

119 Walther, T., Wessel, N., Kang, N., Tschöpe, C., Malberg, H., Bader, M. and Voss, A. (2000) Altered heart rate and blood pressure variability in mice lacking the Mas protooncogene. *Brazilian Journal of Medical and Biological Research*, **33**, 1–9.

120 Wiemer, G., Dobrucki, L.W., Louka, F.R., Malinski, T. and Heitsch, H. (2002) AVE 0991, a nonpeptide mimic of the effects of angiotensin-(1–7) on the endothelium. *Hypertension*, **40**, 847–52.

121 Pinheiro, S.V., Simões e Silva, A.C., Sampaio, W.O., Paula, R.D., Mendes, E.P., Bontempo, E.D., Pesquero, J.B., Walther, T., Alenina, N., Bader, M., Bleich, M. and Santos, R.A.S. (2004) Nonpeptide AVE 0991 is an angiotensin-(1–7) receptor Mas agonist in the mouse kidney. *Hypertension*, **44**, 490–6.

122 Lemos, V.S., Silva, D.M., Walther, T., Alenina, N., Bader, M. and Santos, R.A.S. (2005) The endothelium-dependent vasodilator effect of the nonpeptide Ang-(1–7) mimic AVE 0991 is abolished in the aorta of Mas-knockout mice. *Journal of Cardiovascular Pharmacology*, **46**, 274–9.

123 Faria-Silva, R., Duarte, F.V. and Santos, R.A.S. (2005) Short-term angiotensin (1–7) receptor MAS stimulation improves endothelial function in normotensive rats. *Hypertension*, **46**, 948–52.

124 Ferreira, A.J., Jacoby, B.A., Araujo, C.A., Macedo, F.A., Silva, G.A.B., Almeida, A.P., Caliari, M.V. and Santos, R.A.S. (2007) The nonpeptide angiotensin-(1–7) receptor Mas agonist AVE 0991

attenuates heart failure induced by myocardial infarction. *American Journal of Physiology*, **292**, H1113–19.

125 Ferreira, A.J., Oliveira, T.L., Castro, M.C., Almeida, A.P., Castro, C.H., Caliari, M.V., Gava, E., Kitten, G.T. and Santos, R.A.S. (2007) Isoproterenol-induced impairment of heart function and remodeling are attenuated by the nonpeptide angiotensin-(1–7) analogue AVE 0991. *Life Sciences*, **81**, 916–23.

126 Nadu, A.P., Ferreira, A.J., Reudelhuber, T.L., Bader, M. and Santos, R.A.S. (2007) Reduced isoproterenol-induced collagen III deposition in the left ventricle of rats harboring an angiotensin-(1–7)-producing fusion protein [TGR (A-7), 3292]. *Hypertension*, **50**, e149.

127 Santos, R.A.S., Brosnihan, K.B., Jacobsen, D.W., DiCorleto, P.E. and Ferrario, C.M. (1992) Production of angiotensin-(1–7) by human vascular endothelium. *Hypertension*, **19**, II56–61.

128 Chappell, M.C., Pirro, N.T., Sykes, A. and Ferrario, C.M. (1998) Metabolism of angiotensin-(1–7) by angiotensin-converting enzyme. *Hypertension*, **31**, 362–7.

129 Muthalif, M.M., Benter, I.F., Uddin, M.R., Haper, J.L. and Malik, K.U. (1998) Signal transduction mechanisms involved in angiotensin-(1–7)-stimulated arachidonic acid release and prostanoid synthesis in rabbit aortic smooth muscle cells. *Journal of Pharmacology and Experimental Therapeutics*, **284**, 388–98.

130 Iyer, S.N., Yamada, K., Diz, D.I., Ferrario, C.M., Chappel, M.C. (2000) Evidence that prostaglandins mediate the antihypertensive actions of angiotensin-(1–7) during chronic blockade of the renin–angiotensin system. *Journal of Cardiovascular Pharmacology*, **36**, 109–17.

131 Heitsch, H., Brovkovych, S., Malinsk, T. and Wiemer, G. (2001) Angiotensin-(1–7)-stimulated nitric oxide and superoxide release from endothelial cells. *Hypertension*, **37**, 72–6.

132 Sampaio, W.O., Santos, R.A.S., Faria-Silva, R., Mata Machado, L.T., Schiffrin, E.L. and Touyz, R.M. (2007) Angiotensin-(1–7) through receptor Mas mediates endothelial nitric oxide synthase activation via Akt-dependent pathways. *Hypertension*, **49**, 185–92.

133 Feterik, K., Smith, L. and Katusic, Z.S. (2000) Angiotensin-(1–7) causes endothelium-dependent relaxation in canine middle cerebral artery. *Brain Research*, **873**, 75–82.

134 Meng, W. and Busija, D.W. (1993) Comparative effects of angiotensin-(1–7) and angiotensin II on piglet pial arterioles. *Stroke*, **24**, 2041–4.

135 Osei, S.Y., Ahima, R.S., Minkes, R.K., Weaver, J.P., Khosla, M.C. and Kadowitz, P.J. (1993) Differential responses to angiotensin-(1–7) in the feline mesenteric and hindquarters vascular beds. *European Journal of Pharmacology*, **234**, 35–42.

136 le Tran, Y. and Forster, C. (1997) Angiotensin-(1–7) and the rat aorta: modulation by the endothelium. *Journal of Cardiovascular Pharmacology*, **30**, 676–82.

137 Oliveira, M.A., Fortes, Z.B., Santos, R.A.S., Khosla, M.C. and Carvalho, M.H.C. (1999) Synergistic effect of angiotensin-(1–7) on bradykinin arteriolar dilation *in vivo*. *Peptides*, **20**, 1195–201.

138 Machado, R.D., Ferreira, M.A., Belo, A.V., Santos, R.A.S. and Andrade, S.P. (2002) Vasodilator effect of angiotensin-(1–7) in mature and sponge-induced neovasculature. *Regulatory Peptides*, **107**, 105–13.

139 Sasaki, S., Higashi, Y., Nakagawa, K., Matsuura, H., Kajiyama, G. and Oshima, T. (2001) Effects of angiotensin-(1–7) on forearm circulation in normotensive subjects and patients with essential hypertension. *Hypertension*, **38**, 90–4.

140 Davie, A.P. and McMurray, J.J. (1999) Effect of angiotensin-(1–7) and bradykinin in patients with heart failure treated with an ACE inhibitor. *Hypertension*, **34**, 457–60.

141 Iyer, S.N., Chappell, M.C., Averill, D.B., Diz, D.I. and Ferrario, C.M. (1998) Vasodepressor actions of angiotensin-(1–7) unmasked during combined treatment with lisinopril and losartan. *Hypertension*, **31**, 699–705.

142 Collister, J.P. and Hendel, M.D. (2003) The role of Ang (1–7) in mediating the

143 Soares de Moura, R., Resende, A.C., Emiliano, A.F., Tano, T., Mendes-Ribeiro, A.C., Correia, M.L. and Carvalho, L.C. (2004) The role of bradykinin, AT_2 and angiotensin 1–7 receptors in the EDRF-dependent vasodilator effect of angiotensin II on the isolated mesenteric vascular bed of the rat. *British Journal of Pharmacology*, **141**, 860–6.

144 Paula, R.D., Lima, C.V., Khosla, M.C. and Santos, R.A.S. (1995) Angiotensin-(1–7) potentiates the hypotensive effect of bradykinin in conscious rats. *Hypertension*, **26**, 1154–9.

145 Li, P., Chappell, M.C., Ferrario, C.M. and Brosnihan, K.B. (1997) Angiotensin-(1–7) augments bradykinin-induced vasodilation by competing with ACE and releasing nitric oxide. *Hypertension*, **29**, 394–400.

146 Hecker, M., Blaukat, A., Bara, A.T., Muller-Esterl, W. and Busse, R. (1997) ACE inhibitor potentiation of bradykinin-induced venoconstriction. *British Journal of Pharmacology*, **121**, 1475–81.

147 Ueda, S., Masumori-Maemoto, S., Wada, A., Ishii, M., Brosnihan, K.B. and Umemura, S. (2001) Angiotensin (1–7) potentiates bradykinin-induced vasodilatation in man. *Journal of Hypertension*, **19**, 2001–9.

148 Wilsdorf, T., Gainer, J.V., Murphey L.J. Vaughan, D.E. and Brown, N.J. (2001) Angiotensin-(1–7) does not affect vasodilator or TPA responses to bradykinin in human forearm. *Hypertension*, **37**, 1136–40.

149 Erdos, E.G., Jackman, H.L., Brovkovych, V., Tan, F. and Deddish, P.A. (2002) Products of angiotensin I hydrolysis by human cardiac enzymes potentiate bradykinin. *Journal of Molecular and Cellular Cardiology*, **34**, 1569–76.

150 Maia, L.G., Ramos, M.C., Fernandes, L., Carvalho, M.H., Campagnole-Santos, M.J. and Santos, R.A.S. (2004) Angiotensin-(1–7) antagonist A-779 attenuates the potentiation of brady-kinin by captopril in rats. *Journal of Cardiovascular Pharmacology*, **43**, 685–91.

151 Roks, A.J., van Geel, P.P., Pinto, Y.M., Buikema, H., Henning, R.H., de Zeeuw, D. and van Gilst, W.H. (1999) Angiotensin-(1–7) is a modulator of the human renin–angiotensin system. *Hypertension*, **34**, 296–301.

152 Stegbauer, J., Oberhauser, V., Vonend, O. and Rump, L.C. (2004) Angiotensin-(1–7) modulates vascular resistance and sympathetic neurotransmission in kidneys of spontaneously hypertensive rats. *Cardiovascular Research*, **61**, 352–9.

153 Gallagher, P.E. and Tallant, E.A. (2004) Inhibition of human lung cancer cell growth by angiotensin-(1–7). *Carcinogenesis*, **25**, 2045–52.

154 Freeman, E.J., Chisolm, G.M., Ferrario, C.M. and Tallant, E.A. (1996) Angiotensin-(1–7) inhibits vascular smooth muscle cell growth. *Hypertension*, **28**, 104–8.

155 Machado, R.D., Santos, R.A.S. and Andrade, S.P. (2000) Opposing actions of angiotensins on angiogenesis. *Life Sciences*, **66**, 67–76.

156 Tallant, E.A., Chappell, M.C., Ferrario, C.M. and Gallagher, P.E. (2004) Inhibition of MAP kinase activity by angiotensin-(1–7) in vascular smooth muscle cells is mediated by the Mas receptor. *Hypertension*, **43**, 1348.

157 Langeveld, B., van Gilst, W.H., Tio, R.A., Zijlstra, F. and Roks, A.J. (2005) Angiotensin-(1–7) attenuates neointimal formation after stent implantation in the rat. *Hypertension*, **45**, 1–4.

158 Tallant, E.A. and Clark, M.A. (2003) Molecular mechanisms of inhibition of vascular growth by angiotensin-(1–7). *Hypertension*, **42**, 574–9.

159 Walters, P.E., Gaspari, T.A. and Widdop, R.E. (2005) Angiotensin-(1–7) acts as a vasodepressor agent via angiotensin II type 2 receptors in conscious rats. *Hypertension*, **45**, 960–6.

160 Silva, D.M., Vianna, H.R., Cortes, S.F., Campagnole-Santos, M.J., Santos, R.A.S. and Lemos, V.S. (2006) Evidence for a new angiotensin-(1–7) receptor subtype in the aorta of Sprague-Dawley rats. *Peptides*, **28**, 702–7.

161 Castro, C.H., Santos, R.A.S., Ferreira, A.J., Bader, M., Alenina, N. and Almeida, A.P. (2005) Evidence for a functional interaction of the angiotensin-(1–7) receptor Mas with AT_1 and AT_2 receptors in the mouse heart. *Hypertension*, **46**, 937–42.

162 Kostenis, E., Milligan, G., Christopoulos, A., Sanchez-Ferrer, C.F., Heringer-Walther, S., Sexton, P.M., Gembardt, F., Kellett, E., Martini, L., Vanderheyden, P., Schultheiss, H.P. and Walther, T. (2005) G-protein-coupled receptor Mas is a physiological antagonist of the angiotensin II type 1 receptor. *Circulation*, **111**, 1806–13.

163 Canals, M., Jenkins, L., Kellett, E. and Milligan, G. (2006) Up-regulation of the angiotensin II type 1 receptor by the Mas proto-oncogene is due to constitutive activation of G_q/G^{11} by Mas. *Journal of Biological Chemistry*, **281**, 16757–67.

164 Oro, C., Qian, H. and Thomas, W.G. (2007) Angiotensin type 1 receptor pharmacology: signaling beyond G proteins. *Pharmacology and Therapeutics*, **113**, 210–26.

165 Mostard, G.J.M., Houben, A.J.H.M., Kroon, A.A., van Engelshoven, J.M.A. and de Leeuw, P.W. (2007) Angiotensin 1–7 induces renal vasodilation in hypertensive patients independent of an activated renin–angiotensin system. *Hypertension*, **50**, 804.

166 Botelho-Santos, G.A., Sampaio, W.O., Reudelhuber, T.L., Bader, M., Campagnole-Santos, M.J. and Santos, R.A.S. (2007) Expression of an angiotensin-(1–7)-producing fusion protein in rats induced marked changes in regional vascular resistance. *American Journal of Physiology – Heart and Circulatory Physiology*, **292**, H2485–90.

167 van der Wouden, E.A., Ochodnick, P., van Dokkum, R.P., Roks, A.J., Deelman, L.E., de Zeeuw, D. and Henning, R.H. (2006) The role of angiotensin (1–7) in renal vasculature of the rat. *Journal of Hypertension*, **24**, 1971–8.

168 Lara, L.S., Bica, R.B., Sena, S.L., Correa, J.S., Marques-Fernandes, M.F., Lopes, A.G. and Caruso-Neves, C. (2002) Angiotensin-(1–7) reverts the stimulatory effect of angiotensin II on the proximal tubule Na^+-ATPase activity via a A779-sensitive receptor. *Regulatory Peptides*, **103**, 17–22.

169 Andreatta-van Leyen, S., Romero, M.F., Khosla, M.C. and Douglas, J.G. (1993) Modulation of phospholipase A_2 activity and sodium transport by angiotensin-(1–7). *Kidney International*, **44**, 932–6.

170 Santos, R.A.S., Simões e Silva, A.C., Magaldi, A.J., Khosla, M.C., Cesar, K.R., Passaglio, K.T. and Baracho, N.C. (1996) Evidence for a physiological role of angiotensin-(1–7) in the control of hydroelectrolyte balance. *Hypertension*, **27**, 875–84.

171 Simões e Silva, A.C., Baracho, N.C.V., Passaglio, K.T. and Santos, R.A.S. (1997) Renal actions of angiotensin-(1–7). *Brazilian Journal of Medical and Biological Research*, **30**, 503–13.

172 Magaldi, A.J., Cesar, K.R., Araujo, M., Simões e Silva, A.C. and Santos, R.A.S. (2003) Angiotensin-(1–7) stimulates water transport in rat inner medullary collecting duct: evidence for involvement of vasopressin V2 receptors. *Pflugers Arch*, **447**, 223–30.

173 Simões e Silva, A.C., Bello, A.P.C., Baracho, N.C.V., Khosla, M.C. and Santos, R.A.S. (1998) Diuresis and natriuresis produced by long term administration of a selective angiotensin-(1–7) antagonist in normotensive and hypertensive rats. *Regulatory Peptides*, **74**, 177–84.

174 Vallon, V., Heyne, N., Richter, K., Khosla, M.C. and Fechter, K. (1998) [D-Ala[7]]-angiotensin 1–7 blocks renal actions of angiotensin 1–7 in the anesthetized rat. *Journal of Cardiovascular Pharmacology*, **32**, 164–7.

175 Ferrario, C.M. and Chappell, M.C. (2004) Novel angiotensin peptides. *Cellular and Molecular Life Sciences*, **61**, 2720–7.

176 Campbell, D.J. (2003) The renin angiotensin and kallikrein kinin systems. *International Journal of Biochemistry and Cell Biology*, **35**, 784–91.

177 Luque, M., Martin, P., Martell, N., Fernandez, C., Brosnihan, K.B. and Ferrario, C.M. (1996) Effects of captopril related to increased levels of prostacyclin

178 Ferrario, C.M., Martell, N., Yunis, C., Flack, J.M., Chappell, M.C., Brosnihan, K.B., Dean, R.H., Fernandez, A., Novikov, S.V., Pinillas, C. and Luque, M. (1998) Characterization of angiotensin-(1–7) in the urine of normal and essential hypertensive subjects. *American Journal of Hypertension*, **11**, 137–46.

179 Azizi, M. and Ménard, J. (2004) Combined blockade of the renin angiotensin system with angiotensin-converting enzyme inhibitors and angiotensin II type 1 receptor antagonists. *Circulation*, **109**, 2492–9.

180 Simões e Silva, A.C., Diniz, J.S., Regueira-Filho, A. and Santos, R.A.S. (2004) The renin angiotensin system in childhood hypertension. Selective increase of angiotensin-(1–7) in essential hypertension. *Journal of Pediatrics*, **145**, 93–8.

181 Simões e Silva, A.C., Dini, Z.J., Pereira, R.M., Pinheiro, S.V. and Santos, R.A.S. (2006) Circulating renin angiotensin system in childhood chronic renal failure. Marked increase of angiotensin-(1–7) in end-stage renal disease. *Pediatric Research*, **60**, 734–9.

182 Zini, S., Fournie-Zaluski, M.C., Chauvel, E., Roques, B.P., Corvol, P. and Llorens-Cortes, C. (1996) Identification of metabolic pathways of brain angiotensin II and III using specific aminopeptidase inhibitors: predominant role of angiotensin III in the control of vasopressin release. *Proceedings of the National Academy of Sciences of the United States of America*, **93**, 11968–73.

183 Cesari, M., Rossi, G.P. and Pessina, A.C. (2002) Biological properties of the angiotensin peptides other than angiotensin II. implications for hypertension and cardiovascular diseases. *Journal of Hypertension*, **20**, 793–9.

184 Suzuki, S., Doi, Y., Aoi, W., Kuramochi, M. and Hashiba, K. (1984) Effect of angiotensin III on blood pressure, renin–angiotensin–aldosterone system in normal and hypertensive subjects. *Japanese Heart Journal*, **25**, 75–85.

185 Zager, P.G. and Luetscher, J.A. (1982) Effects of angiotensin III and ACTH on aldosterone secretion. *Clinical and Experimental Hypertension Part A – Theory and Practice*, **4**, 1481–504.

186 Plovsing, R.R., Wamberg, C., Sandgaard, N.C., Simonsen, J.A., Holstein-Rathlou, N.H., Hoilund-Carlsen, P.F. and Bie, P. (2003) Effects of truncated angiotensins in humans after double blockade of the renin system. *American Journal of Physiology – Regulatory Integrative and Comparative Physiology*, **285**, R981–91.

187 Padia, S.H., Howell, N.L., Siragy, H.M. and Carey, R.M. (2006) Renal angiotensin type 2 receptors mediate natriuresis via angiotensin III in the angiotensin II type 1 receptor-blocked rat. *Hypertension*, **47**, 537–44.

188 Ruiz-Ortega, M., Lorenzo, O. and Egido, J. (2000) Angiotensin III increases MCP-1 and activates NF-kappaB and AP-1 in cultured mesangial and mononuclear cells. *Kidney International*, **57**, 2285–98.

189 Padia, S.H., Kemp, B.A., Howell, N.L., Siragy, H.M., Fournie-Zaluski, M.C., Roques, B.P. and Carey, R.M. (2007) Intrarenal aminopeptidase N inhibition augments natriuretic responses to angiotensin III in angiotensin type 1 receptor-blocked rats. *Hypertension*, **49**, 625–30.

190 Wright, J.W., Miller-Wing, A.V., Shaffer, M.J., Higginson, C., Wright, D.E., Hanesworth, J.M. and Harding, J.W. (1993) Angiotensin; II (3–8): (ANG IV) hippocampal binding: potential role in the facilitation of memory. *Brain Research Bulletin*, **32**, 497–502.

191 Pederson, E.S., Harding, J.W. and Wright, J.W. (1998) Attenuation of scopolamine-induced spatial learning impairments by an angiotensin IV analog. *Regulatory Peptides*, **74**, 97–103.

192 Braszko, J.J., Walesiuk, A. and Wielgat, P. (2006) Cognitive effects attributed to angiotensin II may result from its conversion to angiotensin IV. *Journal of the Renin–Angiotensin–Aldosterone System*, **7**, 168–74.

193 Wayner, M.J., Armstrong, D.L., Phelix, C.F., Wright, J.W. and Harding, J.W. (2001) Angiotensin IV enhances LTP in rat dentate gyrus *in vivo*. *Peptides*, **22**, 1403–14.

194 Coleman, J.K., Krebs, L.T., Hamilton, T.A., Ong, B., Lawrence, K.A., Sardinia, M.F., Harding, J.W. and Wright, J.W. (1998) Autoradiographic identification of kidney angiotensin IV binding sites and angiotensin IV-induced renal cortical blood flow changes in rats. *Peptides*, **19**, 269–77.

195 Li, X.C., Campbell, D.J., Ohishi, M., Yuan, S. and Zhuo, J.L. (2006) AT1 receptor-activated signaling mediates angiotensin IV-induced renal cortical vasoconstriction in rats. *American Journal of Physiology Renal Physiol*, **290**, F1024–33.

196 Chen, S., Patel, J.M. and Block, ER. (2000) Angiotensin I.V-mediated pulmonary artery vasorelaxation is due to endothelial intracellular calcium release. *American Journal of Physiology – Lung Cellular and Molecular Physiology*, **279**, L849–56.

197 Patel, J.M., Martens, J.R., Li, Y.D., Gelband, C.H., Raizada, M.K. and Block, E.R. (1998) Angiotensin IV receptor-mediated activation of lung endothelial NOS is associated with vasorelaxation. *American Journal of Physiology*, **275**, L1061–8.

198 Fitzgerald, S.M., Evans, R.G., Bergstrom, G. and Anderson, W.P. (1999) Renal hemodynamic responses to intrarenal infusion of ligands for the putative angiotensin IV receptor in anesthetized rats. *Journal of Cardiovascular Pharmacology*, **34**, 206–11.

199 van Rodijnen, W.F., van Lambalgen, T.A., van Wijhe, M.H., Tangelder, G.J. and Ter Wee, P.M. (2002) Renal microvascular actions of angiotensin II fragments. *American Journal of Physiology – Renal Fluid and Electrolyte Physiology*, **83**, F86–92.

200 Slinker, B.K., Wu, Y., Brennan, A.J., Campbell, K.B. and Harding, J.W. (1999) Angiotensin IV has mixed effects on left ventricle systolic function and speeds relaxation. *Cardiovascular Research*, **42**, 660–9.

201 Albiston, A.L., Pederson, E.S., Burns, P., Purcell, B., Wright, J.W., Harding, J.W., Mendelsohn, F.A., Weisinger, R.S. and Chai, S.Y. (2004) Attenuation of scopolamine-induced learning deficits by LVV-hemorphin-7 in rats in the passive avoidance and water maze paradigms. *Behavioral and Brain Sciences*, **154**, 239–43.

202 Sim, M.K. and Qiu, X.S. (1994) Formation of des-Asp-angiotensin I in the hypothalamic extract of normo- and hypertensive rats. *Blood Press*, **3**, 260–4.

203 Sim, M.K. and Radhakrishnan, R. (1994) Novel central action of des-Asp-angiotensin I. *European Journal of Pharmacology*, **257**, R1–3.

204 Sim, M.K. and Min, L. (1998) Effects of des-Asp-angiotensin I on experimentally-induced cardiac hypertrophy in rats. *International Journal of Cardiology*, **63**, 223–7.

205 Wen, Q., Sim, M.K. and Tang, F.R. (2004) Reduction of infarct size by orally administered des-aspartate-angiotensin I in the ischemic reperfused rat heart. *Regulatory Peptides*, **120**, 149–53.

206 Min, L., Sim, M.K. and Xu, X.G. (2000) Effects of des-aspartate-angiotensin I on angiotensin II-induced incorporation of phenylalanine and thymidine in cultured rat cardiomyocytes and aortic smooth muscle cells. *Regulatory Peptides*, **95**, 93–7.

207 Sim, M.K. and Soh, K.S. (1995) Effects of des-Asp-angiotensin I on the electrically stimulated contraction of the rabbit pulmonary artery. *European Journal of Pharmacology*, **284**, 215–9.

208 Sim, M.K. and Chai, S.K. (1996) Subtypes of losartan-sensitive angiotensin receptor in the rabbit pulmonary artery. *British Journal of Pharmacology*, **117**, 1504–6.

209 Chen, W.S. and Sim, M.K. (2004) Effects of des-aspartate-angiotensin I on the expression of angiotensin AT1 and AT2 receptors in ventricles of hypertrophic rat hearts. *Regulatory Peptides*, **117**, 207–12.

210 Ocaranza, M.P., Godoy, I., Jalil, J.E., Varas, M., Collantes, P., Pinto, M., Roman, M., Ramirez, C., Copaja, M., Diaz-Araya, G., Castro, P. and Lavandero, S. (2006) Enalapril attenuates down-

regulation of Angiotensin-converting enzyme 2 in the late phase of ventricular dysfunction in myocardial infarcted rat. *Hypertension*, **48**, 572–8.

211 Rice, G.I., Thomas, D.A., Grant, P.J., Turner, A.J. and Hooper, N.M. (2004) Evaluation of angiotensin-converting enzyme (ACE), its homologue ACE2 and neprilysin in angiotensin peptide metabolism. *Biochemical Journal*, **383**, 45–51.

212 Jackman, H.L., Massad, M.G., Sekosan, M., Tan, F., Brovkovych, V., Marcic, B.M. and Erdos, E.G. (2002) Angiotensin 1–9 and 1–7 release in human heart: role of cathepsin A. *Hypertension*, **39**, 976–81.

213 Mogielnicki, A., Kramkowski, K., Chabielska, E. and Buczko, W. (2003) Angiotensin 1-9 influences hemodynamics and hemostatics parameters in rats. *Polish Journal of Pharmacology*, **55**, 503–4.

214 Karwowska-Polecka, W., Kulakowska, A., Wisniewski, K. and Braszko, J.J. (1997) Losartan influences behavioural effects of angiotensin II(3–7) in rats. *Pharmacological Research*, **36**, 275–83.

215 Ferreira, P.M., Santos, R.A.S. and Campagnole-Santos, M.J. (2007) Angiotensin-(3–7) pressor effect at the rostral ventrolateral medulla. *Regulatory Peptides*, **141**, 168–74.

216 Handa, R.K. (1999) Angiotensin-(1–7) can interact with the rat proximal tubule AT_4 receptor system. *American Journal of Physiology*, **277**, F75–83.

217 Jankowski, V., Vanholder, R., van der Giet, M., Tolle, M., Karadogan, S., Gobom, J., Furkert, J., Oksche, A., Krause, E., Tran, T.N., Tepel, M., Schuchardt, M., Schluter, H., Wiedon, A., Beyermann, M., Bader, M., Todiras, M., Zidek, W. and Jankowski, J. (2007) Mass-spectrometric identification of a novel angiotensin peptide in human plasma. *Arteriosclerosis Thrombosis and Vascular Biology*, **27**, 297–302.

218 Nagata, S., Kato, J., Sasaki, K., Minamino, N., Eto, T. and Kitamura, K. (2006) Isolation and identification of proangiotensin-12, a possible component of the renin–angiotensin system. *Biochemical and Biophysical Research Communications*, **350**, 1026–31.

4
Kinins

Suzana Macedo de Oliveira, Kely de Picoli Souza, Michael Bader, and João Bosco Pesquero

4.1
Introduction

The kallikrein–kinin system (KKS) is an endogenous metabolic cascade, whose activation results in the production of a small group of vasoactive bradykinin (BK)-related peptides denominated kinins. This complex system comprises different molecules implicated in many physiological and pathological processes (Figure 4.1). Activation of the KKS is initiated when the internal kinin moiety is proteolytically excised from its parent glycoproteins, the high-molecular-weight (HMW) and low-molecular-weight (LMW) kininogens, both of which present at relatively high concentrations in the plasma. The release of kinins from kininogens is undertaken by the action of specialized serine proteases, kininogenases, known as kallikreins. The kinin family is composed by a group of peptides known as "intact kinins" like BK (Arg-Pro-Pro-gly-Phe-Ser-Pro-Phe-Arg), kallidin (Lys-Arg-Pro-Pro-Gly-Phe-Ser-Pro-Phe-Arg) and methionyl-lysyl-BK (Met-Lys-Arg-Pro-Pro-Gly-Phe-Ser-Pro-Phe-Arg). These peptides are either converted in the circulation and tissues into the "des-Arg-kinins", by a group of enzymes denominated kininases I or are rapidly inactivated by circulating kininases. Kinins, often considered as either proinflammatory or cardioprotective, bind and activate two types of G-protein-coupled receptors (GPCRs) – kinin B_1 (B_1 receptor) and B_2 (B_2 receptor) – eliciting a broad range of biological effects, many of them related to cardiovascular homeostasis.

The existence of the KKS was first described almost a century ago when Abelous and Bardier, in 1909, showed the hypotensive effect of human urine in anesthetized dogs [1]. However, it took more than 20 years until kallikrein, the kinin-producing enzyme, was described and the pancreas was assumed to be the principal organ responsible for the production of this hypotensive substance. Therefore, by derivation from the Greek word *kallikreas* (pancreas), the authors named it kallikrein. Subsequent studies showed that kallikrein was a proteolytic enzyme able to cleave a plasma protein releasing a smooth muscle contracting substance called "substance DK", from the German *darmkontrahierenderstoff* (intestine-con-

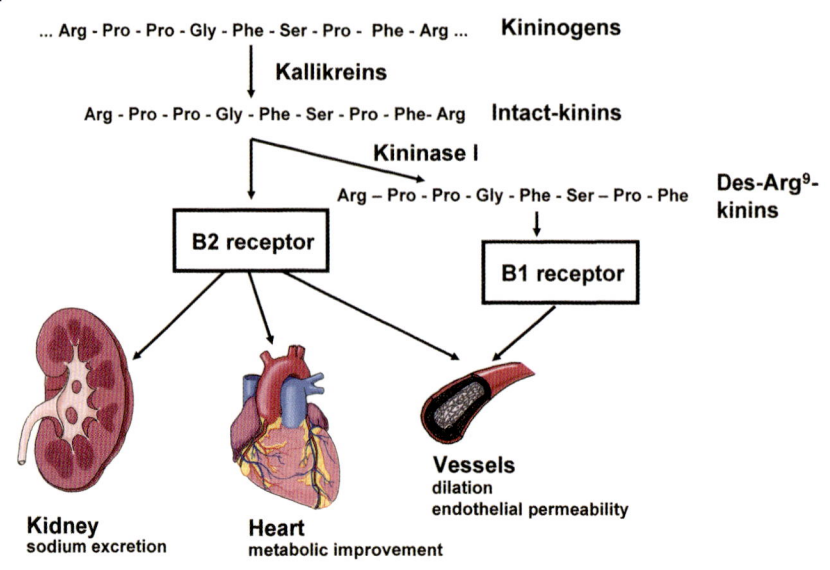

Figure 4.1 Overview of the kallikrein-kinin system. The figure summarises the kallikrein–kinin system showing the precursor molecules, the kininogens; the kinin-forming enzymes, kallikreins; the kininases; the kinin receptors and the main target organs.

tracting substance), later recognized as lysyl-BK (Lys-BK) or kallidin. At the end of 1940s, Rocha e Silva *et al.* [2] reported that the incubation of either *Bothrops jararaca* venom or trypsin with a globulin fraction of bovine plasma resulted in the formation of a substance able to contract isolated guinea pig ileum, rabbit duodenum and rat uterus, as well as to decrease blood pressure when injected intravenously in cats, dogs and rabbits. Using Greek nomenclature, Rocha e Silva *et al.* called this novel substance "bradykinin" (*bradys* meaning slow; *kinein* meaning movement). In 1960, BK was biochemically isolated and its structure defined simultaneously by the group of Elliott [3] and Boissomas [4]. Shortly thereafter, kallidin was defined and shown to be a decapeptide possessing the sequence of BK with an additional lysine residue at the N-terminus [5]. The peptide BK, Lys-BK and Met-Ly-BK (the intact kinins) present similar pharmacological properties acting on the B_2 receptor. On the other hand, when these kinins are hydrolyzed by carboxipeptidases on the C-terminus to des-Arg-kinins, they activate the B_1 receptor. Other kinins such as Ile-Ser-BK (T-kinin), Met-Ile-Ser BK (Met-T-kinin) and Hyp3-kinin were also found in mammals [6–8].

After BK was described by Rocha e Silva *et al.*, the KKS had a strong development and since that time the system has been the subject of intensive research, with more than 20 000 papers referenced in MEDLINE over the last 50 years. The interest in this multiprotein system is explained in part by the plethora of pharmacological activities presented not only by kinins and their receptors, but also by their precursors and their activators, the kinin-forming metallopeptidases (kininogenases) and kinin-degrading (kininases). The relative importance of each component varies from one biological medium to the other, being present in plasma but also in blood cells, in various tissues or their exocrine secretions. Finally, the regulation of this system also by serpins increases the complexity of the system, as well as its multiple relationships with other important metabolic pathways such as the renin–angiotensin, leptin/insulin/glucose, coagulation or complement pathways.

Along with the more than 50 years of BK discovery, a great deal of findings focus on kinins as potential mediators in endogenous cardiovascular protective mechanisms [9–13]. This is due to the fact that the KKS components are localized in the heart and in the vascular tissues [14–19]. In addition, all components of the KKS are also expressed in the kidney, and are important regulators for renal hemodynamics, tubular function, ontogeny and renal remodeling [20–22]. The kidney produces local kinins, like BK, at much higher levels than those present in blood [20], which means that kinins can be locally formed and act as an autocrine/paracrine factor to regulate blood flow perfusion in this organ.

Therefore, based on the above assumptions, an overall view of the KKS will be described in this chapter and the current concepts on the roles of kinins in the cardiovascular system will be discussed.

4.2
Kininogens: Precursors of Kinins

Kininogens are multifunctional proteins derived mainly from α_2 globulins. The existence of kininogen was first postulated in 1937 by Werle *et al.* as a serum-borne substrate for kallikrein, which promoted its hydrolysis liberating a muscle-contracting substance, BK [23]. In most mammals, two types of kininogens, HMW and LMW, have been described. These kininogens differ from each other in molecular weight, susceptibility to plasma and tissue kallikreins, and in their physiological properties [24]. They are synthesized in the liver, and circulate in the plasma and other body fluids. Both kininogen genes code for single-chain glycoproteins consisting of an identical N-terminal part divided into four domains and a divergent C-terminal region consisting of domains 5H and 6 for HMW kininogen, and domain 5L for LMW kininogen [25]. Domains 1, 2 and 3 resemble cystatins, and domains 2 and 3 also exhibit inhibitory effects on cysteine proteinases [26, 27]. Domain 4 contains BK, which is released in a two-step proteolytic cleavage reaction exerted by kallikreins. Subsequent to the release of the kinin moiety, the residual parts of the HMW kininogen molecule, the heavy chain and the light chain, are kept together by an intramolecular cysteine bridge. The two kininogen mRNAs are generated by alternative splicing and polyadenylation from a single kininogen gene in humans, cattle and rats, but not in mice, which have two kininogen genes [28]. The first nine exons and the 5′-end of exon 10 of this gene code for domains 1–4 which are identical in LMW and HMW kininogens. For the generation of HMW kininogen mRNA, a polyadenylation site downstream in exon 10 is used leading to the inclusion of exon 10 sequences encoding domains 5H and 6 in the mature mRNA. For LMW kininogen mRNA, a splice donor site in exon 10, 3′ of the BK coding sequence, is activated skipping the domain 5H and 6 coding part and linking the 5′-end of exon 10 to exon 11 which codes for domain 5L of LMW. The proteins in humans are produced from a gene localized to chromosome 3 [29], consisting of 11 exons encompassing approximately 27 kb [30].

HMW kininogen, an α-globulin secreted by the liver, circulates in plasma as an 88- to 120-kDa single-chain glycoprotein at a concentration of 70–90 µg/ml and exerts an essential function in the contact system of blood coagulation. It binds to plasma prekallikrein and factor XI in the circulation to form the structural basis of a complex that is linked to proteins on the surface of endothelial cells, platelets and neutrophils [31, 32]. Interaction between HMW kininogen and plasma prekallikrein in this complex leads to rapid activation of plasma kallikrein, possibly by prolylcarboxypeptidase activity [33]. Plasma kallikrein proteolytically activates factor XII, which in a positive-feedback loop again activates plasma kallikrein, leading to a burst of kinin release at the site of contact system activation. LMW kininogen is synthesized in most tissues, including liver and kidney, and is mainly metabolized by tissue kallikrein at sites of tissue injury, also yielding kinin peptides [32]. In addition, there is a T-kininogen in the rat plasma considered to be an acute-phase reactant of inflammation [34]. This kininogen releases T-kinin by the enzymatic action of T-kallikrein in rats [35].

4.3
Kinin-Forming Systems

Classically, the synthesis of kinins is determined by the activity of the KKS, which comprises two separate pathways differentiated by their location: the plasma and tissue systems. The plasma KKS, by far the more complex, initiates activation of the intrinsic coagulation pathway. The second and simpler pathway of kinin generation involves tissue kallikrein and its substrate, LMW kininogen. Each of these enzyme systems may play different pathophysiological roles (for recent reviews, see [9–13]).

4.4
Kininases

Kinins are very labile molecules. BK exhibits a remarkably short half-life (10–50 s) in the plasma *in vitro*, depending on species or *in vivo*. On the other hand, the half-life of des-Arg^9-BK is 4- to 12-fold higher than that of BK under comparable experimental conditions and this may explain why the *in vivo* concentration of des-Arg^9-BK is consistently higher than that of BK. The short half-life of these peptides in tissues or biological fluids is due to their hydrolysis by a group of enzymes collectively called kininases, which accounts for their rapid disposal. These enzymes circulate in the plasma, and are expressed in different cells and tissues, regulating the lifetime and function of the kinins in the body. Evidences indicate that the most important kininases are the metallopeptidases angiotensin-converting enzyme (ACE) and neutral endopeptidase (NEP), aminopeptidase P, and the carboxypeptidases M and N [36–38], and the importance of each peptidase depends on the animal species, the biological milieu and the pathophysiological context. Among these activities, the arginine carboxypeptidases N from plasma (EC 3.4.17.3) and membrane-bound carboxypeptidase M remove the Arg^9 residue of the native kinins producing residual des-Arg^9-kinin peptides [39], the physiological agonists of B_1 receptors.

The kininase II group includes ACE, an enzyme located predominantly in luminal membrane of endothelial cells, which explains the extensive pulmonary inactivation of kinins. Another member of the kininase II group of membrane-bound peptidase is NEP (EC 3.4.24.11), which is located at a high concentration in the proximal nephron epithelium, intestine microvilli, lung and skin fibroblasts, and is also present on the membrane of some leukocytes, but barely detectable in the vasculature or plasma [39]. Both kininases inactivate intact kinins by removing the C-terminal Phe8–Arg9 dipeptide, which leads to their complete loss of activity. NEP is also capable of cleaving the Gly4–Phe5 bond in BK. However, the metabolism of des-Arg^9-kinins differs from that of native kinins in several important points. First, des-Arg^9-BK does not react with arginine carboxypeptidases, as it is devoid of the corresponding C-terminal residue. Second, des-Arg^9-BK can be further degraded by a tripeptidyl carboxypeptidase activity of ACE, thus generating the pentapeptide Arg-Pro-Pro-Gly-Phe plus the tripeptide Ser-Pro-Phe, but at a slower rate and with much less affinity than the removal of a C-terminal dipeptide from BK [40].

In addition to kininases I and II, kinins can also be rapidly metabolized by aminopeptidases present in tissues and blood. Aminopeptidase M (EC 3.4.11.2), present in plasma, hydrolyzes Lys-BK into BK and Lys-des-Arg^9-BK into des-Arg^9-BK [41, 42]. This reaction is pharmacologically neutral for the B_2 receptor agonists BK and Lys-BK, because these peptides exhibit similar potencies in this receptor. However, it represents a relative inactivation for the B_1 receptor agonists, as des-Arg^9-BK is an agonist of lesser affinity for the B_1 receptor than Lys-des-Arg^9-BK in some species. An aminopeptidase P (aminoacyl proline aminopeptidase; EC 3.4.11.9), which cleaves the Arg1–Pro2 bond, may also contribute to the pulmonary inactivation of BK [37, 43, 44]. More recently, the emerging role of cathepsin K as a potent kininase has been described [45, 46]. Based on the capacity of this enzyme to attenuate kallikrein-induced decrease of rat blood pressure and to reduce the hypotensive effect of BK in a dose-dependent manner, cathepsin K was described as a functional kininase, a unique property among mammalian cysteine cathepsins, which may act under physiological conditions in circulation [46]. As a result, the peptidase activity of cathepsin K might represent a possible alternative pathway for kinin catabolism, in addition to the other described peptidases.

4.5
Kinin Receptors

The history of kinin receptors started at the beginning of 1970s, when Regoli *et al.* set the early landmarks in the molecular characterization of kinin receptors by pointing to the existence of two types of kinin receptors, B_1 and B_2, differing in their genetic and pharmacological profiles as well as in their expression patterns [47, 48]. The two kinin receptors belong to the family of heptahelical GPCRs and use multiple but mostly overlapping signaling pathways [48]. By acting on both receptors, kinins stimulate cells to liberate autacoids such as nitric oxide (NO) and prostaglandins, which mediate most of their actions. While the B_2 receptor is constitutively expressed in most tissues, the B_1 receptor is peculiar in the family of heptahelical receptors, as it is induced at sites of inflammation by cytokines via activation of NF-κB. In addition, B_2 receptors are immediately desensitized following ligand binding, whilst B_1 receptors are not. Together with their overlapping signaling pathways, these characteristics suggest a partitioning of functions between the two kinin receptors, with B_2 being responsible for the short-term effects of kinins and B_1 for localized long-term actions. Kinin B_2 receptors present a higher affinity to BK and the intact kinins compared with des-Arg^9-BK, while B_1 receptors show a higher affinity to des-Arg^9-BK than to BK. This profile is determined by the discriminating nature of the C-terminal arginine of B_2 receptor-selective agents, which is absent in B_1 receptor-selective compounds. The development of specific agonists and antagonists to the kinin receptors was instrumental in establishing the dualism of kinin receptors in mammals [47]. Antagonists for the B_1 receptor were discovered almost 10 years before the antagonists for the B_2 receptor by Regoli *et al.*, who demonstrated the replacement of Phe8 residue of des-Arg^9-BK by aliphatic amino acid [47]. Later it was demonstrated that Leu8 replacement also resulted in selective B_1 antagonists. Thus, the receptor nomenclature is justified by the fact that it was the first to be pharmacologically fully defined.

Kinin B_1 receptor has been implicated in a number of inflammatory events, including persistent hyperalgesia, fever and increased vascular permeability [49]. These receptors have been identified in a number of arteries (aorta, mesenteric, celiac, pulmonary, renal and basilar), veins (mesenteric and saphena), nonvascular smooth muscle (stomach, colon and duodenum), and isolated cells such as fibroblasts, macrophages, osteoclasts, and vascular, tracheal, mesangeal and endothelial cells. The B_1 receptor has been shown to be an inducible GPCR regulated by organism exposure to cytokines and growth factors. Initially, the human B_1 receptor gene was cloned from human embryonic lung fibroblasts [50] and was shown to be a member of the seven-transmembrane domain GPCR family. Subsequently the mouse [51], rat [52] and rabbit [53] B_1 receptor were also cloned, and shown to display functional and structural homology with the human receptor. The rabbit receptor shows greatest homology with the human receptor whilst homology with the rodent receptors is relatively low. Interestingly this difference in homology is demonstrated by differences in preferred agonists – Lys-des-Arg^9-BK in humans and rabbits and des-Arg^9-BK in rodents [54].

Kinin B_2 receptor is responsible for most of the physiological actions of the KKS, and is widely distributed in mammal tissues such as the intestinal, cardiovascular, genitourinary and respiratory tracts, and ocular and neuronal tissues. Despite the constitutive nature of this receptor, overexpression has been recently found in certain pathologies like diabetes [55] and myocardial infarction [56].

The use of receptor knockout mice has also reinforced that kinins, acting at the B_1 receptor or B_2 receptor, exert a critical role in controlling inflammatory and nociceptive processing mechanisms. B_2 receptor is expressed on nociceptive neurones and mice deficient for the receptor do not respond by hyperalgesia on intrathecal injection of BK as control mice do [57]. Moreover, lack of the B_2 receptor gene led to reduced hyperalgesia in the carrageenan model, while thermal nociception and the nociceptive response to formalin injection were unaltered [58, 59]. The inflammatory response induced by immune complexes in the peritoneal cavity [60] and by closed head trauma in the brain [61] were drastically reduced in B_2-deficient mice, suggesting an important role for this receptor in inflammatory processes.

With respect to the phenotype of kinin B_1 knockout mice, these animals exhibit hypoalgesia in chemical models of nociception, in part by a reduction in the activity-dependent facilitation (wind-up) of a nociceptive spinal reflex, in comparison with wild-type littermates [62]. Later, it was shown that B_1 receptor activation is involved in thermal hyperalgesia in both ipsilateral and contralateral paws induced by intradermal or intrathecal administration of complete Freund's adjuvant or B_1 agonists in mice [57, 63]. Accordingly, the polymorphonuclear leukocyte accumulation into the pleural cavity induced by carrageenan is almost abolished in mice lacking B_1 receptor, while in noninflamed conditions these animals show a hypoalgesic effect in response to noxious stimulus [62, 64]. Similarly, ablation of the B_1 receptor gene in mice (but not of B_2 receptor) reduced the hyperalgesia induced by intraplantar injection of Freund's adjuvant to an extent similar to treating wild-type mice with the B_1 antagonist des-Arg^9-[Leu^8]-BK [57].

4.6
Physiological and Pathological Cardiovascular Roles of Kinins

Evidence suggests that the KKS contributes to maintain homeostasis of arterial pressure and that kinins are potential mediators in endogenous cardiovascular protective mechanisms. The main mediators of the physiologic effects of kinins, kinin B_2 receptors, are expressed in the heart, kidney, vessels and central nervous system – important centers for the control of cardiovascular homeostasis. Apart from the presence of B_2 receptor in endothelial and smooth muscle cells, several pieces of evidence indicate the presence of the whole KKS in blood vessels, such as the expression of kallikrein mRNA in the arteries and veins, endothelial cells in culture present activity similar to kallikrein, and vascular smooth muscle cells in culture release kallikrein and kininogen. BK is known for its multiple effects on the cardiovascular system and systemic administration of BK induces transient hypotension as a consequence of marked systemic vasodilatation. On the other hand, direct administration of this peptide in the central nervous system increases blood pressure. These autacoids evoke endothelium-dependent relaxations and endothelium-independent contractions in isolated vessels *in vitro*. In recent decades, the use of drugs that interfere with the KKS, such as inhibitors of kallikreins and kininase II as well as selective B_1 receptor and B_2 receptor antagonists, and also different pharmacologic and genetic approaches greatly contributed to the understanding of the pathophysiological role of this system in the body. Therefore, disturbances of this system may account for the pathogenesis of arterial hypertension, myocardial ischemia and other cardiovascular disorders. This is due to the fact that the KKS components are localized in the heart and in the vascular tissues [14–19]. In addition, kinins are released during ischemia [65] and cause beneficial cardiac effects [66]. Whereas BK can contribute to the cardioprotective effects of preconditioning by reducing infarct size [67], the use of BK antagonist prevents this beneficial effect and worsens ischemia-induced effects [68]. In rats with hypertension caused by aortic banding, BK can prevent left ventricular hypertrophy at a dose that has no effect on blood pressure [69]. In humans and experimental animals, a reduction in peripheral and cardiac KKS components may also be the cause of developing high blood pressure [70–73]. Inhibitors of kininase II (ACE) have been prescribed successfully to patients with cardiovascular diseases, but there is still a great interest in developing drugs or pharmacological strategies that augment the activity of the KKS in pathological conditions which could serve mostly as potential anti-inflammatory therapeutic agents. Therefore, the use of ACE inhibitors has provided evidence of the role played by kinins in the cardiovascular effects of this class of drugs. Since ACE inhibitors block not only the formation of angiotensin (Ang) II but also inhibit kinin degradation, kinin receptor antagonists were used to prove this assumption. More recently, local gene therapy based on components of the KKS (administration of the kallikrein gene) has been suggested as a pharmacological strategy to treat vascular diseases such as arterial hypertension, cardiac hypertrophy, restenosis and atherosclerosis. In summary, these findings suggest that the KKS participates in the regulation of arterial blood pressure control and circulatory homeostasis. Therefore, based in these assumptions, in the next sections the role played by the different components of the KKS, as well as drugs interfering in the system, in cardiovascular control and in different pathological conditions will be analyzed in detail.

4.7
Kinins in Blood Pressure Control and Hypertension

Kinins exert marked hypotensive effects by inducing relaxation of vascular smooth muscle, which suggests that these peptides may participate in the physiological regulation of organ perfusion and blood pressure. The pharmacological action of BK in the regulation of systemic blood pressure was initially demonstrated as vasodilatation in most areas of the circulation, a reduction of total peripheral vascular resistance and a regulation of sodium excretion from the kidney [74, 75]. This endothelium-dependent phenomenon is mediated by the release of different vasodilators such as prostacyclin (prostaglandin I2), NO and endothelium-derived hyperpolarizing factor. The vasodilatory effect is normally mediated by the kinin B_2 receptor, but under inflammatory conditions, B_1 receptor upregulation mediates kinin-induced vasodilation and hypotension. However, as seen above, kinins are very labile molecules, being rapidly degraded and thereby inactivated by several kininases present in plasma and in the vascular system. Despite that, many findings point to a significant role for kinins in regulating blood pressure.

Much progress in understanding the KKS in physiological and pathological conditions has been made after the development of knockout mice to both of the kinin receptors. Some authors have reported an increase in blood pressure and cardiac weight in B_2 receptor knockout animals at baseline or after a chronic high-salt diet compared to control animals [76, 77]. However, there are also reports of unchanged cardiovascular parameters in these animals, even after high-salt treatment [78–80]. These discrepancies may be partially explained by differences in the background strains used in the experiments [79, 81]. B_2 receptor knockout animals show no difference in cardioprotection induced by ischemic preconditioning or ACE inhibitor treatment, probably be due to the upregulation of B_1 receptor in this model [77, 82, 83], which may take over the protective functions of B_2 receptor. Furthermore, overexpression of B_1 receptor can be observed in B_2 receptor gene knockout mice [77, 82, 83].

Hypertension is a major risk factor for the development of cardiovascular diseases, such as coronary heart disease, congestive heart failure, and peripheral vascular and renal diseases [84]. There is ample evidence documenting the role of KKS in the pathogenesis of hypertension [9–13]. The work of Margolius et al. [85] described for the first time that urinary kallikrein excretion was significantly reduced in hypertensive patients and hypertensive rats, suggesting a correlation between reduced urinary kallikrein excretion and a defect in kinin generation in hypertension. Recently, new findings have attributed a significant effect to the KKS in postexercise hypotension, an important event for blood pressure regulation, especially in hypertensive individuals. Postexercise hypotension is a well-known phenomenon, but the mechanism responsible is still unclear. The work of Moraes et al. [86] suggests that the KKS may be involved in postexercise hypotension in normotensive and hypertensive individuals subjected to exercise training.

ACE (kininase II) inhibitors are currently used in the treatment of both clinical and experimental hypertension [87–89]. However, treatment with a kinin B_2

receptor antagonist can abolish the effect of captopril, an ACE inhibitor, suggesting that the hypotensive action of this class of drugs might be due to the activation of the B_2 receptor. In addition, treatment of spontaneously hypertensive rats with aprotinin is able to suppress the hypotensive responses of ACE inhibitors, highlighting a role of tissue kallikrein in the regulation of blood pressure. Recently, it has been proposed that tissue kallikrein gene delivery into various hypertensive models exhibits protection, such as a reduction in high blood pressure, attenuation of cardiac hypertrophy, and inhibition of renal damage and stenosis [90]. These findings may indicate the prospect of kallikrein gene therapy for cardiovascular and renal pathology.

Recent studies also suggest that the B_1 receptor is involved in regulation of vasodilatation, inflammation, and tissue repair, including myocardial infarction. Lamontagne et al. [91] reported that activation of the B_1 receptor by intravenous infusion of des-Arg^9-BK caused a profound hypotensive response, which was partially blocked by the NO synthase inhibitor N^G-nitro-L-arginine, suggesting a NO-mediated mechanism. Using a combined administration of B_1 receptor and B_2 receptor antagonists in rats, Duka et al. [92] could demonstrate that only with both receptors blockade blood pressure level was significantly increased, leading to an upregulation of genes related to vasoactive systems like endothelial NO synthase, AT_1 receptor, prostaglandin E2 receptor and tissue kallikrein [92].

4.8
Kinins in the Heart

BK has been known for a long time for its multiple effects on the cardiovascular system. The expression and functionality of the KKS has been detected in the cardiac tissue of different species. Kinins are well known to dilate the resistance vessels and to increase the coronary blood flow [93]. Earlier studies using isolated human heart have shown that BK increases the coronary blood flow [12, 94], a finding confirmed later in other species *in vivo* [93, 95, 96]. On the other hand, icatibant (HOE 140) is able to decrease coronary blood flow by blocking the kinin B_2 receptor, evidencing the participation of endogenous BK in the regulation of coronary vascular tone [97, 98]. In addition, Su et al. [99] demonstrated that also intracoronary infusion of des-Arg^9-BK produced dose-dependent coronary vasodilatation, as evidenced by increased coronary diameter and blood flow, an effect that was not altered by a kinin B_2 receptor antagonist, but was attenuated by NO synthase blockade. It is now established that this effect takes place by endothelium-dependent mechanisms through the release of NO and prostacyclin. However, after inhibition of NO synthase and cyclooxygenase, acetylcholine, BK and substance P still relax blood vessels by mechanisms that involve endothelium-dependent hyperpolarization of vascular myocytes [100]. Thus, by stimulating endothelial release of these compounds, kinins may increase coronary perfusion and reduce pre- and afterload, oxidative stress and sympathetic tone – actions that can potentially protect the heart from acute ischemic damage. Apart from increasing coronary flow, BK has been shown to improve the cardiac metabolism. In addition, direct beneficial actions of kinins may also result from the opening of endothelial intermediate- and small-conductance Ca^{2+}-sensitive K^+ channels [101].

Several effects of kinins in the heart are independent of their vasodilator actions, and include the modulation of cell growth and division in the heart and the modulation of myocardial responses to ischemia/reperfusion. Evidence indicates that kinins are beneficial in myocardial ischemia. In animals as well as in humans, myocardial ischemia increased plasma BK levels [9]. In isolated rat hearts under myocardial ischemia, BK reduces both the activity of cytosolic enzymes (lactate dehydrogenase and creatine kinase) and the lactate content, as well as preserves energy-rich phosphates and glycogen stores [102]. Additionally, BK reduces the incidence and duration of ventricular fibrillation in rats, and increases ventricular pressure, myocardial contractility and coronary flow without changing the heart rate [103].

The cardioprotective effects of BK were also evidenced in other studies where the peptide administration was shown to diminish myocardial ischemia and/or to reduce infarct area in experimental animals and in patients undergoing coronary angioplasty [9, 12]. The beneficial effects of BK in myocardial ischemia have been confirmed with the use of ACE inhibitors and selective B_2 receptor antagonists and transgenic animal models. Recent study demonstrated that B_1 receptor is induced during myocardial ischemia where it plays a detrimental role in mice. Infarct size to risk zone ratio was significantly reduced in hearts of B1 knockout mice compared to those of wild-type mice [104]. To address the role of an increase in kinin formation in the protection of cardiac hypertrophy and ischemia, a rat model of overexpression of human tissue kallikrein was used. Induction of cardiac hypertrophy by isoproterenol treatment revealed a marked protective effect of the kallikrein transgene, whereas the cardiac weight of TGR(hKLK1) transgenic rats harboring the human kallikrein gene increased significantly less, and the expression of atrial natriuretic peptide and collagen III as markers for hypertrophy and fibrosis, respectively, were less enhanced. The specific kinin B_2 receptor antagonist, icatibant, abolished this cardioprotective effect. In addition, in the transgenic rats the overflow of nucleotide breakdown products upon reperfusion was significantly less and systolic dP/dt at the end of ischemia was significantly higher in the transgenic rats, despite a similar reduction in total coronary flow during ischemia [105]. These differences were abolished in the rat hearts where icatibant was coinfused. This finding suggests that a chronic increase in tissue kallikrein can increase BK levels sufficiently to protect against cardiac ischemia and that this effect is brought about by a rapid mechanism, rather than by long-term structural alterations. It also exerts antihypertrophic and antifibrotic actions in the heart. Recent data from Koch et al. [106], using the same rat model, showed that BK coronary outflow was 3.5-fold increased in TGR(hKLK1) in ischemic conditions. However, despite similar unchanged infarction sizes, left ventricular function and remodeling improved in TGR(hKLK1) after myocardial infarction. On the other hand, in mice lacking tissue kallikrein, the cardiac function estimated *in vivo* and *in vitro* is decreased both under basal conditions and in response to β-adrenergic stimulation, supporting the idea that a functional KKS is necessary for normal cardiac and arterial function in the mouse [107]. In addition, the same group proposed a key role of tissue kallikrein in the flow-dependent dilation in resistance arteries [108, 109].

In line with these findings, kallikrein gene delivery has been shown to be an important therapeutic tool aimed at enhancing capillarization and arteriogenesis, and decreasing apoptosis [110, 111]. It has been shown that local delivery of an adenovirus carrying the human tissue kallikrein gene accelerates spontaneous angiogenesis and blood flow recovery in a mouse model of peripheral ischemia [112], induces arteriogenesis [113, 114], and inhibits the apoptotic death of vascular cells and skeletal myocytes [113]. More recently, Spillman et al. [115] have shown that human tissue kallikrein delivery to the peri-infarct myocardium is able to prevent postischemic heart failure.

Although cardioprotection given by ACE inhibitors reflects principally inhibition of Ang II formation, many studies have shown that BK accumulation contributes positively to cardiovascular actions of these drugs indirectly via the production of NO and prostacyclin. Accordingly, cardioprotection by ACE inhibitors are unaffected by Ang II receptor antagonists, whereas icatibant usually worsens the acute ischemic effects. Interestingly, in nonischemic hearts in vitro and in toxicological studies in vivo, icatibant alone does not modify basal cardiovascular parameters such as left ventricular pressure, contractility, cardiac output and coronary flow, thus suggesting that these antagonist actions occur only in conditions of acute myocardial ischemia. In experimental models of myocardial infarct by occlusion of coronary artery, ACE inhibitors, including captopril, enalapril and ramipril, also reduce cellular necrosis and infarct size by mechanisms mediated, at least partially, by kinin generation [9, 12]. In isolated hearts of different species subjected to ischemia/reperfusion, BK perfusion resulted in a better recovery of coronary flow and cardiac function during reperfusion, a reduced release of soluble markers of tissue injury, an improvement of metabolic efficiency, and a reduction in infarct size [9, 12].

4.9
Renal Effects of Kinins

As mentioned above, all components of the KKS are also expressed in the kidney, and are important regulators for renal hemodynamics, tubular function, renal ontogeny and remodeling [29, 116, 117]. The fact that kinins, like BK, are produced locally in the kidney at much higher levels than those present in blood [20] means that kinins act as autocrine/paracrine factors to regulate blood flow perfusion of these organs. Among the factors regulating kinin receptors expression are oxidative stress [48], proinflammatory cytokines [118], such as interleukin-1β or tumor necrosis factor-α, as well as other vasoactive peptide systems, such as the renin–angiotensin system (RAS) [119–123].

When BK is injected into the renal artery, it causes diuresis and natriuresis by increasing renal blood flow. These actions of BK have been attributed to prostaglandin release in the renal circulation. The importance of renal KKS is well documented, and generally seen as nephroprotective, natriuretic and diuretic [124, 125]. In this connection, it is suggested that the role of renal KKS is to excrete excess of sodium. Therefore, a reduction in the activation of renal KKS may cause development of hypertension as a result of sodium accumulation in the body. Thus, the development of a compound having renal kallikrein-like activity may serve the purpose of excreting excessive sodium from the kidney. This action may be useful for the treatment of hypertension. Also, it has been demonstrated that transgenic mice overexpressing renal tissue kallikrein were hypotensive and that the administration of aprotinin, a tissue kallikrein inhibitor, restored the blood pressure in the transgenic mice [90]. Kininase II inhibitors could lower blood pressure by inhibiting the biodegradation of kinin as well as blocking the formation of Ang II at the renal site. A calcium channel blocker, nifedipine, used to treat patients with essential hypertension, can normalize the reduced urinary kallikrein excretion [126]. In sodium-depleted dogs, intrarenal administration of kinin receptor antagonist did not affect blood pressure, but reduced renal blood flow and the autoregulation of glomerular filtration [127]. These effects were observed with B_2 receptor antagonists but not with B_1 receptor antagonists, indicating an essential role of B_2 receptor in this setting [128]. Exogenous BK increases renal blood flow without a significant change in glomerular filtration rate or proximal reabsorption, but with an increase in fluid delivery to the distal nephron [129], which contributes to the increased urine volume and sodium excretion. Endogenous BK is involved in the renal hemodynamic regulation. In rats, kinin antagonists decrease renal blood flow [130], papillary blood flow and urine flow [131] as well as volume expansion-induced water and sodium excretion [128, 132].

Tsuchida *et al.* [133], using AT_1 receptor-deficient mice, have demonstrated a potent antihypertrophic effect of the kinin B_2 receptor on the renal vasculature. In the kidney of these animals, the KKS is functionally activated by local suppression of kininase II and extensive redistribution of kallikrein to perivascular areas. Thus, the kinin B_2 antagonist icatibant accelerated significantly the vascular hypertrophy as determined by wall thickness ratio, indicating an antihypertrophic effect of the kinin B_2 receptor activation on the renal vasculature *in vivo* [133]. B_2 receptor-deficient mice show discrete changes in kidney morphology [134], which are aggravated by high-salt treatment during development. Furthermore, two-kidney-one-clip-, desoxycorticosterone acetate- and Ang II-induced hypertension, as well as diabetic and obstructive nephropathy, is worsened in these mice [135, 136]. On the other hand, cyclooxygenase-2 and the RAS are downregulated in the kidney and may not be responsible for the increased sensitivity of the mice to hypertensive stimuli [137].

It has been shown that renal kallikrein production can be reduced in patients with mild renal disease and more markedly in patients with severe renal failure.

Moreover, kallikrein reversed salt-induced renal fibrosis and glomerular hypertrophy, and reduced recruitment and accumulation of inflammatory cells in the tubulo-interstitium and vasculature of hypertensive rats [138]. Malignant hypertension is observed when B_2 receptor knockout mice are overloaded with dietary sodium [139]. These data confirm the long-time suspected role of the renal KKS in handling sodium and water metabolism, in parallel to the renin–angiotensin–aldosterone endocrine axis [125]. In addition, specific polymorphisms of the human kinin receptor locus have been associated with end-stage renal failure (mixed etiology), hypertension and nephropathy in diabetics [140–142]. The protective effect of kinins in renal disease has been extended to the B_1 receptor, as various aspects of the chronic inflammatory response seem to be limited by endogenous ligands of these receptor subtypes in animal models [143].

In addition to their hemodynamic actions and tubular function, kinins may have important anti-inflammatory and antiproliferative properties (e.g. under diabetic conditions). *In vivo* kinin B_2 receptor activation reduces renal fibrosis. In this respect, the renal KKS may belong to an important system that can provide relevant renal protection. In particular, BK has been shown to have nitric oxide-releasing properties, to induce endothelium-dependent vasodilation, and to exert antifibrotic and antihypertrophic actions, as well as to stimulate glucose uptake [144].

Recently, Kakoki *et al.* [145] have described that both kinin receptors are important in protecting the kidney from damage caused by ischemia/reperfusion. Using mice lacking B_1 and B_2 receptors, they have shown reduced DNA damage, apoptosis, morphological and functional kidney changes, and reduced mortality. The results suggest that both receptors, presumably acting in part via NO, mitigate renal ischemia/reperfusion injury by blunting the marked increase in oxidative stress that accompanies the model. Although blockade of the B_2 receptor in streptozotocin-induced diabetic rats by the B_2 receptor antagonist icatibant did not greatly improve diabetic nephropathy [146], recent studies in mice lacking the B_2 receptor implicated the role of the KKS in the development of diabetic nephropathy [147]. In diabetic mice lacking the B_2 receptor, the condition of the diabetic nephropathy worsens, which was associated with a further increase in glomerular mesangial sclerosis. Although these mice also showed a further rise in renal B_1 receptor expression, it is obvious that this increase did not fully compensate for the loss of B_2 receptor in this model.

Apart from the direct actions of kinins in the kidney, the interactions between inhibition of the RAS and activation of the renal KKS are relevant. The therapeutic actions of ACE inhibitors and AT_1 receptor antagonists revealed complex interactions between the RAS and KKS [108]. Several studies [148–150] have demonstrated experimentally that the protective effects of ACE inhibitors are at least partly mediated by a direct potentiation of kinin receptor response to BK stimulation. Kinins are partly involved in the antiproteinuric action of ACE inhibitors in experimental diabetic nephropathy [151]. Such interactions are supported by the fact that ACE efficiently degrades kinins or BK. In addition, angiotensin fragments such as Ang-(1–7) exert kinin-like effects.

References

1. Abelous, J. and Bardier, E. (1909) Les substances hypotensives de l'urine humaine normale. *Comptes Rendus Des Seances De La Societe De Biologie Et De Ses Filiales*, **66**, 511–12.
2. Rocha e Silva, M., Beraldo, W.T. and Rosenfeld, G. (1949) Bradykinin, a hypotensive and smooth muscle stimulating factor released from plasma globulin by snake venoms and by trypsin. *American Journal of Physiology*, **156**, 261–73.
3. Elliott, D.F., Horton, E.W. and Lewis, G.P. (1960) Actions of pure bradykinin. *Journal of Physiology*, **153**, 473–80.
4. Boissonnas, R.A., Guttmann, S. and Jaquenoud, P.A. (1960) Synthèse de la L-arginyl-L-propyl-L-propyl-glycyl-L-phenylalanyl-L-seryl-L-propyl-L-phenylalanyl-L-arginine, un nonapeptide presentant le proprietés de la bradykinine. *Helvetica Chimica Acta*, **43**, 1349–55.
5. Webster, M.E. and Pierce, J.V. (1963) The nature of the kallidins released from human plasma by kallikreins and other enzymes. *Annals of the New York Academy of Sciences*, **104**, 91–107.
6. Kato, H., Matsumura, Y. and Maeda, H. (1988) Isolation and identification of hydroxyproline analogues of bradykinin in human urine. *FEBS Letters*, **232**, 252–4.
7. Maier, M., Reissert, G., Jerabek, I., Lottspeich, F. and Binder, B.R. (1988) Identification of [hydroxyproline3]-lysyl-bradykinin released from human kininogens by human urinary kallikrein. *FEBS Letters*, **232**, 395–8.
8. Sakamoto, W., Satoh, F., Gotoh, K. and Uehara, S. (1987) Ile-Ser-bradykinin (T-kinin) and Met-Ile-Ser-bradykinin (Met-T-kinin) are released from T-kininogen by an acid proteinase of granulomatous tissues in rats. *FEBS Letters*, **219**, 437–40.
9. Su, J.B. (2006) Kinins and cardiovascular diseases. *Current Pharmaceutical Design*, **12**, 3423–35.
10. Sharma, J.N. (2006) Role of tissue kallikrein–kininogen–kinin pathways in the cardiovascular system. *Archives of Medical Research*, **37**, 299–306.
11. Dendorfer, A., Wolfrum, S. and Dominiak, P. (1999) Pharmacology and cardiovascular implications of the kinin–kallikrein system. *Japanese Journal of Pharmacology*, **79**, 403–26.
12. Marcondes, S. and Antunes, E. (2005) The plasma and tissue kininogen–kallikrein–kinin system: role in the cardiovascular system. *Current Medicinal Chemistry – Cardiovascular & Hematological Agents*, **3**, 33–44.
13. Moreau, M.E., Garbacki, N., Molinaro, G., Brown, N.J., Marceau, F. and Adam, A. (2005) The kallikrein–kinin system: current and future pharmacological targets. *Journal of Pharmacy and Pharmaceutical Sciences*, **99**, 6–38.
14. Nolly, H.L. and Brotis, J. (1981) Kinin-forming enzyme in rat cardiac tissue. *American Journal of Physiology*, **265**, H1209–14.
15. Sharma, J.N. and Uma, K. (1996) Cardiac kallikrein in hypertensive and diabetic rats with and without diabetes. *Immunopharmacology*, **33**, 341–3.
16. Sharma, J.N., Uma, K. and Yusof, A.P.M. (1998) Left ventricular hypertrophy and its relation to the cardiac kinin-forming system in hypertensive and diabetic rats. *International Journal of Cardiology*, **63**, 229–35.
17. Sharma, J.N., Uma, K. and Yusof, A.P.M. (1999) Altered cardiac tissue and plasma kininogen levels in hypertensive and diabetic rats. *Immunopharmacology*, **34**, 129–32.
18. Oza, N.B. and Goud, H.D. (1992) Kininogenase of the aortic wall in spontaneously hypertensive rats. *Journal of Cardiovascular Electrophysiology*, **20** (Suppl. 9), 1–3.
19. Nolly, H.L., Carretero, O.A. and Sclicli, A.J. (1993) Kallikrein release by vascular tissue. *American Journal of Physiology*, **265**, H1209–14.
20. Campbell, D.J. (2000) Towards understanding the kallikrein–kinin system: insights from measurement of kinin peptides. *Brazilian Journal of*

Medical and Biological Research, **33**, 665–77.

21 Madeddu, P., Emanueli, C. and El-Dahr, S. (2007) Mechanisms of disease: the tissue kallikrein–kinin system in hypertension and vascular remodelling. *Nature Clinical Practice Nephrology*, **3**, 208–21.

22 El-Dahr, S.S. (2004) Spatial expression of the kallikrein–kinin system during nephrogenesis. *Histology and Histopathology*, **19**, 1301–10.

23 Werle, E., Götze, W. and Keppler, A. (1937) Über die Wirkung des Kallikreins auf den isolierten Darm und über eine neue darmkontrahierende substanz. *Biochemische Zietung*, **289**, 217–33.

24 Muller-Esterl, W., Iwanaga, S. and Nakanishi, S. (1986) Kininogens revisited. *Trends in Biochemical Sciences*, **11**, pp. 336–9.

25 Kitamura, N., Ohkubo, H. and Nakanishi, S. (1987) Molecular biology of the angiotensinogen and kininogen genes. *Journal of Cardiovascular Electrophysiology*, **10** (Suppl. 7), 49–53.

26 Ohkubo, I., Kurachi, K., Takasawa, T., Shiokawa, H. and Sasaki, M. (1984) Isolation of a human cDNA for a2-thiol proteinase inhibitor and its identity with low molecular weight kininogen. *Biochemistry*, **23**, 5691–7.

27 Salvesen, G., Parkes, C., Abrahamson, M., Grubb, A. and Barrett, A.J. (1986) Human low-Mr kininogen contains three copies of a cystatin sequence that are divergent in structure and in inhibitory activity for cysteine proteinases. *Biochemical Journal*, **234**, 429–34.

28 Cardoso, C.C., Garrett, T., Cayla, C., Meneton, P., Pesquero, J.B. and Bader, M. (2004) Structure and expression of two kininogen genes in mice. *Biological Chemistry*, **385**, 295–301.

29 Cheung, P.P., Cannizzaro, L.A. and Colman, R.W. (1992) Chromosomal mapping of human kininogen gene (KNG) to 3q26–;qter. *Cytogenetics and Cell Genetics*, **59**, 24–6.

30 Kitamura, N., Kitagawa, H., Fukushima, D., Takagaki, Y., Miyata, T. and Nakanishi S. (1985) *Journal of Biological Chemistry* **260**, 8610–7.

31 Colman, R.W. and Schmaier, A.H. (1997) Contact system: a vascular biology modulator with anticoagulant, profibrinolytic, antiadhesive, and proinflammatory attributes. *Blood*, **90**, 3819–43.

32 Kaplan, A.P., Joseph, K. and Silverberg, M. (2002) Pathways for bradykinin formation and inflammatory disease. *Journal of Allergy and Clinical Immunology*, **109**, 195–209.

33 Shariat-Madar, Z., Mahdi, F. and Schmaier, A.H. (2002) Identification and characterization of prolylcarboxypeptidase as an endothelial cell prekallikrein activator. *Journal of Biological Chemistry*, **277**, 17962–9.

34 Greenbaum, L.M. (1982) T-kinin and T-kininogen. Children of technology. *Biochemical Pharmacology*, **33**, 2943–4.

35 Okamoto, H. and Greenbaum, L.M. (1983) Pharmacological properties of T-kinin. *Biochemical Pharmacology*, **32**, 2637–8.

36 Dendorfer, A., Wolfrum, S., Wellhöner, P., Korsman, K. and Dominiak, P. (1997) Intravascular and interstitial degradation of bradykinin in isolated perfused rat heart. *British Journal of Pharmacology*, **122**, 1179–87.

37 Pesquero, J.B., Jubilut, G.N., Lindsey, C.J. and Paiva, A.C. (1992) Bradykinin metabolism pathway in the rat pulmonary circulation. *Journal of Hypertension*, **10**, 1471–8.

38 Kokkonen, J.O., Lindstedt, K.A., Kuoppala, A. and Kovanen, P.T. (2000) Kinin-degrading pathways in the human heart. *Trends in Cardiovascular Medicine*, **10**, 42–5.

39 Erdos, E.G. (1990) Some old and some new ideas on kinin metabolism. *Journal of Cardiovascular Electrophysiology*, **15**, S20–24.

40 Drapeau, G., Chow, A. and Ward, P.E. (1991) Metabolism of bradykinin analogs by angiotensin I converting enzyme and carboxypeptidase N. *Peptides*, **12**, 631–8.

41 Proud, D., Baumgarten, C.R., Nacleiro, R.M. and Ward, P.E. (1987) Kinin metabolism in human nasal secretions during experimentally induced allergic

rhinitis. *Journal of Immunology*, **138**, 428–34.

42. Sheikh, I.A. and Kaplan, A.P. (1989) Mechanism of digestion of bradykinin and lysylbradykinin (kallidin) in human serum: Roles of carboxypeptidase, angiotensin-converting enzyme and determination of final degradation products. *Biochemical Pharmacology*, **38**, 993–1000.

43. Ryan, J.W., Berryer, P., Chung, A.Y.K. and Sheffy, D.H. (1994) Characterization of rat pulmonary vascular aminopeptidase P *in vivo*: role in the inactivation of bradykinin. *Journal of Pharmacology and Experimental Therapeutics*, **269**, 941–7.

44. Ward, P.E. (1991) Metabolism of bradykinin and bradykinin analogs, in *Bradykinin Antagonists* (ed. R.M. Burch), Marcel Dekker, New York, pp. 147–70.

45. Alves, M.F., Puzer, L., Cotrin, S.S., Juliano, M.A., Juliano, L., Bromme, D. and Carmona, A.K. (2003) S3 to S3′ subsite specificity of recombinant human cathepsin K and development of selective internally quenched fluorescent substrates. *Biochemical Journal*, **373**, 981–6.

46. Lecaille, F., Vandier, C., Godat, E., Hervé-Grépinet, V., Brömme, D. and Lalmanach, G. (2007) Modulation of hypotensive effects of kinins by cathepsin K. *Archives of Biochemistry and Biophysics*, **459**, 129–36.

47. Regoli, D. and Barabe, J. (1980) Pharmacology of bradykinin and related kinins. *Pharmacological Reviews*, **32**, 1–46.

48. Pesquero, J.B. and Bader, M. (1998) Molecular biology of the kallikrein–kinin system: from structure to function. *Brazilian Journal of Medical and Biological Research*, **31**, 1197–203.

49. Marceau, F. and Hess, J.F. and Bachvarov, D.R. (1998) The B_1 receptors for kinins. *Pharmacological Reviews*, **50**, 357–86.

50. Menke, J.G., Borkowski, J.A., Bierilo, K.K., MacNeil, T., Derrick, A.W., Schneck, K.A., Ransom, R.W. and Strader, C.D. Linemeyer, D.L. and Hess, F.J. (1994) Expression cloning of a human B_1 bradykinin receptor. *Journal of Biological Chemistry*, **269**, 21583–6.

51. Pesquero, J.B., Pesquero, J.L., Oliveira, S.M., Roscher, A.A., Metzger, R. and Ganten, D. and Bader M. (1996) Molecular cloning and functional characterization of a mouse bradykinin B_1 receptor gene. *Biochemical and Biophysical Research Communications*, **220**, 219–25.

52. Ni, A., Chai, K.X., Chao, L. and Chao, J. (1998) Molecular cloning and expression of rat bradykinin B_1 receptor. *Biochimica Et Biophysica Acta*, **1442**, 177–85.

53. MacNeil, T., Bierilo, K.K., Menke, J.G. and Hess, J.F. (1995) Cloning and pharmacological characterization of a rabbit bradykinin B_1 receptor. *Biochimica Et Biophysica Acta*, **1264**, 223–8.

54. Hess, J.F., Derrick, A.W. and MacNeil, T. and Borkowski, J.A. (1996) The agonist selectivity of a mouse B_1 bradykinin receptor differs from human and rabbit B_1 receptors. *Immunopharmacology*, **33**, 1–8.

55. Christopher, J. and Jaffa, A.A. (2002) Diabetes modulates the expression of glomerular kinin receptors. *International Immunopharmacology*, **2**, 1771–9.

56. Tschope, C., Heringer-Walther, S. and Walther, T. (2000) Regulation of the kinin receptors after induction of myocardial infarction: a mini-review. *Brazilian Journal of Medical and Biological Research*, **33**, 701–8.

57. Ferreira, J., Campos, M.M., Araujo, R.C., Bader, M., Pesquero, J.B. and Calixto, J.B. (2002) The use of kinin B_1 and B_2R knockout mice and selective antagonists to characterize the nociceptive responses caused by kinins at the spinal level. *Neuropharmacology*, **43**, 1188–97.

58. Boyce, S., Rupniak, N.M., Carlson, E.J., Webb, J., Borkowski, J.A., Hess, J.F., Strader, C.D. and Hill, R.G. (1996) Nociception and inflammatory hyperalgesia in B_2 bradykinin receptor knockout mice. *Immunopharmacology*, **33**, 333–5.

59. Rupniak, N.M., Boyce, S., Webb, J.K., Williams, A.R., Carlson, E.J., Hill, R.G., Borkowski, J.A. and Hess, J.F. (1997) Effects of the bradykinin B_1 receptor

antagonist des-Arg9[Leu8]bradykinin and genetic disruption of the B_2 receptor on nociception in rats and mice. *Pain*, **71**, 89–97.

60 Samadfam, R., Teixeira, C., Bkaily, G., Sirois, P., Brum-Fernandes, A. and D'Orleans-Juste, P. (2000) Contribution of B_2 receptors for bradykinin in arthus reaction-induced plasma extravasation in wild-type or B_2 transgenic knockout mice. *British Journal of Pharmacology*, **129**, 1732–8.

61 Hellal, F., Pruneau, D., Palmier, B., Faye, P., Croci, N., Plotkine, M. and Marchand-Verrecchia, C. (2003) Detrimental role of bradykinin B_2 receptor in a murine model of diffuse brain injury. *Journal of Neurotrauma*, **20**, 841–51.

62 Pesquero, J.B., Araujo, R.C., Heppenstall, P.A., Stucky, C.L., Silva, J.A. Jr, Walther, T., Oliveira, S.M., Pesquero, J.L., Paiva, A.C., Calixto, J.B., Lewin, G.R. and Bader, M. (2000) Hypoalgesia and altered inflammatory responses in mice lacking kinin B_1 receptors. *Proceedings of the National Academy of Sciences of the United States of America*, **97**, 8140–5.

63 Ferreira, J., Campos, M.M., Pesquero, J.B., Araujo, R.C., Bader, M. and Calixto, J.B. (2001) Evidence for the participation of kinins in Freund's adjuvant-induced inflammatory and nociceptive responses in kinin B_1 and B_2 receptor knockout mice. *Neuropharmacology*, **41**, 1006–12.

64 (a) Araujo, R.C., Kettritz, R., Fichtner, I., Paiva, A.C., Pesquero, J.B. and Bader, M. (2001) Altered neutrophil homeostasis in kinin B_1 receptor-deficient mice. *Biological Chemistry*, **382**, 91–5. (b) Xiong, W., Chao, J. and Chao, L. (1995) Muscle delivery of human tissue kallikrein gene reduces blood pressure in hypertensive rats. *Hypertension*, **25**, 715–19.

65 Vegh, A., Szekeres, L. and Parratt, J.R. (1991) Local intracoronary infusions of bradykinin profoundly reduce the severity of ischaemia-induced arrhythmia in anaesthetized dogs. *British Journal of Pharmacology*, **104**, 294–5.

66 Linz, W., Wiemer, G. and Scholkens, B.A. (1993) Bradykinin prevents left ventricular hypertrophy in rats. *Journal of Hypertension*, **11** (Suppl. 5), S96–97.

67 Vegh, A., Rapp, J.G. and Parratt, J.R. (1994) Attenuation of the antiarrhythmic effects of ischaemia preconditioning by blocked of bradykinin B_2 receptors. *British Journal of Pharmacology*, **107**, 1167–72.

68 Walls, T.M., Sheehy, R. and Hartman, J.C. (1994) Role of bradykinin in myocardial preconditioning. *Journal of Pharmacology and Experimental Therapeutics*, **270**, 681–9.

69 Linz, W.W., Wiemer, G. and Scholkens, B.A. (1993) Contribution of bradykinin to the cardiovascular effects of ramipril. *Journal of Cardiovascular Electrophysiology*, **22** (Suppl. 9), S1–8.

70 Sharma, J.N. (1984) Kinin-forming system in the genesis of hypertension. *Agents Actions*, **14**, 200–5.

71 Sharma, J.N. (1988) Interrelationship between the kallikrein–kinin system and hypertension: a review. *General Pharmacology*, **19**, 177–87.

72 Sharma, J.N. (1989) Contribution of kinin system to the antihypertensive action of angiotensin converting enzyme inhibitors. *Advances in Experimental Medicine and Biology*, **247A**, 197–205.

73 Sharma, J.N., Uma, K.K. and Noor, A.R. (1996) Blood pressure regulation by the kallikrein–kinin system. *General Pharmacology*, **27**, 55–63.

74 De Freitas, F.M., Farraco, E.Z. and de Azevedo, D.F. (1964) General circulatory alterations induced by intravenous infusion of synthetic bradykinin in man. *Circulation*, **29**, 66–70.

75 Webster, M.E. and Gilmore, J.P. (1964) Influence of kallidin-10 on renal function. *American Journal of Physiology*, **206**, 714–18.

76 Alfie, M.E., Yang, X.P., Hess, F. and Carretero, O.A. (1996) Salt sensitive hypertension in bradykinin B_2 receptor knockout mice. *Biochemical and Biophysical Research Communications*, **224**, 625–30.

77 Madeddu, P., Varoni, M.V., Palomba, D., Emanueli, C., Demontis, M.P., Glorioso, N., Dessi Fulgheri, P., Sarzani, R. and

Anania, V. (1997) Cardiovascular phenotype of a mouse strain with disruption of bradykinin B_2-receptor gene. *Circulation*, **96**, 3570–8.

78 Milia, A.F., Gross, V., Plehm, R., Silva, J.A. Jr, Bader, M. and Luft, F.C. (2001) Normal blood pressure and renal function in mice lacking the bradykinin B_2 receptor. *Hypertension*, **37**, 1473–9.

79 Schanstra, J.P., Neau, E., Drogoz, P., Arevalo Gomez, M.A., Lopez Novoa, J.M., Calise, C., Pecher, C., Bader, M., Girolami, J.-P. and Bascands, J.-L. (2002) In vivo bradykinin B_2-receptor activation reduces renal fibrosis. *Journal of Clinical Investigation*, **110**, 371–9.

80 Trabold, F., Pons, S., Hagege, A.A., Bloch-Faure, M., Alhenc-Gelas, F., Giudicelli, J.F., Richer-Giudicelli, C. and Meneton, P. (2002) Cardiovascular phenotypes of kinin B_2 receptor and tissue kallikrein-deficient mice. *Hypertension*, **40**, 90–5.

81 Maestri, R., Milia, A.F., Salis, M.B., Graiani, G., Lagrasta, C., Monica, M., Corradi, D., Emanueli, C. and Madeddu, P. (2003) Cardiac hypertrophy and microvascular deficit in kinin B_2 receptor knockout mice. *Hypertension*, **41**, 1151–5.

82 Duka, I., Shenouda, S., Johns, C., Kintsurashvili, E., Gavras, I. and Gavras, H. (2001) Role of the B_2 receptor of bradykinin in insulin sensitivity. *Hypertension*, **38**, 1355–60.

83 Griol-Charhbili, V., Messadi-Laribi, E., Bascands, J.L., Heudes, D., Meneton, P., Giudicelli, J.F., Alhenc-Gelas, F. and Richer, C. (2005) Role of tissue kallikrein in the cardioprotective effects of ischemic and pharmacological preconditioning in myocardial ischemia. *FASEB Journal*, **19**, 1172–4.

84 Sharma, J.N. (1988) Interrelationship between the kallikrein–kinin system and hypertension: a review. *General Pharmacology*, **19**, 177–87.

85 Margolius, H.S., Geller, R., de Jong, W., Pisano, J.J. and Sjoerdsma, A. (1972) Altered urinary kallikrein excretion in rats with hypertension. *Circulation Research*, **30**, 358–62.

86 Moraes, M.R., Bacurau, R.F., Ramalho, J.D., Reis, F.C., Casarini, D.E., Chagas, J.R., Oliveira, V., Higa, E.M., Abdalla, D.S., Pesquero, J.L., Pesquero, J.B. and Araujo, R.C. (2007) Increase in kinins on post-exercise hypotension in normotensive and hypertensive volunteers. *Biological Chemistry*, **388**, 533–40.

87 Antonaccio, M. (1982) Angiotensin converting enzyme (ACE) inhibitors. *Annual Review of Pharmacology and Toxicology*, **22**, 57–87.

88 Sharma, J.N., Ferandez, P.G., Kim, B.K. and Triggle, C.R. (1984) Systolic blood pressure responses to enalapril maleate (MK 421), an angiotensin converting enzyme inhibitor and hydrochlorothiazide in conscious Dahl salt-sensitive (S) and salt-resistant (R) rats. *Canadian Journal of Physiology and Pharmacology*, **62**, 241–3.

89 Edery, H., Rosenthal, T., Amitzur, G., Rubinstein, A. and Stern, N. (1981) The influence of SQ 20881 on the blood kinin system of renal hypertensive patients. *Drugs under Experimental and Clinical Research*, **VII**, 749–56.

90 Chao, J. and Chao, L. (1997) Kallikrein gene therapy in hypertension, cardiovascular and renal diseases. *Pharmacological Research*, **35**, 517–22.

91 Lamontagne, D., Nakhostine, N., Couture, R. and Nadeau, R. (1996) Mechanisms of kinin B_1-receptor-induced hypotension in the anesthetized dog. *Journal of Cardiovascular Electrophysiology*, **28**, 645–50.

92 Duka, A., Duka, I., Gao, G., Shenouda, S., Gavras, I. and Gavras, H. (2006) Role of bradykinin B_1 and B_2 receptors in normal blood pressure regulation. *American Journal of Physiology-Endocrinology and Metabolism*, **291**, E268–74.

93 Linz, W., Wiemer, G., Gohlke, P., Unger, T. and Scholkens, B.A. (1995) Contribution of kinins to the cardiovascular actions of angiotensin-converting enzyme inhibitors. *Pharmacological Reviews*, **47**, 25–49.

94 Antonio, A. and Rocha e Silva, M. (1962) Coronary vasodilation produced by bradykinin on isolated mammalian heart. *Circulation Research*, **11**, 910–15.

95 Ebrahim, Z., Yellon, D.M. and Baxter, G.F. (2007) Attenuated cardioprotective response to bradykinin, but not classical ischaemic preconditioning, in DOCA-salt hypertensive left ventricular hypertrophy. *Pharmacological Research*, **55**, 42–8.

96 Staszewska-Woolley, J. and Woodman, O.L. (1991) Kinin receptors mediating the effects of bradykinin on the coronary circulation in anaesthetized greyhounds. *European Journal of Pharmacology*, **196**, 9–14.

97 Groves, P., Kurz, S., Just, H. and Drexler, H. (1995) Role of endogenous bradykinin in human coronary vasomotor control. *Circulation*, **92**, 3424–30.

98 Cheng, C-P, Onishi, K., Ohte, N., Suzuki, M. and Little, W.C. (1998) Functional effects of endogenous bradykinin in congestive heart failure. *Journal of the American College of Cardiology*, **31**, 1679–86.

99 Su, J.B., Houel, R., Heloire, F., Barbe, F., Beverelli, F., Sambin, L., Castaigne, A., Berdeaux, A., Crozatier, B. and Hittinger, L. (2000) Stimulation of bradykinin B_1 receptors induces vasodilation in conductance and resistance coronary vessels in conscious dogs: comparison with B_2 receptor stimulation. *Circulation*, **101**, 1848–53.

100 Busse, R., Edwards, G., Félétou, M., Fleming, I., Vanhoutte, P.M. and Weston A.H. (2002) EDHF: bringing the concepts together. *Trends in Pharmacological Science*, **23**, 374–80.

101 Weston, A.H., Félétou, M., Vanhoutte, P.M., Falck, J.R., Campbell, W.B. and Edwards, G. (2005) Bradykinin-induced, endothelium-dependent responses in porcine coronary arteries: involvement of potassium channel activation and epoxyeicosatrienoic acids. and from the increase in myocardial glucose uptake. *British Journal of Pharmacology*, **145**, 775–84.

102 Linz, W., Wiemer, G. and Schölkens, B.A. (1996) Role of kinins in the pathophysiology of myocardial ischemia. *In vitro* and *in vivo* studies. *Diabetes*, **45**, S51.

103 Linz, W., Wiemer, G. and Schölkens, B.A. (1992) ACE-inhibition induces NO-formation in cultured bovine endothelial cells and protects isolated ischemic rat hearts. *Journal of Molecular and Cellular Cardiology*, **24**, 909–19.

104 Lagneux, C., Bader, M., Pesquero, J.B., Demenge, P. and Ribuot, C. (2002) Detrimental implication of B_1 receptors in myocardial ischemia: evidence from pharmacological blockade and gene knockout mice. *International Immuno-pharmacology*, **2**, 815–22.

105 Silva, J.A. Jr, Araujo, R.C., Baltatu, O., Oliveira, S.M., Tschope, C., Fink, E., Hoffmann, S., Plehm, R., Chai, K.X., Chao, L., Chao, J., Ganten, D., Pesquero, J.B. and Bader, M. (2000) Reduced cardiac hypertrophy and altered blood pressure control in transgenic rats with the human tissue kallikrein gene. *FASEB Journal*, **14**, 1858–60.

106 Koch, M., Spillmann, F., Dendorfer, A., Westermann, D., Altmann, C., Sahabi, M., Linthout, S.V., Bader, M., Walther, T., Schultheiss, H.P. and Tschöpe, C. (2006) Cardiac function and remodeling is attenuated in transgenic rats expressing the human kallikrein-1 gene after myocardial infarction. *European Journal of Pharmacology*, **550**, 143–8.

107 Meneton, P., Bloch-Faure, M., Hagege, A.A., Ruetten, H., Huang, W., Bergaya, S., Ceiler, D., Gehring, D., Martins, I., Salmon, G., Boulanger, C.M., Nussberger, J., Crozatier, B., Gasc, J.M., Heudes, D., Bruneval, P., Doetschman, T., Menard, J. and Alhenc-Gelas, F. (2001) Cardiovascular abnormalities with normal blood pressure in tissue kallikrein-deficient mice. *Proceedings of the National Academy of Sciences of the United States of America*, **98**, 2634–9.

108 Bergaya, S., Meneton, P., Bloch-Faure, M., Mathieu, E., Alhenc-Gelas, F., Levy, B.I. and Boulanger, C.M. (2001) Decreased flow-dependent dilation in carotid arteries of tissue kallikrein knockout mice. *Circulation Research*, **88**, 593–9.

109 Bergaya, S., Matrougui, K., Meneton, P., Henrion, D. and Boulanger, C.M. (2004) Role of tissue kallikrein in response to

flow in mouse resistance arteries. *Journal of Hypertension*, **22**, 745–50.

110 Emanueli, C., Bonaria Salis, M., Stacca, T., Pintus, G., Kirchmair, R., Isner, J. M., Pinna, A., Gaspa, L., Regoli, D., Cayla, C., Pesquero, J.B., Bader, M. and Madeddu, P. (2002) Targeting kinin B_1 receptor for therapeutic neovascularization. *Circulation*, **105**, 360–6.

111 Emanueli, C., Salis, M.B., Pinna, A., Stacca, T., Milia, A.F., Spano, A., Chao, J., Chao, L., Sciola, L. and Madeddu, P. (2002) Prevention of diabetes-induced microangiopathy by human tissue kallikrein gene transfer. *Circulation*, **106**, 993–9.

112 Emanueli, C., Minasi, A., Zacheo, A. *et al.* (2001) Local delivery of human tissue kallikrein gene accelerates spontaneous angiogenesis in mouse model of hindlimb ischemia. *Circulation*, **103**, 125–32.

113 Emanueli, C., Graiani, G., Salis, M.B., Gadau, S., Desortes, E. and Madeddu, P. (2004) Prophylactic gene therapy with human tissue kallikrein ameliorates limb ischemia recovery in Type 1 diabetic mice. *Diabetes*, **53**, 1096–103.

114 Emanueli, C., Salis, M.B., Van Linthout, S. *et al.* (2004) Akt/protein kinase B and endothelial nitric oxide synthase mediate muscular neovascularization induced by tissue kallikrein gene transfer. *Circulation*, **110**, 1638–44.

115 Spillmann, F., Graiani, G., Van Linthout, S., Meloni, M., Campesi, I., Lagrasta, C., Westermann, D., Tschope, C., Quaini, F., Emanueli, C. and Madeddu, P. (2006) Regional and global protective effects of tissue kallikrein gene delivery to the peri-infarct myocardium. *Regenerative Medicine*, **1**, 235–54.

116 Madeddu, P., Emanueli, C. and El-Dahr, S. (2007) Mechanisms of disease: the tissue kallikrein–kinin system in hypertension and vascular remodeling. *Nature Clinical Practice Nephrology*, **3**, 208–21.

117 El-Dahr, S.S. (2004) Spatial expression of the kallikrein–kinin system during nephrogenesis. *Histology and Histopathology*, **19**, 1301–10.

118 Phagoo, S.B., Poole, S. and Leeb-Lundberg, L.M. (1999) Autoregulation of bradykinin receptors: agonists in the presence of interleukin-1beta shift the repertoire of receptor subtypes from B_2 to B_1 in human lung fibroblasts. *Molecular Pharmacology*, **56**, 325–33.

119 Tschope, C., Schultheiss, H.P. and Walther, T. (2002) Multiple interactions between the renin-angiotensin and the kallikrein–kinin systems: role of ACE inhibition and AT_1 receptor blockade. *Journal of Cardiovascular Electrophysiology*, **39**, 478–87.

120 Dean, R., Murone, C., Lew, R.A. *et al.* (1997) Localization of bradykinin B_2 binding sites in rat kidney following chronic ACE inhibitor treatment. *Kidney International*, **52**, 1261–70.

121 Marin-Castano, M.E., Schanstra, J.P., Neau, E. *et al.* (2002) Induction of functional bradykinin B_1-receptors in normotensive rats and mice under chronic angiotensin-converting enzyme inhibitor treatment. *Circulation*, **105**, 627–32.

122 Sabourin, T., Morissette, G., Bouthillier, J. *et al.* (2002) Expression of kinin B_1 receptor in fresh or cultured rabbit aortic smooth muscle: role of NF-kappa B. *American Journal of Physiology – Heart and Circulatory Physiology*, **283**, H227–37.

123 Campbell, D.J., Kladis, A. and Duncan, A.M. (1994) Effect of converting enzyme inhibitors on angiotensin and bradykinin peptides. *Hypertension*, **23**, 439–49.

124 Majima, M. and Katori, M. (1995) Approaches to the development of novel antihypertensive drugs: crucial role of the renal kallikrein–kinin system. *Trends in Pharmacological Sciences*, **16**, 239–46.

125 Margolius, H.S. (1995) Kallikreins and kinins: some unanswered questions about system characteristics and roles in human disease. *Hypertension*, **26**, 221–9.

126 Manninen, A., Metsä-Ketelä, T., Tuimala, R. and Vapaatalo, H. (1991) Nifedipine increases urinary excretion of prostacyclin metabolite in hypertensive pregnancy. *Pharmacology & Toxicology*, **69**, 60–3.

127 Roman, R., Kaldunski, M., Scicli, A. and Carretero, O. (1988) Influence of kinins and angiotensin II on the regulation of

papillary blood flow. *American Journal of Physiology*, **255**, F690–8.

128 Madeddu, P., Glorioso, N., Soro, A., Manunta, P., Troffa, C., Tonolo, G. et al. (1990) Effect of a kinin antagonist on renal function and haemodynamics during alterations in sodium balance in conscious normotensive rats. *Clinical Science*, **78**, 165–8.

129 Stein, J.H., Congbalay, R.C., Karsh, D.L., Osgood, R.W. and Ferris, T.F. (1972) The effect of bradykinin on proximal tubular sodium reabsorption in the dog: evidence for functional nephron heterogeneity. *Journal of Clinical Investigation*, **51**, 1709–21.

130 Tornel, J., Madrid, M.I., García-Salom, M. and Wirth, K.J. (2000) Role of kinins in the control of renal papillary blood flow, pressure natriuresis, and arterial pressure. *Circulation Research*, **86**, 589–95.

131 Fenoy, F.J. and Roman, R.J. (1992) Effect of kinin receptor antagonists on renal hemodynamic and natriuretic responses to volume expansion. *American Journal of Physiology*, **263**.R1136–40.

132 Beierwaltes, W.H., Carretero, O.A. and Scicli, A.G. (1988) Renal hemodynamics in response to a kinin analogue antagonist. *American Journal of Physiology*, **255**, F408–14.

133 Tsuchida, S., Miyazaki, Y., Matsusaka, T., Hunley, T.E., Inagami, T., Fogo, A. and Ichikawa, I. (1999) Potent antihypertrophic effect of the bradykinin B_2 receptor system on the renal vasculature. *Kidney International*, **56**, 509–16.

134 Schanstra, J.P., Duchene, J., Praddaude, F., Bruneval, P., Tack, I., Chevalier, J., Girolami, J.P. and Bascands, J.L. (2003) Decreased renal NO excretion and reduced glomerular tuft area in mice lacking the bradykinin B_2 receptor. *American Journal of Physiology – Heart and Circulatory Physiology*, **284**, H1904–8.

135 Kakoki, M., Takahashi, N., Jennette, J.C. and Smithies, O. (2004) Diabetic nephropathy is markedly enhanced in mice lacking the bradykinin B_2 receptor. *Proceedings of the National Academy of Sciences of the United States of America*, **101**, 13302–5.

136 Cervenka, L., Vanecková, I., Malý, J., Horácek, V. and El-Dahr, S.S. (2003) Genetic inactivation of the B_2 receptor in mice worsens two-kidney, one-clip hypertension: role of NO and the AT_2 receptor. *Journal of Hypertension*, **21**, 1531–8.

137 Imig, J.D., Zhao, X., Orengo, S.R., Dipp, S. and El-Dahr, S.S. (2003) The Bradykinin B_2 receptor is required for full expression of renal COX-2 and renin. *Peptides*, **24**, 1141–7.

138 Bledsoe, G., Shen, B., Yao, Y., Zhang, J.J., Chao, L. and Chao, J. (2006) Reversal of renal fibrosis, inflammation, and glomerular hypertrophy by kallikrein gene delivery. *Human Gene Therapy*, **17**, 545–55.

139 Alfie, M.E., Sigmon, D.H., Pomposiello, S.I. and Carretero, O.A. (1997) Effect of high salt intake in mutant mice lacking bradykinin-B_2 receptors. *Hypertension*, **29**, 483–7.

140 Maltais, I., Bachvarova, M., Maheux, P., Perron, P., Marceau, F. and Bachvarov, D. (2002) Bradykinin B_2 receptor gene polymorphism is associated with altered urinary albumin/creatinine values in diabetic patients. *Canadian Journal of Physiology and Pharmacology*, **80**, 323–7.

141 Zychma, M.J., Gumprecht, J., Trautsolt, W., Szydlowska, I. and Grzeszczak, W. (2003) Polymorphic genes for kinin receptors, nephropathy and blood pressure in type 2 diabetic patients. *American Journal of Nephrology*, **23**, 112–16.

142 Bachvarov, D.R., Landry, M., Pelletier, I., Chevrette, M., Betard, C., Houde, I. et al. (1998) Characterization of two polymorphic sites in the human kinin B_1 receptor gene: altered frequency of an allele in patients with a history of end-stage renal failure. *Journal of the American Society of Nephrology*, **9**, 598–604.

143 Couture, R. and Girolami, J.P. (2004) Putative roles of kinin receptors in the therapeutic effects of angiotensin 1-converting enzyme inhibitors in diabetes mellitus. *European Journal of Pharmacology*, **500**, 467–85.

144 Spillmann, F., Van Linthout, S., Schultheiss, H.P. and Tschope, C. (2006) Cardioprotective mechanisms of the kallikrein–kinin system in diabetic cardiopathy. *Current Opinion in Nephrology and Hypertension*, **15**, 22–9.

145 Kakoki, M., McGarrah, R.W., Kim, H.S. and Smithies, O. (2007) Bradykinin B_1 and B_2 receptors both have protective roles in renal ischemia/reperfusion injury. *Proceedings of the National Academy of Sciences of the United States of America*, **104**, 7576–81.

146 Allen, T.J., Cao, Z., Youssef, S., Hulthen, U.L. and Cooper, M.E. (1997) Role of angiotensin II and bradykinin in experimental diabetic nephropathy. Functional and structural studies. *Diabetes*, **46**, 1612–18.

147 Kakoki, M., Kizer, C.M., Yi, X. *et al.* (2006) Senescence-associated phenotypes in Akita diabetic mice are enhanced by absence of bradykinin B_2 receptors. *Journal of Clinical Investigation*, **116**, 1302–9.

148 -Auch-Schwelk, W., Kuchenbuch, C., Claus, M., *et al.* (1993) Local regulation of vascular tone by bradykinin and angiotensin converting enzyme inhibitors. *European Heart Journal*, **14** (Suppl. I), 154–60.

149 Hecker, M., Blaukat, A., Bara, A.T. *et al.* (1997) ACE inhibitor potentiation of bradykinin-induced venoconstriction. *British Journal of Pharmacology*, **121**, 1475–81.

150 Hecker, M., Porsti, I., Bara, A.T. and Busse, R. (1994) Potentiation by ACE inhibitors of the dilator response to bradykinin in the coronary microcirculation: interaction at the receptor level. *British Journal of Pharmacology*, **111**, 238–44.

151 Tschope, C., Seidl, U., Reinecke, A. *et al.* (2003) Kinins are involved in the antiproteinuric effect of angiotensin-converting enzyme inhibition in experimental diabetic nephropathy. *International Immunopharmacology*, **3**, 335–44.

5
Natriuretic Peptides

Paula M. Bryan and Lincoln R. Potter

5.1
History

The history of natriuretic peptides began over 50 years ago with the electron microscopic observation that cardiac atria in mammals are differentiated as both secretory granules and contractile cells [1]. A prescribed function for these "specific atrial granules", considered by some to be an evolutionary remnant, remained a mystery for some time due to a lack of biochemical and morphological techniques.

In the late 1960s, Adolpho J. de Bold began studying the function, if any, of these granules. Over 12 years, de Bold and his colleagues developed biochemical and light microscopy techniques to investigate the nature of the atrial granules. Autoradiographic studies showed that the content of the granules had a high turnover in a manner similar to secretory cells. Furthermore, light microscopy quantitation of granules in atrial tissue revealed changes in the number of granules after procedures known to alter electrolyte and water balance [2].

More convincingly, experiments performed in collaboration with Harald Sonnenberg demonstrated that atrial extracts caused robust natriuresis and diuresis that was immediate and similar to the effects of furosemide, a potent diuretic. The hypothesis that the atrial granules contained a polypeptide hormone was supported by the observation that proteases destroyed this activity. Subsequently, de Bold and colleagues, as well as other groups, isolated, purified and sequenced the agent responsible for the observed physiological responses and named it atrial natriuretic factor (ANF). Over time most groups in the field began calling it atrial natriuretic peptide (ANP), which is the founding member of the natriuretic peptide family that includes brain natriuretic peptide (BNP) and C-type natriuretic peptide (CNP). Together, these three peptides regulate cardiac hypertrophy, blood pressure and long-bone growth.

Cardiovascular Hormone Systems. Edited by Michael Bader
Copyright © 2008 WILEY-VCH Verlag GmbH & Co. KGaA, Weinheim
ISBN: 978-3-527-31920-6

5.2
Natriuretic Peptides: Structure, Processing and Expression

ANP, BNP and CNP share a common 17-amino-acid disulfide-linked ring structure (Figure 5.1).

ANP is synthesized as a 151-amino-acid preprohormone, which is cleaved of its N-terminal signal sequence to a 126-amino-acid propeptide (pro-ANP). This form is stored in atrial granules. Upon secretion, pro-ANP is cleaved by corin – a transmembrane cardiac serine protease [3]. This biologically active 28-amino-acid peptide is released into the cardiac sinus in response to mechanical stimuli such as cardiac distention [4] and is then distributed to various target organs via the circulatory system.

The second member of the family, BNP, was initially isolated from porcine brain, although it is now known to be synthesized predominantly in the ventricle and is released in response to ventricular load-induced stretch [5]. Unlike ANP, BNP is not stored in granules, but its production is transcriptionally regulated by cardiac wall stretch [6]. In humans, BNP is synthesized as a 134-amino-acid preprohormone that is clipped to yield a 108-residue pro-BNP peptide [7]. The protease that cleaves pro-BNP to its mature 32-amino-acid form is unknown.

The third member of the family, CNP, was also initially isolated from porcine brain but has subsequently been found in endothelial cells and chondrocytes [7], and is considered a noncirculatory hormone. In cell culture, its secretion is triggered by sheer stress, tumor necrosis factor-α, transforming growth factor-β, interleukin-I and is suppressed by insulin [8–11]. Pro-CNP is a 103-amino-acid peptide that is possibly clipped by furin, an intracellular endoprotease, to the mature 53-amino-acid form, which is found predominantly in heart [12], endothelial cells [13] and brain [14]. The 53-amino-acid form is further cleaved by an unknown enzyme to yield a 22-amino-acid form, which is the primary form found in body fluids such as cerebral spinal fluid [15] and plasma [13].

5.3
Natriuretic Peptide Receptors

There are three known receptors that bind natriuretic peptides: natriuretic peptide receptor (NPR)-A, NPR-B and the clearance receptor, NPR-C. NPR-A and NPR-B are also called guanylyl cyclase-A and guanylyl cyclase-B, respectively, due to their C-terminal guanylyl cyclase catalytic domains. The topology of both proteins is similar to growth factor receptors, consisting of an extracellular ligand-binding domain of approximately 450 amino acids. A single hydrophobic membrane-spanning domain consists of 20–25 amino acids. The intracellular domain is composed of approximately 570 amino acids (Figure 5.2) and can be further divided into three distinct domains, one of those being the 250-amino-acid kinase homology domain which has sequence similarity to known protein kinases, although NPRs have not been shown to display intrinsic phosphotransferase activity. Carboxyl to the

5.3 Natriuretic Peptide Receptors

TISSUE EXPRESSION	PROTEOLYTIC PROCESSING	MATURE PROTEIN

ANP

Atria
Ventricles
Kidney

BNP

Ventricles
Atria

CNP

Bone
Brain
Endothelium
Heart

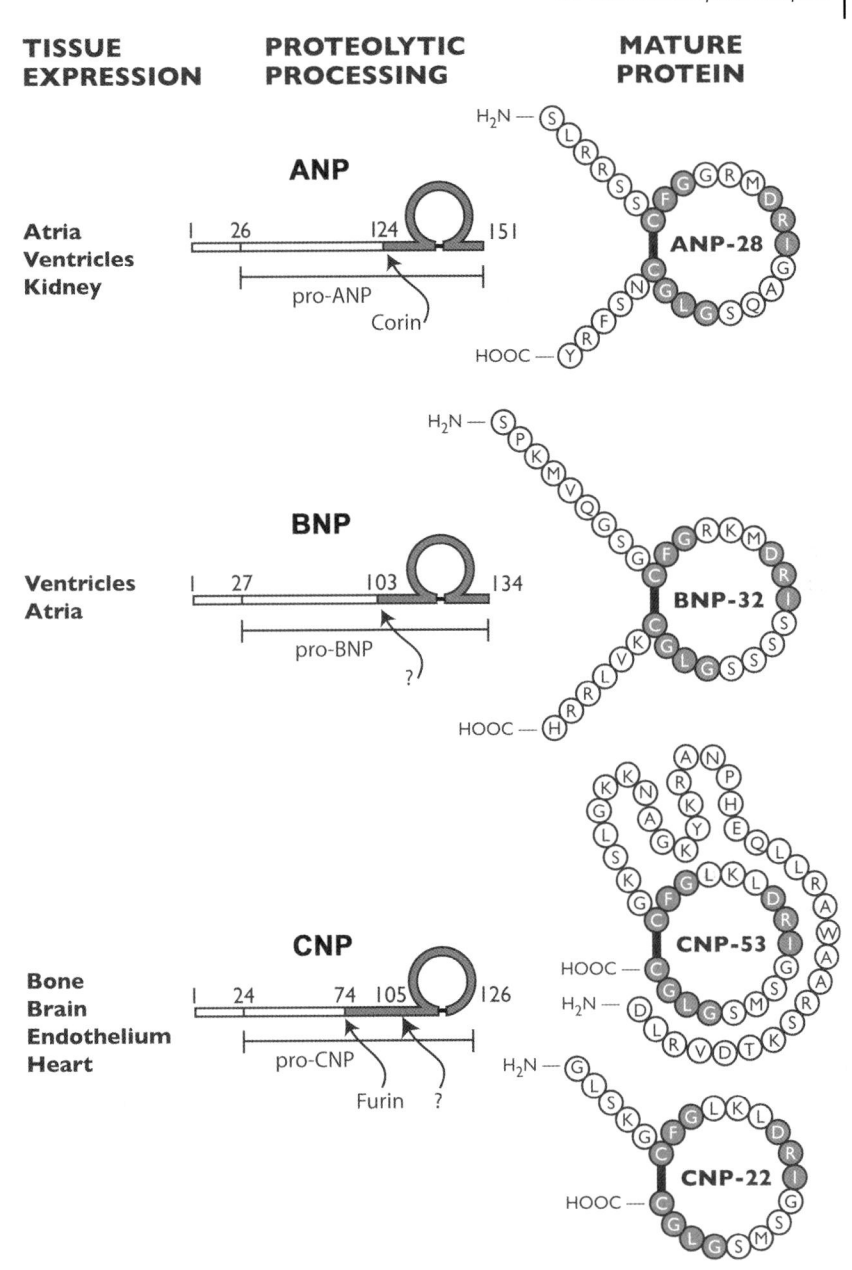

Figure 5.1 Natriuretic peptide structure, tissue expression and processing. ANP is expressed in the indicated tissues as a preprohormone. The propeptide is cleaved by corin to yield a 28-amino-acid mature peptide. BNP is expressed in the ventricles and atria as a preprohormone, and is cleaved by a yet-to-be-identified enzyme to yield a 32-amino-acid human hormone. Pro-CNP is expressed in the indicated tissues and is possibly cleaved by furin to yield a 53-membered peptide. A further cleavage by an unknown enzyme yields a 22-amino-acid peptide.

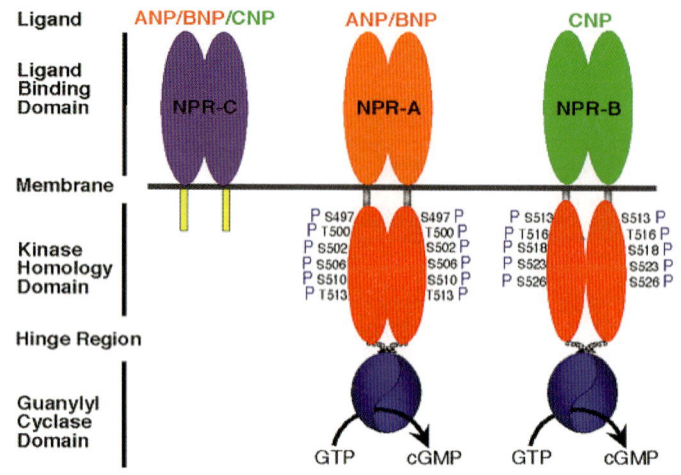

Figure 5.2 The topology of NPRs and ligand preferences. NPR-A and NPR-B are membrane-bound guanylyl cyclases that consist of an extracellular ligand-binding domain, a single hydrophobic transmembrane domain, and intracellular kinase homology domain, dimerization and C-terminal domains. NPR-C only contains 37 intracellular amino acids, and shares approximately 30% sequence identity to NPR-A and NPR-B within its extracellular ligand-binding domain. Known sites of phosphorylation are indicated within the kinase homology domain.

kinase homology domain is an approximately 40-amino-acid coiled-coil dimerization domain and an approximately 250-amino-acid guanylyl cyclase catalytic domain [16].

5.3.1
NPR-A

NPR-A represents one of five transmembrane guanylyl cyclases found in humans [16]. NPR-A is bound and activated by the natriuretic peptides ANP and BNP (Figure 5.2), with ANP displaying a slightly higher affinity for the receptor than BNP [17].

Three intramolecular disulfide bonds are found in the extracellular domain of rat NPR-A [18]. NPR-A displays considerable size heterogeneity as evidenced by the fact that it produces a "hazy" band when fractionated by sodium dodecylsulfate–polyacrylamide gel electrophoresis. This is primarily due to differential N-linked glycosylation [19, 20]. Whether glycosylation affects the ability of NPR-A to bind hormone is controversial [21–24].

Basally, NPR-A is phosphorylated by a yet to be identified kinase(s) on six known residues within its kinase homology domain (Ser497, Thr500, Ser502, Ser506, Ser510 and Thr513). Mutation of any of these single sites to alanine results in reduced receptor phosphorylation, changes in tryptic phosphopeptide mapping patterns and decreased guanylyl cyclase activity. Phosphorylation of NPR-A is absolutely required for hormonal activation because mutation of four or more of the known sites results in a hormonally unresponsive receptor [25]. Indeed, dephosphorylation of NPR-A causes its desensitization [26]. There are two known phosphatase activities associated with NPR-A. One of these is sensitive to the phosphatase inhibitor microcystin, and the other is microcystin-insensitive and Mg^{2+}-dependent [27].

The hypothetical model for NPR-A activation can be separated into three steps. In the first step, NPR-A is in its basal form where it has been shown to exist as a higher-ordered dimer or oligomer and ligand binding does not lead to further oligomerization [23]. In this state, the guanylyl cyclase activity of the receptor is tightly repressed via the kinase homology domain (KHD); receptors that are devoid of the KHD are constitutively active [28, 29]. In the second step, ANP binds to NPR-A in a 1:2 ratio [30] and through a poorly defined mechanism that involves a "tightening" of the juxtamembrane regions of the two monomers, the extracellular hormone-binding signal is transmitted across the plasma membrane. In this "active" state of the receptor, the repression exerted by the KHD is relieved and the guanylyl cyclase domains are hypothesized to come together in a head-to-tail arrangement [31, 32]. Based on the catalytic mechanism for the closely related adenylyl cyclase, it is hypothesized that there are two GTP-binding sites per NPR-A dimer. In the third step, the affinity of the hormone-binding domain for the ligand decreases, increasing the rate of dissociation [33, 34]. Phosphorylated sites within the KHD become dephosphorylated and the receptor is now in a "desensitized" state [26]. In order to go through another round of activation, the receptor must be rephosphorylated by a kinase(s). Whether this takes place at the plasma membrane or whether the receptor must be internalized to be rephosphorylated is not known.

The role of ATP in the regulation of NPR-A is controversial. Original studies showed that ANP could increase guanylyl cyclase activity in crude membrane preparations in the absence of exogenous ATP [35]. However, subsequent studies reported that ATP dramatically increases ANP-dependent guanylyl cyclase activity [36–38]. Kurose et al. suggested that ATP directly binds and activates NPR-A because adenosine 5'-(β,γ-imido)triphosphate (AMPPNP), a nonhydrolyzable ATP analog that presumably cannot be used as a substrate for protein kinases, also increased ANP-dependent guanylyl cyclase activity [37]. Later reports suggested that ATP was absolutely required for NPR-A activation [39–41], which led to a two-stage model for the activation of NPR-A [18]. In this model, ATP binding to the kinase homology domain requires prior natriuretic peptide binding to the extracellular domain and ultimately the catalytic domains are brought together to form an active site. More recently, a portion of the ATP-dependent regulation of NPR-A was shown to involve changes in its phosphorylation state. For example, ATPγS was shown to sensitize NPR-A to activation by ANP and |AMPPNP|. ATPγS can serve as a substrate for protein kinases. However, thiophosphorylated proteins are resistant to phosphatase activity. This suggests that ATP serves as a substrate for the NPR-A kinase(s) [42]. More recently, Antos et al. demonstrated that ATP is not required for activation of NPR-A and NPR-B. Rather, it stabilizes previously activated receptors. The basis for this theory is that the addition of ATP does not significantly increase initial enzymatic rates but does increase activities measured at longer time periods [43]. Whether or not ATP binds directly to NPR-A or NPR-B is unknown.

The expression of NPR-A mRNA is high in kidney, adrenal, terminal ileum, adipose, aortic and lung tissue [44–46]. In rhesus monkeys, NPR-A mRNA is prevalent in the kidney, adrenal glomerulosa, adrenal medulla, pituitary, cerebellum and endocardial endothelial cells [47]. In studies where NPR-A protein levels are measured by Western blot or by immunoprecipitation, NPR-A was detected in rodent lung, adrenal, kidney, testis and liver tissue [48–50]. Finally, primary bovine endothelial cells, human cervical (HeLa) and human embryonic kidney cells express low amounts of NPR-A based on sequential immunoprecipitation–Western blot studies.

NPR-A null mice exhibit cardiac hypertrophy, high blood pressure and ventricular fibrosis [51, 52]. In humans, loss-of-function mutations in NPR-A correlate with disease. Eight individuals have been identified with a single allele mutation in the promoter region of the NPR-A gene, resulting in a 70% decrease in receptor expression; seven of the individuals had hypertension and one had congestive heart failure [53]. To our knowledge, no mutations in the coding region have been identified that modulate NPR-A activity.

5.3.2
NPR-B

NPR-B has a similar topology to NPR-A (Figure 5.2) and preferentially binds and is activated by CNP [17, 54]. Like NPR-A, NPR-B is also glycosylated [55, 56]. There are five known sites of phosphorylation within the kinase homology domain of NPR-B (Ser513, Thr516, Ser518, Ser523 and Ser526) [57]. Phosphorylation of these residues is required for activation of NPR-B because mutation of any of these residues to alanine results in a markedly diminished response to CNP [58].

NPR-B mRNA has been found in brain, adrenal, kidney, lung and ovary tissue [44, 46, 59]. NPR-B mRNA has also been detected via *in situ* hybridization studies in adrenal medulla, cerebellum, pituitary and skin [47]. NPR-B is the predominant NPR in the brain [60], and the protein can also be found at relatively high concentrations in fibroblasts [61–63], chondrocytes, vascular smooth muscle cells, uterus [7], liver, lung and heart [50].

Two loss-of-function mouse models exist for NPR-B. In one model the exons that encode the C-terminal half of the extracellular domain and transmembrane segment were deleted by homologous recombination [64]. These mice have a dwarfism phenotype and the females are sterile. The heterozygous mice were significantly shorter than the wild-type animals. The other mouse model has a substitution of an arginine for a highly conserved leucine in the catalytic domain of NPR-B. This so-called *cn/cn* mouse that contains two defective alleles also displays dwarfism, although female sterility was not noted in this model [65]. In humans, a homozygous loss-of-function mutation has been identified. These patients have acromesomelic dysplasia, type Maroteaux, a rare form of short-limbed dwarfism. These patients are not sterile and like the "knockout" mice, individuals with one normal and one abnormal allele display normal limb proportions but are statistically shorter than the average person from their respective populations [66].

5.3.3
NPR-C

The natriuretic peptide clearance receptor (NPR-C) binds all three known natriuretic peptides. The rank order of affinity is ANP ≥ CNP > BNP, although there is only one order of magnitude difference between the strongest and weakest binders [17, 19]. Its main function is to clear natriuretic peptides from the circulation through receptor-mediated internalization and degradation, as is evidenced by loss-of-function mutations in mice [67, 68]. The internalization of NPR-C is constitutive [69] and its endocytosis is mediated by a clathrin-dependent mechanism [70]. It has been shown that ligand bound to NPR-C undergoes lysosomal hydrolysis followed by NPR-C recycling to the cell surface [69, 71]. NPR-C shares 30% amino acid identity with NPR-A and NPR-B in its extracellular domain [72]. Like NPR-A and NPR-B, the clearance receptor is glycosylated and contains two sets of intramolecular disulfide bonds that are conserved in NPR-A and NPR-B. Human NPR-C also contains two intermolecular bonds, unlike NPR-A and NPR-B [73]. However, NPR-C only contains 37 intracellular amino acids and, therefore, has no known catalytic activity [74].

A number of groups have reported that NPR-C exhibits signaling functions [75]. The synthetic peptide c-ANF (ANP 4–23) binds to NPR-C, but not to NPR-A or NPR-B, and has been shown to reduce cAMP concentrations in whole cells or adenylyl activity in membranes. These effects were inhibited by pertussis toxin [76, 77], consistent with a pathway involving an inhibitory G-protein (G_i or G_o). Furthermore, this c-ANF-dependent inhibition of adenylyl cyclase can be mimicked with small peptide fragments of the NPR-C intracellular domain [78] or blocked with specific antibodies to the intracellular domain of NPR-C [79]. NPR-C also stimulates phospholipase C in a G-protein-dependent manner [80–82] and inhibits catecholamine (neurotransmitter) efflux from pheochromocytoma (PC-12) cells as well [83, 84]. Importantly, however, no physiologic functions have been shown to require NPR-C to date.

NPR-C mRNA has been detected in mesentery, placenta, atrial, lung, venous and kidney tissue [44, 85], aortic endothelial and smooth muscle cells [74]. Through *in situ* hybridization studies, NPR-C mRNA was detected in adrenal, heart, kidney, cerebral cortex and cerebellum tissue [47].

Deletion of the NPR-C gene in mice resulted in a two-thirds longer half-life of [^{125}I]ANP in the circulation. These animals also displayed phenotypes associated with overactivation of NPR-A and NPR-B. For instance, they had a reduced ability to concentrate their urine, were slightly hypotensive and exhibited long bone overgrowth [68]. In addition, three different mouse models, resulting from a chemical mutagenesis screen, contain recessive loss-of-function mutations in NPR-C. All mutations were found within the extracellular domain and presumably disrupt ligand binding. These mice also display skeletal overgrowth and are exceptionally thin when they are older and have frequent tail and/or sacral kinks [67].

5.4
Physiological Effects of the Natriuretic Peptide System

There are many physiological effects of natriuretic peptides. Stimulation of NPR-A or NPR-B converts GTP to the intracellular second messenger cGMP. The physiological effects of NPR signaling are mediated through three known cGMP-binding proteins: cGMP-binding phosphodiesterases (PDEs), cGMP-dependent protein kinases (cGKs) and cyclic nucleotide-gated (CNG) ion channels.

PDEs degrade cyclic nucleotides into inactive 5′-nucleotide monophosphates and thus regulate intracellular second messenger concentrations. There are 11 different families of PDEs consisting of at least 25 different proteins [86, 87]. Some PDEs degrade only cGMP or cAMP while others degrade both cyclic nucleotides [88].

cGKs are serine/threonine protein kinases that are bound and activated by cGMP [89, 90]. There are two cGK genes; cGK-I, which can be alternatively spliced to form α and β isoforms, and cGK-II. cGK-I is predominantly found in the cytosol, whereas cGK-II is myristolated and therefore a membrane resident. cGK-I is required for natriuretic peptide dependent vasorelaxation, whereas cGK-II is required for ANP-dependent renin release and CNP-dependent long bone growth.

CNG channels are a family of nonselective nucleotide-gated ion channels that bind cGMP or cAMP within their C-terminals [91]. To date, there has not been any data linking specific natriuretic peptide functions to CNG channels.

5.4.1
Effects on Blood Pressure

The role of ANP/NPR-A in the regulation of basal blood pressure has been elucidated, in part, through the use of knockout mice. Deletion of ANP or NPR-A genes results in mice with blood pressures 20–40 mmHg higher than normal [51, 52, 92]. Conversely, mice that overexpress ANP or BNP have blood pressures that are 20–30 mmHg lower than normal [93, 94]. In a gene dosage study in which mice were produced containing zero to four alleles, blood pressures and ANP-dependent guanylyl cyclase activities were found to be directly proportional to gene number [95].

5.4.2
Effects of ANP/NPR-A on Intravascular Volume and Endothelium Permeability

Transvascular fluid balance is regulated via microvascular permeability. Interestingly, the initial report on ANP indicated that it increases hematocrit levels, consistent with a role in vascular volume contraction [2]. Later studies showed that ANP increases vascular permeability as evidenced by increased endothelial permeability to albumin [96]. Further evidence of ANP playing a role in endothelial permeability is found in knockout mice that lack NPR-A in their vascular endothelium [97]. These mice have blood pressures that are 10–15 mmHg higher than normal and are volume-expanded by 11–13%. This suggests that ANP-regulated endothelial permeability accounts for around 30% of the hypotensive effects of ANP since the NPR-A knockout animals are volume-expanded by 30% and have blood pressures that are 20–30 mmHg above normal.

5.4.3
Effects of ANP and BNP on Cardiac Hypertrophy and Fibrosis

ANP and BNP have been shown to have direct effects on the size and morphology of the heart. Animals overexpressing ANP have smaller hearts than normal [94, 98], and ANP and NPR-A knockout mice have hypertrophied hearts [52, 92, 99]. NPR-A knockout mice treated with hypertensive drugs from birth do not have hypertension, but have cardiac hypertrophy [100]. Conversely, NPR-A knockout mice that have specific transgenic replacement of NPR-A in the heart have reduced cardiomyocyte size while still being hypertensive [101]. This suggests that the antihypertrophic effects of NPR-A and ANP are due to a local inhibitory effect on heart growth and inhibition of systemic hypertension. In culture, BNP inhibits the growth of cardiac fibroblasts [102]. This has also been observed *in vivo* in BNP knockout mice that exhibit pressure-sensitive ventricular fibrosis [103].

5.4.4
Effects of ANP and CNP on Vascular Relaxation and Remodeling

NPR-A-stimulated vasorelaxation is important for acute, but not chronic blood pressure regulation. This was determined by a knockout mouse study where NPR-A was selectively removed from vascular smooth muscle cells using Cre–lox technology [104]. When a bolus injection of ANP was administered, the knockout animals did not undergo an acute reduction in blood pressure. However, the resting blood pressures of these animals did not differ from wild-type littermates, suggesting that the vasorelaxing properties of NPR-A do not mediate the chronic blood pressure lowering effects of the ANP/NPR-A pathway.

CNP, which is released in response to vascular injury [9], is also a vasodilator, but its effects are mediated through NPR-B that is present in aortic vascular smooth muscle cells and has been shown to mediate the relaxation of precontracted rat aorta [105]. CNP also inhibits vascular smooth muscle proliferation in a cGMP-dependent manner [106].

5.4.5
Effects of ANP on Natriuresis and Diuresis

In the kidney, ANP functions to inhibit water and sodium reabsorption, increase glomerular filtration rate, and reduce renin secretion. No ANP-dependent natriuresis and diuresis is exhibited in NPR-A knockout mice, suggesting that the effects of ANP are mediated exclusively through NPR-A [107]. ANP causes a coordinated constriction of efferent arteriolar glomerular capillaries and dilation of afferent arteriolar glomerular capillaries; the net effect is increased glomerular filtration rate [108]. In the proximal tubule of the kidney, ANP inhibits angiotensin (Ang) II-stimulated water and sodium transport [109]. In the collecting ducts, it inhibits an amiloride-sensitive cation channel and reduces sodium transport [110].

5.4.6
Natriuretic Peptides and Renal Function

The enzyme renin is a protease secreted from juxtaglomerular cells in the kidney. Angiotensinogen is cleaved to Ang I by renin, and a subsequent cleavage of Ang I by angiotensin-converting enzyme (ACE) yields Ang II [111]. This leads to the release of aldosterone – the major hormone responsible for regulating sodium reabsorption in the cortical collecting duct. Ang II is also a potent vasoconstrictor. Both of these actions result in an increase in blood pressure. In the kidney, ANP regulates blood pressure, in part, by inhibiting the renin–Ang II–aldosterone system. In humans, physiological doses of ANP suppress both aldosterone and renin levels [112].

ANP also directly inhibits aldosterone production in the adrenal gland. The involvement of cGMP in this process is controversial because one group showed that a cell-permeable cGMP analog mimicked the effects of ANP [113] while another group showed that it had no effect [114]. Other studies reported that the ANP-induced reduction in aldosterone could be mimicked by an NPR-C-specific ligand and blocked by pertussis toxin, a G_i/G_o inhibitor [79]. NPR-A knockout mice have 2-fold higher aldosterone levels that their wild-type littermates, suggesting that NPR-A mediates the ANP-dependent reduction in aldosterone levels by increasing cGMP levels [115]. The mechanism for ANP-dependent aldosterone inhibition may include PDE2 [116], which is activated by cGMP and degrades cAMP, the major determinant for aldosterone synthesis.

5.5
Natriuretic Peptides as Diagnostic Indicators of Heart Failure

Elevated circulating concentrations of natriuretic peptides are clinical hallmarks of cardiac dysfunction. The serum levels of both ANP and BNP are elevated in heart failure. However, the differential between the BNP levels in normal patients and patients with congestive heart failure is larger than the ANP differential. Therefore, plasma BNP concentrations are the better diagnostic indicator of congestive heart failure. The normal upper limit of BNP levels in young, healthy patients is around 20 pg/ml [117]. In patients with congestive heart failure, serum BNP levels vary according to age, renal function, sex, cardiac rhythm, drug therapy and body mass index [118]. However, determination of serum BNP levels remains a strong predictor of congestive heart failure. In a recent study, the N-terminal BNP (NT-pro-BNP) Investigation of Dyspnea in the Emergency Department, NT-pro-BNP levels were determined in 600 patients [119]. NT-pro-BNP is present in higher levels *in vivo* and has a longer half-life than BNP (60–120 versus 15–20 min, respectively) [120]. The study demonstrated that a cutpoint of 300 pg/ml was best for ruling out acute decompensated heart failure (ADHF), for example patients with BNP levels higher than 300 pg/ml are very likely to have heart failure. For the diagnosis of ADHF in patients less than 50 years of age, the cutpoint for NT-BNP was 450 pg/ml and for those greater than 50 years of age, 900 pg/ml [119].

High BNP levels have been correlated with a poor prognosis in many diseases. BNP levels have been used successfully to estimate the risk of death for heart failure patients [121] as well as predicting poststroke mortality [122].

Paradoxically, it has been suggested that patients with a poor prognosis and end-stage heart failure may actually have lower levels of circulating BNP [123]. It has been hypothesized that during the progression of heart failure there is a point where the neurohormonal systems are no longer capable of providing sufficient levels of natriuretic peptides as compensation. The mechanism behind this shift is not yet clearly understood [124]. Therefore, although increased circulating levels of BNP are a good indicator of acute heart failure and may aid in the short-term prognosis of patients at the acute stage, low BNP levels may serve as a prognostic indicator of advanced or chronic heart failure.

5.6
NPRs and Heart Failure

Evidence that receptor levels change in response to heart failure has been demonstrated using a transaortic banding model of congestive heart failure in mice [125]. After 8 weeks of transaortic banding, the banded mice exhibited cardiac hypertrophy, increased left ventricular end-diastolic and systolic diameters, and decreased left ventricular ejection fractions. ANP-dependent activity was reduced by 44% in the kidneys of these animals compared to sham operated control mice. NPR-A phosphorylation was reduced by 25% with a concomitant 30% decrease in NPR-A protein levels, suggesting that downregulation, not dephosphorylation, of NPR-A contributes to the renal insensitivity to natriuretic peptides in congestive heart failure.

Another study using transaortic banded mice demonstrated that ANP-dependent activity in the ventricle is depressed in congestive heart failure. Interestingly, maximal CNP-dependent activity was not decreased and was nearly 2-fold higher than ANP-dependent activity in the failed heart. Hence, NPR-B, not NPR-A, accounts for the majority of natriuretic peptide-dependent activity in the failed mouse heart [126].

5.7
Therapeutic Applications and Future Directions

Synthetic ANP and human recombinant BNP have been used therapeutically to treat heart failure and hypertension. Infusion of ANP, clinically known as carperitide causes a decrease in blood pressure, and increases in sodium and water excretion when given to patients with chronic heart failure [127, 128] and hypertension [129]. BNP, known clinically as nesiritide or by its trade name Natrecor™, causes potent vasorelaxation and increased natriuresis and a decrease in plasma aldosterone and endothelin levels in patients with acute heart failure [130].

Studies showing increased CNP levels in patients with chronic heart failure in a manner that positively correlates with mean pulmonary capillary wedge pressure or degree of failure suggest a regulatory role for CNP in the heart [131, 132]. This proposition is bolstered by the aforementioned study that demonstrated for the first time that NPR-B is responsible for a previously unappreciated and significant amount of natriuretic peptide dependent guanylyl cyclase activity in the nonfailed heart and the majority of natriuretic peptide-dependent guanylyl cyclase activity in the failed heart. Thus, NPR-B represents an exciting new potential drug target for the treatment of congestive heart failure.

References

1 Kisch, B. (1956) *Experimental Medicine and Surgery*, **14**, 99–112.
2 de Bold, A.J., Borenstein, H.B., Veress, A.T. and Sonnenberg, H. (1981) *Life Sciences*, **28**, 89–94.
3 Yan, W., Wu, F., Morser, J. and Wu, Q. (2000) *Proceedings of the National Academy of Sciences of the United States of America*, **97**, 8525–9.
4 Brenner, B.M., Ballermann, B.J., Gunning, M.E. and Zeidel, M.L. (1990) *Physiological Reviews*, **70**, 665–99.
5 Pandey, K.N. (2005) *Peptides*, **26**, 901–32.
6 Vanderheyden, M., Goethals, M., Verstreken, S., De Bruyne, B., Muller, K., Van Schuerbeeck, E. and Bartunek, J. (2004) *Journal of the American College of Cardiology*, **44**, 2349–54.
7 Potter, L.R., Abbey-Hosch, S. and Dickey, D.M. (2006) *Endocrine Reviews*, **27**, 47–72.
8 Suga, S., Nakao, K., Itoh, H., Komatsu, Y., Ogawa, Y., Hama, N. and Imura, H. (1992) *Journal of Clinical Investigation*, **90**, 1145–9.
9 Suga, S., Itoh, H., Komatsu, Y., Ogawa, Y., Hama, N., Yoshimasa, T. and

Nakao, K. (1993) *Endocrinology*, **133**, 3038–41.
10 Chun, T.H., Itoh, H., Ogawa, Y., Tamura, N., Takaya, K., Igaki, T., Yamashita, J., Doi, K., Inoue, M., Masatsugu, K., Korenaga, R., Ando, J. and Nakao, K. (1997) *Hypertension*, **29**, 1296–302.
11 Igaki, T., Itoh, H., Suga, S., Komatsu, Y., Ogawa, Y., Doi, K., Yoshimasa, T. and Nakao, K. (1996) *Diabetes*, **45** (Suppl. 3), S62–64.
12 Minamino, N., Makino, Y., Tateyama, H., Kangawa, K. and Matsuo, H. (1991) *Biochemical and Biophysical Research Communications*, **179**, 535–42.
13 Stingo, A.J., Clavell, A.L., Heublein, D.M., Wei, C.M., Pittelkow, M.R. and Burnett, J.C. Jr (1992) *American Journal of Physiology*, **263** (4 Pt 2), H1318–21.
14 Totsune, K., Takahashi, K., Ohneda, M., Itoi, K., Murakami, O. and Mouri, T. (1994) *Peptides*, **15**, 37–40.
15 Togashi, K., Kameya, T., Kurosawa, T., Hasegawa, N. and Kawakami, M. (1992) *Clinical Chemistry*, **38**, 2136–9.
16 Potter, L.R. (2005) *Frontiers in Bioscience*, **10**, 1205–20.
17 Suga, S., Nakao, K., Hosoda, K., Mukoyama, M., Ogawa, Y., Shirakami, G., Arai, H., Saito, Y., Kambayashi, Y., Inouye, K. and Imura, H. (1992) *Endocrinology*, **130**, 229–39.
18 Potter, L.R. and Hunter, T. (2001) *Journal of Biological Chemistry*, **276**, 6057–60.
19 Bennett, B.D., Bennett, G.L., Vitangcol, R.V., Jewett, J.R., Burnier, J., Henzel, W. and Lowe, D.G. (1991) *Journal of Biological Chemistry*, **266**, 23060–7.
20 Miyagi, M., Zhang, X. and Misono, K.S. (2000) *European Journal of Biochemistry*, **267**, 5758–68.
21 Heim, J.M., Singh, S. and Gerzer, R. (1996) *Life Sciences*, **59**, PL61–8.
22 Koller, K.J., Lipari, M.T. and Goeddel, D.V. (1993) *Journal of Biological Chemistry*, **268**, 5997–6003.
23 Lowe, D.G. (1992) *Biochemistry*, **31**, 10421–5.
24 Muller, D., Middendorff, R., Olcese, J. and Mukhopadhyay, A.K. (2002) *Endocrinology*, **143**, 23–9.
25 Potter, L.R. and Hunter, T. (1998) *Molecular and Cellular Biology*, **18**, 2164–72.
26 Potter, L.R. and Garbers, D.L. (1992) *Journal of Biological Chemistry*, **267**, 14531–4.
27 Bryan, P.M. and Potter, L.R. (2002) *Journal of Biological Chemistry*, **277**, 16041–7.
28 Chinkers, M. and Garbers, D.L. (1989) *Science*, **245**, 1392–4.
29 Koller, K.J., de Sauvage, F.J., Lowe, D.G. and Goeddel, D.V. (1992) *Molecular and Cellular Biology*, **12**, 2581–90.
30 Ogawa, H., Qiu, Y., Ogata, C.M. and Misono, K.S. (2004) *Journal of Biological Chemistry*, **279**, 28625–31.
31 Sunahara, R.K., Beuve, A., Tesmer, J.J., Sprang, S.R., Garbers, D.L. and Gilman, A.G. (1998) *Journal of Biological Chemistry*, **273**, 16332–8.
32 Tucker, C.L., Hurley, J.H., Miller, T.R. and Hurley, J.B. (1998) *Proceedings of the National Academy of Sciences of the United States of America*, 95, 5993–7.
33 Jewett, J.R., Koller, K.J., Goeddel, D.V. and Lowe, D.G. (1993) *EMBO Journal*, **12**, 769–77.
34 Koh, G.Y., Nussenzveig, D.R., Okolicany, J., Price, D.A. and Maack, T. (1992) *Journal of Biological Chemistry*, **267**, 11987–94.
35 Waldman, S.A., Rapoport, R.M. and Murad, F. (1984) *Journal of Biological Chemistry*, **259**, 14332–4.
36 Chang, C.H., Kohse, K.P., Chang, B., Hirata, M., Jiang, B., Douglas, J.E. and Murad, F. (1990) *Biochimica et Biophysica Acta*, **1052**, 159–65.
37 Kurose, H., Inagami, T. and Ui, M. (1987) *FEBS Letters*, **219**, 375–9.
38 Song, D.L., Kohse, K.P. and Murad, F. (1988) *FEBS Letters*, **232**, 125–9.
39 Chinkers, M., Singh, S. and Garbers, D.L. (1991) *Journal of Biological Chemistry*, **266**, 4088–93.
40 Marala, R.B., Sitaramayya, A. and Sharma, R.K. (1991) *FEBS Letters*, **281**, 73–6.
41 Wong, S.K., Ma, C.P., Foster, D.C., Chen, A.Y. and Garbers, D.L. (1995) *Journal of Biological Chemistry*, **270**, 30818–22.

42 Foster, D.C. and Garbers, D.L. (1998) *Journal of Biological Chemistry*, **273**, 16311–18.

43 Antos, L.K., Abbey-Hosch, S.E., Flora, D.R. and Potter, L.R. (2005) *Journal of Biological Chemistry*, **280**, 26928–32.

44 Nagase, M., Katafuchi, T., Hirose, S. and Fujita, T. (1997) *Journal of Hypertension*, **15**, 1235–43.

45 Lowe, D.G., Chang, M.S., Hellmiss, R., Chen, E., Singh, S., Garbers, D.L. and Goeddel, D.V. (1989) *EMBO Journal*, **8**, 1377–84.

46 Schulz, S., Singh, S., Bellet, R.A., Singh, G., Tubb, D.J., Chin, H. and Garbers, D.L. (1989) *Cell*, **58**, 1155–62.

47 Wilcox, J.N., Augustine, A., Goeddel, D.V. and Lowe, D.G. (1991) *Molecular and Cellular Biology*, **11**, 3454–62.

48 Goy, M.F., Oliver, P.M., Purdy, K.E., Knowles, J.W., Fox, J.E., Mohler, P.J., Qian, X., Smithies, O. and Maeda, N. (2001) *Biochemical Journal*, **358**, 379–87.

49 Muller, D., Mukhopadhyay, A.K., Speth, R.C., Guidone, G., Potthast, R., Potter, L.R. and Middendorff, R. (2004) *Endocrinology*, **145**, 1392–401.

50 Bryan, P.M., Smirnov, D., Smolenski, A., Feil, S., Feil, R., Hofmann, F., Lohmann, S. and Potter, L.R. (2006) *Biochemistry*, **45**, 1295–303.

51 Lopez, M.J., Wong, S.K., Kishimoto, I., Dubois, S., Mach, V., Friesen, J., Garbers, D.L. and Beuve, A. (1995) *Nature*, **378**, 65–8.

52 Oliver, P.M., Fox, J.E., Kim, R., Rockman, H.A., Kim, H.S., Reddick, R.L., Pandey, K.N., Milgram, S.L., Smithies, O. and Maeda, N. (1997) *Proceedings of the National Academy of Sciences of the United States of America*, **94**, 14730–5.

53 Nakayama, T., Soma, M., Takahashi, Y., Rehemudula, D., Kanmatsuse, K. and Furuya, K. (2000) *Circulation Research*, **86**, 841–5.

54 Koller, K.J., Lowe, D.G., Bennett, G.L., Minamino, N., Kangawa, K., Matsuo, H. and Goeddel, D.V. (1991) *Science*, **252**, 120–3.

55 Fenrick, R., McNicoll, N. and De Lean, A. (1996) *Molecular and Cellular Biochemistry*, **165**, 103–9.

56 Fenrick, R., Bouchard, N., McNicoll, N. and De Lean, A. (1997) *Molecular and Cellular Biochemistry*, **173**, 25–32.

57 Potter, L.R. and Hunter, T. (1998) *Journal of Biological Chemistry*, **273**, 15533–9.

58 Potter, L.R. and Hunter, T. (1999) *Methods*, **19**, 506–20.

59 Chrisman, T.D., Schulz, S., Potter, L.R. and Garbers, D.L. (1993) *Journal of Biological Chemistry*, **268**, 3698–703.

60 Herman, J.P., Dolgas, C.M., Rucker, D. and Langub, M.C. Jr (1996) *Journal of Comparative Neurology*, **369**, 165–87.

61 Abbey, S.E. and Potter, L.R. (2002) *Journal of Biological Chemistry*, **277**, 42423–30.

62 Abbey, S.E. and Potter, L.R. (2003) *Endocrinology*, **144**, 240–6.

63 Chrisman, T.D. and Garbers, D.L. (1999) *Journal of Biological Chemistry*, **274**, 4293–9.

64 Tamura, N., Doolittle, L.K., Hammer, R.E., Shelton, J.M., Richardson, J.A. and Garbers, D.L. (2004) *Proceedings of the National Academy of Sciences of the United States of America*, **101**, 17300–5.

65 Tsuji, T. and Kunieda, T. (2005) *Journal of Biological Chemistry*, **280**, 14288–92.

66 Bartels, C.F., Bukulmez, H., Padayatti, P., Rhee, D.K., van Ravenswaaij-Arts, C., Pauli, R.M., Mundlos, S., Chitayat, D., Shih, L.Y., Al-Gazali, L.I., Kant, S., Cole, T., Morton, J., Cormier-Daire, V., Faivre, L., Lees, M., Kirk, J., Mortier, G.R., Leroy, J., Zabel, B., Kim, C.A., Crow, Y., Braverman, N.E., van den Akker, F. and Warman, M.L.a. (2004) *American Journal of Human Genetics*, **75**, 27–34.

67 Jaubert, J., Jaubert, F., Martin, N., Washburn, L.L., Lee, B.K., Eicher, E.M. and Guenet, J.L. (1999) *Proceedings of the National Academy of Sciences of the United States of America*, **96**, 10278–83.

68 Matsukawa, N., Grzesik, W.J., Takahashi, N., Pandey, K.N., Pang, S., Yamauchi, M. and Smithies, O. (1999) *Proceedings of the National Academy of Sciences of the United States of America*, **96**, 7403–8.

69 Nussenzveig, D.R., Lewicki, J.A. and Maack, T. (1990) *Journal of Biological Chemistry*, **265**, 20952–8.

70 Cohen, D., Koh, G.Y., Nikonova, L.N., Porter, J.G. and Maack, T. (1996) *Journal of Biological Chemistry*, **271**, 9863–9.

71 Fan, D., Bryan, P.M., Antos, L.K., Potthast, R.J. and Potter, L.R. (2005) *Molecular Pharmacology*, **67**, 174–83.
72 van den Akker, F. (2001) *Journal of Molecular Biology*, **311**, 923–37.
73 Stults, J.T., O'Connell, K.L., Garcia, C., Wong, S., Engel, A.M., Garbers, D.L. and Lowe, D.G. (1994) *Biochemistry*, **33**, 11372–81.
74 Fuller, F., Porter, J.G., Arfsten, A.E., Miller, J., Schilling, J.W., Scarborough, R.M., Lewicki, J.A. and Schenk, D.B. (1988) *Journal of Biological Chemistry*, **263**, 9395–401.
75 Anand-Srivastava, M.B. and Trachte, G.J. (1993) *Pharmacological Reviews*, **45**, 455–97.
76 Anand-Srivastava, M.B., Sairam, M.R. and Cantin, M. (1990) *Journal of Biological Chemistry*, **265**, 8566–72.
77 Anand, S.M., Srivastava, A.K. and Cantin, M. (1987) *Journal of Biological Chemistry*, **262**, 4931–4.
78 Pagano, M. and Anand-Srivastava, M.B. (2001) *Journal of Biological Chemistry*, **276**, 22064–70.
79 Anand-Srivastava, M.B., Sehl, P.D. and Lowe, D.G. (1996) *Journal of Biological Chemistry*, **271**, 19324–9.
80 Berl, T., Mansour, J. and Teitelbaum, I. (1991) *American Journal of Physiology*, **260** (4 Pt 2), F590–5.
81 Resink, T.J., Scott, B.T., Baur, U., Jones, C.R. and Buhler, F.R. (1988) *European Journal of Biochemistry*, **172**, 499–505.
82 Murthy, K.S., Teng, B.Q., Zhou, H., Jin, J.G., Grider, J.R. and Makhlouf, G.M. (2000) *American Journal of Physiology– Gastrointestinal and Liver Physiology*, **278**, G974–80.
83 Trachte, G.J. (2000) *Journal of Pharmacology and Experimental Therapeutics*, **294**, 210–15.
84 Trachte, G.J. (2003) *Endocrinology*, **144**, 94–100.
85 Porter, J.G., Arfsten, A., Fuller, F., Miller, J.A., Gregory, L.C. and Lewicki, J.A. (1990) *Biochemical and Biophysical Research Communications*, **171**, 796–803.
86 Rybalkin, S.D., Yan, C., Bornfeldt, K.E. and Beavo, J.A. (2003) *Circulation Research*, **93**, 280–91.
87 Beavo, J.A. (1995) *Physiological Reviews*, **75**, 725–48.
88 Maurice, D.H., Palmer, D., Tilley, D.G., Dunkerley, H.A., Netherton, S.J., Raymond, D.R., Elbatarny, H.S. and Jimmo, S.L. (2003) *Molecular Pharmacology*, **64**, 533–46.
89 Lohmann, S.M., Vaandrager, A.B., Smolenski, A., Walter, U. and De Jonge, H.R. (1997) *Trends in Biochemical Sciences*, **22**, 307–12.
90 Schlossmann, J., Feil, R. and Hofmann, F. (2005) *Frontiers in Bioscience*, **10**, 1279–89.
91 Kaupp, U.B. and Seifert, R. (2002) *Physiological Reviews*, **82**, 769–824.
92 John, S.W., Krege, J.H., Oliver, P.M., Hagaman, J.R., Hodgin, J.B., Pang, S.C., Flynn, T.G. and Smithies, O. (1995) *Science*, **267**, 679–81.
93 Ogawa, Y., Itoh, H., Tamura, N., Suga, S., Yoshimasa, T., Uehira, M., Matsuda, S., Shiono, S., Nishimoto, H. and Nakao, K. (1994) *Journal of Clinical Investigation*, **93**, 1911–21.
94 Steinhelper, M.E., Cochrane, K.L. and Field, L.J. (1990) *Hypertension*, **16**, 301–7.
95 Oliver, P.M., John, S.W., Purdy, K.E., Kim, R., Maeda, N., Goy, M.F. and Smithies, O. (1998) *Proceedings of the National Academy of Sciences of the United States of America*, **95**, 2547–51.
96 McKay, M.K. and Huxley, V.H. (1995) *American Journal of Physiology*, **268** (3 Pt 2), H1139–48.
97 Sabrane, K., Kruse, M.N., Fabritz, L., Zetsche, B., Mitko, D., Skryabin, B.V., Zwiener, M., Baba, H.A., Yanagisawa, M. and Kuhn, M. (2005) *Journal of Clinical Investigation*, **115**, 1666–74.
98 Barbee, R.W., Perry, B.D., Re, R.N., Murgo, J.P. and Field, L.J. (1994) *Circulation Research*, **74**, 747–51.
99 Franco, F., Dubois, S.K., Peshock, R.M. and Shohet, R.V. (1998) *American Journal of Physiology*, **274** (2 Pt 2), H679–83.
100 Knowles, J.W., Esposito, G., Mao, L., Hagaman, J.R., Fox, J.E., Smithies, O., Rockman, H.A. and Maeda, N. (2001) *Journal of Clinical Investigation*, **107**, 975–84.
101 Kishimoto, I., Rossi, K. and Garbers, D.L. (2001) *Proceedings of the National Academy of Sciences of the United States of America*, **98**, 2703–6.

102 Cao, L. and Gardner, D.G. (1995) *Hypertension*, **25**, 227–34.
103 Tamura, N., Ogawa, Y., Chusho, H., Nakamura, K., Nakao, K., Suda, M., Kasahara, M., Hashimoto, R., Katsuura, G., Mukoyama, M., Itoh, H., Saito, Y., Tanaka, I., Otani, H. and Katsuki, M. (2000) *Proceedings of the National Academy of Sciences of the United States of America*, **97**, 4239–44.
104 Holtwick, R., Gotthardt, M., Skryabin, B., Steinmetz, M., Potthast, R., Zetsche, B., Hammer, R.E., Herz, J. and Kuhn, M. (2002) *Proceedings of the National Academy of Sciences of the United States of America*, **99**, 7142–7.
105 Drewett, J.G., Fendly, B.M., Garbers, D.L. and Lowe, D.G. (1995) *Journal of Biological Chemistry*, **270**, 4668–74.
106 Furuya, M., Yoshida, M., Hayashi, Y., Ohnuma, N., Minamino, N., Kangawa, K. and Matsuo, H. (1991) *Biochemical and Biophysical Research Communications*, **177**, 927–31.
107 Kishimoto, I., Dubois, S.K. and Garbers, D.L. (1996) *Proceedings of the National Academy of Sciences of the United States of America*, **93**, 6215–19.
108 Marin-Grez, M., Fleming, J.T. and Steinhausen, M. (1986) *Nature*, **324**, 473–6.
109 Harris, P.J., Thomas, D. and Morgan, T.O. (1987) *Nature*, **326**, 697–8.
110 Light, D.B., Corbin, J.D. and Stanton, B.A. (1990) *Nature*, **344**, 336–9.
111 Erdos, E.G. (1990) *Hypertension*, **16**, 363–70.
112 Richards, A.M., McDonald, D., Fitzpatrick, M.A., Nicholls, M.G., Espiner, E.A., Ikram, H., Jans, S., Grant, S. and Yandle, T. (1988) *Journal of Clinical Endocrinology and Metabolism*, **67**, 1134–9.
113 Barrett, P.Q. and Isales, C.M. (1988) *Endocrinology*, **122**, 799–808.
114 Ganguly, A., Chiou, S., West, L.A. and Davis, J.S. (1989) *Biochemical and Biophysical Research Communications*, **159**, 148–54.
115 Shi, S.J., Nguyen, H.T., Sharma, G.D., Navar, L.G. and Pandey, K.N. (2001) *American Journal of Physiology – Renal Fluid and Electrolyte Physiology*, **281**, F665–73.
116 MacFarland, R.T., Zelus, B.D. and Beavo, J.A. (1991) *Journal of Biological Chemistry*, **266**, 136–42.
117 Burke, M.A. and Cotts, W.G. (2007) *Heart Failure Reviews*, **12**, 23–36.
118 Mark, P.B., Petrie, C.J. and Jardine, A.G. (2007) *Seminars in Dialysis*, **20**, 40–9.
119 Januzzi, J.L. Jr, Camargo, C.A., Anwaruddin, S., Baggish, A.L., Chen, A.A., Krauser, D.G., Tung, R., Cameron, R., Nagurney, J.T., Chae, C.U., Lloyd-Jones, D.M., Brown, D.F., Foran-Melanson, S., Sluss, P.M., Lee-Lewandrowski, E. and Lewandrowski, K.B. (2005) *American Journal of Cardiology*, **95**, 948–54.
120 de Denus, S., Pharand, C. and Williamson, D.R. (2004) *Chest*, **125**, 652–68.
121 Doust, J.A., Pietrzak, E., Dobson, A. and Glasziou, P. (2005) *British Medical Journal*, **330**, 625.
122 Makikallio, A.M., Makikallio, T.H., Korpelainen, J.T., Vuolteenaho, O., Tapanainen, J.M., Ylitalo, K., Sotaniemi, K.A., Huikuri, H.V. and Myllyla, V.V. (2005) *Stroke*, **36**, 1016–20.
123 Beck-da-Silva, L., de Bold, A., Fraser, M., Williams, K. and Haddad, H. (2005) *Congest Heart Fail*, **11**, 248–53, quiz 254–5.
124 Sun, T., Wang, L. and Zhang, Y. (2007) *Journal of Internal Medicine*, **37**, 168–71.
125 Bryan, P.M., Xu, X., Dickey, D.M., Chen, Y. and Potter, L.R. (2007) *American Journal of Physiology – Renal Fluid and Electrolyte Physiology*, **292**, F1636–44.
126 Dickey, D.M., Flora, D.R., Bryan, P.M., Xu, X., Chen, Y. and Potter, L.R. (2007) *Endocrinology*, **148**, 3518–22.
127 Cody, R.J., Atlas, S.A., Laragh, J.H., Kubo, S.H., Covit, A.B., Ryman, K.S., Shaknovich, A., Pondolfino, K., Clark, M., Camargo, M.J., Scarborough, R.M. and Lewicki, J.A. (1986) *Journal of Clinical Investigation*, **78**, 1362–74.,
128 Fifer, M.A., Molina, C.R., Quiroz, A.C., Giles, T.D., Herrmann, H.C., De Scheerder, I.R., Clement, D.L., Kubo, S., Cody, R.J., Cohn, J.N. and Fowler, M.B. (1990) *American Journal of Cardiology*, **65**, 211–16.

129 Weder, A.B., Sekkarie, M.A., Takiyyuddin, M., Schork, N.J. and Julius, S. (1987) *Hypertension*, **10**, 582–9.
130 Fonarow, G.C. (2003) *Heart Failure Reviews*, **8**, 321–5.
131 Kalra, P.R., Clague, J.R., Bolger, A.P., Anker, S.D., Poole-Wilson, P.A., Struthers, A.D. and Coats, A.J. (2003) *Circulation*, **107**, 571–3.
132 Del Ry, S., Passino, C., Maltinti, M., Emdin, M. and Giannessi, D. (2005) *European Journal of Heart Failure*, **7**, 1145–8.

6
Endothelins
Gian Paolo Rossi and Teresa M. Seccia

6.1
Endothelin System

Endothelin (ET) was isolated and chemically identified by Masaki's group in March 1988. The peptide, then described as one of the most potent vasoconstrictors known, has a strong resemblance to sarafotoxin peptides present in the venom of snakes of the *Atractapis* family. Four ET prototypes exist, ET-1, ET-2, ET-3 and ET-4 (vasoactive intestinal contractor), which exhibit differential tissue distribution [1, 2]. The most widely studied and predominant isoform, particularly in the cardiovascular system, is ET-1 [3].

ET-1 is a 21-amino-acid peptide predominantly synthesized by endothelial cells from which it is released abluminally toward the vascular smooth muscle cells, where it acts in a paracrine fashion [4]. Under disease conditions other cell types implicated in cardiovascular disease, such as smooth muscle cells, leukocytes, macrophages, cardiac myocytes and mesangial cells, can produce ET-1 [5, 6]. The regulation and synthesis of ET is shown in Figure 6.1. Once formed, prepro-ET-1 is cleaved by a furin-like convertase to the 38-amino-acid precursor big (pro)-ET-1. Endothelin-converting enzyme (ECE)-1 (or ECE-2) then converts this peptide into ET-1. ECE-3 selectively converts big ET-1 to ET-3. ECE-independent pathways also play a major role ET-1 production, as in ECE-1 knockout mice in which tissue levels of ET-1 are reduced by only 30% [7]. Importantly, chymase, a major angiotensin (Ang) II-forming enzyme in the human cardiovascular system, also catalyses the conversion of big ET-1 to ET-1 (1-31) [8, 9]. Hence, the synthesis of the peptide is not isolated, but closely integrated with the renin–angiotensin system (RAS).

The biological action of ET-1 occurs via activation of specific G-protein-coupled receptors. Two ET receptor subtypes, defined as ET_A and ET_B, have been identified in mammalians (Table 6.1). In the vasculature, ET_A subtypes are found in smooth muscle cells, whilst ET_B subtypes are located on endothelial cells and sometimes in smooth muscle cells (see below). ET-1 exerts its deleterious effects, including proliferation, vasoconstriction, fibrosis, hypertrophy and inflammation, via both

Cardiovascular Hormone Systems. Edited by Michael Bader
Copyright © 2008 WILEY-VCH Verlag GmbH & Co. KGaA, Weinheim
ISBN: 978-3-527-31920-6

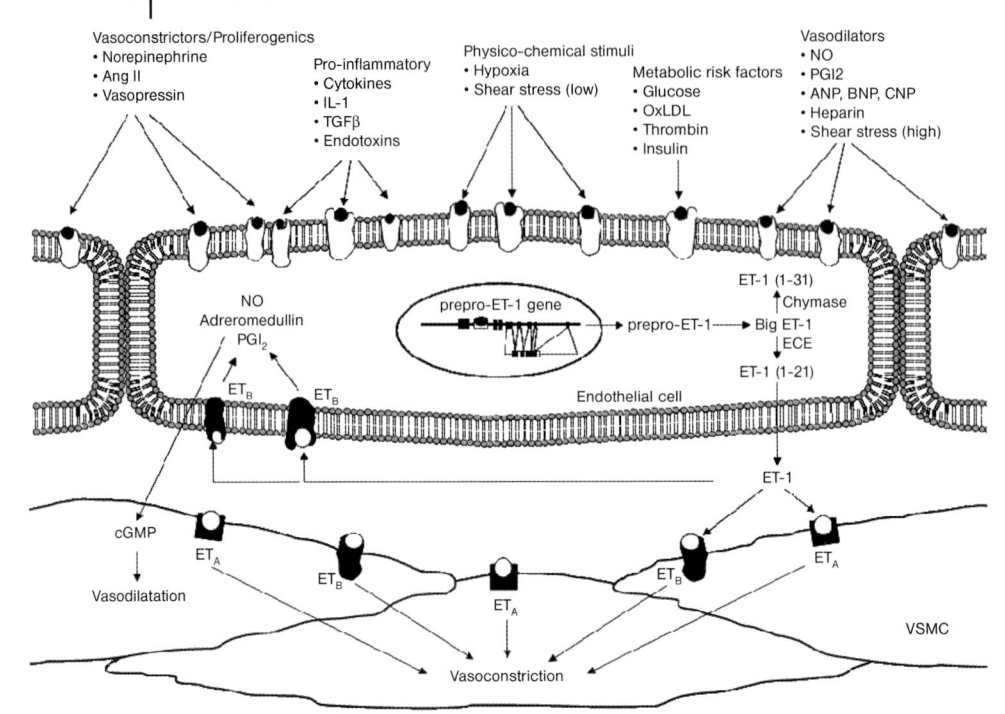

Figure 6.1 Overview of the ET system. The cartoon summarizes the synthesis of ET-1, and illustrates the factors affecting ET-1 synthesis and related intracellular signaling pathways in the endothelial cell. Exons and consensus sequences in the promoter of prepro-ET-1 are schematically depicted in the nucleus. After conversion of prepro-ET-1 in big ET-1, chymase or ECE generates ET-1 (1–31) or ET-1 (1–21), respectively. ET-1 interacts with ET_A and ET_B receptors located on the surface of the vascular muscle cells (VSMC), thereby inducing vasoconstriction. ET-1 may also interact with ET_B receptors, which are located on the surface of the endothelial cells. Interaction with these receptors stimulate the synthesis of NO, prostaglandin I_2 (PGI_2) and adrenomedullin that, in turn, induce vasodilatation via cGMP increase. IL-1, Interleukin-1; oxLDL, oxidized low-density lipoprotein; ANP, atrial natriuretic peptide; BNP, brain natriuretic peptide; CNP, C-type natriuretic peptide.

receptor subtypes albeit predominantly via ET_A [10]. Activation of the ET_A receptor evokes a long-lasting vasoconstriction. Activation of the ET_B receptor, on the other hand, induces vasodilatation via the release of nitric oxide (NO), prostacyclin and adrenomedullin. ET_B receptors also mediate the pulmonary clearance of circulating ET-1, the reuptake of ET-1 by endothelial cells and the release of aldosterone [11]. Under disease conditions, however, ET_B receptors can also be expressed in vascular smooth muscle cells and mediate vasoconstriction [12, 13].

ETs play an important role in growth and physiology, and are considered growth factors during embryonic development. ET-1 and ET_A receptor null mice die

Table 6.1 The biological effects of endothelin receptor activation.

ET_A receptor	ET_B receptor
Vasoconstriction	Release of NO, prostacyclin, adrenomedullin
Cell growth	ET-1 clearance
Stimulation of synthesis of cytokines and growth factors: vascular endothelial growth factor and basic fibroblast growth factor	Inhibition of ECE-1
TGF-β- and PDGF-mediated effects	Migration and proliferation of endothelial cells
Neutrophil adhesion	Release of aldosterone
Platelet adhesion	
Macrophages chemiotaxis	

shortly following birth from respiratory failure and cardiac abnormalities; moreover, these animals have hypoplasia of the facial bones [14, 15].

The ET gene appears to be regulated in a highly complex manner, but comprises only five exons, of which exon 2 codes for mature ET-1. The synthesis of ET-1 is regulated by many different factors (Figure 6.1), including physicochemical factors, such as pulsatile stretch, shear stress, hypoxia and pH [16–18]. Importantly, ET-1 production is stimulated by cardiovascular risk factors such as increased levels of oxidized low-density lipoprotein and glucose, estrogen deficiency, obesity, aging, and procoagulants (e.g. thrombin) [19–23]. Vasoconstrictors like norepinephrine, and Ang II, growth factors and cytokines also stimulate ET production [24–30]. By contrast, NO, prostacyclin and natriuretic peptides can inhibit ET synthesis [23, 31, 32].

6.2
ET and Cardiovascular Disease

ET-1 is a potent vasoconstrictor that interacts closely with other peptide systems to produce complex effects. ET-1 stimulates the production of cytokines and growth factors and exerts mitogenic effects [33, 34]. It also promotes the production of reactive oxygen species and the deposition of extracellular matrix proteins partly through upregulation of collagen 1 gene activity. ET-1 promotes expression of fibronectin and exacerbates the effects of transforming growth factor (TGF)-β and platelet-derived growth factor (PDGF) [35]. ET-1 also stimulates neutrophil adhesion and platelet aggregation, and is a chemotactic factor for macrophages. Most if not all of these factors influenced by ET-1 are involved in cardiovascular disease. Therefore, the therapeutic applicability of endothelin receptor antagonists (ERAs) has been and is currently being pursued in many cardiovascular pathologies, including pulmonary hypertension (for which an ERA, bosentan, is already approved), renal insufficiency, systemic arterial hypertension, heart failure, pulmonary fibrosis and scleroderma, among others.

6.3
Assessment of the ET System in Disease States

The ET-1 secretion from endothelial cells occurs abluminally, toward the smooth muscle of the media and only about one-fifth of the peptide is secreted in the vessel lumen, where its half-life is in the range of tenths of a second. Therefore, the plasma levels of the peptide provide only a rough estimate of the degree of activation of the ET system. Notwithstanding this, increased ET-1 levels have been shown in a number of cardiovascular diseases (Table 6.2).

Table 6.2 Cardiovascular diseases with increased ET-1 levels.

• Pulmonary hypertension	• Hemangioendothelioma
• Congestive heart failure	• Acute myocardial ischemia
• Systemic arterial hypertension	• Atherosclerosis
• Renal insufficiency	• Primary aldosteronism
• Scleroderma	

In human hypertension, although some studies have demonstrated increased plasma levels of ET-1, others have not shown elevated levels of the peptide. This is not surprising for the aforementioned reasons [4]. Additionally, most human hypertension studies had a statistical power inadequate to provide conclusive data [36]. Moreover, given the pathophysiologic diversity of essential hypertension, it is likely naive to presume that ET-1 is consistently increased in this condition. More robust results were obtained when the synthesis of ET is examined at the tissue level. We examined immunoreactive (ir) ET-1 levels in the different layers of the wall of human arteries, obtained *ex vivo* from patients with coronary artery disease (CAD) and/or hypertension undergoing surgery, as well as from organ donors. In organ donors, irET-1 was found mainly in endothelial cells, whereas in CAD and/or hypertension patients, irET-1 was detectable in the tunica media of different types of arteries. In CAD patients irET-1 was detected in the endothelium of all arteries and in the tunica media of internal thoracic artery from most CAD patients. We also found significant correlations between the amount of irET-1 in the tunica media and mean blood pressure, total serum cholesterol and number of atherosclerotic sites. Hence, these findings indicate that endothelial damage occurs in atherosclerosis and/or hypertension and that in these patients ET-1 is synthesized in vascular smooth muscle cells – a cell phenotype which is not usually involved in ET-1 synthesis [5].

Plasma ET concentrations are elevated in pulmonary arterial hypertension (PAH) of various etiologies, including primary PAH, scleroderma-associated PAH and congenital heart disease-associated PAH [37–39]. These augmented ET-1 levels, which are even more remarkable based on the considerations made above, likely indicate that the ET system is markedly activated at the tissue level in these conditions and the leakage of ET-1 is sufficient to consistently raise plasma levels to such an extent.

ET-1 is difficult to measure in plasma due to its short half-life instability, low levels and its binding to plasma proteins and receptors. A novel assay technique to measure ET-1 has been recently developed [40]. This assay quantifies ET-1 levels by measuring the C-terminal ET-1 precursor fragment (CT-pro-ET-1), which is more stable than big-ET-1 or the mature peptide found in the circulation (Figure 6.2). CT-pro-ET-1 values follow a normal distribution in healthy subjects, without

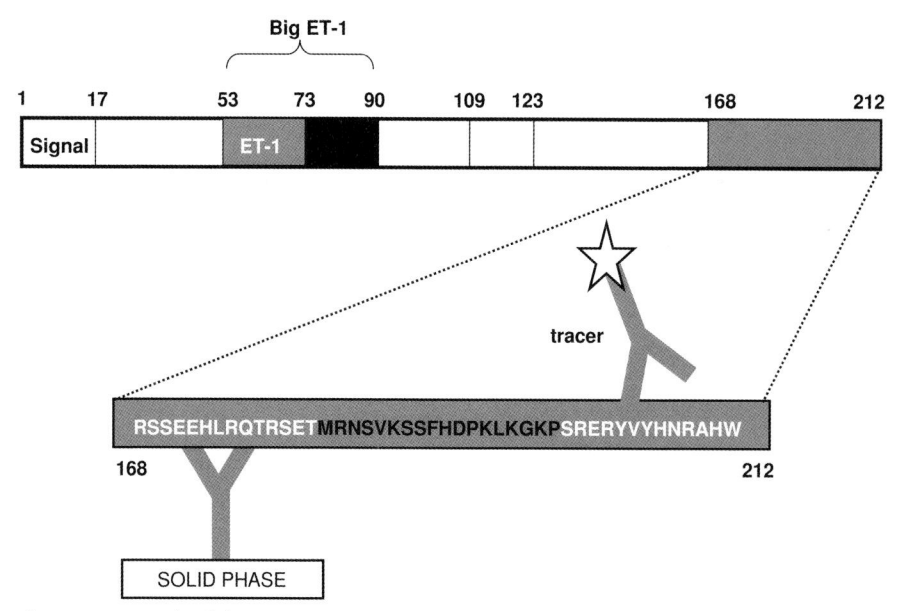

Figure 6.2 Principle of the assay measuring CT-pro-ET-1. The magnified portion of prepro-ET-1 represents the C-terminal precursor fragment (CT-pro-ET-1). Numbers indicate amino acids and white letters indicate antibody epitopes. Signal, signal peptide; tracer, labeled antibody; solid phase, antibody coated on tubes.

differences between genders. They were found to be significantly increased in two pathological states associated with enhanced ET-1 production, including heart failure and sepsis [40]. This assay technique therefore entails a promising tool for indirectly assessing ET-1 levels.

6.4
ET in PAH

PAH reflects a multifactorial process with complex evolution that involves dysfunction of underlying cellular pathways and mediators [41]. The vasculature of the lung is a site of intense synthesis and clearance of ET-1 and NO, which can be disturbed in lung disease and PAH [42]. An imbalance in the synthesis of NO and ET-1 is a hallmark of both primary and secondary PAH, which implies that the ET system plays an important pathogenic role in this disease [42, 43].

In hypoxia-induced pulmonary hypertension pretreatment with bosentan prevented the pulmonary vasoconstrictor response to acute hypoxia and vascular remodeling in the rat, which is consistent with the notion that hypoxia potently evokes activation of ET-1 synthesis (Figure 6.1) [44]. Chronic bosentan treatment initiated 48 h prior to hypoxia, prevented the development of pulmonary hypertension, reduced right heart hypertrophy and prevented the remodeling of small pulmonary arteries. Institution of bosentan treatment 2 weeks following hypoxia produced a significant reversal of pulmonary hypertension, right heart hypertrophy, and pulmonary vascular remodeling [44].

In humans, substantial evidence from clinical trials and long-term data indicate that monotherapy with an ERA is a beneficial therapeutic approach in PAH [45]. In a pilot study of a few patients with primary and scleroderma-induced PAH, bosentan induced a dose-dependent fall in total pulmonary resistance and mean pulmonary artery pressure [46]. A subsequent double-blind, placebo-controlled study, multicenter study, the Bosentan Randomized Trial of Endothelin Antagonist Therapy (BREATHE-1), confirmed the efficacy of bosentan for this indication: 213 PAH patients were randomly assigned placebo or 62.5 mg bosentan twice daily for 4 weeks, followed by 125.0 or 250.0 mg bosentan twice daily for a minimum of 12 weeks [47]. The primary endpoint was the degree of change in exercise capacity (6-min walking test), and secondary endpoints included change in the Borg dyspnea index and World Health Organization functional class, and the time to clinical worsening [47]. At 16 weeks all these features had improved in the bosentan group. These findings led to approval of bosentan in both the US and Europe for the treatment of pulmonary hypertension in patients in NYHA (New York Heart Association) class III–IV.

Results of a Cochrane database meta-analysis of randomized and quasirandomized trials of ERAs versus placebo in PAH showed a consistent and significant 37 m increase in distance walked during a treadmill exercise test, a significant improvement in NYHA functional class, a lowering of Borg dyspnea index scores, a significant decrease in mean pulmonary artery pressure, a significant decrease of pulmonary vascular resistance and a significant increase of cardiac index [45].

Even despite these impressive results, no significant differences were seen for mortality between the ERA treatment and placebo groups. However, the number of deaths recorded was very small and, therefore, the lack of effect of ERAs on survival should be taken with caution as most likely it reflects only the lack of statistical power.

Moreover, as mentioned, PAH reflects a multifactorial process that involves dysfunction of several cellular pathways and mediators. Therefore, the strategy of combining ERAs with a prostanoid or phosphodiesterase-5 inhibitor is conceptually appealing. The findings from pilot studies suggest this approach is useful and safe, but experimental and clinical investigation is ongoing [44, 48, 49]. Hence, whether combination therapy will be better than monotherapy with any of the drugs in these three classes remains to be proven.

6.5
ET in Systemic Arterial Hypertension

6.5.1
Animal Studies

The role of ET-1 differs markedly across experimental models of hypertension. In Ang II-induced and in salt-sensitive hypertension, as in deoxycorticosterone acetate-salt-treated, stroke-prone spontaneously hypertensive rats (SHRs), Dahl-salt sensitive rats, salt-depleted squirrel monkeys, rats on a high fructose diet and dogs with renal hypertension, the ET system is upregulated and chronic ERA lowers blood pressure [50, 51]. In contrast, no blood pressure lowering was seen in SHRs, renovascular hypertension (2K/1C), N^G-nitro-L-arginine (L-NAME)-induced hypertension and transgenic TGRen2 rats [44, 52].

Irrespective of ET-1 importance in the various models of hypertension, the peptide is believed to be implicated in cardiovascular hypertrophy and damage associated with hypertension. ET is a proinflammatory agent, and exerts potent mitogenic actions that contribute to vascular smooth muscle proliferation and end-organ damage by extracellular matrix deposition. Thus, it is reasonable to assume that ET blockade should ameliorate hypertension-induced end-organ damage (Figure 6.3). In keeping with this, Barton et al. studied the effects of darusentan in salt-induced Dahl hypertension [53]. Salt-sensitive and salt-resistant Dahl rats were treated with a high sodium diet with or without darusentan for 2 months. Despite only partially reducing systolic blood pressure, darusentan prevented increased tissue ET-1 content and vascular hypertrophy. None of these effects were found in salt-resistant Dahl rats [53]. These findings imply that ET-1 acts as a local mediator of vascular dysfunction and hypertrophy in Dahl-salt-induced hypertension.

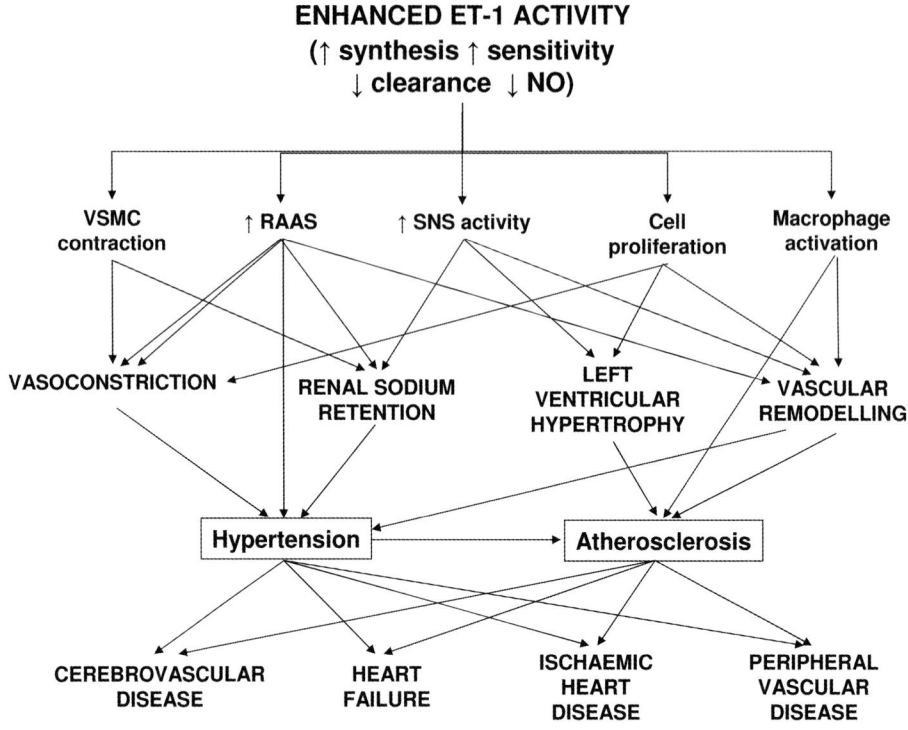

Figure 6.3 Mechanisms by which an increased activity of the ET system contributes to triggering and progression of CV diseases. An enhanced ET-1 synthesis and sensitivity, combined with a decreased ET-1 clearance and NO production, activates the renin–angiotensin–aldosterone system (RAAS) and the sympathetic nervous system (SNS), as well as promotes cell proliferation and inflammation, thereby favoring development and progression of hypertension and atherosclerosis. Both are risk factors for cerebrovascular disease, heart failure, ischemic heart disease and peripheral vascular disease.

Seccia et al. evaluated the role of ET-1 and the renin–angiotensin–aldosterone system in cardiac fibrosis [54]. Transgenic (mRen2)27 rats were treated with placebo, bosentan, irbesartan (angiotensin receptor blocker), BMS-182874 (ET$_A$ selective antagonist) or a combination of irbesartan and BMS-182874. In the placebo group, hypertension was associated with left ventricular hypertrophy and cardiac fibrosis, which were prevented with irbesartan, but not BMS-182874. Bosentan prevented fibrosis, but not hypertension and left ventricular hypertrophy. Combined irbesartan and BMS-182874 prevented left ventricular hypertrophy, but not fibrosis. Thus, in a model of Ang II-dependent hypertension, fibrosis and left ventricular hypertrophy were reduced by both mixed ET$_A$/ET$_B$ blockade and AT$_1$ blockade. However, only the latter treatment prevented both hypertension and left ventricular hypertrophy [54]. Hence, there is dissociation between the mechanisms of cardiac fibrosis and hypertension, which do and do not implicate ET-1, respectively, in this model of hypertension.

In summary, overall experimental studies indicate an involvement of ET-1 in the pathogenesis of hypertension and target organ damage.

6.5.2
Human Studies

ET-1 contributes to vascular tone via the activation of both ET_A and ET_B receptors: under physiological conditions, the ET system is carefully balanced so that the vasoconstrictor effect of the peptide is counterbalanced via ET_B-mediated NO and prostacyclin release. Thus, in humans, studies have unequivocally demonstrated the vasoconstrictor effects of ET-1: the ET-1-induced vasoconstriction was abolished by a selective ET_A antagonist, BQ-123. BQ-123 and ET-1 coinfusion not only reversed the vasoconstrictor effect of ET-1, but resulted in vasodilatation, likely via ET_B receptor activation.

Vasoconstrictors such as norepinephrine and Ang II, growth factors, cytokines and thrombin enhance ET-1 secretion. In contrast, mediators such as NO, natriuretic peptides and prostacyclin inhibit ET-1 release (Figure 6.1) [23–27, 29, 31, 32]. Hence, under normal conditions, the effects of ET-1 are carefully regulated through inhibition or stimulation of its release from the endothelium [55]. This balance becomes deranged under pathological conditions due to multiple potential mechanisms, including an increase in ET-1 synthesis and/or release, and/or a decrease in its clearance, an increase in ET receptor sensitivity or a decrease in the bioavailability of NO. Thus, an imbalance in the production of vasodilator and the enhanced activity of the ET system may contribute to the initiation and maintenance of high blood pressure. Consistently with this hypothesis, by infusing a very low dose of BQ-123, alone or combined with BQ-788, an ET_B-selective antagonist, at very low doses that competes only with endogenous ET-1, and using ambulatory blood pressure monitoring, Rossi *et al.* documented a lowering of blood pressure in patients with a high- or low-renin hypertension [56]. On the other hand, Cardillo *et al.* reported an increased vasoconstrictor response to both exogenous and endogenous ET-1 in hypertensive patients compared with normotensive subjects [57]. They also assessed the vascular response to intrabrachial administration of BQ-123 and BQ-788 separately and in combination. In normal subjects, BQ-123 alone or with BQ-788 did not significantly affect forearm blood flow. In contrast, in hypertensive patients, BQ-123 increased forearm blood flow, and the combination of BQ-123 and BQ-788 resulted in a greater vasodilator effect. BQ-788 produced a decrease in forearm blood flow in control subjects and transient vasodilatation in hypertensive patients. Thus, the vasoconstrictor effect of endogenous ET is increased in patients with primary hypertension. Moreover, the increase in vascular resistance was normalized by nonselective ET receptor blockade, suggesting that ET plays a role in the pathophysiology of hypertension. The increased vasoconstrictor effect of exogenous ET in hypertensive patients may thus be attributed to ET receptor upregulation, postreceptor sensitization and/or impaired ET_B receptor vasodilatation.

Enhanced vasoconstriction to endogenous ET-1 in primary hypertension has also been reported by Taddei *et al.* [58]. In hypertensive patients, TAK-044 (mixed

ET_A/ET_B antagonist) caused significantly greater vasodilatation than in normotensive subjects [58]. Additionally, vasoconstriction due to NO synthase inhibition with L-N^G-monomethyl-arginine (L-NMMA) was significantly decreased in hypertensive patients compared with controls. Thus, (i) patients with primary hypertension have a greater endogenous ET-1-mediated vasoconstrictor activity; (ii) the greater vasoconstrictor effect of L-NMMA in normotensives demonstrates the higher bioactivity of NO in these subjects as compared with hypertensive patients; and (iii) the inverse relationship between the degree of relaxation induced by TAK-044 and contraction due to L-NMMA suggests that the greater vasoconstriction response exerted by endogenous ET-1 in primary hypertensive patients might be partially related to reduced NO bioactivity.

Compelling evidence for a role of ET-1 in human hypertension derives also from two large clinical trials. Krum *et al.* reported that bosentan lowered both systolic and diastolic blood pressure similar to enalapril in mild-to-moderate hypertensive patients without inducing reflex tachycardia or neurohormonal activation [59]. Another study, named the Heart Failure ET_A Receptor Blockade Trial (HEAT), demonstrated that the ET_A-selective agent, darusentan, dose-dependently lowered in blood pressure in hypertensive patients [60].

Furthermore, ERAs have demonstrated antihypertensive effects in special hypertensive populations, including blacks with primary hypertension, cyclosporine-induced hypertension and erythropoietin-induced hypertension [61].

We will herein briefly mention only obstructive sleep apnea syndrome, which is an emerging clinical problem in hypertension. Chronic hypoxia may lead to hypertension in obstructive sleep apnea syndrome through numerous mechanisms, including vasoconstriction, sympathetic activation of the inflammatory system, cytokine production as well as endothelial damage [62].

Changes in ET-1 concentrations were associated with changes in both mean arterial blood pressure and oxygen saturation, thereby suggesting that ERAs may ameliorate the increased cardiovascular disease risk associated with the obstructive sleep apnea syndrome [63].

6.5.3
ET-1 in Renal Disease

The global incidence of end-stage renal disease (ESRD) and diabetic nephropathy (DN), which is the leading cause of ESRD, is increasing even despite the availability of a vast array of antidiabetic and antihypertensive drugs, likely because of the longer survival of type 2 diabetic patients, the increasing number of diabetics, a lag phase between improvement in management and a decline in end-stage complications, and an expansion in the aging population [64]. However, the deficiencies in current treatment regimes might also account for this worrisome trend.

Available data indicate that both baseline proteinuria and albuminuria are strong risk factors for the progression to ESRD [65]. Hence, proteinuria *per se* is a risk factor for renal disease progression, which implies that the greater the reduction of albuminuria, the greater the renoprotection. Even though drugs available to treat DN are beneficial, they do not provide optimal reduction of proteinuria and do not arrest disease progression, thus raising the question as to whether additional clinically relevant improvements can be obtained with further novel treatments, including ERAs.

ET and RAS "cross-talk", and closely interact under physiological and pathophysiologic conditions, which may provide opportunities for targeting both systems to obtain more effective inhibition of the cascade of events leading to glomerulosclerosis and thus slow down further or even prevent ESRD [9]. Both ET-1 and Ang II are potent vasoconstrictors where they stimulate protein kinase C, TGF-β1 expression and extracellular matrix accumulation. Ang II modulates ET-1 production, and RAS inhibition has been shown to reduce ET-1 production in blood vessels and in glomeruli, and ET-1 excretion in urine. Ang II-induced vasoconstriction and extracellular matrix expressions are at least partially mediated by ET-1. Additionally, ET-1 enhances conversion of Ang I to Ang II [66]. ERAs could therefore exert beneficial effects via their inhibitory effects on the RAS, which suggest angiotensin-converting enzyme (ACE) inhibitors/angiotensin receptor blockers and ERAs may act "synergistically".

In the kidney, ET-1 regulates extracellular volume by modulating sodium and water excretion and hemodynamics, including intraglomerular pressure, via its vasoactive effects. ET-1 is a potent constrictor of both pre- and postglomerular arterioles, and it decreases renal blood flow and glomerular filtration rate, leading to a reduced urine flow and sodium excretion [67–69]. As it plays a role in fibrosis, cell proliferation and matrix formation, ET-1 can be implicated in chronic renal failure, as well as the increase in proteinuria of renal disease [70].

Of the two receptor subtypes, in the kidney, the ET_B receptor is predominantly expressed in the collecting system, and the ET_A receptor in the glomeruli and vascular structures [71]. Plasma ET-1 levels are raised in both diabetes and kidney disease [72, 73]. In patients with ESRD on hemodialysis, an increase in plasma ET-1 concentrations has been noted [72]. Additionally, urinary excretion of ET-1 is also increased in patients with chronic glomerulonephritis [74]. In animal models of renal disease, ERAs were shown to exert renoprotective effects [75, 76]. In a NO-deficient rat model of hypertension induced by L-NAME mRNA expression of prepro-ET-1 and ET-1 content increased in renal microvessels [77]. Coadministration of bosentan with L-NAME abolished the increased mRNA expression and synthesis of collagen 1, and reduced the severity of renal vascular lesions without lowering blood pressure. Thus, when NO production is reduced, synthesis of ET-1 in renal microvessels is increased. In this particular model of hypertension, therefore, ET-1 is involved in the development of renal fibrosis independently of hypertension.

Nephroprotective effects of selective ET_A blockade were demonstrated also in uninephrectomized, stroke-prone, SHRs receiving a high-salt diet, development of glomerulosclerosis and tubulointerstitial and vascular damage was completely prevented by darusentan [47]. Thus, ET-1 plays a role via ET_A receptors in the development of progressive renal injury in salt-loaded uninephrectomized, stroke-prone, SHRs.

In diabetic animals both ET_A receptor-selective and nonselective antagonists were shown to reduce renal damage. Bosentan prevented the increase in blood pressure in streptozotocin-induced diabetic rats as potently as enalapril treatment, while combination treatment of bosentan with enalapril had no further blood pressure lowering effects [78].

An antiproteinuric effects of bosentan in diabetic rats has been also demonstrated, independent of blood pressure lowering [79].

The selective ET_A antagonist, YM598 dose-dependently reduced proteinuria in Long-Evans Tokushima Fatty rats, with a high dose providing a similar reduction to that seen with enalapril [80]. Furthermore, urinary excretions of heparan sulfate and type IV collagen were also decreased with YM598 [80]. In another study, darusentan decreased glomerular hyaline deposits in streptozotocin-induced diabetes in rats [81].

Both TGF-β1 and podocyte loss can play central roles in DN; however, available data suggest that in DN the renoprotective effects of ET receptor blockade does not involve preservation of podocyte structure, but can operate through mechanisms which may be complementary to these ACE inhibitors [82].

In a rat model of age-related glomerulosclerosis ERAs may improve podocyte dysfunction and structural injury in glomeruli and podocytes [83]. Plasma protein overload in podocytes led to an increase in ET-1 gene expression, with ensuing increased generation of the peptide, suggesting that glomerular epithelial cells exposed to high concentration of protein are a significant source of ET-1. Enhanced ET-1 generation may alter glomerular perm selectivity and exaggerate the toxic effect of protein overload on dysfunctional podocytes; ERAs may limit this harmful effect [84]. Whether this applies to ET receptor blockade in renal disease in humans remains to be proven.

6.5.3.1 Blood Pressure, Proteinuria and ERAs

Whether the renal protective action of ERAs is specific to their mechanism of action or depends on their antihypertensive properties remains, as far all antihypertensive agents with renal protective action, controversial (see below).

As already alluded to, the greater the reduction of proteinuria, the greater the reduction in renal disease. ERAs have been consistently shown to reduce proteinuria, which may explain their renoprotective actions.

6.6
ERAs in Human Renal Diseases

The effects of BQ-123 and BQ-788, either alone or in combination, were studied in hypertensive chronic renal failure patients [85]. BQ-123, alone and in combination with BQ-788, lowered blood pressure in chronic renal failure. BQ-123 significantly increased renal blood flow and reduced renal vascular resistance when given alone, but not when combined with BQ-788. However, combined ET_A/ET_B receptor antagonism, while lowering blood pressure, did not confer renal benefits, presumably due to the blockade of ET_B-mediated vasodilator actions. These data confirm a role for ET-1 in chronic renal failure and the potential therapeutic applicability of selective ET_A receptor antagonism in this pathology. This suggests that ERAs in hypertensive patients with chronic renal failure may lead to a reduction in end-organ damage, which along with normalizing blood pressure, is the ultimate therapeutic goal.

A randomized, double-blind, placebo- controlled, dose-range study evaluated the efficacy of avosentan (SPP301), a selective ET_A receptor antagonist belonging to a class of functionalized pyridylsulfonamido pyrimidines, in 286 patients with DN [86]. This phase IIb study ascertained the effects of four doses of avosentan (5, 10, 25 and 50 mg) or placebo administered once daily for 12 weeks on urinary albumin excretion rate (UAER). Avosentan was administered on top of RAS inhibitors, therefore suggesting that as maximal blood pressure reduction was already achieved with RAS inhibition no further reduction with an ERA was not possible. Compared to placebo, all four doses of avosentan led to a significant decrease in UAER [87]. Creatinine clearance remained unchanged following 12 weeks. Avosentan produced minor, nonsignificant antihypertensive effects. The main adverse effects were edema and headache, which were more commonly observed with the highest dose of avosentan.

Thus, a larger phase III outcome study (Avosentan on doubling of Serum Creatinine, ENd stage renal disease or death in Diabetic nephropathy; ASCEND), began in July 2005 to investigate the effects of avosentan (25 and 50 mg) administered on top of RAS blockade in more than 2000 type 2 diabetes patients with DN. Primary endpoints of the study included time to doubling of serum creatinine, time to ESRD or death. The study was stopped in December 2006 because patients taking avosentan developed higher levels of fluid retention than those on placebo.

In summary, animal data and a proof of concept study in human DN have shown the renoprotective potential of ERAs. They were shown to increase survival, ameliorate proteinuria and normalize renal structural injury in diabetic animals. Most importantly, an ET_A selective antagonist, avosentan has demonstrated the potential for renoprotection in type 2 diabetic patients. However, an ongoing phase III outcome study involving avosentan did not fulfill the expectations in terms of improving the course of diabetes mellitus because it was prematurely stopped due to excess fluid retention. Thus, further investigation is clearly needed to explore the potential usefulness of ERA for renal protection.

6.7
ET and Heart Failure

Despite much progress in the treatment of heart failure, mortality in these patients remains substantial, which justifies the search for novel therapeutic approaches

[88]. Among the neurohormonal systems implicated in the pathogenesis of heart failure, the ET system can play a substantial role in heart failure which is characterized by increased peripheral resistance and volume retention.

6.8
Role of ET in Heart Failure

Increased afterload is a key feature of heart failure, making ET-1 a pathological suspect inasmuch as it contributes to systemic vasoconstriction, decreased left ventricular function and fluid retention in heart failure [89]. ET-1 also plays a role in increasing pulmonary artery pressure, which is important because pulmonary hypertension is a secondary manifestation of heart failure. Moreover, ET influences the inotropic state of the myocardium, the central and autonomic nervous system, and the baroreflex [90–92]. As mentioned, it also plays a key role in the kidney, where it influences Na^+ and water excretion partly through enhanced aldosterone production (Figure 6.4) [93].

6.8.1
ET-1 Plasma Levels in Heart Failure

In rats with heart failure the production of ET-1 is increased in the myocardium and in the peripheral circulation, particularly in the lung [94–98]. The plasma ET-1 levels are also enhanced in experimental heart failure, along with the density of myocardial ET receptor [99, 100]. An upregulation of ET_A as well as ET_B receptor mRNA levels in the left ventricle of rats with heart failure has also been shown [98].

The hemodynamic severity of heart failure correlates with plasma ET-1 levels, and an inverse relationship between stroke volume and ET-1 exists, whereas peripheral vascular resistance is positively associated with circulating ET-1 [101]. Furthermore, ET-1 levels correlated positively with pulmonary capillary wedge pressure (PCWP), mean arterial pulmonary artery pressure (MPAP), but negatively with cardiac index (CI). Tissue levels of ET-1 are also increased in the failing human heart, furthermore, plasma big ET-1 levels independently predicted death [102–107].

Endogenous ET exerts differential effects in the normal and failing human heart: infusion into the left coronary artery of BQ123 in normal subjects significantly reduced left ventricular dP/dt_{max} and left ventricular dP/dt, suggesting that endogenous ET-1 exerts a tonic inotropic effect [108]. By contrast, in patients with nonischemic dilated cardiomyopathy, BQ123 caused no reductions in left ventricular dP/dt_{max} or left ventricular dP/dt_{40}. BQ123 did not affect heart rate, left ventricular relaxation, left ventricular end-diastolic pressure, right arterial pressure or pulmonary pressure in either group. Thus, endogenous ET-1 has a positive inotropic effect in normal myocardium, but this effect of ET-1 is absent in the failing

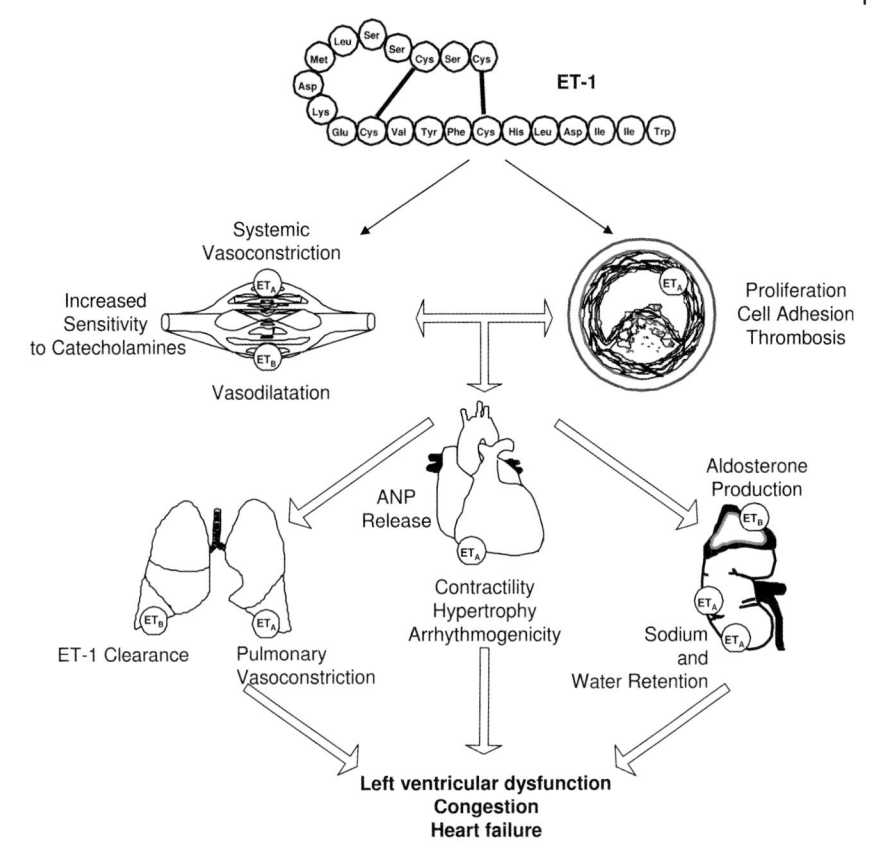

Figure 6.4 Potential role of ET-1 in hypertension and its complications. ET-1, by inducing systemic vasoconstriction and atherosclerosis, facilitates the development of left ventricular hypertrophy and dysfunction, sodium and water retention, and pulmonary hypertension. The progressive ET-1 receptor desensitization and downregulation in the myocardium causes a loss of the inotropic effect played by ET-1 under physiological conditions, thereby favoring the development of heart failure.

human heart. This loss of inotropic effect could be due to ET receptor downregulation or desensitization and may contribute to contractile left ventricular dysfunction [109].

6.8.2
ERAs in Heart Failure

Both ET_A-selective and mixed ERAs are in clinical development for acute and chronic blockade. Selective ET_A blockade is of more interest as ET_B receptor activation induces vasodilatation via NO and prostacyclin release; hence, blockade of this receptor may be theoretically detrimental in heart failure.

The acute hemodynamic effects of a single oral dose of darusentan (1, 10, 30, 100 or 300 mg) were investigated in a multicenter study of 95 patients with heart failure (NYHA II–III) with an ejection fraction of 35% or below [101].

Darusentan treatment dose-dependently increased plasma ET-1 levels. A substantial surge in ET plasma concentrations was observed at the highest dose of 300 mg, which may result from ET displacement from the ET_A receptor or from reduced ET clearance via ET_B receptors, as darusentan may loose some specificity at higher doses. Darusentan also dose-dependently increased CI and decreased mean arterial pressure and systemic vascular resistance. Pulmonary vascular resistance, PCWP, MPAP and right atrial pressure also decreased significantly. Neurohormonal systems were not affected following ET_A blockade [101]. Therefore, in heart failure patients, acute administration of an ET_A receptor-selective antagonist dose-dependently improves hemodynamics without affecting other neurohormonal systems.

Even more marked and persistent hemodynamic improvement was subsequently reported in the Heart Failure ET_A Receptor Blockade Trial, which investigated the effects of darusentan in heart failure patients [110]. Double-blind treatment with placebo or darusentan (30, 100 or 300 mg/day) in addition to standard therapy for 3 weeks was randomly assigned in 157 patients with heart failure (present or recent NYHA class III of at least 3 months duration). Darusentan dose-dependently increased plasma ET-1 levels. However, it led to reductions in pulmonary vascular resistance versus placebo – the greatest reduction being observed in the 300 mg darusentan group. A similar trend was seen with CI. At variance with results from the acute study, darusentan increased CI, significantly versus placebo, following 3 weeks of treatment [110]. Thus, the effects of darusentan are superior when given chronically compared with acutely. The effects of acute and chronic darusentan treatment on PCWP were less pronounced. Importantly, there were no significant alterations in catecholamine levels, plasma natriuretic peptide levels and/or cGMP, and heart rate following darusentan treatment. Thus, selective ET_A receptor blockade for 3 weeks improves hemodynamics in an heart failure population in addition to current standard therapy, implying that ERAs may have a role in heart failure in addition to currently prescribed drugs.

6.8.3
Selective or Nonselective ERAs in Heart Failure?

The blockade of ET_B receptors can impair the pulmonary clearance of ET-1 and reduce NO-induced vasodilatation. Moreover, ET_B receptor deficiency was associated with hypertension in mice and selective ET_B ERAs were found to cause peripheral vasoconstriction in healthy subjects, suggesting that in heart failure, it may be more appropriate to use ET_A-selective antagonists [111, 112]. However, ET_B receptor can be expressed in vascular smooth muscle cells, where they mediate vasoconstriction, under disease conditions. Thus, the short-term effects of a nonselective ERA, bosentan, were examined in 36 men with NYHA class III heart failure who received either bosentan or placebo over 2 weeks on top of a treatment with diuretics, digoxin and ACE inhibitors [113]. Compared to placebo, bosentan led to significant decreases in mean arterial pressure, MPAP, PCWP and right arterial pressure on day 1. Cardiac output increased, but HR remained unchanged. Following 2 weeks treatment, CI further increased, and systemic and pulmonary vascular resistances further decreased compared with day 1. Marked bosentan-related increases in plasma ET-1 levels were observed. Hence, nonselective antagonists also appear to exert beneficial hemodynamic effects in heart failure, indicating that therapeutic effects can be attained provided that the ET_A receptors are blocked, irrespective of concomitant ET_B receptor blockade.

6.9
Long-Term Effects of ERAs in Heart Failure

Experimental studies have demonstrated beneficial effects of ET receptor blockade in heart failure on left ventricular dysfunction, prevention of ventricular remodeling and on survival; thus, the short-term success with ERAs in heart failure prompted the initiation of studies designed to determine their long-term effects [114]. The Endothelin A Receptor Antagonist Trial in Heart Failure (EARTH) was designed to determine the long-term effects of darusentan on left ventricular remodeling and clinical outcomes in heart failure patients [115]. In this trial, 642 heart failure patients were assigned to darusentan treatment at 10, 25, 50, 100 or 300 mg daily or placebo for 24 weeks in addition to standard therapy, in a randomized, double-blind fashion. The primary endpoint was change in left ventricular end-systolic volume at 24 weeks from baseline, measured by mitral regurgitation. Although well tolerated, darusentan treatment did not significantly changed left ventricular end-systolic volume compared with placebo at any dose. Therefore, selective ET_A receptor blockade does not improve cardiac remodeling or outcomes in heart failure patients who were already receiving an ACE inhibitor, β-blocker or aldosterone antagonist [115]. Reasons for these negative findings remain unknown.

Further human studies also failed to demonstrate any benefits of ERAs in heart failure. In the still unpublished Endothelin Antagonist Bosentan for Lowering Cardiac Events in Heart Failure (ENABLE) study, the long-term effects of ET receptor blockade in heart failure patients were evaluated by using low-dose bosentan in patients with severe heart failure (left ventricular ejection fraction below 35%, NYHA class IIIb–IV) [116]. In total, 1613 patients were randomized to either bosentan (62.5 mg b.i.d. later up titrated to 125 mg b.i.d.) or placebo. The primary endpoint of all-cause mortality or hospitalization for heart failure was reached in 321/808 patients given placebo and 312/805 receiving bosentan. Bosentan did not affect the primary endpoint and in fact was associated with an early risk of worsening heart failure requiring hospitalization, due to fluid retention [116]. Thus, this study again could not fulfill expectations on the benefit of ERA on mortality and remodeling in human heart failure, despite significant improvement in hemodynamics in these patients.

In another still unpublished study, the Enrasentan Cooperative Randomized Evaluation (ENCOR) study, 419 patients with heart failure NYHA class II or III were randomized to enrasentan or placebo over a 9-month period, on top on conventional heart failure therapy [87]. Enrasentan did not show any benefit in a composite endpoint made of NYHA class, hospitalization rate and global assessment [87].

In summary, it is uncertain as to why the benefits of ERAs observed in experimental heart failure cannot be reproduced in human heart failure. However, not only is there the issue of significant species variation, but the additional difficulty of replicating heart failure in animals. Animal models of heart failure do not reproduce the exact course of the disease, which can follow a heterogeneous path in humans, and occurs over many decades through the interaction of physiological systems and organs. Furthermore, meta-analysis of experimental studies have revealed that ERAs have no overall significant benefit on mortality in animal models of experimental heart failure [117]. Clearly, the long-term effects of ET receptor blockade in heart failure remain unresolved and merit further investigation.

6.10
Conclusions

ET-1, the most potent vasoconstrictor known so far, plays a relevant physiological role throughout life, first in growth development, and then in the regulation of the vascular tone, extracellular matrix deposition, aldosterone synthesis and inflammation, thereby controlling most systems in the organism. ET-1 also plays an important mechanistic role in the development of diseases, as in PAH. Hence, the ERA bosentan has been approved for treatment of PAH. Compelling evidence for a role of ET-1 in systemic arterial hypertension, heart failure and renal insufficiency exists. Nonetheless, despite the promising results obtained from the experimental studies, no benefits of ERAs on mortality were found in long-term studies in patients with heart failure. Moreover, it is controversial whether the protective action of ERAs in the kidney of hypertensives depends on their specific mechanism of action or, rather, on the antihypertensive effects.

Thus, many issues concerning the role of ET-1 and use of ERAs in kidney and cardiovascular diseases still remain unclear. As the enhanced production of ET-1 may be one major, but not the only, abnormality occurring in these diseases and, furthermore, the pathways controlling the vascular tone, growth, salt and water balance are redundantly regulated, it is conceivable that treatment with ERAs alone could not completely reverse the pathologies associated with these conditions. Hence, further research is ongoing to better define the place of ERAs in the treatment of cardiovascular and some noncardiovascular human diseases.

References

1 Saida, K., Mitsui, Y. and Ishida, N. (1989) A novel peptide, vasoactive intestinal contractor, of a new (endothelin) peptide family. *Journal of Biological Chemistry*, **264**, 14613–16.

2 Masaki, T., Yanagisawa, M. and Goto, K. (1992) Physiology and pharmacology of endothelins. *Medicinal Research Reviews*, **12**, 391–421.

3 Rossi, G.P. and Pessina, A.C. (2006) Endothelins: molecular mechanisms in hypertension and cardiovascular diseases, in *Textbook of Molecular Mechanisms of Hypertension* (ed. E.L. Schiffrin), Taylor & Francis, London, pp. 141–53.

4 Wagner, O.F., Christ, G., Wojta, J., Vierhapper, H., Parzer, S., Nowotny, P.J. et al. (1992) Polar secretion of endothelin-1 by cultured endothelial cells. *Journal of Biological Chemistry*, **267**, 16066–8.

5 Rossi, G.P., Colonna, S., Pavan, E., Albertin, G., Della, R.F., Gerosa, G. et al. (1999) Endothelin-1 and its mRNA in the wall layers of human arteries *ex vivo*. *Circulation*, **99**, 1147–55.

6 Rubanyi, G.M. and Polokoff, M.A. (1994) Endothelins: molecular biology, biochemistry, pharmacology, physiology, and pathophysiology. *Pharmacological Reviews*, **46**, 325–415.

7 Yanagisawa, H., Yanagisawa, M., Kapur, R.P., Richardson, J.A., Williams, S.C., Clouthier, D.E. et al. (1998) Dual genetic pathways of endothelin-mediated intercellular signaling revealed by targeted disruption of endothelin converting enzyme-1 gene. *Development*, **125**, 825–36.

8 Urata, H., Kinoshita, A., Misono, K.S., Bumpus, F.M. and Husain, A. (1990) Identification of a highly specific chymase as the major angiotensin II-forming enzyme in the human heart. *Journal of Biological Chemistry*, **265**, 22348–57.

9 Rossi, G.P., Sacchetto, A., Cesari, M. and Pessina, A.C. (1999) Interactions between endothelin-1 and the renin–angiotensin–aldosterone system. *Cardiovascular Research*, **43**, 300–7.

10 Rubin, L.J. and Roux, S. (2002) Bosentan: a dual endothelin receptor antagonist. *Expert Opinion on Investigational Drugs*, **11**, 991–1002.

11 Rossi, G.P., Albertin, G., Belloni, A., Zanin, L., Biasolo, M.A., Prayer Galetti, T. et al. (1994) Gene expression, localization, and characterization of endothelin A and B receptors in the human adrenal cortex. *Journal of Clinical Investigation*, **94**, 1226–34.

12 Dagassan, P.H., Breu, V., Clozel, M., Kunzli, A., Vogt, P., Turina, M. et al. (1996) Up-regulation of endothelin-B receptors in atherosclerotic human coronary arteries. *Journal of Cardiovascular Pharmacology*, **27**, 147–53.

13 Teerlink, J.R., Breu, V., Sprecher, U., Clozel, M. and Clozel, J.P. (1994) Potent vasoconstriction mediated by endothelin ETB receptors in canine coronary arteries. *Circulation Research*, **74**, 105–14.

14 Clouthier, D.E., Williams, S.C., Yanagisawa, H., Wieduwilt, M., Richardson, J.A. and Yanagisawa, M. (2000) Signaling pathways crucial for craniofacial development revealed by endothelin-A receptor-deficient mice. *Developmental Biology*, **217**, 10–24.

15 Kurihara, Y., Kurihara, H., Suzuki, H., Kodama, T., Maemura, K., Nagai, R. et al. (1994) Elevated blood pressure and craniofacial abnormalities in mice deficient in endothelin-1. *Nature*, **368**, 703–10.

16 Macarthur, H., Warner, T.D., Wood, E.G., Corder, R. and Vane, J.R. (1994) Endothelin-1 release from endothelial cells in culture is elevated both acutely and chronically by short periods of mechanical stretch. *Biochemical and Biophysical Research Communications*, **200**, 395–400.

17 Malek, A. and Izumo, S. (1992) Physiological fluid shear stress causes downregulation of endothelin-1 mRNA in bovine aortic endothelium. *American Journal of Physiology*, **263**, C389–96.

18 Wesson, D.E., Simoni, J. and Green, D.F. (1998) Reduced extracellular pH increases endothelin-1 secretion by human renal microvascular endothelial cells. *Journal of Clinical Investigation*, **101**, 578–83.

19 Boulanger, C.M., Tanner, F.C., Bea, M.L., Hahn, A.W., Werner, A. and Luscher, T.F. (1992) Oxidized low density lipoproteins induce mRNA expression and release of endothelin from human and porcine endothelium. *Circulation Research*, **70**, 1191–7.

20 Yamauchi, T., Ohnaka, K., Takayanagi, R., Umeda, F. and Nawata, H. (1990) Enhanced secretion of endothelin-1 by elevated glucose levels from cultured bovine aortic endothelial cells. *FEBS Letters*, **267**, 16–18.

21 Akishita, M., Ouchi, Y., Miyoshi, H., Orimo, A., Kozaki, K., Eto, M. et al. (1996) Estrogen inhibits endothelin-1 production and c-*fos* gene expression in rat aorta. *Atherosclerosis*, **125**, 27–38.

22 Barton, M., Carmona, R., Morawietz, H., d'Uscio, L.V., Goettsch, W., Hillen, H. et al. (2000) Obesity is associated with tissue-specific activation of renal angiotensin-converting enzyme *in vivo*: evidence for a regulatory role of endothelin. *Hypertension*, **35**, 329–36.

23 Boulanger, C. and Luscher, T.F. (1990) Release of endothelin from the porcine aorta. Inhibition by endothelium-derived nitric oxide. *Journal of Clinical Investigation*, **85**, 587–90.

24 Ito, H., Hirata, Y., Adachi, S., Tanaka, M., Tsujino, M., Koike, A. et al. (1993) Endothelin-1 is an autocrine/paracrine factor in the mechanism of angiotensin II-induced hypertrophy in cultured rat cardiomyocytes. *Journal of Clinical Investigation*, **92**, 398–403.

25 Barton, M., Shaw, S., Uscio, L.V., Moreau, P. and Luscher, T.F. (1997) Angiotensin II increases vascular and renal endothelin-1 and functional endothelin converting enzyme activity *in vivo*: role of ETA receptors for endothelin regulation. *Biochemical and Biophysical Research Communications*, **238**, 861–5.

26 Imai, T., Hirata, Y., Emori, T., Yanagisawa, M., Masaki, T. and Marumo, F. (1992) Induction of endothelin-1 gene by angiotensin and vasopressin in endothelial cells. *Hypertension*, **19**, 753–7.

27 Boulanger, C.M. and Luscher, T.F. (1991) Hirudin and nitrates inhibit the thrombin-induced release of endothelin from the intact porcine aorta. *Circulation Research*, **68**, 1768–72.

28 Matsuura, A., Yamochi, W., Hirata, K., Kawashima, S. and Yokoyama, M. (1998) Stimulatory interaction between vascular endothelial growth factor and endothelin-1 on each gene expression. *Hypertension*, **32**, 89–95.

29 Bodin, P., Milner, P., Marshall, J. and Burnstock, G. (1995) Cytokines suppress the shear stress-stimulated release of vasoactive peptides from human endothelial cells. *Peptides*, **16**, 1433–8.

30 Corder, R., Carrier, M., Khan, N., Klemm, P. and Vane, J.R. (1995) Cytokine regulation of endothelin-1 release from bovine aortic endothelial cells. *Journal of Cardiovascular Pharmacology*, **26** (Suppl. 3), S56–8.

31 Stewart, D.J., Cernacek, P., Mohamed, F., Blais, D., Cianflone, K. and Monge, J.C. (1994) Role of cyclic nucleotides in the regulation of endothelin-1 production by human endothelial cells. *American Journal of Physiology*, **266**, H944–51.

32 Fujisaki, H., Ito, H., Hirata, Y., Tanaka, M., Hata, M., Lin, M. et al. (1995) Natriuretic peptides inhibit angiotensin II-induced proliferation of rat cardiac fibroblasts by blocking endothelin-1 gene expression. *Journal of Clinical Investigation*, **96**, 1059–65.

33 Alberts, G.F., Peifley, K.A., Johns, A., Kleha, J.F. and Winkles, J.A. (1994) Constitutive endothelin-1 overexpression promotes smooth muscle cell proliferation via an external autocrine loop. *Journal of Biological Chemistry*, **269**, 10112–18.

34 Agui, T., Xin, X., Cai, Y., Sakai, T. and Matsumoto, K. (1994) Stimulation of interleukin-6 production by endothelin in rat bone marrow-derived stromal cells. *Blood*, **84**, 2531–8.

35 Yang, Z., Krasnici, N. and Luscher, T.F. (1999) Endothelin-1 potentiates human smooth muscle cell growth to PDGF:

effects of ETA and ETB receptor blockade. *Circulation*, **100**, 5–8.

36 Rossi, G.P., Seccia, T.M., Albertin, G. and Pessina, A.C. (2000) Measurement of endothelin: clinical and research use. *Annals of Clinical Biochemistry*, **37**, 608–26.

37 Stewart, D.J., Levy, R.D., Cernacek, P. and Langleben, D. (1991) Increased plasma endothelin-1 in pulmonary hypertension: marker or mediator of disease?. *Annals of Internal Medicine*, **114**, 464–9.

38 Vancheeswaran, R., Magoulas, T., Efrat, G., Wheeler Jones, C., Olsen, I., Penny, R. et al. (1994) Circulating endothelin-1 levels in systemic sclerosis subsets – a marker of fibrosis or vascular dysfunction? *Journal of Rheumatology*, **21**, 1838–44.

39 Yoshibayashi, M., Nishioka, K., Nakao, K., Saito, Y., Matsumura, M., Ueda, T. et al. (1991) Plasma endothelin concentrations in patients with pulmonary hypertension associated with congenital heart defects. Evidence for increased production of endothelin in pulmonary circulation. *Circulation*, **84**, 2280–5.

40 Papassotiriou, J., Morgenthaler, N.G., Struck, J., Alonso, C. and Bergmann, A. (2006) Immunoluminometric assay for measurement of the C-terminal endothelin-1 precursor fragment in human plasma. *Clinical Chemistry*, **52**, 1144–51.

41 Dupuis, J., Cernacek, P., Tardif, J.C., Stewart, D.J., Gosselin, G., Dyrda, I. et al. (1998) Reduced pulmonary clearance of endothelin-1 in pulmonary hypertension. *American Heart Journal*, **135**, 614–20.

42 Giaid, A., Michel, R.P., Stewart, D.J., Sheppard, M., Corrin, B. and Hamid, Q. (1993) Expression of endothelin-1 in lungs of patients with cryptogenic fibrosing alveolitis. *Lancet*, **341**, 1550–4.

43 Williamson, D.J., Wallman, L.L., Jones, R., Keogh, A.M., Scroope, F., Penny, R. et al. (2000) Hemodynamic effects of Bosentan, an endothelin receptor antagonist, in patients with pulmonary hypertension. *Circulation*, **102**, 411–18.

44 Chen, S.J., Chen, Y.F., Meng, Q.C., Durand, J., Dicarlo, V.S. and Oparil, S. (1995) Endothelin-receptor antagonist bosentan prevents and reverses hypoxic pulmonary hypertension in rats. *Journal of Applied Physiology*, **79**, 2122–31.

45 Rubin, L.J., Badesch, D.B., Barst, R.J., Galie, N., Black, C.M., Keogh, A. et al. (2002) Bosentan therapy for pulmonary arterial hypertension. *New England Journal of Medicine*, **346**, 896–903.

46 Tharaux, P.L., Chatziantoniou, C., Casellas, D., Fouassier, L., Ardaillou, R. and Dussaule, J.C. (1999) Vascular endothelin-1 gene expression and synthesis and effect on renal type I collagen synthesis and nephroangiosclerosis during nitric oxide synthase inhibition in rats. *Circulation*, **99**, 2185–91.

47 Orth, S.R., Esslinger, J.P., Amann, K., Schwarz, U., Raschack, M. and Ritz, E. (1998) Nephroprotection of an ETA-receptor blocker (LU 135252) in salt-loaded uninephrectomized stroke-prone spontaneously hypertensive rats. *Hypertension*, **31**, 995–1001.

48 McLaughlin, V.V., Oudiz, R.J., Frost, A., Tapson, V.F., Murali, S., Channick, R.N. et al. (2006) Randomized study of adding inhaled iloprost to existing bosentan in pulmonary arterial hypertension. *American Journal of Respiratory and Critical Care Medicine*, **174**, 1257–63.

49 Mathai, S.C., Girgis, R.E., Fisher, M.R., Champion, H.C., Housten-Harris, T., Zaiman, A. et al. (2007) Addition of sildenafil to bosentan monotherapy in pulmonary arterial hypertension. *European Respiratory Journal*, **29**, 469–75.

50 Schiffrin, E.L. (1999) State-of-the-art lecture. Role of endothelin-1 in hypertension. *Hypertension*, **34**, 876–81.

51 Lariviere, R., Day, R. and Schiffrin, E.L. (1993) Increased expression of endothelin-1 gene in blood vessels of deoxycorticosterone acetate-salt hypertensive rats. *Hypertension*, **21**, 916–20.

52 Rossi, G.P., Sacchetto, A., Rizzoni, D., Bova, S., Porteri, E., Mazzocchi, G. et al. (2000) Blockade of angiotensin II type 1 receptor and not of endothelin receptor prevents hypertension and cardiovascular

disease in transgenic TGR(mRen2)27 rats via adrenocortical steroid-independent mechanisms. *Arteriosclerosis Thrombosis and Vascular Biology*, **20**, 949–56.

53 Barton, M., d'Uscio, L.V., Shaw, S., Meyer, P., Moreau, P. and Luscher, T.F. (1998) ET_A receptor blockade prevents increased tissue endothelin-1, vascular hypertrophy, and endothelial dysfunction in salt-sensitive hypertension. *Hypertension*, **31**, 499–504.

54 Seccia, T.M., Belloni, A.S., Kreutz, R., Paul, M., Nussdorfer, G.G., Pessina, A.C. et al. (2003) Cardiac fibrosis occurs early and involves endothelin and AT-1 receptors in hypertension due to endogenous angiotensin II. *Journal of the American College of Cardiology*, **41**, 666–73.

55 Taddei, S., Virdis, A., Ghiadoni, L., Sudano, I., Magagna, A. and Salvetti, A. (2001) Role of endothelin in the control of peripheral vascular tone in human hypertension. *Heart Failure Reviews*, **6**, 277–85.

56 Rossi, G.P., Ganzaroli, C., Cesari, M., Maresca, A., Plebani, M., Nussdorfer, G.G. et al. (2003) Endothelin receptor blockade lowers plasma aldosterone levels via different mechanisms in primary aldosteronism and high-to-normal renin hypertension. *Cardiovascular Research*, **57**, 277–83.

57 Cardillo, C., Kilcoyne, C.M., Waclawiw, M., Cannon, R.O. and Panza, J.A. (1999) Role of endothelin in the increased vascular tone of patients with essential hypertension. *Hypertension*, **33**, 753–8.

58 Taddei, S., Virdis, A., Ghiadoni, L., Sudano, I., Notari, M. and Salvetti, A. (1999) Vasoconstriction to endogenous endothelin-1 is increased in the peripheral circulation of patients with essential hypertension. *Circulation*, **100**, 1680–3.

59 Krum, H., Viskoper, R.J., Lacourciere, Y. Budde, M. and Charlon, V. for the Bosentan Hypertension Investigators. (1998) The effect of an endothelin receptor antagonist, Bosentan, on blood pressure in patients with essential hypertension. *New England Journal of Medicine*, **338**, 784–90.

60 Nakov, R., Pfarr, E. and Eberle, S. (2002) Darusentan: an effective endothelinA receptor antagonist for treatment of hypertension. *American Journal of Hypertension*, **15**, 583–9.

61 Rich, S. and McLaughlin, V.V. (2003) Endothelin receptor blockers in cardiovascular disease. *Circulation*, **108**, 2184–90.

62 Quan, S.F. and Gersh, B.J. (2004) Cardiovascular consequences of sleep-disordered breathing: past, present and future: report of a workshop from the National Center on Sleep Disorders Research and the National Heart, Lung, and Blood Institute. *Circulation*, **109**, 951–7.

63 Phillips, B.G., Narkiewicz, K., Pesek, C.A., Haynes, W.G., Dyken, M.E. and Somers, V.K. (1999) Effects of obstructive sleep apnea on endothelin-1 and blood pressure. *Journal of Hypertension*, **17**, 61–6.

64 Rossing, P. (2005) The changing epidemiology of diabetic microangiopathy in type 1 diabetes. *Diabetologia*, **48**, 1439–44.

65 de Zeeuw, D., Remuzzi, G., Parving, H.H., Keane, W.F., Zhang, Z., Shahinfar, S. et al. (2004) Proteinuria, a target for renoprotection in patients with type 2 diabetic nephropathy: lessons from RENAAL. *Kidney International*, **65**, 2309–20.

66 Jandeleit-Dahm, K., Allen, T.J., Youssef, S., Gilbert, R.E. and Cooper, M.E. (2000) Is there a role for endothelin antagonists in diabetic renal disease? *Diabetes Obesity and Metabolism*, **2**, 15–24.

67 Haynes, W.G. and Webb, D.J. (1998) Endothelin as a regulator of cardio-vascular function in health and disease. *Journal of Hypertension*, **16**, 1081–98.

68 Pollock, D.M. and Opgenorth, T.J. (1993) Evidence for endothelin-induced renal vasoconstriction independent of ETA receptor activation. *American Journal of Physiology*, **264**, R222–6.

69 Kohan, D.E. (1997) Endothelins in the normal and diseased kidney. *American Journal of Kidney Diseases*, **29**, 2–26.

70 Dhaun, N., Goddard, J. and Webb, D.J. (2006) The endothelin system and its antagonism in chronic kidney disease. *Journal of the American Society of Nephrology*, **17**, 943–55.

71 Karet, F.E. and Davenport, A.P. (1996) Localization of endothelin peptides in human kidney. *Kidney International*, **49**, 382–7.

72 Koyama, H., Tabata, T., Nishzawa, Y., Inoue, T., Morii, H. and Yamaji, T. (1989) Plasma endothelin levels in patients with uraemia. *Lancet*, **1**, 991–2.

73 Bruno, C.M., Meli, S., Marcinno, M., Ierna, D., Sciacca, C. and Neri, S. (2002) Plasma endothelin-1 levels and albumin excretion rate in normotensive, microalbuminuric type 2 diabetic patients. *Journal of Biological Regulators and Homeostatic Agents*, **16**, 114–17.

74 Ohta, K., Hirata, Y., Shichiri, M., Kanno, K., Emori, T., Tomita, K. *et al.* (1991) Urinary excretion of endothelin-1 in normal subjects and patients with renal disease. *Kidney International*, **39**, 307–11.

75 Pollock, D.M. (2001) Endothelin antagonists in the treatment of renal failure. *Current Opinion in Investigational Drugs*, **2**, 513–20.

76 Knoll, T., Schaub, M., Birck, R., Braun, C., Juenemann, K.P. and Rohmeiss, P. (2000) The renoprotective potential of endothelin receptor antagonists. *Expert Opinion on Investigational Drugs*, **9**, 1041–52.

77 Tharaux, P.L., Chatziantoniou, C., Casellas, D., Fouassier, L., Ardaillou, R. and Dussaule, J.C. (1999) Vascular endothelin-1 gene expression and synthesis and effect on renal type I collagen synthesis and nephro-angiosclerosis during nitric oxide synthase inhibition in rats. *Circulation*, **99**, 2185–91.

78 Ding, S.S., Qiu, C., Hess, P., Xi, J.F., Zheng, N. and Clozel, M. (2003) Chronic endothelin receptor blockade prevents both early hyperfiltration and late overt diabetic nephropathy in the rat. *Journal of Cardiovascular Pharmacology*, **42**, 48–54.

79 Cosenzi, A., Bernobich, E., Trevisan, R., Milutinovic, N., Borri, A. and Bellini, G. (2003) Nephroprotective effect of bosentan in diabetic rats. *Journal of Cardiovascular Pharmacology*, **42**, 752–6.

80 Sugimoto, K., Tsuruoka, S. and Fujimura, A. (2002) Renal protective effect of YM598, a selective endothelin ET_A receptor antagonist, against diabetic nephropathy in OLETF rats. *European Journal of Pharmacology*, **450**, 183–9.

81 Dhein, S., Hochreuther, S., Aus Dem, S.C., Bollig, K., Hufnagel, C. and Raschack, M. (2000) Long-term effects of the endothelin$_A$ receptor antagonist LU 135252 and the angiotensin-converting enzyme inhibitor trandolapril on diabetic angiopathy and nephropathy in a chronic type I diabetes mellitus rat model. *Journal of Pharmacology and Experimental Therapeutics*, **293**, 351–9.

82 Gross, M.L., El Shakmak, A., Szabo, A., Koch, A., Kuhlmann, A., Munter, K. *et al.* (2003) ACE-inhibitors but not endothelin receptor blockers prevent podocyte loss in early diabetic nephropathy. *Diabetologia*, **46**, 856–68.

83 Ortmann, J., Amann, K., Brandes, R.P., Kretzler, M., Munter, K., Parekh, N. *et al.* (2004) Role of podocytes for reversal of glomerulosclerosis and proteinuria in the aging kidney after endothelin inhibition. *Hypertension*, **44**, 974–81.

84 Morigi, M., Buelli, S., Angioletti, S., Zanchi, C., Longaretti, L., Zoja, C. *et al.* (2005) In response to protein load podocytes reorganize cytoskeleton and modulate endothelin-1 gene: implication for permselective dysfunction of chronic nephropathies. *American Journal of Pathology*, **166**, 1309–20.

85 Goddard, J., Johnston, N.R., Hand, M.F., Cumming, A.D., Rabelink, T.J., Rankin, A.J. *et al.* (2004) Endothelin-A receptor antagonism reduces blood pressure and increases renal blood flow in hypertensive patients with chronic renal failure: a comparison of selective and combined endothelin receptor blockade. *Circulation*, **109**, 1186–93.

86 Dieterle, W., Mann, J. and Kutz, K. (2005) Multiple-dose pharmacokinetics, pharmacodynamics and tolerability of the oral ET_A endothelin-receptor antagonist SPP301 in man. *International Journal of*

Clinical Pharmacology and Therapeutics, **43**, 178–86.

87 Battistini, B., Berthiaume, N., Kelland, N.F., Webb, D.J. and Kohan, D.E. (2006) Profile of past and current clinical trials involving endothelin receptor antagonists: the novel "-sentan" class of drug. *Experimental Biology and Medicine of Maywood*, **231**, 653–95.

88 Hurlimann, D., Enseleit, F., Noll, G., Luscher, T.F. and Ruschitzka, F. (2002) Endothelin antagonists and heart failure. *Current Hypertension Reports*, **4**, 85–92.

89 Spieker, L.E., Noll, G., Ruschitzka, F.T. and Luscher, T.F. (2001) Endothelin receptor antagonists in congestive heart failure: a new therapeutic principle for the future? *Journal of the American College of Cardiology*, **37**, 1493–505.

90 Yang, Z., Bauer, E., von Segesser, L., Stulz, P., Turina, M. and Luscher, T.F. (1990) Different mobilization of calcium in endothelin-1-induced contractions in human arteries and veins: effects of calcium antagonists. *Journal of Cardiovascular Pharmacology*, **16**, 654–60.

91 Yang, Z.H., Richard, V., von Segesser, L., Bauer, E., Stulz, P., Turina, M. et al. (1990) Threshold concentrations of endothelin-1 potentiate contractions to norepinephrine and serotonin in human arteries. A new mechanism of vasospasm? *Circulation*, **82**, 188–95.

92 Knuepfer, M.M., Han, S.P., Trapani, A.J., Fok, K.F. and Westfall, T.C. (1989) Regional hemodynamic and baroreflex effects of endothelin in rats. *American Journal of Physiology*, **257**, H918–26.

93 Sorensen, S.S., Madsen, J.K. and Pedersen, E.B. (1994) Systemic and renal effect of intravenous infusion of endothelin-1 in healthy human volunteers. *American Journal of Physiology*, **266**, F411–18.

94 Sakai, S., Yorikane, R., Miyauchi, T., Sakurai, T., Kasuya, Y., Yamaguchi, I. et al. (1995) Altered production of endothelin-1 in the hypertrophied rat heart. *Journal of Cardiovascular Pharmacology*, **26** (Suppl. 3), S452–5.

95 Tonnessen, T., Christensen, G., Oie, E., Holt, E., Kjekshus, H., Smiseth, O.A. et al. (1997) Increased cardiac expression of endothelin-1 mRNA in ischemic heart failure in rats. *Cardiovascular Research*, **33**, 601–10.

96 Sakai, S., Miyauchi, T., Sakurai, T., Yamaguchi, I., Kobayashi, M., Goto, K. et al. (1996) Pulmonary hypertension caused by congestive heart failure is ameliorated by long-term application of an endothelin receptor antagonist. Increased expression of endothelin-1 messenger ribonucleic acid and endothelin-1-like immunoreactivity in the lung in congestive heart failure in rats. *Journal of the American College of Cardiology*, **28**, 1580–8.

97 Huntington, K., Picard, P., Moe, G., Stewart, D.J., Albernaz, A. and Monge, J.C. (1998) Increased cardiac and pulmonary endothelin-1 mRNA expression in canine pacing-induced heart failure. *Journal of Cardiovascular Pharmacology*, **31** (Suppl. 1), S424–6.

98 Picard, P., Smith, P.J., Monge, J.C., Rouleau, J.L., Nguyen, Q.T., Calderone, A. et al. (1998) Coordinated upregulation of the cardiac endothelin system in a rat model of heart failure. *Journal of Cardiovascular Pharmacology*, **31** (Suppl. 1), S294–7.

99 Margulies, K.B., Hildebrand, F.L. Jr, Lerman, A., Perrella, M.A. and Burnett, J.C. Jr (1990) Increased endothelin in experimental heart failure. *Circulation*, **82**, 2226–30.

100 Sakai, S., Miyauchi, T., Sakurai, T., Kasuya, Y., Ihara, M., Yamaguchi, I. et al. (1996) Endogenous endothelin-1 participates in the maintenance of cardiac function in rats with congestive heart failure. Marked increase in endothelin-1 production in the failing heart. *Circulation*, **93**, 1214–22.

101 Spieker, L.E., Mitrovic, V., Noll, G., Pacher, R., Schulze, M.R., Muntwyler, J. et al. (2000) Acute hemodynamic and neurohumoral effects of selective ET_A receptor blockade in patients with congestive heart failure. ET 003. Investigators. *Journal of the American College of Cardiology*, **35**, 1745–52.

102 Fukuchi, M. and Giaid, A. (1998) Expression of endothelin-1 and endothelin-converting enzyme-1 mRNAs and proteins in failing human hearts.

Journal of Cardiovascular Pharmacology, **31** (Suppl. 1), S421–3.

103 Zolk, O., Quattek, J., Sitzler, G., Schrader, T., Nickenig, G., Schnabel, P. et al. (1999) Expression of endothelin-1, endothelin-converting enzyme, and endothelin receptors in chronic heart failure. *Circulation*, **99**, 2118–23.

104 Pacher, R., Stanek, B., Hulsmann, M., Koller-Strametz, J., Berger, R., Schuller, M. et al. (1996) Prognostic impact of big endothelin-1 plasma concentrations compared with invasive hemodynamic evaluation in severe heart failure. *Journal of the American College of Cardiology*, **27**, 633–41.

105 Pacher, R., Bergler Klein, J., Globits, S., Teufelsbauer, H., Schuller, M., Krauter, A. et al. (1993) Plasma big endothelin-1 concentrations in congestive heart failure patients with or without systemic hypertension. *American Journal of Cardiology*, **71**, 1293–9.

106 Pousset, F., Isnard, R., Lechat, P., Kalotka, H., Carayon, A., Maistre, G. et al. (1997) Prognostic value of plasma endothelin-1 in patients with chronic heart failure. *European Heart Journal*, **18**, 254–8.

107 Hulsmann, M., Stanek, B., Frey, B., Sturm, B., Putz, D., Kos, T. et al. (1998) Value of cardiopulmonary exercise testing and big endothelin plasma levels to predict short-term prognosis of patients with chronic heart failure. *Journal of the American College of Cardiology*, **32**, 1695–700.

108 MacCarthy, P.A., Grocott-Mason, R., Prendergast, B.D. and Shah, A.M. (2000) Contrasting inotropic effects of endogenous endothelin in the normal and failing human heart: studies with an intracoronary ET_A receptor antagonist. *Circulation*, **101**, 142–7.

109 Ponicke, K., Vogelsang, M., Heinroth, M., Becker, K., Zolk, O., Bohm, M. et al. (1998) Endothelin receptors in the failing and nonfailing human heart. *Circulation*, **97**, 744–51.

110 Luscher, T.F., Enseleit, F., Pacher, R., Mitrovic, V., Schulze, M.R., Willenbrock, R. et al. (2002) Hemodynamic and neurohumoral effects of selective endothelin A (ET_A) receptor blockade in chronic heart failure: the Heart Failure ET_A Receptor Blockade Trial (HEAT). *Circulation*, **106**, 2666–72.

111 Strachan, F.E., Spratt, J.C., Wilkinson, I.B. and Webb, D.J. (1999) Systemic blockade of the ETB receptor increases peripheral vascular resistance in healthy volunteers *in vivo*. *Hypertension*, **33**, 581–5.

112 Ohuchi, T., Kuwaki, T., Ling, G.Y., deWit, D., Ju, K.H., Onodera, M. et al. (1999) Elevation of blood pressure by genetic and pharmacological disruption of the ETB receptor in mice. *American Journal of Physiology*, **276**, R1071–7.

113 Sutsch, G., Kiowski, W., Yan, X.W., Hunziker, P., Christen, S., Strobel, W. et al. (1998) Short-term oral endothelin-receptor antagonist therapy in conventionally treated patients with symptomatic severe chronic heart failure. *Circulation*, **98**, 2262–8.

114 Yamauchi-Kohno, R., Miyauchi, T., Hoshino, T., Kobayashi, T., Aihara, H., Sakai, S. et al. (1999) Role of endothelin in deterioration of heart failure due to cardiomyopathy in hamsters: increase in endothelin-1 production in the heart and beneficial effect of endothelin-A receptor antagonist on survival and cardiac function. *Circulation*, **99**, 2171–6.

115 Anand, I., McMurray, J., Cohn, J.N., Konstam, M.A., Notter, T., Quitzau, K. et al. (2004) Long-term effects of darusentan on left-ventricular remodelling and clinical outcomes in the EndothelinA Receptor Antagonist Trial in Heart Failure (EARTH): randomised, double-blind, placebo-controlled trial. *Lancet*, **364**, 347–54.

116 Kalra, P.R., Moon, J.C. and Coats, A.J. (2002) Do results of the ENABLE (Endothelin Antagonist Bosentan for Lowering Cardiac Events in Heart Failure) study spell the end for non-selective endothelin antagonism in heart failure?. *International Journal of Cardiology*, **85**, 195–7.

117 Lee, D.S., Nguyen, Q.T., Lapointe, N., Austin, P.C., Ohlsson, A., Tu, J.V. et al. (2003) Meta-analysis of the effects of endothelin receptor blockade on survival in experimental heart failure. *Journal of Cardiac Failure*, **9**, 368–74.

7
Adrenomedullin

István Szokodi and Heikki Ruskoaho

In 1993, Kitamura *et al.* isolated a novel peptide, adrenomedullin (AM), from human pheochromocytoma extracts that induced long-lasting hypotensive effects *in vivo* [1]. Since its discovery, tremendous efforts have been made to reveal the exact biological role of AM. Now, substantial evidence supports the concept that AM is an important regulator of the cardiovascular system in health and disease, and may offer a new therapeutic tool for the treatment of acute and chronic heart failure [2–12].

7.1
Molecular Aspects of AM

7.1.1
Structure and Synthesis of AM

Human AM consists of 52 amino acids with an intramolecular disulfide bridge forming a ring structure of six residues with an amidated tyrosine at the C-terminus [1]. Due to moderate sequence homology with calcitonin/calcitonin gene-related peptide (CGRP) and amylin, AM is classified into the CGRP/amylin peptide family. Compared with the human peptide, rat AM has six substitutions and two deletions, thus consisting of 50 amino acids [13]. Human AM is synthesized as a 185-amino-acid precursor. The cleavage of a 21-amino-acid signal peptide from the N-terminus of this prepro-AM leads to formation of a 164-amino-acid prohormone [14]. The prohormone is further cleaved to liberate a 20-amino-acid proadrenomedullin N-terminal 20-peptide (PAMP), which has been found to elicit biological effects independent of those of AM [15]. The remaining prohormone is finally cleaved between positions 93 and 94 as well as 148 and 149, resulting in the formation of human AM [14] (Figure 7.1).

Cardiovascular Hormone Systems. Edited by Michael Bader
Copyright © 2008 WILEY-VCH Verlag GmbH & Co. KGaA, Weinheim
ISBN: 978-3-527-31920-6

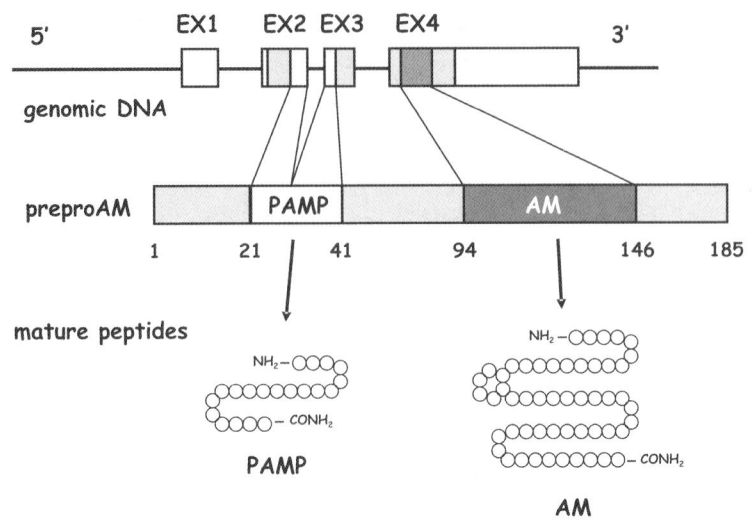

Figure 7.1 Schematic presentations of the AM gene, prepro-AM and biosynthesis of the derivative peptides. EX, exon.

7.1.2
Distribution and Sites of AM Production

Immunoreactive AM and AM mRNA have been found to be ubiquitously expressed in peripheral tissues as well as various regions of the central nervous system. The adrenal gland contains proportionally the largest quantity of AM mRNA and immunoreactive AM in both man and rat, but substantial amounts are also found in the cardiovascular system [13, 14, 16–18]. Generally, AM expression shows close correlation with the degree of tissue vascularization, as both endothelial cells and vascular smooth muscle cells (VSMCs) synthesize and secrete AM in large quantities [19, 20]. Indeed, these cells appear to be the major source of circulating AM in the plasma. Immunohistochemical staining showed the presence of AM in atria, ventricles and the muscular layer of the aorta in dog heart [21]. In neonatal rats, both cardiac myocytes and nonmyocytes secrete AM, the basal rate of secretion being higher in nonmyocytes [22]. AM is present in a variety of embryonic tissues, particularly in the heart [23], and this cardiac expression is developmentally regulated [24].

7.1.3
Regulators of AM Gene Expression

The gene encoding prepro-AM is located at a single locus on chromosome 11 [25], and consists of four exons and three introns. The complete nucleotide sequence of the mature AM protein is encoded by the fourth exon, while the PAMP protein is encoded by portions of the second and third exons. The 5′-flanking region of the AM gene contains TATA, CAAT and GC boxes. There are also several binding sites for activator protein-2, cAMP-regulated enhancer [25], nuclear factor for interleukin-6 and hypoxia inducible factor [24, 26]. A variety of cytokines, growth factors and hormones can modify AM gene expression and AM synthesis. Vasoactive substances, such as angiotensin (Ang) II and endothelin (ET)-1, have been shown to increase AM mRNA levels and secretion of AM from cardiac myocytes [27–30] and VSMCs [31]. Inflammatory cytokines interleukin-1α and -1β, tumor necrosis factor α and β, and a bacterial endotoxin, lipopolysaccharide, induce upregulation of AM mRNA and peptide levels in cultured rat VSMCs and endothelial cells [32, 33]. Mechanical stretch appears to regulate AM secretion and gene expression. In human umbilical vein endothelial cells (HUVECs) shear stress augmented the gene expression of AM [34], whereas in human aortic endothelial cells AM mRNA and peptide levels were downregulated in response to shear stress [35]. In cultured cardiac myocytes Tsuruda et al. reported enhanced gene expression and production of AM in response to static stretch [36]. Notably, Luodonpää et al. showed that cyclic stretch significantly attenuated AM mRNA levels in cultured cardiomyocytes, whereas direct left ventricular wall stretch in isolated, perfused rat heart preparations activated AM gene expression [37]. These results suggest that cardiac nonmyocytes may be important in mediating the upregulation of AM gene expression in the heart in response to mechanical overload, such as during pressure overload *in vivo* [37–41]. Hypoxia is a potent stimulus for AM expression and secretion in cultured cardiac myocytes [24, 42, 43], and in endothelial cells from human coronary artery [44]. Oxidative stress has also been shown to enhance AM production in various cell types including rat VSMCs [45], bovine carotid artery endothelial cells [46] and rat cardiac myocytes [43].

7.1.4
AM Receptors

The presence of specific binding sites for AM has been shown in various tissues including the myocardium [47]. However, the attempts to characterize a specific and functional AM receptor failed until McLatchie et al. described a family of receptor-activity modifying proteins (RAMPs) [48], which indeed represents a new paradigm in G-protein-coupled receptor signaling [12]. RAMPs are single-transmembrane-domain proteins that are associated with calcitonin receptor-like receptor (CLR) to direct its ligand binding specificity and affinity. Coexpression of RAMP1 with CLR results in a functional CGRP receptor, whereas association of CLR with RAMP2 or RAMP3 confers preferential AM binding [48].

7.2
Functional Role of AM in the Heart

7.2.1
AM and Myocardial Contractility

7.2.1.1 Effect of AM on Cardiac Contractility

Considerable evidence suggests that AM acts as an autocrine or paracrine factor in the regulation of cardiac function. Intracoronary infusion of AM (10 pM to 1 nM) induced a potent, dose-dependent positive inotropic effect *in vitro* in isolated perfused rat heart preparations [49–51]. On a molar basis, ET-1 has been considered previously to be the most potent stimulator of cardiac contractility [52]. As AM was active in the subnanomolar range, with an EC_{50} value of 50 pM (similar to that of ET-1), AM appears to be among the most potent endogenous positive inotropic substances yet identified. Moreover, the AM-induced increase in developed tension was approximately 60% of the maximal inotropic response to the β-adrenergic agonist isoproterenol [53] and it was comparable to the effect of ET-1 [51, 53]. Of particular interest, AM increased the contraction force gradually with a mean time of 25–30 min to reach its maximum effect. This unique, slowly developing but sustained inotropic response clearly differs from the classical β-adrenergic effect, which develops rapidly, usually over a matter of seconds, suggesting that AM may modulate the inotropic responsiveness of the heart over a different time frame (e.g. from minutes to hours) [50, 51]. In line with these findings, AM has been reported to increase isometric tension in isolated rat papillary muscles [54]. However, other studies suggested a distinct effect of AM on cardiac contractility. A dual inotropic effect has been observed in isolated adult rat ventricular myocytes, as AM produced an initial increase in cell shortening followed by a negative inotropic effect on prolonged incubation (greater than 1 h) [55, 56]. Moreover, AM reduced cell shortening in isolated rabbit cardiomyocytes [57] and human ventricular myocytes [58]. Finally, others failed to detect any effect of AM on cardiac contractility in rat papillary muscles [59] or isolated trabeculae from nonfailing human hearts [60].

Systemic administration of AM has been shown to induce a marked decrease of peripheral resistance accompanied by a fall in blood pressure, and increases in cardiac output and heart rate in rats [61], sheep [62–64], dogs [65] and humans [66, 67]. However, the pronounced effects of AM on pre- and afterload make assessment of its direct cardiac effects difficult.

7.2.1.2 Signaling Mechanisms

An intimate relation has been suggested between AM-induced biological effects and the increased cAMP levels in various cell types, including VSMCs [68, 69], endothelial cells [70] and glomerular mesangial cells [71]. Similarly, AM has been reported to stimulate cAMP formation in cultured neonatal rat cardiac myocytes [72, 73]. These observations suggest that activation of the adenylate cyclase/cAMP system, which is one of the major pathways for the regulation of cardiac contractility in the mammalian heart [74], may also mediate the cardiac effects of AM. Indeed, AM has been reported to augment cardiac contractility in association with increased cAMP formation at high doses (10–100 nM) and inhibition of protein kinase A (PKA) could abolish the positive inotropic effect of the peptide in rat papillary muscles [54]. Moreover, increases in cell shortening and intracellular Ca^{2+} transients induced by AM (100 nM) were partially attenuated by inhibition of adenylate cyclase and PKA in isolated adult ventricular myocytes [56]. However, substantial evidence indicates that cAMP is not the major second messenger of the inotropic effect of AM at physiologically more relevant concentrations (0.1–1 nM) [50]. First, AM failed to increase left ventricular cAMP content in perfused rat hearts. Second, a PKA inhibitor did not reduce the positive inotropic effect of AM. Finally, the response to AM could not be enhanced in the presence of a phosphodiesterase inhibitor [50]. In contrast to the adenylate cyclase/cAMP/PKA pathway, it has been suggested that activation of protein kinase C (PKC) is involved in the positive inotropic effect of AM, because the AM-induced increase in developed tension was markedly attenuated by pharmacological inhibition of PKC [50]. Activated PKC can phosphorylate a wide spectrum of cellular proteins including the sarcolemmal L-type voltage-dependent Ca^{2+} channels [75, 76]. During depolarization extracellular Ca^{2+} enters the cell through the L-type channels. This inward Ca^{2+} current triggers the release of large amount of Ca^{2+} from the sarcoplasmic reticulum (SR) by opening the SR Ca^{2+} release channels (ryanodine receptors), the phenomenon known as Ca^{2+}-induced Ca^{2+} release. The combination of Ca^{2+} influx and release raises the free intracellular Ca^{2+} concentration, allowing Ca^{2+} to bind to troponin C, which then switches on the contractile machinery [74]. AM has been suggested to influence the intracellular Ca^{2+} homeostasis. In isolated rat hearts an L-type Ca^{2+} channel blocker suppressed the inotropic effect of AM. Superfusion of atrial myocytes with AM in isolated left atrial preparations resulted in a marked prolongation of action potential duration between 10 and −50 mV transmembrane voltage, consistent with an increase in L-type Ca^{2+} channel current during the plateau [50]. Moreover, administration of ryanodine, at a low concentration that increases the open probability of ryanodine receptors, caused a persistent leak of Ca^{2+} from the SR, significantly reducing the positive inotropic effect of AM. Furthermore, thapsigargin, which inhibits SR Ca^{2+} ATPase and depletes intracellular Ca^{2+} stores, attenuated the increase in developed tension produced by AM. These results suggest that AM-induced positive inotropic action may involve enhanced Ca^{2+} influx through L-type Ca^{2+} channels, which could then enhance contractility further by increasing Ca^{2+} release from the SR [50] (Figure 7.2).

7.2.2
AM and Coronary Blood Flow

7.2.2.1 Effect of AM on Coronary Vascular Tone
Initial studies characterized AM as a potent vasodilator, as intravenous administration of AM induced a long-lasting blood pressure-lowering effect accompanied by a decrease in total peripheral resistance [1, 77]. Now, the systemic hypotensive

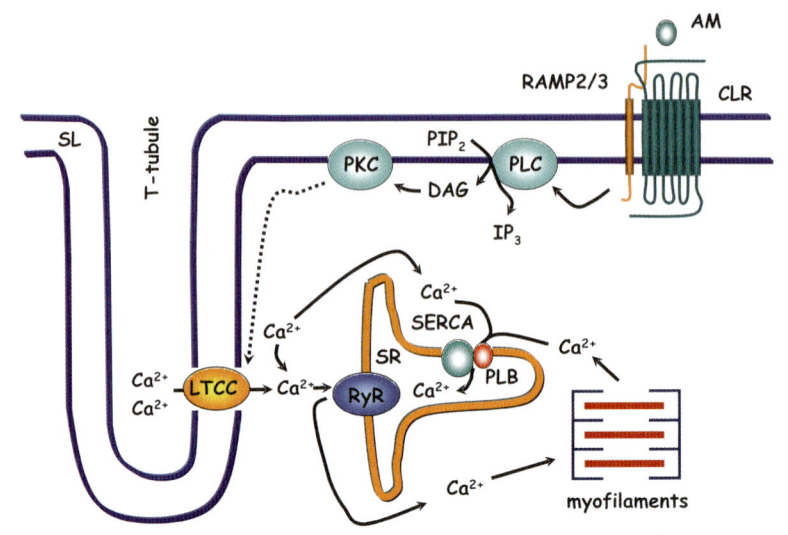

Figure 7.2 Proposed mechanism of action of adrenomedullin regulating cardiac contractility. AM, adrenomedullin; CLR, calcitonin receptor-like receptor; DAG, diacylglycerol; IP$_3$, inositol (1,4,5)-trisphosphate; LTCC, L-type Ca^{2+} channel; PIP$_2$, phosphatidylinositol 4,5-bisphosphate; PLB, phospholamban; PLC, phospholipase C; PKC, protein kinase C; RAMP, receptor-activity modifying protein; RYR, ryanodine-receptor; SERCA, sarcoplasmic reticulum Ca^{2+} ATPase; SL, sarcolemma; SR, sarcoplasmic reticulum.

effect of AM is well established in a range of species including rats [1, 61, 77, 78], rabbits [79, 80], cats [81], sheep [62–64], dogs [65] and humans [66, 67, 82]. Dilatation of resistance vessels has been observed in various regional vascular beds, such as the coronary, renal, cerebral and pulmonary circulation [83–88]. In fact, the coronary relaxant effect of AM has been shown to be comparable to that of CGRP [84] – the most potent vasodilator known so far [89]. The presence of specific AM receptors on VSMCs [68], combined with findings of synthesis and secretion of AM in vascular endothelial cells [19, 33] and smooth muscle cells [20], suggests that AM may act predominantly in a paracrine–autocrine way.

Endothelium-dependent as well as endothelium-independent mechanisms are involved in the vasodilator effect of AM. Nitric oxide (NO) is generally considered to be the primary endothelium-derived relaxing factor regulating vascular resistance [90]. AM has been reported to trigger NO synthesis via Ca^{2+}-dependent activation of endothelial NO synthase in endothelial cells [70]. Furthermore, the stimulation of NO release appears to be the major mechanism mediating the vasorelaxant effect of AM in various rat vascular preparations, including thoracic aorta, renal arteries and pulmonary arteries [91–94]. In contrast, inhibition of NO synthase produced only a slight attenuation of the vasodilator response to AM in hindquarters [83]. In line, AM induced significant vasodilation in the presence of a NO synthase inhibitor in the rat coronary circulation [95]. Denudation of the endothelium did not modulate the relaxant effect of AM in porcine coronary artery rings [85]. In human coronary arterioles, AM elicited vasodilation in part through production of NO [96]. Thus, the mechanism by which AM induces vasorelaxation may differ depending on species, caliber of blood vessels or regions where the vessels were harvested.

Considerable evidence suggests that AM is able to function as a physiological antagonist of the most potent vasoconstrictor, ET-1 [97], in the coronary vasculature. It has been reported that administration of ET-1 *in vivo* at low doses mimicking pathophysiological concentrations induces a coronary constrictor effect predominantly via ET_A receptors [98]. However, simultaneous activation of ET_B receptors, triggering the release of NO from endothelial cells [99, 100], may limit the constrictor effect of ET-1. Accordingly, removal of this negative feedback by the blockade of NO formation has been shown to augment the effect of ET-1 on vascular tone [95, 101, 102]. A cross-talk between ET-1 and AM has been suggested by the demonstration that ET-1 enhances the production of AM in cultured VSMCs [31] and the stimulation of ET_B receptors increases the secretion of AM in vascular endothelial cells [103], suggesting that the peptide (in a fashion similar to NO) may function to buffer the vasoconstrictor effect of ET-1. In agreement with this hypothesis, AM has been shown to markedly attenuate the coronary vasoconstriction induced by ET-1 [95]. The existence of a paracrine–autocrine regulatory loop between ET-1 and AM is further supported by the finding that administration of ET-1 significantly increases the release of AM into the coronary effluent of the perfused rat heart [95]. Taken together, these data indicate an intimate, specific relationship between ET-1 and AM. Since AM could reverse the coronary vasoconstriction induced by ET-1 even under the blockade of NO synthesis in the rat coronary circulation [95], AM may represent an alternative pathway distinct from NO to counteract the pressor response to ET-1 in the rat.

7.2.2.2 Signaling Mechanisms

There is limited information available on the signaling mechanisms of the vasorelaxant effect of AM in the coronary circulation. Detailed analysis of the AM-induced endothelium-dependent response has been provided using rat thoracic aorta [91]. AM-induced vasodilation in aortic rings with intact endothelium was inhibited by pretreatment with phosphatidylinositol 3-kinase (PI3K) inhibitors to the same level as that in endothelium-denuded aorta. AM elicited Akt phosphorylation, which was attenuated by pharmacological inhibition of calmodulin-dependent protein kinase and PI3K. AM-induced vasodilation was diminished by expressing an adenovirus construct containing a dominant-negative Akt mutant in the endothelium. Furthermore, AM-induced cGMP production, which was used as an indicator for NO production, was suppressed by PI3K inhibition or by a dominant-negative Akt mutant. These results suggest that AM may dilate rat aorta by activating PI3K and Akt via the Ca^{2+}/calmodulin-dependent pathway, which leads to increased production of NO through phosphorylation of endothelial NO synthase [91] (Figure 7.3).

The AM receptor is coupled with adenyl cyclase in VSMCs [69] and increased cAMP generation is likely to play an important role in the endothelium-

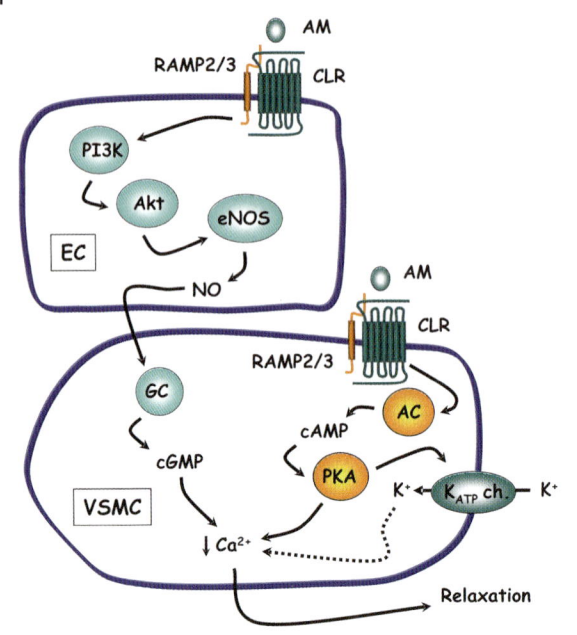

Figure 7.3 Proposed mechanism of endothelium-dependent and -independent vasodilatation to adrenomedullin. AC indicates adenylate cyclase; AM, adrenomedullin; cAMP, adenosine 3'5'-cyclic monophosphate; cGMP, guanosine 3'5'-cyclic monophosphate; CLR, calcitonin receptor-like receptor; EC, endothelial cell; eNOS, endothelial nitric oxide synthase; GC, guanylate cyclase; K_{ATP} ch, ATP-sensitive potassium channel; PI3K, phosphatidylinositol 3-kinase; PKA, protein kinase A; RAMP, receptor-activity modifying protein; VSMC, vascular smooth muscle cell.

independent vasodilator action of AM. Phosphorylation of PKA by cAMP can activate adenosine triphosphate-sensitive potassium (K_{ATP}) channels, which in turn evoke smooth muscle cell hyperpolarization and decrease the open probability of voltage-sensitive Ca^{2+} channels [87]. Indeed, activation of K_{ATP} channels has been implicated in the coronary vasorelaxant effect of AM [85, 104] (Figure 7.3). Furthermore, the functional antagonism between AM and ET-1 in the coronaries [95] may occur at the level of K_{ATP} channels, as inhibition of these channels occupies central role in the vasoconstrictor effect of ET-1 [105].

7.2.3
AM in Myocardial Ischemia

7.2.3.1 AM Production in Myocardial Ischemia
In a rat model of myocardial infarction, left ventricular AM gene expression and peptide synthesis were upregulated in both the infarcted region and the noninfarcted region as early as 6 h after coronary ligation, suggesting a potential role of AM in ischemia/reperfusion injury [106]. Clinical studies have shown that plasma

AM levels increase in the acute phase of myocardial infarction in proportion with clinical severity, reaching a maximum after 2–3 days and returning to baseline after 3 weeks [107–109].

7.2.3.2 Myocardial Cytoprotection by AM

There is mounting data suggesting that AM is a critical factor regulating myocardial tolerance to ischemia/reperfusion injury. The first evidence for the potential cardioprotective effect of AM has been provided using a systemic gene transfer approach in rats. Intravenous injection of adenovirus containing human AM cDNA resulted in the expression of human AM mRNA in various rat tissues including the heart, kidney, adrenal gland, aorta, lung and liver [110, 111]. One week after AM gene delivery, rats were subjected to 30 min of coronary occlusion, followed by 2 h of reperfusion. AM gene transfer significantly reduced the ratio of infarct size to ischemic area at risk and attenuated cardiomyocyte apoptosis [112]. In a different *in vivo* study, rats were exposed to 30 min of coronary artery occlusion followed by 24 h of reperfusion. Animals receiving a 60-min intravenous AM infusion (0.05 µg/kg/min) after coronary ligation displayed reduced myocardial infarct size, attenuated left ventricular end-diastolic pressure and decreased number of apoptotic nuclei in myocytes, whereas left ventricular systolic function tended to improve [113]. Notably, in these models one cannot discriminate between the direct and the systemic effects of AM. Recently, AM has been reported to have direct cardioprotective effects. Local AM gene transfer using a catheter-based technique resulted in highly efficient and specific expression of recombinant human AM in the left ventricle. Five days after AM gene delivery, rats were exposed to a 30 min of regional myocardial ischemia and 2 h of reperfusion. AM gene transfer significantly attenuated ischemia/reperfusion-induced cardiomyocyte apoptosis [114]. Further evidence of the direct infarct-limiting effect of AM has been provided in a rat perfused heart model. Isolated rat hearts underwent 35 min of regional ischemia, followed by 2 h of reperfusion. Administration of AM (1 nM) for 20 min during the early reperfusion period produced a marked limitation of infarct size with improvements in coronary flow and cardiac contractility. In contrast, AM had no effect on irreversible tissue injury during the early ischemic period [115]. The development of mouse strains with targeted deletion of the AM gene [116–119] offered the possibility to study the pathophysiological importance of endogenous AM in the development of myocardial injury. The null mutation of the AM gene ($AM^{-/-}$) is lethal *in utero*. In contrast, heterozygous mice ($AM^{+/-}$), with AM peptide levels 50% of wild-type, survive until adulthood and exhibit only mild changes in the cardiovascular phenotype under baseline conditions [116–119]. When wild-type and $AM^{+/-}$ mice underwent 30 min of regional myocardial ischemia and 2 h of reperfusion, $AM^{+/-}$ mice had higher mortality rates and the surviving animals showed larger infarct size than their wild-type littermates [120]. These findings indicate that innate AM deficiency profoundly affects propensity toward myocardial ischemia/reperfusion injury. Although various endogenous factors can increase the tolerance of the heart to myocardial infarction, it appears that these cardioprotective factors could not counterbalance AM deficiency. Thus, AM may play a mandatory role in determining myocardial resistance to injury. Of note, hearts from AM$^{+/-}$ mice displayed mild left ventricular hypertrophy, and hypertrophy itself can increase the susceptibility to myocardial infarction.

7.2.3.3 Signaling Mechanisms

Juhaszova et al. were the first to propose that the general mechanism of cardioprotection is the convergence of diverse upstream signaling pathways via inhibition of the master switch kinase, glycogen synthase kinase (GSK)-3β, on the end-effector, the permeability transition pore complex, to limit mitochondrial permeability transition induction [121]. Recently, it has been reported that local AM gene transfer can increase the phosphorylation of GSK-3β, thereby reducing its activity in the myocardium upon ischemia/reperfusion insult. Moreover, pharmacological inhibition of GSK-3β and AM gene delivery were equally effective in reducing apoptotic cell death [114]. PI3K and its downstream effector Akt are important regulators of GSK-3β activity [122]. There is a consensus that activation of the prosurvival kinase Akt plays a vital role in AM-induced myocardial protection. AM gene transfer or administration of exogenous AM increased the phosphorylation of Akt in the myocardium exposed to ischemia/reperfusion *in vivo* as well as *in vitro* [112–115, 120]. In agreement, pharmacological inhibition of PI3K abolished AM-induced cardioprotection, such as reduction of infarct size, hemodynamic improvements and inhibition of apoptosis [113]. Moreover, adenoviruses containing a dominant-negative mutant of Akt blocked AM-induced phosphorylation of GSK-3β [114]. Thus, Akt-dependent inhibition of GSK-3β activity appears to be the upstream signaling mechanism regulating the cytoprotective effect of AM.

AM may also affect the central death machinery via the members of the Bcl-2 family. The Bcl-2 family of proteins regulates the intrinsic (or mitochondrial) death pathway by modulating mitochondrial permeability and the release of cytochrome *c* [123]. The antiapoptotic protein Bcl-2 resides in the outer mitochondrial wall and inhibits cytochrome *c* release. The proapoptotic proteins, such as Bad and Bax, are located in the cytosol, but translocate to and insert into the outer mitochondrial membrane and form a proapoptotic complex with Bcl-2 upon stimulation. Phosphorylation of Bad prevents the translocation of this protein to the mitochondria, thereby promoting cell survival [124, 125]. The finding that AM gene delivery significantly increased phospho-Bad and Bcl-2, and decreased Bax in the myocardium [112] suggests that Bcl-2 family members may play a crucial role in the antiapoptotic effect of AM. Although Akt can phosphorylate various Bcl-2 proteins, it is not known currently if the effect of AM is mediated by Akt. Likewise, further studies are required to determine the connection between GSK-3β and Bcl-2 family members.

Ischemia/reperfusion-induced increases in mitochondrial permeability lead to cytochrome *c* release and activation of caspase-9, which cleaves and activates the downstream procaspase-3 resulting in apoptosis [123]. Accordingly, broad-spectrum caspase inhibitors have been shown to reduce infarct size and decrease cardiac dysfunction following ischemia/reperfusion. These effects showed a close correlation with reduction of cardiomyocyte death, suggesting that the benefit of caspase inhibition is attributable to its antiapoptotic property [126, 127]. AM gene transfer has been reported to attenuate caspase-3 activity by an Akt-dependent mechanism. Moreover, caspase-3 activation was reduced by pharmacological inhibition of GSK-3β [114]. Collectively, these data suggest that AM may suppress ischemia/reperfusion-induced apoptosis through an Akt/GSK-3β/caspase-3-dependent pathway.

7.2.4
AM and Angiogenesis

7.2.4.1 Angiogenic Effect of AM

Generation of AM gene knockout mice has revealed that AM is indispensable for vascular morphogenesis during embryonic development. As mentioned earlier, targeted null mutation of the AM gene is lethal *in utero* [116, 117]. The mortality rate among $AM^{-/-}$ embryos was greater than 80% at embryonic day 13.5. Although no obvious abnormality was detected in the vascular structure before embryonic day 10.5, the surviving $AM^{-/-}$ embryos at embryonic day 13.5–14.0 displayed severe hemorrhage, readily observable under the skin and in visceral organs [117]. Electron microscopic examination of $AM^{-/-}$ embryos revealed severe abnormalities in basement membrane structures, which may disrupt the stable adhesion of endothelial cells to the basement membrane resulting in increased permeability of the vasculature [117]. In agreement, a distinct strain of $AM^{-/-}$ mice showed defects in VSMC development [116].

Recent studies have suggested that AM may be an important angiogenic factor in pathological conditions during adulthood [10, 128]. AM has been reported to increase DNA synthesis, migration, and tube formation in HUVECs *in vitro* [129]. The angiogenic activity of AM has been confirmed *in vivo* using the mouse Matrigel plug assay and the chick embryo chorioallantoic membrane assay [129, 130]. Endothelial cells spread and aligned with each other to form branching anastomosing tubes with multicentric junctions that gave rise to a meshwork of capillary-like structures in Matrigel [130]. Using laser Doppler perfusion imaging, AM has been reported to stimulate recovery of blood flow to the affected limb in a mouse hindlimb ischemia model. Accordingly, $AM^{+/-}$ mice showed significantly less blood flow recovery with less collateral capillary development than their wild-type littermates. Importantly, AM administration was able to rescue these abnormalities [131].

AM has been shown to enhance the angiogenic potency of bone marrow-derived mononuclear cell (MNC) transplantation. MNCs consist of a variety of stem and progenitor cells, such as endothelial progenitor cells, and contribute to pathological neovascularization [132–134]. In a rat model of hindlimb ischemia, administration of AM enhanced MNC transplantation-induced angiogenesis. AM promoted the differentiation of MNCs into endothelial cells, and increased the number of α-smooth muscle actin-positive cells involved in the formation of vascular structures. These morphological changes were accompanied by significant improvement in blood perfusion of the ischemic hindlimb [135]. In a rat model of myocardial infarction, the combination of AM infusion and MNC transplantation significantly increased capillary density of the ischemic myocardium compared with MNC transplantation alone. Although MNC apoptosis was frequently observed after transplantation, AM markedly decreased the number of apoptotic cells among the transplanted MNCs [136]. Therefore, it has been suggested that the beneficial effect may be mediated partly by the angiogenic action of AM itself and by its antiapoptotic effect on MNCs [136].

Vascular endothelial growth factor (VEGF) has been considered as the most potent inducer of the angiogenic process [128, 137]. Preclinical data indicate that the vessels resulting from delivery of VEGF are stable for some time, but eventually resemble immature tumor vessels [138] associated with bleeding and microvascular leakage [139, 140]. Clearly, formation of a mature vascular network requires additional mediators acting in concert with VEGF. AM may be a particularly attractive candidate for such a role. AM has been shown to enhance VEGF-induced capillary formation [131, 141]. On the other hand, blocking antibodies to VEGF did not attenuate AM-induced angiogenic-related effects [141]. Thus, AM seems to be an independent mediator of angiogenesis, which can act synergistically with VEGF. Further studies are warranted to clarify whether delivery of VEGF and AM in combination leads to vessels that appear more mature than those formed using VEGF alone.

7.2.4.2 Signaling Mechanisms

Recent studies have suggested that Akt plays a crucial role in the angiogenic effect of AM. Miyashita *et al.* have reported that AM increased the phosphorylation of Akt in HUVECs and this effect was attenuated by inhibition of PKA or PI3K. Similarly, AM-induced re-endothelialization *in vitro* of wounded monolayer of HUVECs and neovascularization *in vivo* in murine gel plugs were also inhibited by the inhibitors for PKA or PI3K [142]. Concurrently, Kim *et al.* showed that AM can phosphorylate Akt, extracellular signal-regulated kinase (ERK) 1/2 and focal adhesion kinase (p125^{FAK}) in HUVECs. No cross-talk was observed between Akt and ERK1/2, whereas the activation of p125^{FAK} appeared to be PI3K/Akt-dependent. Pharmacological inhibition of PI3K or ERK1/2 suppressed AM-induced endothelial tube formation *in vitro* as well as neovessel formation *in vivo* [129]. *In vitro*, AM enhanced smooth muscle cell migration, which was inhibited by a PI3K inhibitor [135]. These findings suggest that AM exerts angiogenic activities through activation of PI3K/Akt/p125^{FAK} and Ras/Raf/MEK/ERK1/2 pathways.

7.2.5
AM in Heart Failure

7.2.5.1 Cardiac AM Production in Heart Failure

It is well established that the levels of circulating AM are increased in response to hemodynamic overload in various experimental models, such as hypertensive

Dahl salt-sensitive rats [143], aortic banding [39], vasopressin-induced hypertension [40], deoxycorticosterone acetate (DOCA)-salt hypertensive rats [144], and in stroke-prone spontaneously hypertensive rats [145]. A number of studies have revealed an increase in plasma AM in experimental heart failure [39, 146, 147]. In agreement with these findings, AM immunoreactivity and gene expression is augmented in the failing left ventricle [106, 147–150]. However, there is no consensus regarding the exact source of increased AM production in the heart. Øie et al. have reported a substantial increase of immunoreactive AM in microvascular endothelium and perivascular interstitial cells in the left ventricular tissue in ischemic heart failure [150]. In contrast, using the same experimental model, Nagaya et al. have shown that intense immunostaining for AM was limited to myocytes in both the infarcted and noninfarcted regions [106].

Plasma concentration of circulating AM has been shown to be increased in patients with congestive heart failure in proportion to the symptomatic and hemodynamic severity of the syndrome [151–160]. Immunohistochemical staining for AM is significantly increased in the failing human ventricular myocardium compared with the normal human ventricle [151]. Moreover, there was a significant step-up in plasma AM between aorta and anterior interventricular vein, and between aorta and coronary sinus, suggesting that the failing human heart secretes AM and contributes to the increase in plasma AM [154]. Furthermore, plasma AM appears to be an independent predictor of mortality and disease progression in heart failure [159, 160].

7.2.5.2 Acute Hemodynamic Effects of AM in Heart Failure

Rademaker et al. presented the first evidence for the beneficial hemodynamic, renal, and neurohormonal effects of AM in heart failure [161]. Intravenous infusion of AM (10–100 ng/kg/min) for 90 min dose-dependently decreased peripheral resistance, arterial pressure and left atrial pressure with concomitant vigorous increases in cardiac output in sheep with pacing-induced heart failure. Despite a marked decline in arterial pressure, glomerular filtration rate and urine sodium excretion increased, and urine output was maintained. Regardless of brisk falls in atrial and left ventricular pressures, circulating levels of norepinephrine, epinephrine and renin were not stimulated, and atrial natriuretic peptide (ANP) and brain natriuretic peptide (BNP) concentrations were maintained, whereas aldosterone levels were reduced [161]. In a rat model of chronic heart failure produced by left coronary artery ligation, 20-min intravenous administration of AM (10 ng/kg/min) exerted diuresis and natriuresis without inducing hypotension. AM at a higher dose (50 ng/kg/min) markedly increased cardiac output with a slight decrease in mean arterial pressure. In addition, AM significantly increased glomerular filtration rate as well as urine flow and urinary sodium excretion [162].

Short-term clinical studies have confirmed the favorable effects of AM in the setting of chronic heart failure. Eight patients with definitive heart failure (left ventricular ejection fraction below 35%) received AM infusion (16 and 32 ng/kg/min for 2 h at each dose) to increase plasma AM levels within the pathophysiological range [163]. AM administration reduced blood pressure without significant effect on cardiac output. Despite the hypotensive effect, urine volume and sodium excretion were maintained. Although plasma renin activity was more than doubled, AM infusion eventually suppressed plasma aldosterone levels [163]. In a different, placebo-controlled study, AM induced more pronounced hemodynamic and renal effects in patients with chronic heart failure [67]. A 30-min AM (50 ng/kg/min) infusion markedly increased cardiac index with a small decrease in mean arterial pressure. Moreover, AM increased urine volume and urinary sodium excretion. AM significantly reduced plasma aldosterone levels, whereas renin and norepinephrine levels were not altered [67]. Similarly, AM induced marked hemodynamic responses in patients with more preserved cardiac function (left ventricular ejection fraction 40–50%) [164]. Short-term intravenous AM administration (50 ng/kg/min) decreased left ventricle systolic pressure and left ventricular end-diastolic pressure with a striking increase in cardiac index in patients with ischemic heart disease. Left ventricular pressure-volume analysis revealed that AM increases left ventricular end-systolic elastance, a load-independent index of cardiac contractility. Despite a similar reduction of left ventricular systolic pressure, ANP administration had negligible effects on cardiac index and left ventricular end-systolic elastance. These data indicate that both afterload reduction and enhanced cardiac contractility are likely to contribute to the AM-induced increase in cardiac index. Moreover, AM significantly shortened the time constant of left ventricular pressure decay, Tau, suggesting that AM may improve diastolic function as well. Importantly, AM did not increase myocardial oxygen consumption as the peptide increased coronary blood flow and decreased arterial–coronary sinus oxygen saturation difference [164].

7.2.5.3 Chronic Effects of AM on Heart Failure Progression

Experimental data unequivocally indicate that chronic AM delivery attenuates left ventricular remodeling in various heart failure models. Chronic infusion of human recombinant AM (0.5 µg/h) for 7 weeks in Dahl salt-sensitive rats attenuated the transition from left ventricular hypertrophy (LVH) to heart failure [165]. Administration of AM significantly improved hemodynamic parameters, as systemic vascular resistance decreased and cardiac output increased without changes in mean arterial pressure. Analysis of the pressure–volume relationship revealed that AM can increase left ventricular end-systolic elastance – a load-insensitive measure of cardiac contractility. These data suggest that improvements in systolic function cannot be attributed solely to the reduction of afterload, but AM can directly increase cardiac contractility in failing hearts. Furthermore, AM treatment decreased left ventricular weight to body weight ratio and myocardial levels of Ang II, ANP and BNP. Overall, the favorable AM-induced changes resulted in improved survival in Dahl salt-sensitive rats [165]. In a different study, rats were infused with AM (1.0 µg/h) for 7 days immediately after coronary ligation and were examined 9 weeks later. AM infusion significantly improved survival and reduced heart weight, lung weight, left ventricular end-diastolic pressure and collagen volume fraction of noninfarcted left ventricle without affecting infarct size [166]. In agreement, systemic AM gene transfer has been reported to confer cardiorenal protection in DOCA-salt hypertensive rats [167], in Dahl salt-sensitive rats [110] and in 2K1C hypertensive rats [111]. Intravenous injection of adenovirus containing human AM cDNA resulted in the expression of human AM mRNA in various rat tissues including the heart, kidney, adrenal gland, aorta, lung and liver [110, 111, 167]. Human AM gene delivery resulted in a prolonged reduction of blood pressure, which was associated with significant decreases in left ventricular weight and cardiomyocyte diameter, and reduced interstitial fibrosis and extracellular matrix formation within the heart. Furthermore, AM gene transfer attenuated glomerular sclerosis, tubular injuries, and protein casts in the kidney of DOCA-salt, Dahl salt-sensitive and 2K1C rats [110, 111, 167]. Of note, it is impossible to separate the direct and the systemic effects of AM in these models.

Recently, more direct evidence for the putative antihypertrophic effect of endogenous AM has been provided using AM knockout mice. Compared with wild-type littermates, pressure overload induced by aortic constriction resulted in more pronounced LVH with impaired left ventricular function in heterozygous mice ($AM^{+/-}$) [119]. Notably, $AM^{+/-}$ mice showed further downregulation of the SERCA2 gene, which is associated with cardiac dysfunction [168]. Furthermore, $AM^{+/-}$ mice subjected to pressure overload exhibited lower survival rates [119] reinforcing the significance of endogenous AM signaling in stress adaptation. However, important gender differences have been observed regarding the protective effect of AM [169]. Male mice heterozygous for AM gene deletion ($AM^{+/-}$) and heterozygous for the presence of a renin transgene (RenTgMK) displayed greater degree of LVH compared with mice with two copies of the AM gene ($AM^{+/+}$) and the RenTgMK transgene. In contrast, cardiac hypertrophy induced by the RenTgMK transgene was not exacerbated by the reduction of AM levels in female mice [169].

It is an intriguing question how AM can alleviate the progression of left ventricular remodeling and heart failure. Although it has been reported that administration of AM is frequently associated with decreases in preload and/or afterload in a number of studies [110, 111, 165–167], it is unlikely that the reduction in load can exclusively explain the beneficial effects of AM. Nishikimi *et al.* showed that AM and diuretic treatment similarly decreased systolic blood pressure and increased sodium excretion in Dahl salt-sensitive rats; however, the favorable effects on left ventricular remodeling were greater in AM than in the diuretic group [165]. Caron *et al.* reported that even though renin transgene increased blood pressure to a similar extent in male mice with one copy or two copies of the AM gene, $AM^{+/-}$ mice displayed more pronounced LVH [169]. In contrast to hemodynamic effects, it is more likely that AM could directly attenuate the growth processes in the myocardium. AM has been shown to attenuate the growth-promoting effects of Ang II *in vitro* in cultured cardiac myocytes, as AM inhibited Ang II-induced *de novo* protein synthesis [27, 29], ANP and BNP gene expression as well as assembly of sarcomeric structures [29]. Furthermore, Ang II induced more prominent hypertrophic response in cultured myocytes from $AM^{+/-}$ mice than in wild-type myocytes [119]. AM has also been shown to inhibit proliferation and protein synthesis of cardiac fibroblasts in response to Ang II and ET-1 [170]. In line, fibroblasts isolated from $AM^{+/-}$ mice showed faster proliferation and enhanced collagen type I gene expression [119]. In AM$^{+/-}$ mice, the exacerbated hypertrophic phenotype was associated with super-induction of the local renin–angiotensin system including the genes encoding angiotensinogen and angiotensin converting enzyme [119]. Furthermore, AM infusion reduced left ventricular mRNA levels of angiotensin converting enzyme and p22^{phox}, an essential component of NADH/NADPH oxidase, suggesting that AM administration might alleviate the progression of left ventricular remodeling and heart failure by blocking the upregulation of local renin-angiotensin system as well as by reducing oxidative stress [166]. These findings indicate that AM may play a role in left ventricular remodeling by suppressing cardiomyocyte hypertrophy and proliferation of fibroblasts. Further studies are required to clarify whether antiapoptotic and angiogenic effects of AM may also contribute to the overall beneficial action of AM in heart failure. Taking into account the finding that AM administration enhanced cardiac contractility without increasing myocardial oxygen consumption [164], one may speculate that the cAMP-independent positive inotropic effect of AM [50] may not exhaust the contractile reserve of the failing heart as β-adrenergic agonists and phosphodiesterase inhibitors do. Collectively, AM may be a new therapeutic approach for the treatment of heart failure as it reduces pre- and afterload, increases coronary blood flow, and improves cardiac contractility in conjunction with inhibition of cardiac hypertrophy and interstitial fibrosis.

Acknowledgments

This work was supported by the Academy of Finland, the Sigrid Juselius Foundation, the Finnish Foundation for Cardiovascular Research and the Finnish Medical Foundation. I.S. was supported by the János Bolyai Research Fellowship (Hungarian Academy of Sciences).

References

1 Kitamura, K., Kangawa, K., Kawamoto, M., Ichiki, Y., Nakamara, S., Matsuo, H. and Eto, T. (1993) *Biochemical and Biophysical Research Communications*, **192**, 553–60.
2 Samson, W.K. (1999) *Annual Review of Physiology*, **61**, 363–89.
3 Bunton, D.C., Petrie, M.C., Hillier, C., Johnston, F. and McMurray, J.J. (2004) *Pharmacology and Therapeutics*, **103**, 179–201.
4 Hinson, J.P., Kapas, S. and Smith, D.M. (2000) *Endocrine Reviews*, **21**, 138–67.
5 Eto, T. (2001) *Peptides*, **22**, 1693–711.
6 Nicholls, M.G., Lainchbury, J.G., Lewis, L.K., McGregor, D.O., Richards, A.M., Troughton, R.W. and Yandle, T.G. (2001) *Peptides*, **22**, 1745–52.
7 Kato, J., Tsuruda, T., Kita, T., Kitamura, K. and Eto, T. (2005) *Arteriosclerosis, Thrombosis, and Vascular Biology*, **25**, 2480–7.
8 Brain, S.D. and Grant, A.D. (2004) *Physiological Reviews*, **84**, 903–34.
9 Hamid, S.A. and Baxter, G.F. (2005) *Pharmacology and Therapeutics*, **105**, 95–112.
10 Nagaya, N., Mori, H., Murakami, S., Kangawa, K. and Kitamura, S. (2005)

American Journal of Physiology – Regulatory Integrative and Comparative Physiology, **288**, R1432–7.
11. Ishimitsu, T., Ono, H., Minami, J. and Matsuoka, H. (2006) *Pharmacology and Therapeutics*, **111**, 909–27.
12. Gibbons, C., Dackor, R., Dunworth, W., Fritz-Six, K. and Caron, K.M. (2007) *Molecular Endocrinology*, **21**, 783–96.
13. Sakata, J., Shimokubo, T., Kitamura, K., Nakamura, S., Kangawa, K., Matsuo, H. and Eto, T. (1993) *Biochemical and Biophysical Research Communications*, **195**, 921–7.
14. Kitamura, K., Sakata, J., Kangawa, K., Kojima, M., Matsuo, H. and Eto, T. (1993) *Biochemical and Biophysical Research Communications*, **194**, 720–5.
15. Kitamura, K., Kangawa, K., Ishiyama, Y., Washimine, H., Ichiki, Y., Kawamoto, M., Minamino, N., Matsuo, H. and Eto, T. (1994) *FEBS Letters*, **351**, 35–7.
16. Kitamura, K., Kangawa, K., Kojima, M., Ichiki, Y., Matsuo, H. and Eto, T. (1994) *FEBS Letters*, **338**, 306–10.
17. Ichiki, Y., Kitamura, K., Kangawa, K., Kawamoto, M., Matsuo, H. and Eto, T. (1994) *FEBS Letters*, **338**, 6–10.
18. Sakata, J., Shimokubo, T., Kitamura, K., Nishizono, M., Ichiki, Y., Kangawa, K., Matsuo H. and Eto, T. (1994) *FEBS Letters*, **352**, 105–8.
19. Sugo, S., Minamino, N., Kangawa, K., Miyamoto, K., Kitamura, K., Sakata, J., Eto, T. and Matsuo, H. (1994) *Biochemical and Biophysical Research Communications*, **201**, 1160–6.
20. Sugo, S., Minamino, N., Shoji, H., Kangawa, K., Kitamura, K., Eto, T. and Matsuo, H. (1994) *Biochemical and Biophysical Research Communications*, **203**, 719–26.
21. Jougasaki, M., Wei, C.-M., Heublein, D.M., Sandberg, S.M. and Burnett, J.C. Jr (1995) *Peptides*, **16**, 773–5.
22. Tomoda, Y., Kikumoto, K., Isumi, Y., Katafuchi, T., Tanaka, A., Kangawa, K., Dohi, K. and Minamino, N. (2001) *Cardiovascular Research*, **49**, 721–30.
23. Montuenga, L.M., Martínez, A., Miller, M.J., Unsworth, E.J. and Cuttitta, F. (1997) *Endocrinology*, **138**, 440–51.
24. Cormier-Regard, S., Nguyen, S.V. and Claycomb, W.C. (1998) *Journal of Biological Chemistry*, **273**, 17787–92.
25. Ishimitsu, T., Kojima, M., Kangawa, K., Hino, J., Matsuoka, H., Kitamura, K., Eto, T. and Matsuo, H. (1994) *Biochemical and Biophysical Research Communications*, **203**, 631–9.
26. Ishimitsu, T., Miyata, A., Matsuoka, H. and Kangawa, K. (1998) *Biochemical and Biophysical Research Communications*, **243**, 463–70.
27. Tsuruda, T., Kato, J., Kitamura, K., Kuwasako, K., Imamura, T., Koiwaya, Y., Tsuji, T., Kangawa, K. and Eto, T. (1998) *Hypertension*, **31**, 505–10.
28. Mishima, K., Kato, J., Kuwasako, K., Ito, K., Imamura, T., Kitamura, K. and Eto, T. (2001) *Biochemical and Biophysical Research Communications*, **287**, 264–9.
29. Luodonpää, M., Vuolteenaho, O., Eskelinen, S. and Ruskoaho, H. (2001) *Peptides*, **22**, 1859–66.
30. Mishima, K., Kato, J., Kuwasako, K., Imamura, T., Kitamura, K. and Eto, T. (2003) *Life Sciences*, **73**, 1629–35.
31. Sugo, S., Minamino, N., Shoji, H., Kangawa, K. and Matsuo, H. (1995) *FEBS Letters*, **369**, 311–4.
32. Sugo, S., Minamino, N., Shoji, H., Kangawa, K., Kitamura, K., Eto, T. and Matsuo, H. (1995) *Biochemical and Biophysical Research Communications*, **207**, 25–32.
33. Isumi, Y., Shoji, H., Sugo, S., Tochimoto, T., Yoshioka, M., Kangawa, K., Matsuo, H. and Minamino, N. (1998) *Endocrinology*, **139**, 838–46.
34. Chun, T.-H., Itoh, H., Ogawa, Y., Tamura, N., Takaya, K., Igaki, T., Yamashita, J., Doi, K., Inoue, M., Masatsugu, K., Korenaga, R., Ando, J. and Nakao, K. (1997) *Hypertension*, **29**, 1296–302.
35. Shinoki, N., Kawasaki, T., Minamino, N., Okahara, K., Ogawa, A., Ariyoshi, H., Sakon, M., Kambayashi, J., Kangawa, K. and Monden, M. (1998) *Journal of Cellular Biochemistry*, **71**, 109–15.
36. Tsuruda, T., Kato, J., Kitamura, K., Imamura, T., Koiwaya, Y., Kangawa, K., Komuro, I., Yazaki, Y. and Eto, T. (2000) *Hypertension*, **35**, 1210–14.

37 Luodonpää, M., Rysä, J., Pikkarainen, S., Tenhunen, O., Tokola, H., Puhakka, J., Marttila, M., Vuolteenaho, O. and Ruskoaho, H. (2003) *Regulatory Peptides*, **112**, 153–9.
38 Romppanen, H., Marttila, M., Magga, J., Vuolteenaho, O., Kinnunen, P., Szokodi, I. and Ruskoaho, H. (1997) *Endocrinology*, **138**, 2636–9.
39 Morimoto, A., Nishikimi, T., Yoshihara, F., Horio, T., Nagaya, N., Matsuo, H., Dohi, K. and Kangawa, K. (1999) *Hypertension*, **33**, 1146–52.
40 Romppanen, H., Puhakka, J., Földes, G., Szokodi, I., Vuolteenaho, O., Tokola, H., Tóth, M. and Ruskoaho, H. (2001) *Hypertension*, **37**, 84–90.
41 Földes, G., Suo, M., Szokodi, I., Lakó-Futó, Z., deChâtel, R., Vuolteenaho, O., Huttunen, P., Ruskoaho, H. and Tóth, M. (2001) *Endocrinology*, **142**, 4256–63.
42 Nguyen, S.V. and Claycomb, W.C. (1999) *Biochemical and Biophysical Research Communications*, **265**, 382–6.
43 Yoshihara, F., Horio, T., Nishikimi, T., Matsuo, H. and Kangawa, K. (2002) *European Journal of Pharmacology*, **436**, 1–6.
44 Nakayama, M., Takahashi, K., Murakami, O., Shirato, K. and Shibahara, S. (1999) *Peptides*, **20**, 769–72.
45 Ando, K., Ito, Y., Kumada, M. and Fujita, T. (1998) *Hypertension Research*, **21**, 187–91.
46 Chun, T.-H., Itoh, H., Saito, T., Yamahara, K., Doi, K., Mori, Y., Ogawa, Y., Yamashita, J., Tanaka, T., Inoue, M., Masatsugu, K., Sawada, N., Fukunaga, Y. and Nakao, K. (2000) *Journal of Hypertension*, **18**, 575–80.
47 Owji, A.A., Smith, D.M., Coppock, H.A., Morgan, D.G.A., Bhogal, R., Ghatel, M.A. and Bloom, S.R. (1995) *Endocrinology*, **136**, 2127–34.
48 McLatchie, L.M., Fraser, N.J., Main, M.J., Wise, A., Brown, J., Thompson, N., Solari, R., Lee, M.G. and Foord, S.M. (1998) *Nature*, **393**, 333–9.
49 Szokodi, I., Kinnunen, P. and Ruskoaho, H. (1996) *Acta Physiologica Scandinavica*, **156**, 151–2.
50 Szokodi, I., Kinnunen, P., Tavi, P., Weckström, M., Toth, M. and Ruskoaho, H. (1998) *Circulation*, **97**, 1062–70.
51 Kinnunen, P., Szokodi, I., Nicholls, M.G. and Ruskoaho, H. (2000) *American Journal of Physiology – Regulatory Integrative and Comparative Physiology*, **279**, R569–75.
52 Kelly, R.A., Eid, H., Krämer, B.K., O'Neill, M., Liang, B.T., Reers, M. and Smith, T.W. (1990) *Journal of Clinical Investigation*, **86**, 1164–71.
53 Szokodi, I., Tavi, P., Földes, G., Voutilainen-Myllylä, S., Ilves, M., Tokola, H., Pikkarainen, S., Piuhola, J., Rysä, J., Tóth, M. and Ruskoaho, H. (2002) *Circulation Research*, **91**, 434–40.
54 Ihara, T., Ikeda, U., Tate, Y., Ishibashi, S. and Shimada, K. (2000) *European Journal of Pharmacology*, **390**, 167–72.
55 Mittra, S., Hyvelin, J.M., Shan, Q., Tang, F. and Bourreau, J.P. (2004) *American Journal of Physiology, Heart and Circulatory Physiology*, **286**, H1034–42.
56 Mittra, S. and Bourreau, J.P. (2006) *American Journal of Physiology, Heart and Circulatory Physiology*, **290**, H1842–7.
57 Ikenouchi, H., Kangawa, K., Matsuo, H. and Hirata, Y. (1997) *Circulation*, **95**, 2318–24.
58 Mukherjee, R., Multani, M.M., Sample, J.A., Dowdy, K.B., Zellner, J.L., Hoover, D.B. and Spinale, F.G. (2002) *Journal of Cardiovascular Pharmacology and Therapeutics*, **7**, 235–40.
59 Stangl, V., Dschietzig, T., Bramlage, P., Boye, P., Kinkel, H.T., Staudt, A. and Baumann, G. (2000) *European Journal of Pharmacology*, **408**, 83–9.
60 Saetrum Opgaard, O., Hasbak, P., de Vries, R., Saxena, P.R. and Edvinsson, L. (2000) *European Journal of Pharmacology*, **397**, 373–82.
61 He, H., Bessho, H., Fujisawa, Y., Horiuchi, K., Tomohiro, A., Kita, T., Aki, Y., Kimura, S., Tamaki, T. and Abe, Y. (1995) *European Journal of Pharmacology*, **273**, 209–14.
62 Parkes, D.G. (1995) *American Journal of Physiology, Heart and Circulatory Physiology*, **37**, H2574–8.
63 Parkes, D.G. and May, C.N. (1997) *British Journal of Pharmacology*, **120**, 1179–85.
64 Charles, C.J., Rademaker, M.T., Richards, A.M., Cooper, G.J.S., Coy, D.H., Jing,

N.-Y. and Nicholls, M.G. (1997) *American Journal of Physiology*, **272**, R2040–7.
65. Lainchbury, J.G., Meyer, D.M., Jougasaki, M., Burnett, J.C. Jr and Redfield, M.M. (2000) *American Journal of Physiology, Heart and Circulatory Physiology*, **279**, H1000–6.
66. Lainchbury, J.G., Troughton, R.W., Lewis, L.K., Yandle, T.G., Richards, A.M. and Nicholls, M.G. (2000) *Journal of Clinical Endocrinology and Metabolism*, **85**, 1016–20.
67. Nagaya, N., Satoh, T., Nishikimi, T., Uematsu, M., Furuichi, S., Sakamaki, F., Oya, H., Kyotani, S., Nakanishi, N., Goto, Y., Masuda, Y., Miyatake, K. and Kangawa, K. (2000) *Circulation*, **101**, 498–503.
68. Eguchi, S., Hirata, Y., Kano, H., Sato, K., Watanabe, Y., Watanabe, T.X., Nakajima, K., Sakakibara, S. and Marumo, F. (1994) *FEBS Letters*, **340**, 226–30.
69. Ishizaka, Y., Tanaka, M., Kitamura, K., Kangawa, K., Minamino, N., Matsuo, H. and Eto, T. (1994) *Biochemical and Biophysical Research Communications*, **200**, 642–6.
70. Shimekake, Y., Nagata, K., Ohta, S., Kambayashi, Y., Teraoka, H., Kitamura, K., Eto, T., Kangawa, K. and Matsuo, H. (1995) *Journal of Biological Chemistry*, **270**, 4412–7.
71. Chini, E.N., Choi, E., Grande, J.P., Burnett, J.C. and Dousa, T.P. (1995) *Biochemical and Biophysical Research Communications*, **215**, 868–73.
72. Ikeda, U., Kanbe, T., Kawahara, Y., Yokoyama, M. and Shimada, K. (1996) *Circulation*, **94**, 2560–5.
73. Sato, A., Canny, B.J. and Autelitano, D.L. (1997) *Biochemical and Biophysical Research Communications*, **230**, 311–14.
74. Bers, D.M. (2002) *Nature*, **415**, 198–205.
75. Vlahos, C.J., McDowell, S.A. and Clerk, A. (2003) *Nature Reviews Drug Discovery*, **2**, 99–113.
76. Kamp, T.J. and Hell, J.W. (2000) *Circulation Research*, **87**, 1095–102.
77. Ishiyama, Y., Kitamura, K., Ichiki, Y., Nakamura, S., Kida, O., Kangawa, K. and Eto, T. (1993) *European Journal of Pharmacology*, **241**, 271–3.
78. Khan, A.I., Kato, J., Kitamura, K., Kangawa, K. and Eto, T. (1997) *Clinical and Experimental Pharmacology and Physiology*, **24**, 139–42.
79. Fukuhara, M., Tsuchihashi, T., Abe, I. and Fujishima, M. (1995) *American Journal of Physiology*, **269**, R1289–93.
80. Hjelmquist, H., Keil, R., Mathai, M., Hubschle, T. and Gerstberger, R. (1997) *American Journal of Physiology*, **273**, R716–24.
81. Hao, Q., Chang, J.K., Gharavi, H., Fortenberry, Y., Hyman, A. and Lippton, H. (1994) *Life Sciences*, **54**, PL265–70.
82. Lainchbury, J.G., Cooper, G.J., Coy, D.H., Jiang, N.Y., Lewis, L.K., Yandle, T.G., Richards, A.M. and Nicholls, M.G. (1997) *Clinical Science*, **92**, 467–72.
83. Gardiner, S.M., Kemp, P.A., March, J.E. and Bennett, T. (1995) *British Journal of Pharmacology*, **114**, 584–91.
84. Entzeroth, M., Doods, H.N., Wieland, H.A. and Wienen, W. (1995) *Life Sciences*, **56**, L19–25.
85. Yoshimoto, R., Mitsui-Saito, M., Ozaki, H. and Karaki, H. (1998) *British Journal of Pharmacology*, **123**, 1645–54.
86. Hirata, Y., Hayakawa, H., Suzuki, Y., Suzuki, E., Ikenouchi, H., Kohmoto, O., Kimura, K., Kitamura, K., Eto, T., Kangawa, K., Matsuo, H. and Omata, M. (1995) *Hypertension*, **25**, 790–5.
87. Lang, M.G., Paterno, R., Faraci, F.M. and Heistad, D.D. (1997) *Stroke*, **28**, 181–5.
88. Lippton, H., Chang, J.K., Hao, Q., Summer, W. and Hyman, A.L. (1994) *Journal of Applied Physiology*, **76**, 2154–6.
89. Brain, S.D., Williams, T.J., Tippins, J.R., Morris, H.R. and MacIntyre, I. (1985) *Nature*, **313**, 54–6.
90. Furchgott, R.F. (1993) *Circulation*, **87**, V3–8.
91. Nishimatsu, H., Suzuki, E., Nagata, D., Moriyama, N., Satonaka, H., Walsh, K., Sata, M., Kangawa, K., Matsuo, H., Goto, A., Kitamura, T. and Hirata, Y. (2001) *Circulation Research*, **89**, 63–70.
92. Hayakawa, H., Hirata, Y., Kakoki, M., Suzuki, Y., Nishimatsu, H., Nagata, D., Suzuki, E., Kikuchi, K., Nagano, T., Kangawa, K., Matsuo, H., Sugimoto, T. and Omata, M. (1999) *Hypertension*, **33**, 689–93.

93 Nossaman, B.D., Feng, C.J., Kaye, A.D., DeWit, B., Coy, D.H., Murphy, W.A. and Kadowitz, P.J. (1996) *American Journal Physiology – Lung, Cell and Physiology*, **270**, L782–89.

94 Feng, C.J., Kang, B., Kaye, A.D., Kadowitz, P.J. and Nossaman, B.D. (1994) *Life Sciences*, **55**, L433–8.

95 Kinnunen, P., Piuhola, J., Ruskoaho, H. and Szokodi, I. (2001) *American Journal of Physiology, Heart and Circulatory Physiology*, **281**, H1178–83.

96 Terata, K., Miura, H., Liu, Y., Loberiza, F. and Gutterman, D.D. (2000) *American Journal of Physiology, Heart and Circulatory Physiology*, **279**, H2620–6.

97 Yanagisawa, M., Kurihara, H., Kimura, S., Tomobe, Y., Kobayashi, M., Mitsui, Y., Yazaki, Y., Goto, K. and Masaki, T. (1988) *Nature*, **332**, 411–5.

98 Cannan, C.R., Brandt, J.C. Jr, Burnett R.R. and Lerman, A. (1995) *Circulation*, **92**, 3312–7.

99 Hirata, Y., Emori, T., Eguchi, S., Kanno, K., Imai, T., Ohta, K. and Marumo, F. (1993) *Journal of Clinical Investigation*, **91**, 1367–73.

100 Tsukahara, H., Ende, H., Magazine, H.I., Bahou, W.F. and Goligorsky, M.S. (1994) *Journal of Biological Chemistry*, **269**, 21778–85.

101 Lerman, A., Sandok, E.K., Hildebrand, F.L. Jr and Burnett, J.C. Jr (1992) *Circulation*, **85**, 1894–8.

102 Wang, Q.D., Li, X.S. and Pernow, J. (1994) *European Journal of Pharmacology*, **271**, 25–30.

103 Jougasaki, M., Schirger, J.A., Simari, R.D. and Burnett J.C. Jr (1998) *Hypertension*, **32**, 917–22.

104 Sabates, B.L., Pigott, J.D., Choe, E.U., Cruz, M.P., Lippton, H.L., Hyman, A.L., Flint, L.M. and Ferrara, J.J. (1997) *Journal of Surgical Research*, **67**, 163–8.

105 Miyoshi, Y., Nakaya, Y., Wakatsuki, T., Nakaya, S., Fujino, K., Saito, K. and Inoue, I. (1992) *Circulation Research*, **70**, 612–6.

106 Nagaya, N., Nishikimi, T., Yoshihara, F., Horio, T., Morimoto, A. and Kangawa, K. (2000) *American Journal of Physiology – Regulatory Integrative and Comparative Physiology*, **278**, R1019–26.

107 Kobayashi, K., Kitamura, K., Hirayama, N., Date, H., Kashiwagi, T., Ikushima, I., Hanada, Y., Nagatomo, Y., Takenaga, M., Ishikawa, T., Imamura, T., Koiwaya, Y. and Eto, T. (1996) *American Heart Journal*, **131**, 676–80.

108 Miyao, Y., Nishikimi, T., Goto, Y., Miyazaki, S., Daikoku, S., Morii, I., Matsumoto, T., Takishita, S., Miyata, A., Matsuo, H., Kangawa, K. and Nonogi, H. (1998) *Heart*, **79**, 39–44.

109 Yoshitomi, Y., Nishikimi, T., Kojima, S., Kuramochi, M., Takishita, S., Matsuoka, H., Miyata, A., Matsuo, H. and Kangawa, K. (1998) *Clinical Science*, **94**, 135–9.

110 Zhang, J.J., Yoshida, H., Chao, L. and Chao, J. (2000) *Human Gene Therapy*, **11**, 1817–27.

111 Wang, C., Dobrzynski, E., Chao, J. and Chao, L. (2001) *American Journal of Physiology – Renal Physiology*, **280**, F934–71.

112 Kato, K., Yin, H., Agata, J., Yoshida, H., Chao, L. and Chao, J. (2003) *American Journal of Physiology, Heart and Circulatory Physiology*, **285**, H1506–14.

113 Okumura, H., Nagaya, N., Itoh, T., Okano, I., Hino, J., Mori, K., Tsukamoto, Y., Ishibashi-Ueda, H., Miwa, S., Tambara, K., Toyokuni, S., Yutani, C. and Kangawa, K. (2004) *Circulation*, **109**, 242–8.

114 Yin, H., Chao, L. and Chao, J. (2004) *Hypertension*, **43**, 109–16.

115 Hamid, S.A. and Baxter, G.F. (2005) *Basic Research in Cardiology*, **100**, 387–96.

116 Caron, K.M. and Smithies, O. (2001) *Proceedings of the National Academy of Sciences of the United States of America*, **98**, 615–9.

117 Shindo, T., Kurihara, Y., Nishimatsu, H., Moriyama, N., Kakoki, M., Wang, Y., Imai, Y., Ebihara, A., Kuwaki, T., Ju, K.H., Minamino, N., Kangawa, K., Ishikawa, T., Fukuda, M., Akimoto, Y., Kawakami, H., Imai, T., Morita, H., Yazaki, Y., Nagai, R., Hirata, Y. and Kurihara, H. (2001) *Circulation*, **104**, 1964–71.

118 Shimosawa, T., Shibagaki, Y., Ishibashi, K., Kitamura, K., Kangawa, K., Kato, S., Ando, K. and Fujita, T. (2002) *Circulation*, **105**, 106–11.

119 Niu, P., Shindo, T., Iwata, H., Iimuro, S., Takeda, N., Zhang, Y., Ebihara, A., Suematsu, Y., Kangawa, K., Hirata, Y. and Nagai, R. (2004) *Circulation*, **109**, 1789–94.

120 Hamid, S.A. and Baxter, G.F. (2006) *Journal of Molecular and Cellular Cardiology*, **41**, 360–3.

121 Juhaszova, M., Zorov, D.B., Kim, S.H., Pepe, S., Fu, Q. Fishbein, K.W., Ziman, B.D., Wang, S., Ytrehus, K., Antos, C.L., Olson, E.N. and Sollott, S.J. (2004) *Journal of Clinical Investigation*, **113**, 1535–49.

122 Tong, H., Imahashi, K., Steenbergen, C. and Murphy, E. (2002) *Circulation Research*, **90**, 377–9.

123 Crow, M.T., Mani, K., Nam, Y.J. and Kitsis, R.N. (2004) *Circulation Research*, **95**, 957–70.

124 Zhao, Z.Q. and Vinten-Johansen, J. (2002) *Cardiovascular Research*, **55**, 438–55.

125 Valen, G. (2003) *Annals of Thoracic Surgery*, **75**, S656–60.

126 Yaoita, H., Ogawa, K., Maehara, K. and Maruyama, Y. (1998) *Circulation*, **97**, 276–81.

127 Holly, T.A., Drincic, A., Byun, Y., Nakamura, S., Harris, K., Klocke, F.J. and Cryns, V.L. (1999) *Journal of Molecular and Cellular Cardiology*, **31**, 1709–15.

128 Ribatti, D., Conconi, M.T. and Nussdorfer, G.G. (2007) *Pharmacological Reviews*, **59**, 185–205.

129 Kim, W., Moon, S.O., Sung, M.J., Kim, S.H., Lee, S., So, J.N. and Park, S.K. (2003) *FASEB Journal*, **17**, 1937–9.

130 Ribatti, D., Guidolin, D., Conconi, M.T., Nico, B., Baiguera, S., Parnigotto, P.P., Vacca, A. and Nussdorfer, G.G. (2003) *Oncogene*, **22**, 6458–61.

131 Iimuro, S., Shindo, T., Moriyama, N., Amaki, T., Niu, P., Takeda, N., Iwata, H., Zhang, Y., Ebihara, A. and Nagai, R. (2004) *Circulation Research*, **95**, 415–23.

132 Asahara, T., Murohara, T., Sullivan, A., Silver, M., van der Zee, R., Li, T., Witzenbichler, B., Schatteman, G. and Isner, J.M. (1997) *Science*, **275**, 964–7.

133 Takahashi, T., Kalka, C., Masuda, H., Chen, D., Silver, M., Kearney, M., Magner, M., Isner, J.M. and Asahara, T. (1999) *Nature Medicine*, **5**, 434–8.

134 Kawamoto, A., Gwon, H.C., Iwaguro, H., Yamaguchi, J.I., Uchida, S., Masuda, H., Silver, M., Ma, H., Kearney, M., Isner, J.M. and Asahara, T. (2001) *Circulation*, **103**, 634–7.

135 Iwase, T., Nagaya, N., Fujii, T., Itoh, T., Ishibashi-Ueda, H., Yamagishi, M., Miyatake, K., Matsumoto, T., Kitamura, S. and Kangawa, K. (2005) *Circulation*, **111**, 356–62.

136 Fujii, T., Nagaya, N., Iwase, T., Murakami, S., Miyahara, Y., Nishigami, K., Ishibashi-Ueda, H., Shirai, M., Itoh, T., Ishino, K., Sano, S., Kangawa, K. and Mori, H. (2005) *American Journal of Physiology, Heart and Circulatory Physiology*, **288**, H1444–50.

137 Jain, R.K. (2003) *Nature Medicine*, **9**, 685–93.

138 Dellian, M., Witwer, B.P., Salehi, H.A., Yuan, F. and Jain, R.K. (1996) *American Journal of Pathology*, **149**, 59–71.

139 Thurston, G., Suri, C., Smith, K., McClain, J., Sato, T.N., Yancopoulos, G.D. and McDonald, D.M. (1999) *Science*, **286**, 2511–4.

140 Lee, R.J., Springer, M.L., Blanco-Bose, W.E., Shaw, R., Ursell, P.C. and Blau, H.M. (2000) *Circulation*, **102**, 898–901.

141 Fernandez-Sauze, S., Delfino, C., Mabrouk, K., Dussert, C., Chinot, O., Martin, P.M., Grisoli, F., Ouafik, L. and Boudouresque, F. (2004) *International Journal of Cancer*, **108**, 797–804.

142 Miyashita, K., Itoh, H., Sawada, N., Fukunaga, Y., Sone, M., Yamahara, K., Yurugi-Kobayashi, T., Park, K. and Nakao, K. (2003) *FEBS Letters*, **544**, 86–92.

143 Shimokubo, T., Sakata, J., Kitamura, K., Kangawa, K., Matsuo, H. and Eto, T. (1996) *Clinical and Experimental Hypertension*, **18**, 949–61.

144 Nishikimi, T., Yoshihara, F., Kanazawa, A., Okano, I., Horio, T., Nagaya, N., Yutani, C., Matsuo, H., Matsuoka, H. and Kangawa, K. (2001) Role of increased circulating and renal adrenomedullin in

145 Wang, X., Nishikimi, T., Akimoto, K., Tadokoro, K., Mori, Y. and Minamino, N. (2003) *Journal of Hypertension*, **21**, 1171–81.
146 Nishikimi, T., Horio, T., Sasaki, T., Yoshihara, F., Takishita, S., Miyata, A., Matsuo, H. and Kangawa, K. (1997) *Hypertension*, **30**, 1369–75.
147 Nishikimi, T., Tadokoro, K., Mori, Y., Wang, X., Akimoto, K., Yoshihara, F., Minamino, N., Kangawa, K. and Matsuoka, H. (2003) *Hypertension*, **41**, 512–8.
148 Willenbrock, R., Langenickel, T., Knecht, M., Pagel, I., Höhnel, K., Philipp, S. and Dietz, R. (1999) *Life Sciences*, **21**, 2241–9.
149 Totsune, K., Takahashi, K., Mackenzie, H.S., Murakami, O., Arihara, Z., Sone, M., Mouris, T., Brenner, B.M. and Ito, S. (2000) *Clinical Science*, **99**, 541–6.
150 Vinge, E., Øie, L.E., Yndestad, A., Sandbreg, C., Grøgaard, H. and Attramadal, H. (2000) *Circulation*, **101**, 415–22.
151 Jougasaki, M., Wei, C.M., McKinley, L.J. and Burnett, J.C. Jr (1995) *Circulation*, **92**, 286–9.
152 Nishikimi, T., Saito, Y., Kitamura, K., Ishimitsu, T., Eto, T., Kangawa, K., Matsuo, H., Omae, T. and Matsuoka, H. (1995) *Journal of the American College of Cardiology*, **26**, 1424–31.
153 Kato, J., Kobayashi, K., Etoh, T., Tanaka, M., Kitamura, K., Imamura, T., Koiwaya, Y., Kangawa, K. and Eto, T. (1996) *Journal of Clinical Endocrinology and Metabolism*, **81**, 180–3.
154 Jougasaki, M., Rodeheffer, R.J., Redfield, M.M., Yamamoto, K., Wei, C.M., McKinley, L.J. and Burnett, J.C. Jr (1996) *Journal of Clinical Investigation*, **97**, 2370–6.
155 Kobayashi, K., Kitamura, K., Etoh, T., Nagatomo, Y., Takenaga, M., Ishikawa, T., Imamura, T., Koiwaya, Y. and Eto, T. (1996) *American Heart Journal*, **131**, 994–8.
156 Cheung, B. and Leung, R. (1997) *Clinical Science*, **92**, 59–62.
157 Etoh, T., Kato, J., Washimine, H., Imamura, T., Kitamura, K., Koiwaya, Y., Kangawa, K. and Eto, T. (1997) *Hormone and Metabolic Research*, **29**, 46–7.
158 Etoh, T., Kato, J., Takenaga, M., Imamura, T., Kitamura, K., Koiwaya, Y. and Eto, T. (1999) *Clinical Cardiology*, **22**, 113–7.
159 Pousset, F., Masson, F., Chavirovskia, O., Isnard, R., Carayon, A., Golmard, J.L., Lechat, P., Thomas, D. and Komajda, M. (2000) *European Heart Journal*, **21**, 1009–104.
160 Richards, A.M., Doughty, R., Nicholls, M.G., MacMahon, S., Sharpe, N., Murphy, J., Espiner, E.A., Frampton, C. and Yandle, T.G. (2001) *Journal of the American College of Cardiology*, **37**, 1781–7.
161 Rademaker, M.T., Charles, C.J., Lewis, L.K., Yandle, T.G., Cooper, G.S., Coy, D.H., Richards, A.M. and Nicholls, M.G. (1997) *Circulation*, **96**, 1983–90.
162 Nagaya, N., Nishikimi, T., Horio, T., Yoshihara, F., Kanazawa, A., Matsuo, H. and Kangawa, K. (1999) *American Journal of Physiology*, **276**, R213–18.
163 Lainchbury, J.G., Nicholls, M.G., Espiner, E.A., Yandle, T.G., Lewis, L.K. and Richards, A.M. (1999) *Hypertension*, **34**, 70–5.
164 Nagaya, N., Goto, Y., Satoh, T., Sumida, H., Kojima, S., Miyatake, K. and Kangawa, K. (2002) *Journal of Cardiovascular Pharmacology*, **39**, 754–60.
165 Nishikimi, T., Yoshihara, F., Horinaka, S., Kobayashi, N., Mori, Y., Tadokoro, K., Akimoto, K., Minamino, N., Kangawa, K. and Matsuoka, H. (2003) *Hypertension*, **42**, 1034–42.
166 Nakamura, R., Kato, J., Kitamura, K., Onitsuka, H., Imamura, T., Cao, Y., Marutsuka, K., Asada, Y., Kangawa, K. and Eto, T. (2004) *Circulation*, **110**, 426–31.
167 Dobrzynski, E., Wang, C., Chao, J. and Chao, L. (2000) *Hypertension*, **36**, 995–1001.
168 Frank, K.F., Bolck, B., Erdmann, E. and Schwinger, R.H.G. (2003) *Cardiovascular Research*, **57**, 20–7.

169 Caron, K., Hagaman, J., Nishikimi, T., Kim, H.S. and Smithies, O. (2007) *Proceedings of the National Academy of Sciences of the United States of America*, **104**, 3420–5.

170 Tsuruda, T., Kato, J., Kitamura, K., Kawamoto, M., Kuwasako, K., Imamura, T., Koiwaya, Y., Tsuji, T., Kangawa, K. and Eto, T. (1999) *Cardiovascular Research*, **43**, 958–67.

8
Apelin and Vasopressin

Xavier Iturrioz, Annabelle Reaux-Le Goazigo, Françoise Moos, and Catherine Llorens-Cortes

8.1
Discovery of Apelin

The apelin story began in 1993 with the cloning of the cDNA for the APJ receptor (putative receptor protein related to the type 1 angiotensin receptor) from a human genomic library by O'Dowd *et al.* [1]. This receptor was identified as a member of the orphan seven-transmembrane domain G protein-coupled receptor (GPCR) family. In parallel, we initially searched for another angiotensin receptor subtype specific for angiotensin (Ang) III, as we had previously shown that Ang III was one of the major effector peptides of the brain renin–angiotensin system, exerting tonic stimulatory control over arterial blood pressure in hypertensive rats [2], triggering arginine vasopressin (AVP) release and activating magnocellular vasopressinergic neurons in the supraoptic nucleus (SON) [3, 4]. However, these effects in the SON are not mediated by the conventional Ang II type 1 (AT_1) or type 2 (AT_2) receptors, as no mRNA expression or binding sites corresponding to AT_1 or AT_2 receptors have been found in the SON. Ang III-induced AVP stimulation may therefore involve another angiotensin receptor subtype. In an attempt to isolate a putative angiotensin receptor subtype specific for Ang III, we cloned a partial GPCR sequence displaying 35% amino acid sequence identity to the rat AT_1 receptor from rat SON by a homology cloning method. This sequence was subsequently used to screen a rat brain cDNA library. We then isolated a seven-transmembrane domain GPCR composed of 377 amino acids [5] which exhibited a high level of amino acid sequence identity (90–96%) to the sequences of mouse [6] and human [1] orphan APJ receptors and only 31% identity to the rat AT_{1A} receptor sequence [7]. Despite the low amino acid sequence identity (32%) between AT_1 and AT_2 receptors, these receptors bind Ang II and Ang III with similar affinity. It was therefore possible that the receptor we had isolated might bind an angiotensin peptide. However, this receptor, when stably expressed in Chinese hamster ovary (CHO) cells, was unable to bind iodinated angiotensin fragments such as Ang II, Ang III and Ang IV, thus demonstrating that this receptor did not

Cardiovascular Hormone Systems. Edited by Michael Bader
Copyright © 2008 WILEY-VCH Verlag GmbH & Co. KGaA, Weinheim
ISBN: 978-3-527-31920-6

correspond to a new angiotensin receptor subtype [5]. It therefore remained an orphan receptor for which the endogenous ligand had to be isolated. For this purpose, we developed a screening process for identifying the ligands of orphan GPCRs based on the property of GPCRs to internalize following agonist ligand exposure [8]. It involves expressing the orphan GPCR, tagged at its C-terminus with enhanced green fluorescent protein in eukaryotic cells, incubating the cells with high-performance liquid chromatography fractions purified from frog brain extracts and identifying the fraction inducing receptor internalization. Positive fractions, which contain the endogenous ligand, are then subjected to successive purification steps and internalization tests, to isolate a fraction containing only a few peptides, the sequencing of which by matrix-assisted laser desorption ionization–time-of-flight analysis, reveals the sequence of the endogenous ligand. We were close to completing the identification of the endogenous ligand of the orphan GPCR we had cloned when Tatemoto *et al.* isolated, from bovine stomach tissue extracts, the endogenous ligand of the human APJ receptor, using the Cytosensor microphysiometer method. They named this peptide apelin, for APJ endogenous ligand [9].

8.2
Structure and Processing of the Apelin Precursor

Apelin is a 36-amino-acid peptide (apelin 36) generated from a larger precursor, the 77-amino-acid proapelin (Figure 8.1A). This precursor has been isolated from various species [9–11]. The human proapelin gene is located on chromosome X at locus Xq25–q26.1 and contains three exons, with the coding region spanning exons 1 and 2. The 3'-untranslated region also spans two exons (2 and 3). This may account for the presence of transcripts of two different sizes (around 3 and 3.6 kb) in various tissues [11, 12]. The alignment of proapelin amino acid sequences from cattle, humans, rats and mice has demonstrated strict conservation of the C-terminal 17 amino acids, known as apelin 17 or K17F (Figure 8.1A). *In vivo*, various molecular forms of apelin are present, differing only in length (36, 17 or 13 amino acids) (Figure 8.1B). The occurrence of two internal dibasic motifs (Arg59–Arg60 and Arg63–Arg64) within the cattle, human rat and mouse proapelin sequences suggests that K17F and apelin 13 (pE13F) may be processed by prohormone convertases. For apelin 36, the maturation mechanism remains to be defined. In rat brain and plasma, the predominant forms of apelin are the pyroglutamyl form of pE13F and, to a lesser extent, K17F (Figure 8.1B) [12]. In rat lung, testis, and uterus and bovine colostrum, apelin 36 predominates, whereas both apelin 36 and pE13F have been detected in rat mammary gland (Figure 8.1B) [14, 15].

(A) Proapelin

```
Cattle   M N L R R C V Q A L L L L L W L C L S A V C G G P L L Q T S D  30
Humans   M N L R L C V Q A L L L L L W L S L T A V C G G S L M P L P D  30
Rats     M N L S F C V Q A L L L L L W L S L T A V C G V P L M L P P D  30
Mice     M N L R L C V Q A L L L L L W L S L T A V C G V P L M L P P D  30

         ▼
Cattle   G K E M E E G T I R Y L V Q P R G P R S G P G P W Q G G R R  60
Humans   G N G L E D G N V R H L V Q P R G S R N G P G P W Q G G R R  60
Rats     G K G L E E G N M R Y L V K P R T S R T G P G A W Q G G R R  60
Mice     G T G L E E G S M R Y L V K P R T S R T G P G A W Q G G R R  60

Cattle   │K F R R Q R P R L S H K G P M P F│  77
Humans   │K F R R Q R P R L S H K G P M P F│  77
Rats     │K F R R Q R P R L S H K G P M P F│  77
Mice     │K F R R Q R P R L S H K G P M P F│  77
```

Apelin 17 (K17F)

(B) Apelin fragments detected *in vivo* in mammals and tissular distribution

Apelin 36 R R G G Q W A G P G T R S T R P K V L -NH2   *Testis, Lung*
           K F R R Q R P R L S H G P M P F -COOH   *Colostrum, Uterus*
                                                         Mammary gland

Apelin 17
           NH2- K F R R Q R P R L S H G P M P F -COOH   *Brain, Plasma*

Pyroglutamyl form
of apelin 13 (pE13F)   pE R P R L S H G P M P F -COOH   *Brain, Plasma*
                                                         Mammary gland

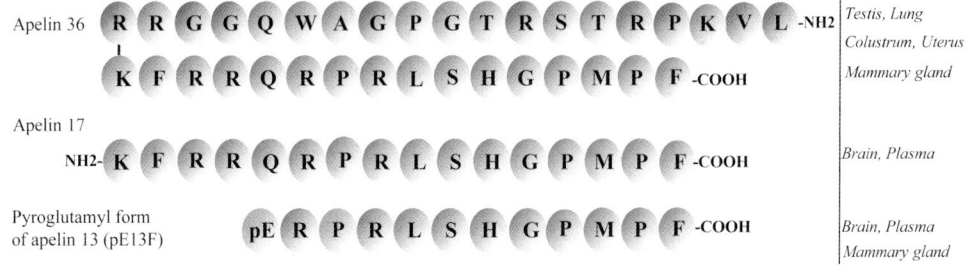

Figure 8.1 (A) Amino acid sequence of the apelin precursor, proapelin, in cattle, humans, rats, and mice. The first amino acid of apelin 36 is indicated by an arrowhead and the apelin 17 (K17F) sequence is indicated by a gray box (see [9, 10]). (B) Apelin fragments detected *in vivo* in mammals and their tissue distribution: amino acid sequence of apelin 36, apelin 17 and the pyroglutamyl form of apelin 13 (pE13F). In rat brain and plasma, the predominant forms of apelin are pE13F and, to a lesser extent, K17F [13]. In rat lung, testis and uterus, and bovine colostrum, apelin 36 predominates, whereas both apelin 36 and pE13F have been detected in rat mammary gland [14, 15].

8.3
Apelin Receptor Signaling and Internalization

Activation of the apelin receptor induces a number of effects involving multiple signaling pathways. Rat and human apelin receptors stably expressed in CHO or human embryonic kidney cells are negatively coupled to adenylate cyclase activity, and both are activated by several apelin fragments [5, 10, 16]. The most potent inhibitors of forskolin-induced cAMP production were found to be apelin 36, K17F, apelin 13 (Q13F) and pE13F, whereas apelin fragments R10F and G5F were inactive [10, 16, 17]. An Ala scan of pE13F [16] and N- or C-terminal deletions of K17F [17] have shown that the arginine residues in positions 2 and 4 and the leucine residue in position 5 in pE13F play a critical role in binding affinity and in the inhibition of cAMP production. Apelin 36, K17F and pE13F also increase intracellular calcium mobilization in both NTera 2 human teratocarcinoma cells and RBL-2H3 cells derived from rat basophils stably expressing the human apelin receptor [16, 18]. Apelin 36, K17F and pE13F induce internalization of the rat and human apelin receptors [17, 19, 20], whereas deletion of the C-terminal phenylalanine of K17F abolishes internalization without affecting the adenylate cyclase coupling of the apelin receptor [17]. This interesting feature revealed that the apelin receptor G-protein coupling and internalization processes are dissociated, possibly reflecting the existence of several conformational states of this receptor, stabilized by the binding of different apelin fragments to the receptor [17]. Moreover, Masri et al. have also shown that pE13F activates extracellular signal-regulated kinases (ERKs) via a pertussis toxin-sensitive G-protein and Ras-independent pathway [21]. pE13F also activates p70 S6 kinase in human umbilical vein endothelial cells and in CHO cells expressing the mouse apelin receptor, via both the ERK and phosphoinositide 3-kinase pathways [22]. Apelin activates the sarcolemmal Na^+/H^+ exchanger and increases intracellular pH in cardiomyocytes in culture, certainly in a phospholipase C- and protein kinase C-dependent manner [23, 24]. Apelin was also recently shown to phosphorylate Akt in aorta, inducing the activation of endothelial nitric oxide (NO) synthase, leading to NO release and vasodilation [25].

8.4
Distribution of Apelin and Its Receptor within the Adult Rat Brain

8.4.1
Topographical Distribution of Apelin

The production and characterization of a polyclonal antibody with high affinity and selectivity for K17F [13] has made it possible, for the first time, to visualize apelin neurons in the rat central nervous system. Apelin cell bodies are present in the hypothalamus, pons and medulla oblongata, regions involved in neuroendocrine control, drinking behavior, and the regulation of arterial blood pressure. Apelin cell bodies are particularly abundant in the SON and the magnocellular part of the paraventricular nucleus (PVN), the arcuate nucleus, the lateral reticular nucleus and the nucleus ambiguus [26]. Conversely, apelin nerve fibers are much more widely distributed in many brain regions than neuronal apelin cell bodies. The density of apelin nerve fibers and apelin nerve endings is highest in the inner layer of the median eminence and in the posterior pituitary gland [19, 27], suggesting that the apelin neurons of the SON and PVN, like the magnocellular AVP and ocytocin neurons, project into the posterior pituitary gland. Apelin nerve fibers also innervate the mesencephalon, the pons, the medulla oblongata and several circumventricular organs, such as the vascular organ of the lamina terminalis, the subfornical organ, the subcommissural organ and the area postrema [26]. Consistent with these data, an overlap between the distribution of the apelin precursor mRNA [11] and that of apelin cell bodies and/or fibers is seen in the septum, the preoptic region, the amygdala, the hypothalamus at the level of the periventricular, supraoptic, paraventricular and ventromedial nuclei, the periaqueductal gray matter, and the parabrachial and spinal trigeminal nuclei.

8.4.2
Distribution of Apelin Receptor mRNA Expression

Like apelin, the apelin receptor is widely distributed throughout the rat central nervous system [5, 11, 12]. *In situ* hybridization has shown that apelin receptor mRNA is present in the piriform and enthorinal cortices, the septum, the hippocampus, the central gray matter, and structures containing monoaminergic neuronal cell bodies (pars compacta of the substantia nigra, dorsal raphe nucleus and locus coeruleus). The apelin receptor is particularly abundant in the apelin-rich hypothalamic nuclei, including the SON, PVN and the arcuate nucleus, and in the pineal gland and the anterior and intermediate lobes of the pituitary gland [5, 11, 12]. Several areas with high levels of apelin receptor mRNA expression [5, 11, 12], including the septum, the supraoptic, paraventricular and arcuate nuclei, the central gray matter, and the dorsal raphe nucleus, also contain apelin fibers. In the absence of data concerning the location of apelin receptor-binding sites, we can hypothesize that, in these nuclei, locally synthesized apelin receptors are present on the perikarya or dendrites, serving as postsynaptic receptors for neurotransmission. In many brain structures, the distribution of apelin cell bodies and/or fibers and that of mRNAs encoding preproapelin and the apelin precursor [5, 11, 12] show considerable overlap with those of angiotensin and natriuretic peptides and their receptors [28–30]. These observations suggest that apelin, like these vasoactive peptides, may be involved in the central control of body fluid homeostasis and cardiovascular functions and pituitary hormone release.

8.5
Apelin: Physiological Actions within the Brain and Anterior Pituitary Gland

8.5.1
Involvement of Vasopressin and Apelin in the Maintenance of Water Balance

8.5.1.1 Vasopressinergic System
The neuropeptide AVP is synthesized and packaged in large dense core vesicles within around 9000 magnocellular neurons of the SON and PVN nuclei of the hypothalamus, the axons of which project to the neurohypophysis via the internal zone of median eminence. In these vesicles, AVP is tightly associated to neurophysin, a chaperone protein involved in hormone biosynthesis. The hormone domain of the AVP prohormone is crucial for correct trafficking of the prohor-

mone through the secretory pathway and AVP–neurophysin association ensures correct prohormone folding in the endoplasmic reticulum [31]. Vesicles are targeted to the different compartments of the AVP neurons, including the perikarya, dendrites and neurohypophyseal nerve endings. At this later level, vesicles are densely stored, exocytosis occurring in response to neuronal action potentials. Vesicle content is then directly released into the systemic circulation where AVP exerts its various hormonal effects. This release is triggered by various stimuli, the most potent being dehydration, osmotic challenge and change in volemia (e.g. during hemorrhage). Other stimuli also affect the release of AVP, such as pain, cold and stress. The amount of released AVP depends on the characteristics of phasic bursting activity (alternation of periods of stable electrical activity and silence), which mainly emerge when the demand for AVP release is increased. The activation is translated by an increased burst frequency or increased rates within bursts or both (see references in the review by Hatton [32]). Phasic bursting is more effective on AVP release per action potential basis and, during a burst, the release per pulse is greatest early in the burst when the rate is high.

The hypothalamo-neurohypophyseal complex is not, however a simple output system releasing AVP into blood circulation, as AVP peptide is also released by soma and dendrites in the SON/PVN under many physiological conditions (for review, see [33]). This dendritic exocytosis [34, 35] induced by many transmitters and neuropeptides present in nerve fibers terminating in the SON/PVN [36] does not occur in conjunction with the release from axon terminals (for review, see [33]). It first requires the relocation of peptide pools closer to sites of secretion, a mechanism selectively regulated through activation of intracellular calcium stores [37]. After priming, vesicles can subsequently be released by electrical and depolarization-dependent activation [38, 39]. This dendritic release has a key functional significance, as it regulates in turn the electrical activity of AVP neurons by favoring the expression of the phasic pattern known to be the most efficient for hormone release [40]. These data led to the new concept of modulation of neuronal activity by centrally released peptides: the optimization of a neuronal discharge in accordance with the physiological demand. Optimization of neuronal firing is reached through an autocrine mechanism since AVP neurons express V_{1a} and V_{1b} autoreceptors [40]. The fact that these receptors are colocalized with AVP in the same vesicles [41] would facilitate binding of the peptide to its receptors when they are simultaneously released during exocytosis. Such a binding results in both an influx of calcium entry and a mobilization of intracellular calcium stores (see references in the review by Ludwig [33]). This "postsynaptic" control is complemented by a presynaptic action of AVP on afferent inputs [42–44], in particular the γ-aminobutyric acid ones via the V_{1a} receptors [45]. These modulatory actions would be particularly prominent in physiological situation increasing local AVP release.

However, the process of auto-retrocontrol is much more complex since it involves co-stored and co-released peptides such as neuropeptide Y [46], galanin [47] or dynorphin [48, 49], peptides that are also likely to regulate the neuronal activity in a complementary way to AVP. This has been clearly demonstrated for dynorphin which terminates spontaneous phasic bursts in magnocellular neurosecretory cells by autocrine inhibition of plateau potentials via the activation of κ-opioid receptors [50, 51]. The temporal dissociation of the effects by dynorphin and AVP attests for a complex and exquisite feedback regulation of rhythmogenesis in AVP cells by these two dendritically copackaged and coreleased peptides. A final feedback regulation by colocalized opioid peptides is exerted in the neurohypophysis on AVP release. AVP neurons also express peptides, which are packaged in granules distinct from those containing AVP, and these peptides control different autocrine/paracrine and endocrine functions. This is the case for apelin, which behaves the opposite way to AVP (see Section 8.5.1.2).

8.5.1.2 Apelinergic System

The neurosecretory neurons release AVP into the fenestrated capillaries of the posterior pituitary gland in response to changes in plasma osmolality and volemia [52, 53]. Dual immunofluorescence labeling confocal microscopy studies have shown that apelin colocalizes with AVP [13, 54] and ocytocin [27] in magnocellular hypothalamic neurons. Furthermore, double-labeling studies combining immunohistochemistry and *in situ* hybridization have demonstrated that, in the SON and PVN, apelin receptors [19, 55] are synthesized by magnocellular AVP neurons. Thus, AVP neurons express apelin, in addition to apelin receptors and V_{1a} and V_{1b} vasopressinergic receptors [56], suggesting a direct autocrine feedback regulation of these neurons by the two neuropeptides.

In response to osmotic stimulation, there is an increase in the somatodendritic release of AVP from SON and PVN cell bodies [33, 40], enhancing, via V_1 autoreceptors the phasic activity pattern of AVP neurons [40] and thereby facilitating systemic AVP release. Similarly, apelin may be involved in the autocrine somatodendritic feedback regulation of neurohypophyseal AVP neurons. This hypothesis has been tested in lactating rats exhibiting a reinforced phasic pattern of AVP neurons during lactation, thereby facilitating systemic AVP release to maintain body water content for optimal milk production. In this model, the intracerebroventricular injection of K17F inhibits the phasic firing activity of AVP neurons, thereby decreasing AVP release into the bloodstream, leading to aqueous diuresis [13] (Figure 8.2). Similarly, a marked decrease in systemic AVP release is observed following the intracerebroventricular injection of K17F or pE13F in mice deprived of water for 24 h [19], a condition known to increase AVP neuron activity. These data suggest that apelin is probably released from the SON and PVN AVP cell bodies to inhibit both AVP neuron activity and AVP release by acting directly on the apelin autoreceptors expressed by AVP/apelin-containing neurons. This mechanism probably involves apelin acting as a natural inhibitor of the antidiuretic effect of AVP. The colocalization and opposite biological actions of these two peptides raise questions concerning how these two peptides are regulated to maintain body fluid homeostasis.

This led to the investigation of the effect of water deprivation on the neuronal content and release of both apelin and AVP. Water deprivation increases somatodendritic and systemic AVP release, thereby causing the depletion of

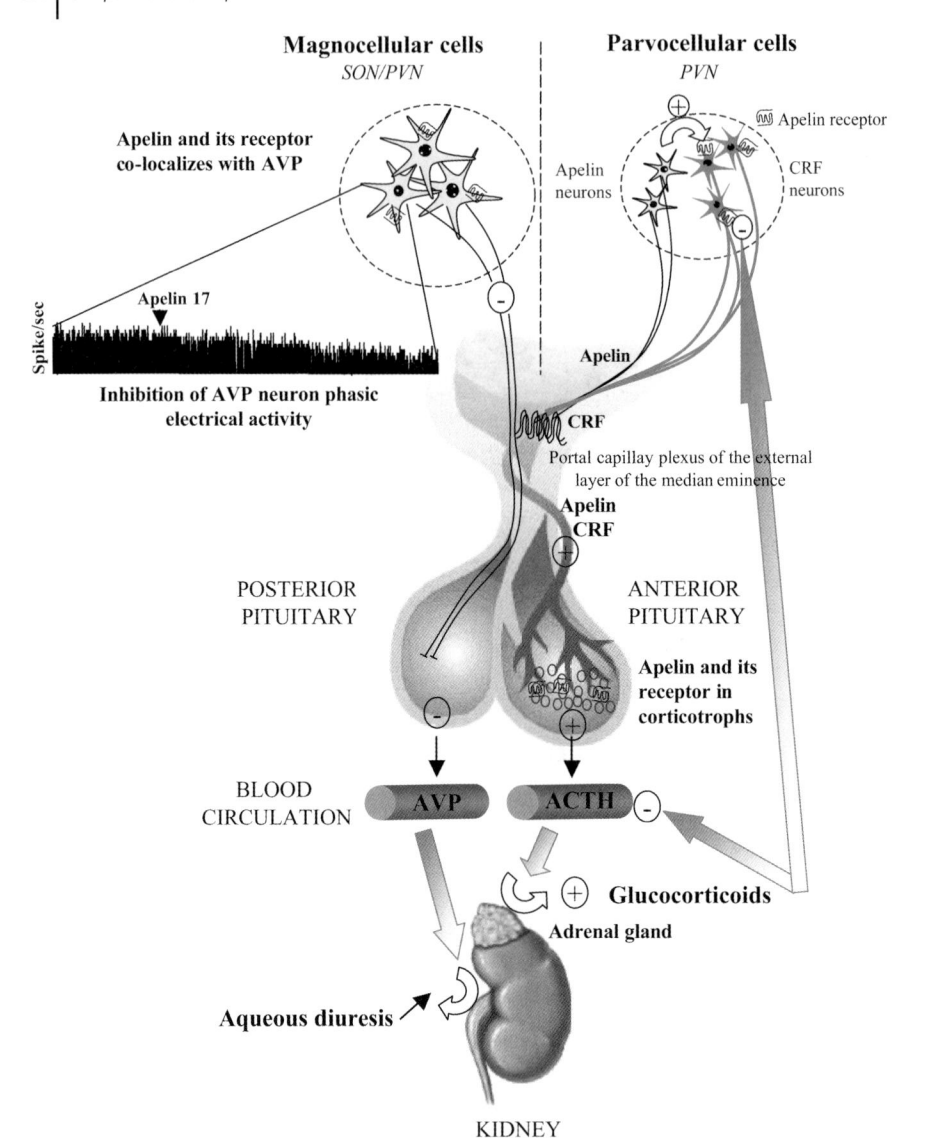

Figure 8.2 Schematic diagram illustrating regulation of the vasopressinergic magnocellular system and the hypothalamic–adrenal–pituitary axis by apelin and its receptor. Within the magnocellular neurons, in rodents, apelin and its receptor are colocalized with AVP in the SON and PVN magnocellular neurons. In lactating animals, the central injection of apelin 17 induces gradual and sustained inhibition of the phasic electrical activity of AVP neurons, thereby decreasing systemic AVP secretion and increasing aqueous diuresis. Within the hypothalamic–adrenal–pituitary axis, apelin stimulates ACTH secretion *indirectly* by activating CRF parvocellular neurons in the hypothalamus or *directly* in the anterior pituitary gland, in an autocrine/paracrine fashion. Within the adrenal gland, ACTH then induces the secretion of glucocorticoids, which regulate, via a negative-feedback loop, the secretion of CRF, and apelin receptor mRNA expression in the hypothalamus and ACTH secretion in the pituitary gland.

magnocellular neuronal AVP. It also decreases plasma apelin concentrations and induces a large increase in hypothalamic apelin neuronal content, especially that of the PVN and SON magnocellular neurons [13, 54] (Figure 8.3), suggesting that, under these conditions, apelin accumulates within AVP neurons rather than being released. The dehydration-induced increase in hypothalamic neuron apelin content is markedly diminished by the central injection of a selective V_1 receptor antagonist and is mimicked by the central infusion of AVP. This suggests that the dehydration-induced accumulation of apelin in hypothalamic neurons is caused by the somatodendritic release of AVP, which, by acting on the V_1 receptor, inhibits apelin release [54]. The apelin response to dehydration is therefore the opposite of that of AVP, which is released faster than it is synthesized [13, 57]. This interpretation implies that apelin and AVP are released differentially by the

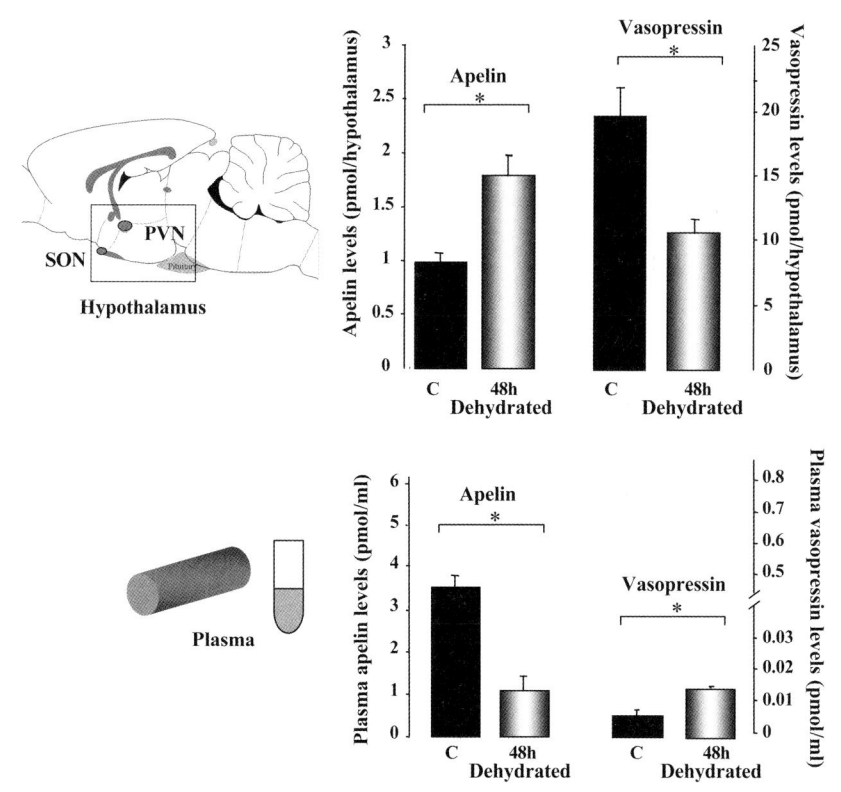

Figure 8.3 Apelin and vasopressin are inversely regulated during water deprivation in the rat hypothalamus and plasma. In rats deprived of water for 48 h, a large increase in hypothalamic apelin content (especially in that of the PVN and SON magnocellular neurons) is mirrored by a decrease in plasma apelin levels, suggesting that, under these conditions, apelin accumulates within AVP neurons rather than being released. Conversely, AVP levels decrease within the hypothalamus and increase in the bloodstream. Plasma (pmol/ml) and hypothalamic (pmol/hypothalamus) apelin concentrations and AVP levels were measured by radio-immunoassay.

magnocellular AVP neurons in which they are produced. Consistent with this hypothesis, double-labeling and confocal microscopy studies have demonstrated that AVP and apelin are mainly present in populations of vesicles differing in size and distribution in magnocellular neurons [13, 54].

These opposite regulatory patterns of apelin and AVP suggest that these peptides act in concert to maintain body fluid homeostasis. During dehydration, increases in the somatodendritic release of AVP optimize the phasic activity of AVP neurons [33, 40], facilitating the release of AVP into the bloodstream, whereas apelin accumulates in these neurons rather than being released into the bloodstream and, probably, into the nuclei. Thus, decreases in the local supply of apelin to the SON and PVN AVP cell bodies may facilitate the expression by AVP neurons of optimized phasic activity, by decreasing the inhibitory effects of apelin on these neurons. This concerted regulation by apelin and AVP has a biological purpose, making it possible to maintain the water balance of the organism by preventing additional water loss via the kidneys. This reveals a new physiological concept of dual potentiality for endocrine neurones that, according to the degree of their activation/inhibition, will dynamically ensure opposite physiological functions in perfect accordance with the hormonal demand, owing to the selective release of one of their coexpressed peptides.

It remains unclear whether the aquaretic effect of apelin involves only a central effect on systemic AVP release or whether it also involves a renal effect mediated by the activation of intrarenal apelin receptors. Consistent with a renal action of apelin, APJ receptor, preproapelin mRNA and apelin peptide have been detected in the rat and human kidney [12, 16, 58], and APJ receptor mRNA has been shown to be particularly abundant in the vasa recta in the inner stripe of the outer medulla, a region known to play a key role in water and sodium balance [12]. These studies highlight the crucial role played by apelin in the maintenance of body fluid homeostasis, by counteracting AVP actions.

Moreover, further evidence for a role for apelin in the control of water balance, depending not only on water loss but also on water and salt intake, is provided by the significant decrease in water intake following the intracerebroventricular administration of apelin in rats deprived of water for 24 h [19].

It would be of interest to investigate plasma apelin levels in parallel with plasma AVP concentration in various pathological states of euvolemic and hypervolemic hyponatremia. Analyses of the relationship between apelin concentration, AVP concentration and osmolality in plasma may reveal new classifications of the multiple etiologies of hypoosmolar states of impaired urinary dilution or concentration in patients.

8.5.2
Apelin, Like Vasopressin, is Involved in Regulating the Hypothalamic–Adrenal–Pituitary Axis

The existence of an apelinergic system within the adult male rat anterior pituitary gland has recently been reported. Numerous apelin-immunoreactive cells have been visualized within the anterior pituitary gland. Double-fluorescence immunohistochemistry studies have shown that apelin is highly expressed by corticotrophs and, to a lesser extent, by somatotrophs [59]. *In situ* hybridization combined with immunohistochemistry has shown that apelin receptor mRNA is also highly expressed in corticotrophs [59], like V_{1b} [60], AT_{1B} [61] and corticotropin-releasing factor (CRF) type 1 [62] receptors, suggesting a local interaction between apelin and adrenocorticotrophic hormone (ACTH). This was confirmed by *ex vivo* anterior pituitary gland perfusion studies, which showed that apelin increased basal or K^+-evoked ACTH release. Together, these data indicate that apelin in corticotrophs or somatotrophs exerts a direct stimulatory effect in the anterior pituitary gland on ACTH release in an autocrine/paracrine fashion.

Apelin nerve fibers and apelin receptor mRNA expression were also visualized in the parvocellular part of the PVN [55]. The central administration of apelin significantly increases plasma ACTH [63] and corticosterone [63, 64] release, at least in part by stimulating CRF release [63]. Apelin has also been shown to stimulate significantly the release of CRF and AVP from hypothalamic explants *in vitro* [9]. These data suggest that apelin stimulates ACTH release not only by direct effects in the anterior pituitary, but also by indirect effects involving the stimulation of CRH release in the hypothalamus (Figure 8.2). Consistent with this model, apelin receptor mRNA expression was found to be increased in the parvocellular division of the PVN in rats subjected to acute stress (restraint stress), which is known to increase the activity of the hypothalamic–adrenal–pituitary axis [55]. More recently, Li Wei *et al.* reported that dexamethasone, a glucocorticoid agonist, strongly decreases apelin mRNA levels in 3T3-L1 mouse adipocytes [65], whereas it increases apelin receptor mRNA levels in adrenalectomized rats [55], suggesting that glucocorticoids downregulate the expression of apelin and its receptor (Figure 8.2).

8.6
Peripheral Cardiovascular Actions

Apelin and its receptor are present in the cardiovascular system. Apelin has been detected in blood vessels, in the human heart and in kidney [58]. Apelin receptor mRNA expression has been detected in the heart and kidney in the adult rat [12], and in vascular endothelial cells of mouse embryos [6]. Apelin receptor-binding sites have been visualized in rat and human myocardium, and in human coronary artery, aorta and saphenous vein [66]. In addition, apelin and its receptor have been shown to play an important role in the development of the cardiovascular system, contributing to endothelium and cardiac differentiation in *Xenopus laevis*, and to myocardial cell specification and heart development in zebrafish [67–69].

The injection of apelin into the bloodstream decreases arterial blood pressure in normotensive Wistar-Kyoto and spontaneously hypertensive rats [11, 17, 19, 70, 71], via a mechanism dependent on NO production [71]. Baseline mean arterial blood pressure is similar in wild-type and apelin receptor-deficient mice, indicating that the apelin system might not be actively involved in maintaining blood pressure. However, apelin receptor-deficient mice have an enhanced vasopressor response to systemic Ang II, suggesting a counterregulatory action of apelin on Ang II [72]. In addition, apelin also modulates abnormal aortic vascular tone in response to Ang II and acetylcholine, by activating endothelial NO synthase in diabetic mice, providing further support for a role for apelin in vascular function [69].

Apelin has been shown *ex vivo* to be one of the most potent endogenous positive inotropic substances when applied to isolated rat hearts [24]. This action may be mediated by direct activation of the sarcolemmal Na^+/H^+ exchanger present in cardiomyocytes, leading to an increase in myofilament sensitivity to Ca^{2+}, as recently shown *in vitro* in cardiomyocytes in culture [23].

Positive inotropic effects of apelin have been also observed *in vivo* in both normal rat hearts and in rats with heart failure after myocardial infarction [73–75]. In addition, the acute administration of apelin *in vivo* reduces left ventricular pre- and afterload, and increases cardiac contractility. These effects are accompanied by a slight decrease in cardiac output. Conversely, chronic apelin infusion increases cardiac output without causing hypertrophy [76]. Apelin immunoreactivity has been found to increase in the plasma of patients in the early stages of heart failure and then to decrease during later, more severe stages of heart failure [77]. Together, these results indicate that apelin system may play a major role in the regulation of cardiovascular functions.

8.7
Conclusions and Pathophysiological Implications

The identification of apelin as the endogenous ligand of the orphan APJ receptor constitutes a major advance, both for fundamental research and, potentially, for clinical practice. It demonstrates the validity of the "deorphanization" approach to orphan receptors for the identification of new bioactive peptides and new therapeutic targets. The experimental data obtained to date in lactating rats demonstrate that apelin, by inhibiting the phasic electrical activity of AVP neurons and the systemic secretion of AVP, induces water diuresis. In the periphery, apelin decreases arterial blood pressure and increases the contractile force of the myocardium. Overall, these data show that this newly identified vasoactive peptide may play a key role in the maintenance of water balance and cardiovascular functions. The development of nonpeptide agonists of the apelin receptor, based on knowledge of the structures of apelin and its receptor, may provide alternative therapeutic approaches, or approaches complementary to the use of V_2 receptor antagonists [78] for the treatment of patients with water retention and/or hyponatremic disorders. These compounds also may constitute promising potential treatments for cardiovascular disorders. In particular, their administration to patients with heart failure might improve the contractile performance of the myocardium while reducing cardiac loading and increasing aqueous diuresis.

References

1. O'Dowd, B.F. et al. (1993) A human gene that shows identity with the gene encoding the angiotensin receptor is located on chromosome 11. *Gene*, **136**, 355–60.
2. Reaux, A. et al. (1999) Aminopeptidase A inhibitors as potential central antihypertensive agents. *Proceedings of the National Academy of Sciences of the United States of America*, 96, 13415–20.
3. Zini, S. et al. (1998) Inhibition of vasopressinergic neurons by central injection of a specific aminopeptidase A inhibitor. *Neuroreport*, **9**, 825–8.
4. Zini, S. et al. (1996) Identification of metabolic pathways of brain angiotensin II and III using specific aminopeptidase inhibitors: predominant role of angiotensin III in the control of vasopressin release. *Proceedings of the National Academy of Sciences of the United States of America*, 93, 11968–73.
5. De Mota, N., Lenkei, Z. and Llorens-Cortes, C. (2000) Cloning, pharmacological characterization and brain distribution of the rat apelin receptor. *Neuroendocrinology*, **72**, 400–7.
6. Devic, E. et al. (1999) Amino acid sequence and embryonic expression of *msr/apj*, the mouse homolog of *Xenopus X-msr* and human *APJ*. *Mechanisms of Development*, **84**, 199–203
7. Murphy, T.J. et al. (1991) Isolation of a cDNA encoding the vascular type-1 angiotensin II receptor. *Nature*, **351**, 233–6.
8. Lenkei, Z. et al. (2000) A highly sensitive quantitative cytosensor technique for the identification of receptor ligands in tissue extracts. *Journal of Histochemistry and Cytochemistry*, **48**, 1553–64.
9. Tatemoto, K. et al. (1998) Isolation and characterization of a novel endogenous peptide ligand for the human APJ receptor. *Biochemical and Biophysical Research Communications*, **251**, 471–6.
10. Habata, Y. et al. (1999) Apelin, the natural ligand of the orphan receptor APJ, is abundantly secreted in the colostrum. *Biochimica et Biophysica Acta*, **13**, 25–35.
11. Lee, D.K. et al. (2000) Characterization of apelin, the ligand for the APJ receptor. *Journal of Neurochemistry*, **74**, 34–41.
12. O'Carroll, A.M. et al. (2000) Distribution of mRNA encoding *B78/apj*, the rat homologue of the human *APJ* receptor, and its endogenous ligand apelin in brain and peripheral tissues. *Biochimica et Biophysica Acta*, **21**, 72–80.
13. De Mota, N. et al. (2004) Apelin, a potent diuretic neuropeptide counteracting vasopressin actions through inhibition of vasopressin neuron activity and vasopressin release. *Proceedings of the National Academy of Sciences of the United States of America*, 101, 10464–9.
14. Hosoya, M. et al. (2000) Molecular and functional characteristics of APJ. Tissue distribution of mRNA and interaction with the endogenous ligand apelin. *Journal of Biological Chemistry*, **275**, 21061–7.
15. Kawamata, Y. et al. (2001) Molecular properties of apelin: tissue distribution and receptor binding. *Biochimica et Biophysica Acta*, **23**, 2–3.
16. Medhurst, A.D. et al. (2003) Pharmacological and immunohistochemical characterization of the APJ receptor and its endogenous ligand apelin. *Journal of Neurochemistry*, **84**, 1162–72.
17. El Messari, S. et al. (2004) Functional dissociation of apelin receptor signaling and endocytosis: implications for the effects of apelin on arterial blood pressure. *Journal of Neurochemistry*, **90**, 1290–301.
18. Choe, W. et al. (2000) Functional expression of the seven-transmembrane HIV-1 co-receptor APJ in neural cells. *Journal of Neurovirology*, **6**, S61–9.
19. Reaux, A. et al. (2001) Physiological role of a novel neuropeptide, apelin, and its receptor in the rat brain. *Journal of Neurochemistry*, **77**, 1085–96.
20. Zhou, N. et al. (2003) Cell-cell fusion and internalization of the CNS-based, HIV-1 co-receptor, APJ. *Virology*, **307**, 22–36.
21. Masri, B. et al. (2002) Apelin (65–77) activates extracellular signal-regulated kinases via a PTX-sensitive G protein.

Biochemical and Biophysical Research Communications, **290**, 539–45.
22 Masri, B. et al. (2004) Apelin (65–77) activates p70 S6 kinase and is mitogenic for umbilical endothelial cells. *FASEB Journal*, **22**, 22.
23 Farkasfalvi, K. et al. (2007) Direct effects of apelin on cardiomyocyte contractility and electrophysiology. *Biochemical and Biophysical Research Communications*, **357**, 889–95.
24 Szokodi, I. et al. (2002) Apelin, the novel endogenous ligand of the orphan receptor APJ, regulates cardiac contractility. *Circulation Research*, **91**, 434–40.
25 Zhong, J.C. et al. (2007) Apelin modulates aortic vascular tone via endothelial nitric oxide synthase phosphorylation pathway in diabetic mice. *Cardiovascular Research*, **74**,, 388–95.
26 Reaux, A. et al. (2002) Distribution of apelin-synthesizing neurons in the adult rat brain. *Neuroscience*, **113**, 653–62.
27 Brailoiu, G.C. et al. (2002) Apelin-immunoreactivity in the rat hypothalamus and pituitary. *Neuroscience Letters*, **327**, 193–7.
28 Imura, H. , Nakao, K. and Itoh, H. (1992) The natriuretic peptide system in the brain: implications in the central control of cardiovascular and neuroendocrine functions. *Frontiers in Neuroendocrinology*, **13**, 217–49.
29 Lenkei, Z. et al. (1997) Expression of angiotensin type-1 (AT_1) and type-2 (AT_2) receptor mRNAs in the adult rat brain: a functional neuroanatomical review. *Frontiers in Neuroendocrinology*, **18**, 383–439.
30 Saavedra, J.M. (1992) Brain and pituitary angiotensin. *Endocrine Reviews*, **13**, 329–80.
31 De Bree, F.M. et al. (2003) The hormone domain of the vasopressin prohormone is required for the correct prohormone trafficking through the secretory pathway. *Journal of Neuroendocrinology*, **15**, 1156–63.
32 Hatton, G.I. (1990) Emerging concepts of structure–function dynamics in adult brain: the hypothalamo-neurohypophysial system. *Progress in Neurobiology*, **34**, 437–504.
33 Ludwig, M. (1998) Dendritic release of vasopressin and oxytocin. *Journal of Neuroendocrinology*, **10**, 881–95.
34 Morris, J.F. and Pow, D.V. (1988) Capturing and quantifying the exocytotic event. *Journal of Experimental Biology*, **139**, 81–103.
35 Pow, D.V. and Morris, J.F. (1989) Dendrites of hypothalamic magnocellular neurons release neurohypophysial peptides by exocytosis. *Neuroscience*, **32**, 435–9.
36 Ludwig, M. et al. (1997) Direct hypertonic stimulation of the rat supraoptic nucleus increases c-fos expressionin glial cells rather than magnocellular neurones. *Cell and Tissue Research*, **287**, 79–90.
37 Tobin, V.A. et al. (2004) Thapsigargin-induced mobilization of dendritic dense-cored vesicles in rat supraoptic neurons. *European Journal of Neuroscience*, **19**, 2909–12.
38 Ludwig, M. et al. (2002) Intracellular calcium stores regulate activity-dependent neuropeptide release from dendrites. *Nature*, **418**, 85–9.
39 Ludwig, M. et al. (2005) Regulation of activity-dependent dendritic vasopressin release from rat supraoptic neurones. *Journal of Physiology*, **564**, 515–22.
40 Gouzenes, L. et al. (1998) Vasopressin regularizes the phasic firing pattern of rat hypothalamic magnocellular vasopressin neurons. *Journal of Neuroscience*, **18**, 1879–85.
41 Hurbin, A. et al. (2002) The vasopressin receptors colocalize with vasopressin in the magnocellular neurons of the rat supraoptic nucleus and are modulated by water balance. *Endocrinology*, **143**, 456–66.
42 Kombian, S.B. , Mouginot, D. and Pittman, Q.J. (1997) Dendritically released peptides act as retrograde modulators of afferent excitation in the supraoptic nucleus *in vitro*. *Neuron*, **19**, 903–12.
43 Pittman, Q.J. et al. (2000) Neurohypophysial peptides as retrograde transmitters in the supraoptic nucleus of the rat. *Experimental Physiology*, **85**, 139S–43S.
44 Kombian, S.B. et al. (2002) Modulation of synaptic transmission by oxytocin and

vasopressin in the supraoptic nucleus. *Progress in Brain Research*, **139**, 235–46.
45. Hermes, M.L. et al. (2000) Vasopressin increases GABAergic inhibition of rat hypothalamic paraventricular nucleus neurons *in vitro*. *Journal of Neurophysiology*, **83**, 705–11.
46. Kagotani, Y. et al. (1990) Intragranular co-storage of neuropeptide Y and arginine vasopressin in the paraventricular magnocellular neurons of the rat hypothalamus. *Cell and Tissue Research*, **262**, 47–52.
47. Landry, M. et al. (2003) Differential routing of coexisting neuropeptides in vasopressin neurons. *European Journal of Neuroscience*, **17**, 579–89.
48. Watson, S.J. et al. (1982) Dynorphin and vasopressin: common localization in magnocellular neurons. *Science*, **216**, 85–7.
49. Whitnall, M.H. et al. (1983) Dynorphin-A-(1–8) is contained within vasopressin neurosecretory vesicles in rat pituitary. *Science*, **222**, 1137–9.
50. Brown, C.H. and Bourque, C.W. (2004) Autocrine feedback inhibition of plateau potentials terminates phasic bursts in magnocellular neurosecretory cells of the rat supraoptic nucleus. *Journal of Physiology*, **557**, 949–60.
51. Brown, C.H. et al. (2006) Endogenous activation of supraoptic nucleus kappa-opioid receptors terminates spontaneous phasic bursts in rat magnocellular neurosecretory cells. *Journal of Neurophysiology*, **95**, 3235–44.
52. Brownstein, M.J., Russell, J.T. and Gainer, H. (1980) Synthesis, transport, and release of posterior pituitary hormones. *Science*, **207**, 373–8.
53. Manning, M. et al. (1977) Design of neurohypophyseal peptides that exhibit selective agonistic and antagonistic properties. *Federation Proceedings*, **36**, 1848–52.
54. Reaux-Le Goazigo, A. et al. (2004) Dehydration-induced cross-regulation of apelin and vasopressin immunoreactivity levels in magnocellular hypothalamic neurons. *Endocrinology*, **145**, 4392–400.
55. O'Carroll, A.M., Don, A.L. and Lolait, S.J. (2003) APJ receptor mRNA expression in the rat hypothalamic paraventricular nucleus: regulation by stress and glucocorticoids. *Journal of Neuroendocrinology*, **15**, 1095–101.
56. Hurbin, A. et al. (1998) The V_{1a} and V_{1b}, but not V_2, vasopressin receptor genes are expressed in the supraoptic nucleus of the rat hypothalamus, and the transcripts are essentially colocalized in the vasopressinergic magnocellular neurons. *Endocrinology*, **139**, 4701–7.
57. Zingg, H.H., Lefebvre, D. and Almazan, G. (1986) Regulation of vasopressin gene expression in rat hypothalamic neurons. Response to osmotic stimulation. *Journal of Biological Chemistry*, **261**, 12956–9.
58. Kleinz, M.J. and Davenport, A.P. (2004) Immunocytochemical localization of the endogenous vasoactive peptide apelin to human vascular and endocardial endothelial cells. *Regulatory Peptides*, **118**, 119–25.
59. Reaux-Le Goazigo, A. et al. (2007) Cellular localization of apelin and its receptor in the anterior pituitary: evidence for a direct stimulatory action of apelin on ACTH release. *American Journal of Physiology. Endocrinology and Metabolism*, **292**, E7–15.
60. Lolait, S.J. et al. (1995) Extrapituitary expression of the rat V_{1b} vasopressin receptor gene. *Proceedings of the National Academy of Sciences of the United States of America*, 92, 6783–7.
61. Lenkei, Z. et al. (1999) Identification of endocrine cell populations expressing the AT_{1B} subtype of angiotensin II receptors in the anterior pituitary. *Endocrinology*, **140**, 472–7.
62. Potter, E. et al. (1994) Distribution of corticotropin-releasing factor receptor mRNA expression in the rat brain and pituitary. *Proceedings of the National Academy of Sciences of the United States of America*, 91, 8777–81.
63. Taheri, S. et al. (2002) The effects of centrally administered apelin-13 on food intake, water intake and pituitary hormone release in rats. *Biochemical and Biophysical Research Communications*, **291**, 1208–12.
64. Jaszberenyi, M., Bujdoso, E. and Telegdy, G. (2004) Behavioral, neuroendocrine and thermoregulatory actions of apelin-13. *Neuroscience*, **129**, 811–6.
65. Wei, L., Hou, X. and Tatemoto, K. (2005) Regulation of apelin mRNA expression by

insulin and glucocorticoids in mouse 3T3-L1 adipocytes. *Regulatory Peptides*, **132**, 27–32.

66 Katugampola, S.D. et al. (2001) [^{125}I]-(Pyr^1)Apelin-13 is a novel radioligand for localizing the APJ orphan receptor in human and rat tissues with evidence for a vasoconstrictor role in man. *British Journal of Pharmacology*, **132**, 1255–60.

67 Inui, M. et al. (2006) Xapelin and Xmsr are required for cardiovascular development in Xenopus laevis. *Developmental Biology*, **298**, 188–200.

68 Scott, I.C. et al. (2007) The G protein-coupled receptor Agtrl1b regulates early development of myocardial progenitors. *Developmental Cell*, **12**, 403–13.

69 Zeng, X.X. et al. (2007) Apelin and its receptor control heart field formation during zebrafish gastrulation. *Developmental Cell*, **12**, 391–402.

70 Lee, D.K. et al. (2005) Modification of the terminal residue of apelin-13 antagonizes its hypotensive action. *Endocrinology*, **146**, 231–6.

71 Tatemoto, K. et al. (2001) The novel peptide apelin lowers blood pressure via a nitric oxide-dependent mechanism. *Regulatory Peptides*, **99**, 87–92.

72 Ishida, J. et al. (2004) Regulatory roles for APJ, a seven-transmembrane receptor related to angiotensin-type 1 receptor in blood pressure *in vivo*. *Journal of Biological Chemistry*, **279**, 26274–9.

73 Berry, M.F. et al. (2004) Apelin has *in vivo* inotropic effects on normal and failing hearts. *Circulation*, **110** (Suppl 1), II187–93.

74 Chen, M.M. et al. (2003) Novel role for the potent endogenous inotrope apelin in human cardiac dysfunction. *Circulation*, **108**, 1432–9.

75 Dai, T., Ramirez-Correa, G. and Gao, W.D. (2006) Apelin increases contractility in failing cardiac muscle. *European Journal of Pharmacology*, **553**, 222–8.

76 Ashley, E.A. et al. (2005) The endogenous peptide apelin potently improves cardiac contractility and reduces cardiac loading *in vivo*. *Cardiovascular Research*, **65**, 73–82.

77 Foldes, G. et al. (2003) Circulating and cardiac levels of apelin, the novel ligand of the orphan receptor APJ, in patients with heart failure. *Biochemical and Biophysical Research Communications*, **308**, 480–5.

78 Schrier, R.W. et al. (2006) Tolvaptan, a selective oral vasopressin V$_2$-receptor antagonist, for hyponatremia. *New England Journal of Medicine*, **355**, 2099–112.

Part Three Amines

9
Serotonin

Michael Bader

9.1
Introduction

At the beginning of the twentieth century it was already known that serum contains a substance that constricts blood vessels [1]. In the late 1940s, Rapport and Page isolated this vasoconstricting substance and named it serotonin according to its origin and its tonic effect on vessels [2]. In 1949, Rapport purified and crystallized serotonin, and uncovered its chemical identity as the indoleamine 5-hydroxytryptophan (5-HT) [3]. More than 10 years earlier, Vittorio Erspamer had also extracted a substance from the rabbit gastric mucosa that induced smooth muscle contraction and had named it enteramine [4]. Erspamer also purified it and found it to be identical to serotonin in 1952 [5]. One year later, Twarog and Page discovered serotonin in the brain of vertebrates [6]. This was the beginning of its career as neurotransmitter, which left behind the analysis as a cardiovascular hormone. Nevertheless, more than 90% of the 5-HT of a mammal is generated in the enterochromaffin cells in the gut and secreted for the use by platelets distributing it into the whole body. Some serotonin is also produced in the pineal gland as a precursor of melatonin. Serotonin from these sources cannot reach central sites since its entry in the brain is blocked by the blood–brain barrier. Thus, the peripheral and the central serotonin system are strictly separated (Figure 9.1). The neurotransmitter serotonin is generated locally, in the raphe nuclei of the brain stem. In these dispersed nuclei, there are only about 20 000 neurons that produce nearly all brain 5-HT and innervate the rest of the central nervous system (CNS) [7]. There, serotonin influences a multitude of behaviors such as food intake, anxiety, aggression, sexuality and mood. Therefore, it is also the target of most drugs of abuse, and also of the majority of psychotropic drugs to treat diseases such as depression, and bipolar, panic and eating disorders.

Cardiovascular Hormone Systems. Edited by Michael Bader
Copyright © 2008 WILEY-VCH Verlag GmbH & Co. KGaA, Weinheim
ISBN: 978-3-527-31920-6

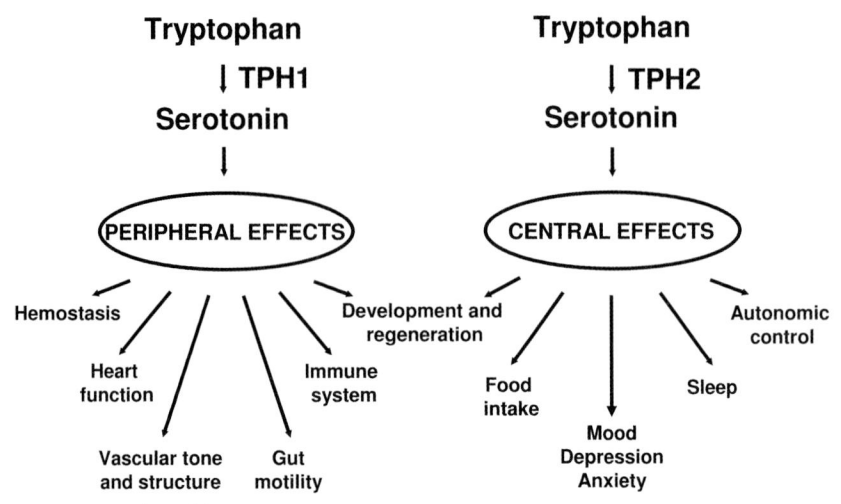

```
       Tryptophan                    Tryptophan
         ↓ TPH1                        ↓ TPH2
        Serotonin                     Serotonin
            ↓                             ↓
   ( PERIPHERAL EFFECTS )         ( CENTRAL EFFECTS )
```

```
Hemostasis     Development and              Autonomic
               regeneration                  control
  Heart        Immune          Food
 function      system         intake        Sleep
                                             ↓
 Vascular tone   Gut           Mood
 and structure  motility    Depression
                              Anxiety
```

Figure 9.1 Duality of the serotonin system. Serotonin is formed by two distinct TPH isoforms in the periphery (mainly in the gut) and the brain, and has a multitude of differential actions in physiology and pathophysiology.

From an evolutionary perspective, serotonin is a very old hormone. It is found in high amounts in plants, where it functions as a trophic factor and antioxidant [8], and in probably all animals. In worms it has important hormonal functions regulating food intake and lifespan [9, 10]. In insects, serotonin already works as cardiovascular hormone by influencing heart function [11].

Extracellular levels of serotonin are minute under normal conditions. It is sequestered very efficiently into intracellular granules in neurons and platelets and other cells by two transporter systems (see Section 9.2.2), and only released locally on demand (Figure 9.2). Interacting with at least 14 different receptors (see Section 9.2.3), it elicits its numerous effects in the CNS and in the periphery (Table 9.1). Immediately after release it is again taken up by the releasing and neighboring cells, where it is recycled and/or degraded. Thus, serotonin is more an autacoid than a hormone since it is locally released in high concentrations, exerts effects for seconds to minutes and then disappears again.

The literature about serotonin is overwhelming and the amount of PubMed entries has just jumped over the 100 000 publications mark. Thus, it is impossible to summarize all of our knowledge about 5-HT. This chapter will therefore focus on the findings about cardiovascular effects of the hormone in the CNS and in the periphery. Even these effects of serotonin cannot be comprehensively listed and the reader is referred to previous, more detailed reviews of the subject [24–32].

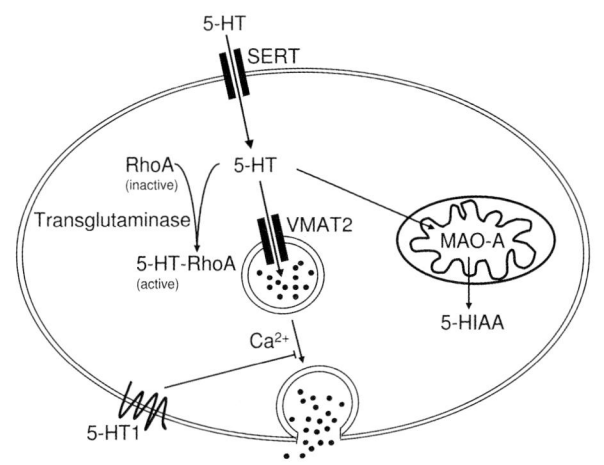

Figure 9.2 Fates of serotonin in neurons and platelets. 5-HT is taken up by SERT into the cytoplasm. There, it can be used for signaling by covalent linkage to small G-proteins (e.g. RhoA) by transglutaminases (shown for platelets and smooth muscle cells), it can be degraded in mitochondria by MAO-A, and it can be further transported and stored in vesicles by VMAT2. The vesicles release their content after activation of the cell eliciting an increase in intracellular Ca^{2+}. $5\text{-}HT_1$ receptors function as negative feedback autoreceptors on neurons inhibiting serotonin secretion. HIAA, hydroxyindoleacetic acid.

9.2 Components of the Serotonin System

9.2.1 Enzymes

9.2.1.1 Tryptophan Hydroxylases

Tryptophan hydroxylases (TPHs) are the rate-limiting enzymes in serotonin synthesis and catalyze the first step of tryptophan metabolism, the formation of 5-hydroxytryptophan (5-HTP) (Figure 9.3). TPHs belong to a superfamily of aromatic amino acid hydroxylases, together with phenylalanine hydroxylase and tyrosine hydroxylase [33]. The enzymatic reaction employs tetrahydrobiopterin and oxygen as cosubstrates, and iron as cofactor. We have recently discovered that two different genes code for TPH enzymes, *TPH1* and *TPH2* [34, 35]. TPH1 is expressed in enterochromaffin cells in the gut and in the pineal gland, and TPH2 is nearly exclusively expressed in the raphe nuclei of the brain stem. Thus, TPH1 is at the basis of the peripheral serotonin system and TPH2 initiates central serotonin generation (Figure 9.1). Mice lacking TPH1 have been generated by two groups [34, 36]. We found that these animals show an impaired thrombogenesis due to a defect in the secretion of von Willebrand factor (vWF) from platelets [37]. Furthermore, with the help of these mice an essential role of peripheral serotonin could be established in mammary gland plasticity and liver regeneration [38, 39]. TPH1-

Table 9.1 Serotonin receptors: the signaling pathways, knockout mouse phenotypes (if existing), and the drugs in current clinical use are shown for all seven families of 5-HT receptors.

Family	Subtype	Main signalling pathway	Main expression sites	Main knockout phenotype	Clinically used drugs
5-HT$_1$	5-HT$_{1A}$	G$_{i/o}$ adenylate cyclase inhibition	raphe nuclei, hippocampus, cortex	increased anxiety [12–14]	"-pirones" agonist
	5-HT$_{1B}$	G$_{i/o}$ adenylate cyclase inhibition	widespread in brain, endothelial cells, smooth muscle cells	increased aggression [15]	"-triptans" agonist
	5-HT$_{1D}$	G$_{i/o}$ adenylate cyclase inhibition	widespread in brain, endothelial cells, smooth muscle cells		"-triptans" agonist
	5-HT$_{1E}$	G$_{i/o}$ adenylate cyclase inhibition	widespread in brain	(not existing in mice)	
	5-HT$_{1F}$	G$_{i/o}$ adenylate cyclase inhibition	widespread in brain		"-triptans" agonist
5-HT$_2$	5-HT$_{2A}$	G$_{q/11}$ phospholipase C activation	platelets, smooth muscle cells, cortex	decreased anxiety [16]	sarpogrelate antagonist
	5-HT$_{2B}$	G$_{q/11}$ phospholipase C activation	cardiac cells, smooth muscle cells	cardiomyopathy [17]	
	5-HT$_{2C}$	G$_{q/11}$ phospholipase C activation	widespread in brain	increased appetite, overweight [18]	
5-HT$_3$	pentamer of 5-HT$_{3A}$ with 5-HT$_{3B}$, 5-HT$_{3C}$, 5-HT$_{3D}$ and 5-HT$_{3E}$	ion channel	peripheral nervous system and brain	reduced pain perception (5-HT$_{3A}$) [19]	"-setrons" antagonist
5-HT$_4$	5-HT$_4$	G$_s$ adenylate cyclase activation	cardiomyocytes, smooth muscle cells, widespread in brain	decreased stress response [20]	"-serods" agonist; "-saprides" agonist
5-HT$_5$	5-HT$_{5A}$	G$_{i/o}$ adenylate cyclase inhibition	widespread in brain	decreased anxiety [21]	
	5-HT$_{5B}$	?	habenula, raphe nuclei, hippocampus; pseudogene in humans		
5-HT$_6$	5-HT$_6$	G$_s$ adenylate cyclase activation	widespread in brain	altered alcohol response [22]	
5-HT$_7$	5-HT$_7$	G$_s$ adenylate cyclase activation	smooth muscle cells, widespread in brain	disturbed thermoregulation [23]	

deficient mice are also less affected by pulmonary hypertension ([40], see Section 9.3.2). Cote et al. describe an impaired fetal growth and lethal dilated cardiomyopathy in their independently established colony of TPH1-deficient animals [36, 41]. These phenotypes have not been observed in our animals (unpublished results).

9.2.1.2 Aromatic Amino Acid Decarboxylase

Aromatic amino acid decarboxylase catalyzes the second step of serotonin synthesis, the decarboxylation of 5-HTP (Figure 9.3). The enzyme is expressed in a multitude of tissues and has additional substrates such as L-3,4-dihydroxyphenylalanine, making it also essential for dopamine generation (see also Chapter 11). Thus, any interference with this enzyme (e.g. inhibition by carbidopa) will not be specific for the serotonin system and therefore the enzyme is not mentioned further in this chapter.

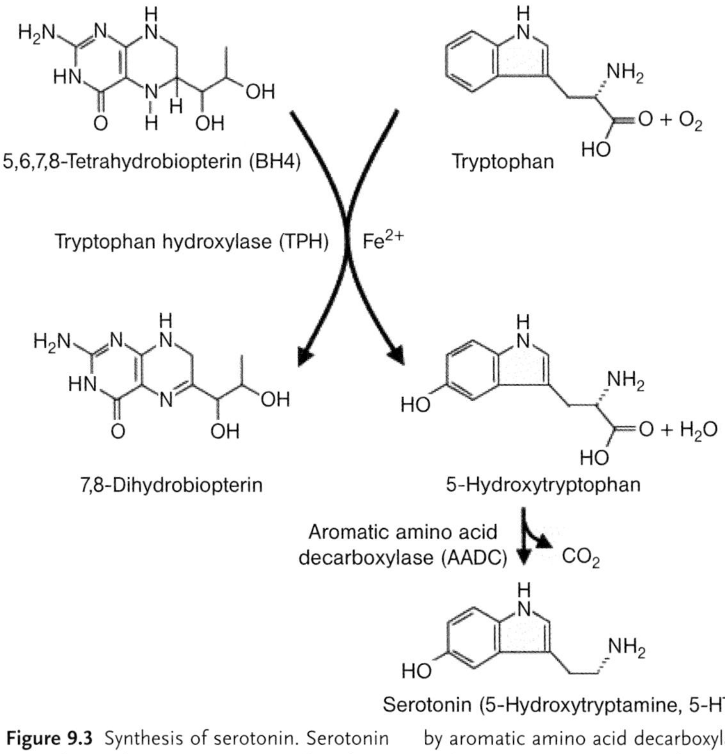

Figure 9.3 Synthesis of serotonin. Serotonin is synthesized in a two-step procedure from the essential amino acid tryptophan. In the rate-limiting step, TPH generates 5-hydroxytryptophan, which is further decarboxylated by aromatic amino acid decarboxylase (AADC) to the active product 5-HT. The TPH reaction needs O_2 and tetrahydrobiopterin (BH_4) as cosubstrates, and iron as cofactor and AADC releases CO_2 in the second step.

9.2.1.3 Monoamine Oxidases

Monoamine oxidase (MAO)-A and -B degrade biogenic amines and MAO-A is the major serotonin degrading enzyme. MAOs are located in mitochondria. Thus, serotonin needs first to be taken up by the cell via the serotonin transporter (SERT, see Section 9.2.2.1) before being metabolized to 5-hydroxyindoleacetic acid (Figure 9.2). Both enzymes are found in variable amounts in nearly all tissues including the brain. MAO inhibitors (e.g. tranylcypromine, moclobemide, selegiline) are used as psychotropic drugs since they increase the lifetime of serotonin. Knockout mice for each of the two and for both enzymes have been reported, and mainly show behavioral alterations [42–44]. However, again, interference with MAOs will affect several aminergic systems and is, thus, not mentioned further in this chapter.

9.2.2
Transporters

9.2.2.1 Serotonin Transporter

At neutral pH, serotonin is positively charged and cannot cross biological membranes. Therefore, it needs a transporter in the plasma membrane to be taken up by cells (Figure 9.2). Serotonin transporter (SERT) (5-HT transporter; other name: SLC6A4) is specific for 5-HT and transports most of the hormone. There is some residual transport activity observed in SERT-deficient animals, which is probably due to other transporters with a low affinity for serotonin [45–47].

SERT is expressed mainly on presynaptic endings of serotonergic neurons, in platelets and a multitude of other cell types. For platelets, the most important peripheral source of serotonin, uptake by SERT is the only supply of the hormone since they lack the 5-HT synthesizing machinery. SERT is extremely effective; the serotonin concentration inside platelets is more than a million times higher than in the plasma. The catalytic mechanism exploits ion gradients on the plasma membrane, and cotransports one Na^+ and one Cl^- ion with 5-HT in exchange for one K^+ ion [48]. SERT blockers, the selective serotonin reuptake inhibitors (SSRIs), such as fluoxetine (Prozac), paroxetine, citalopram, fluvoxamine and sertraline, are classical antidepressive drugs since they sustain extracellular serotonin concentrations and thereby the activation of 5-HT receptors. They also deplete platelets from serotonin, which may be the reason for a rare side-effect of the drugs, an increased propensity for bleeding ([49–51], see Section 9.3.1). Another class of drugs, the serotonin releasers including fenfluramine, reverses the transport when interacting with SERT and allows serotonin to leave the cell through SERT [52]. At the same time, some of them are countertransported into the cell. These drugs have been used as appetite suppressants, but major cardiovascular side-effects such as primary pulmonary hypertension (PPH) and valvular heart disease has stopped their clinical use about 10 years ago (see Sections 9.3.2 and 9.3.3).

In general, the actions of SERT in the cardiovascular system are rarely studied, and also the reports about the established knockout mouse and rat models deal nearly exclusively with behavioral phenotypes [47, 53]. The only cardiovascular phenotypes reported for mice with genetically modified SERT expression are changed susceptibilities for PPH and valvular heart disease ([54–56], see Sections 9.3.2 and 9.3.3).

9.2.2.2 Vesicular Monoamine Transporters

Serotonin, to reach storage granules, again needs to cross a lipid bilayer. Here, vesicular monoamine transporter (VMAT) 1 and 2 are instrumental (Figure 9.2) [57]. VMAT1 (SLC18A1) is present in the gut, and VMAT2 (SLC18A2) in neurons and platelets. The transport utilizes an electrochemical gradient across the vesicular membrane established by the vacuolar H^+-ATPase. Serotonin enters the vesicle by an exchange against two protons. The filling state of the vesicle regulates the activity of VMAT2 in neurons by a G-protein-mediated mechanism [58]. Some serotonin-releasing drugs, such as fenfluramine, also reverse the VMAT-induced transport [52]. VMATs do not only transport serotonin but also dopamine, catecholamines and histamine. Due to this multiple functions, knockout mice for VMAT2 die 1 week after birth and are, therefore, not informative for the cardiovascular actions of serotonin [59–61]. Thus, VMATs are also not mentioned further in this review.

9.2.3
5-HT Receptors

There is no other hormone in mammals which has more different receptors than serotonin. As yet 14 different 5-HT receptors are known, but only 13 have been detected in rodents (5-HT_{1E} does not exist in rats and mice) and also 13 in man (5-HT_{5B} is a pseudogene in humans). They are distributed in seven families by homology and signaling pathways, named 5-HT_{1-7} [24–26]. All of them are G-protein-coupled receptors with seven-transmembrane domains. The only exception is the 5-HT_3 receptor which is a ligand-gated ion channel.

Also except the 5-HT_3 receptor, most 5-HT receptors show ligand-independent constitutive activity [62–66]. Thus, the serotonin system may be partially independent of serotonin itself and serotonin probably just augments the signaling of its receptors. In the following and in Table 9.1 the families and the single receptors are described in more detail, in particular concerning their expression and function in the cardiovascular system.

9.2.3.1 5-HT_1 Family

The 5-HT_1 family of serotonin receptors has five members, 5-HT_{1A}, 5-HT_{1B}, 5-HT_{1D}, 5-HT_{1E} and 5-HT_{1F} (5-HT_{1C} was renamed 5-HT_{2C}). In mice and rats, the 5-HT_{1E} receptor does not exist. Their main signaling pathway comprises $G_{i/o}$-proteins inhibiting adenylate cyclase which reduces intracellular levels of cAMP. Since cAMP relaxes smooth muscles its reduction after 5-HT_1 receptor activation leads to vasoconstriction. However, only 5-HT_{1B} and 5-HT_{1D} are expressed in vessels, 5-HT_{1A}, 5-HT_{1E} and 5-HT_{1F} are centrally expressed receptors with mainly neuronal functions. 5-HT$_{1A}$ is the somatodendritic and 5-HT$_{1B}$ and 5-HT$_{1D}$ are the presynaptic autoreceptors on raphe neurons, all eliciting a negative feedback on serotonin release by hyperpolarizing the expressing neurons (Figure 9.2). 5-HT$_{1B}$ and 5-HT$_{1D}$ are expressed on cerebral vessels and, thus, agonists for 5-HT$_1$ receptors (-triptans) are important anti-migraine drugs (see Section 9.3.4).

9.2.3.2 5-HT$_2$ Family

The 5-HT$_2$ family of serotonin receptors has three members, 5-HT$_{2A}$, 5-HT$_{2B}$ and 5-HT$_{2C}$. There are several isoforms of 5-HT$_{2C}$ receptors due to alternative splicing and editing of its mRNA [67]; however, this receptor is not found in the cardiovascular system and therefore not dealt with in this chapter. The 5-HT$_{2B}$ receptor is hardly expressed in the brain, but mainly in liver, kidney, heart and smooth muscle cells. The 5-HT$_{2A}$ receptor is expressed in the brain and in the periphery. It is the most important receptor for the cardiovascular actions of serotonin since it is expressed in high amounts on platelets and smooth muscle cells. 5-HT$_2$ receptors couple primarily to G$_{q/11}$-proteins activating phospholipase C, which leads to increased diacylglycerol and inositol-1,4,5-triphosphate levels in cells causing Ca^{2+} entry and vasoconstriction. Therefore, antagonists for these receptors such as ketanserin have been used for the treatment of hypertension. However, cardiac and central side-effects have reduced the suitability of this drug. A novel and more specific 5-HT$_{2A}$ antagonist, sarprogrelate, may be useful as antithrombotic and antianginal drug [27].

9.2.3.3 5-HT$_3$ Family

The 5-HT$_3$ family of serotonin receptors has one member, 5-HT$_3$. However this ligand gated ion channel is a homo- or hetero-pentamer of at least five subunits, 5-HT$_{3A}$, 5-HT$_{3B}$, 5-HT$_{3C}$, 5-HT$_{3D}$ and 5-HT$_{3E}$, encoded by different genes [68]. The presence of the 5-HT$_{3A}$ subunit is essential for the function of the pentamer; the other subunits may vary and modulate the activity of the receptor [68, 19]. After binding of serotonin, a cation-driven current rapidly depolarizes 5-HT$_3$-expressing cells. 5-HT$_3$ receptors are mainly found on neurons in the peripheral and, to a lesser extent, CNS. Since they are also on neurons in the area postrema, where the vomiting reflex is relayed, antagonists for this subtype, the "-setrons", are well-established antiemetics. The cardiovascular effects of 5-HT$_3$ receptors are mainly due to their expression on nerve endings in the autonomic nervous system regulating cardiac function (see Section 9.3.3).

9.2.3.4 5-HT$_4$ Family

The 5-HT$_4$ family of serotonin receptors has one member, 5-HT$_4$. However, due to alternative splicing, several proteins with differential tissue distributions are encoded by the 5-HT$_4$ gene, differing in the length of their intracellular C-termini [69]. The 5-HT$_4$ receptor is coupled to G$_s$-proteins causing activation of adenylate cyclase and an increase in intracellular cAMP. It is expressed in the brain, heart and in several other peripheral organs, with high levels in the gut. There it seems to be a major regulator of peristalsis. Therefore, 5-HT$_4$ agonists, such as the

"-serods" and "-saprides", are successfully employed to treat gastrointestinal disorders. The most important site of expression in the cardiovascular system are cardiomyocytes, where 5-HT$_4$ activation improves contraction (see Section 9.3.3).

9.2.3.5 5-HT$_5$ Family

The 5-HT$_5$ family of serotonin receptors has two members, 5-HT$_{5A}$ and 5-HT$_{5B}$ (however, in humans, 5-HT$_{5B}$ is a silent pseudogene). These receptors are the least characterized in the serotonin system. There is evidence that 5-HT$_{5A}$ employs multiple signaling pathways but mainly couples to G$_{i/o}$-proteins inhibiting adenylate cyclase [70]. However, it is not yet clear what functions 5-HT$_5$ receptors have. Since they are mostly expressed in the brain, their relevance for the cardiovascular system may be anyhow limited.

9.2.3.6 5-HT$_6$ Family

The 5-HT$_6$ family of serotonin receptors has one member, 5-HT$_6$. This receptor also increases intracellular cAMP levels by coupling to G$_s$-proteins. It is expressed in the brain and probably of limited importance for the cardiovascular system.

9.2.3.7 5-HT$_7$ Family

The 5-HT$_7$ family of serotonin receptors has one member, 5-HT$_7$. Like the 5-HT$_4$ receptor its mRNA is alternatively spliced leading to the expression of several different proteins. Also like the 5-HT$_4$ and 5-HT$_6$ receptors, this subtype couples to G$_s$-proteins and increases cAMP. Since it is expressed, apart from in the brain, in smooth muscle cells it induces vasodilation. Its importance for the cardiovascular system may not yet be fully estimated since specific agonists and antagonists as well as gene-ablated mice became available only very recently.

9.3
Cardiovascular Actions

When serotonin is given intravenously in an animal a complicated blood pressure response is observed consisting of at least three phases: first a strong blood pressure fall, than an overshooting counterregulation and, finally, a persistent hypotension. These effects are caused by the interaction of the hormone with different receptors on different cardiovascular cell types and will be described below. In a normal rodent, the dominating long-term effect of circulating serotonin may be a blood pressure decrease since TPH1-deficient mice are slightly hypertensive [40].

9.3.1
Platelets

Serotonin is a major constituent of the dense granules in platelets. It is accumulated there by the consecutive actions of SERT and VMAT2 (Figure 9.2). Platelets act like sinks for serotonin and empty the blood from this hormone. The residual free 5-HT concentration in the plasma is far below 1 nmol/l [71], and cannot activate receptors on vascular and circulating cells. When platelets are activated (e.g. by contact with collagen in the subendothelial basal membrane after endothelial injury) dense granules with serotonin are released, increasing the local concentrations of the hormone several thousand-fold. At these concentrations, serotonin activates 5-HT$_{2A}$ receptors on neighboring smooth muscle cells and platelets, and induces a cascade of events including platelet aggregation and vasoconstriction which are important steps in wound healing. Additionally and mechanistically less well defined it also induces long-term effects in tissues which promote regeneration [72].

We have studied in detail the role of serotonin in platelet aggregation and could show that after its release from dense granules it interacts not only with 5-HT$_{2A}$ receptors on the platelet membrane to increase intracellular Ca^{2+}, but is also transported into the platelet cytoplasm by SERT. There it is covalently linked to small G-proteins such as RhoA and Rab4 by Ca^{2+}-activated transglutaminases (Figure 9.2) [37]. This newly discovered signaling pathway activates the G-proteins and was termed "serotonylation" (Figure 9.4). Activated RhoA and Rab4 in turn induce the release of α-granules containing essential factors for platelet aggregation and adhesion, such as vWF, fibronectin and P-selectin. These factors link platelets among each other and with the subendothelial matrix. Serotonin has a second role in this process, at least under certain conditions. In so-called COAT platelets, extracellular transglutaminases use serotonin to further cross-link the α-granular factors and thereby the platelets with each other [73].

Thus, platelet serotonin is an essential factor in thrombogenesis. Accordingly, TPH1-deficient mice with less than 10% serotonin left in platelets are severely deficient in hemostasis, but also protected from thrombosis [37]. SSRIs which also

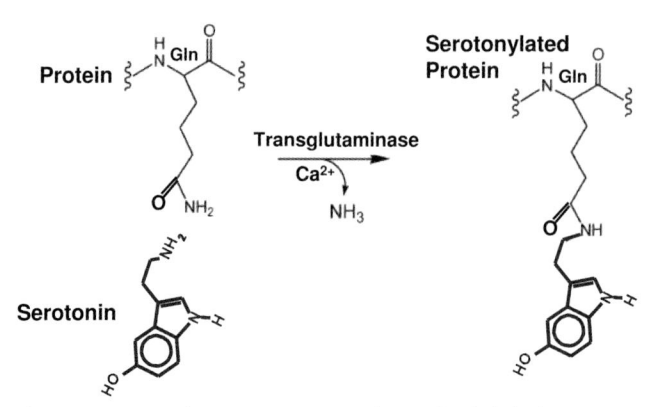

Figure 9.4 Serotonylation. Serotonin can be covalently linked to glutamine residues of proteins by transglutaminases [73, 74]. In the case of small G-proteins this leads to their irreversible activation [75, 74]

drastically reduce platelet serotonin content decrease the risk for myocardial infarction [76, 77] and, thus, may have prospect as antithrombotic drugs [49]. Bleeding has already been reported as a rare side-effect of these drugs [50, 51].

9.3.2
Vessels

When serotonin is released in vessels, it depends on the type of vessel and the presence of the endothelium whether vasoconstriction or vasodilatation is observed. As mentioned above, in case of endothelial injury, platelets are activated and release serotonin that can directly interact with the smooth muscle layer. There, 5-HT$_{2A}$ receptors cause an immediate contraction in nearly all vascular beds. The biological rationale behind this effect is the protection against blood loss by the constriction of an injured vessel.

The situation is more complicated if serotonin hits an intact vessel wall. First it will interact with the endothelium. There, 5-HT$_{1B}$ and 5-HT$_{2B}$ receptors [78], if present, cause an intracellular Ca^{2+} increase and thereby the activation of nitric oxide (NO) synthase and NO release. Consequently, the vessel will relax. Furthermore, 5-HT$_7$ receptors in smooth muscle cells may directly relax vessels by an increase in cAMP [79]. Despite the common view of serotonin as vasoconstrictor, these relaxing mechanisms seem to be dominating. When during extracorporeal circulation serotonin is slowly released from platelets, hypotension is observed in the treated patient [80]. Furthermore, these mechanisms may also be involved in the increased blood pressure in TPH1-deficient mice [40], in which insufficient serotonin leaking from platelets may occur in resistance vessels, keeping them in a permanently constricted state. There is evolutionary evidence that serotonin-induced vasodilation may also be employed in the cutaneous microcirculation of mammals in response to heat. Vertebrates that show this mechanism of thermoregulation such as mammals and birds carry serotonin in platelets; cold-blooded animals such as amphibia and reptiles do not [81]. However, there is also ample evidence that the effect of serotonin on skin perfusion is mainly if not exclusively regulated by central mechanisms involving the sympathetic nervous system (see Section 9.3.4).

On the other hand, a local serotonin-mediated mechanism in arterioles is discussed for Raynaud's phenomenon [26]. This is a vasospasm in the finger induced by cold or mechanical stimulation. The receptors and cell types involved are not yet clear, but possibly the release of serotonin from locally activated platelets may play a role in the pathogenesis of this disease.

The vessel with the highest free serotonin concentration is the portal vein carrying the blood from the gut to the liver. It is the place of uptake of the hormone into platelets, but this uptake will not reach completeness before the blood reaches the liver. The portal vein expresses 5-HT$_{2A}$ receptor and is contracted by serotonin. Thus, it is not surprising that this receptor has been reported to be pathophysiologically important in portal hypertension – a disorder often associated with liver cirrhosis [26].

Apart from $5\text{-}HT_{2A}$, $5\text{-}HT_{1B}$ receptors on vascular smooth muscle cells are also involved in vasoconstrictive actions of serotonin and the relative contribution of the two receptor subtypes depend on the type of vessel studied [26]. While most arteries and veins have both receptors, intracranial vessels seem to be only constricted by the $5\text{-}HT_{1B}$ subtype, rendering triptans, which are agonists at this receptor, suitable drugs for the treatment of migraine accompanied by inappropriate intracranial vasodilatation (see also Section 9.3.4).

A special, but clinically relevant, interaction of serotonin with a vessel occurs in the lung. Serotonin may be the most important etiological factor in PPH. Animals lacking TPH1 [40] or SERT [54] are resistant to this disease, and transgenic mice overexpressing SERT develop it spontaneously [55, 82]. PPH cannot be inhibited by 5-HT receptor antagonist but by SSRIs [83]. Obviously SERT has to transport serotonin into pulmonary arterial smooth muscle cells to induce their growth and thereby PPH. Until recently the mechanisms involved were enigmatic. However, in 2007 it was shown that serotonylation (see Section 9.3.1; Figures 9.2 and 9.4) happens also in the cytosol of this cell type again activating RhoA and that RhoA activation is essential for the growth response to serotonin [75, 84]. On the first sight this hypothesis seems to be contradicted by the fact that PPH is also a major side-effect of drugs that release serotonin from cells, such as fenfluramine [52]. However, first these substances also release 5-HT from intracellular vesicles, thereby increasing the cytosolic serotonin levels. Secondly, the drugs are transported into the cytosol and may be themselves coupled to G-proteins eliciting PPH. This hypothesis is supported by the fact that mainly serotonin releasers, which are countertransported into the cell, induce PPH. The source of serotonin in PPH may not only be the platelets, but also the pulmonary endothelial cells. These cells can be stimulated to express TPH1 and then secrete 5-HT and induce the proliferation of the neighboring smooth muscle cells [85, 86]. Apart from the major importance of SERT, there is also evidence for a role of serotonin receptors, namely $5\text{-}HT_{1B}$ and $5\text{-}HT_{2B}$, in the pathogenesis of PPH. Mice lacking these proteins are also protected from hypoxia-induced PPH [87, 88]. Recent evidence even argues for a cooperativity between $5\text{-}HT_{1B}$ receptors and SERT on pulmonary smooth muscle cells which is necessary to induce PPH employing reactive oxygen species [89]. Thus, the relevance of serotonin in PPH is unequivocal, but its exact role still remains to be determined.

9.3.3
Heart

Serotonin has multiple functions in the heart mediated by several receptors including $5\text{-}HT_{2B}$, $5\text{-}HT_3$ and $5\text{-}HT_4$ [26, 90]. $5\text{-}HT_{2B}$ receptors are already essential for cardiac development. Knockout mice for these receptors show severe cardiac abnormalities such as a lack of trabeculation and most of them die at midgestation or shortly after birth [17]. Since $5\text{-}HT_{2B}$ receptors have also been implicated in the differentiation and correct migration of neural crest cells, and since the heart partially consists of neural crest derivatives, $5\text{-}HT_{2B}$ receptors on neural crest cells may be crucial for correct heart development [91]. It is not clear whether serotonin is necessary for this action of its receptor, since TPH1-deficient mice, at least in our hands, have completely normal hearts. Moreover, TPH1 is only expressed in the embryo after the crucial phase of cardiac development [41]. Thus, the source of the 5-HT$_{2B}$ ligand during early development remains unclear. Possibly, the constitutive activity of 5-HT$_{2B}$ receptors [66] plays a major role in cardiogenesis.

The 5-HT$_{2B}$ receptor is also involved in cardiac hypertrophy. Transgenic mice overexpressing 5-HT$_{2B}$ in cardiomyocytes develop myocardial hypertrophy [92] and knockout mice, if they survive, are resistant to it [93]. These data provide evidence for a role of the 5-HT$_{2B}$ receptor on both cardiac myocytes and fibroblasts in the regulation of cardiac size. While in myocytes the inhibition of apoptosis may be important [94], fibroblasts release prohypertrophic cytokines after 5-HT$_{2B}$ stimulation [93].

Serotonin can cause a disease of heart valves, via its 5-HT$_{2B}$ receptor also expressed there [26]. This has been reported in patients with a carcinoid tumor. Such tumors release high amounts of 5-HT into the circulation which can not be completely cleared any more by the platelets and by MAOs in liver and lung. Valvular heart disease was also observed in mice lacking SERT and thereby not able to clear the plasma from serotonin [56] and in rats chronically treated with serotonin [95]. Furthermore, it occurred in patients treated with fenfluramine. Interaction of this drug or of free serotonin with 5-HT$_{2B}$ probably induces fibrotic and hypertrophic changes in the valves, which impair their functions.

5-HT$_3$ receptors mediate the most dramatic cardiovascular effect of serotonin – the von Bezold Jarisch reflex. Seconds after intravenous administration of 5-HT a very strong bradycardia is observed followed by a blood pressure drop and an immediate recovery. This effect is induced by depolarization and activation of afferent vagal nerve endings in the heart carrying 5-HT$_3$ receptors, since it is blunted by vagotomy and 5-HT$_3$ blockers. The physiological meaning of this reflex, already discovered in 1867 [96, 97], is unclear but it may be of pathophysiological relevance in certain disease states [98].

5-HT$_4$ receptors are expressed on cardiomyocytes in the atria and to a lower extent also in ventricles. When activated by serotonin they increase intracellular cAMP, and thereby induce positive lusitropic, chronotropic and inotropic effects on the heart. These effects can also become proarrhythmic, maybe explaining the reported cardiac side-effects of some serotonergic drugs. However, the physiological role of this serotonergic control of the heart is unclear, since 5-HT$_4$-deficient mice show no cardiac abnormalities [20], and also the source of the serotonin acting on cardiac 5-HT$_4$ receptors under normal conditions remains enigmatic.

9.3.4
Brain

In the brain, serotonin is of major importance for cardiovascular control as has previously been summarized in comprehensive reviews [29, 99, 100]. However, the exact mechanisms are far from being clear and the literature is full of contradicting results. The reasons for this are the differential expression of nearly all 5-HT receptors in different parts of the brain (Table 9.1), and the use of only partially specific agonists and antagonists. A huge amount of evidence supports a role of 5-HT_{1A} receptors in the brain stem, in particular in the nucleus tractus solitarii (NTS), the rostral ventral lateral medulla and other centers involved in sympathetic regulation, which get strong serotonergic innervation from the dorsal group of raphe nuclei B1–B4. When these 5-HT_{1A} receptors are activated a fall in blood pressure is observed accompanied by a decrease in sympathetic nerve activity. Furthermore, the same receptors on the nucleus ambiguus and the dorsal nucleus of the vagus seem to elicit vagal activation further aggravating the hypotensive and bradycardiac effect. Accordingly, 5-HT_{1A} antagonists and serotonin depletion attenuate cardiovascular reflexes, such as the baroreflex, which work via the autonomic nervous system [101]. Thus, serotonin seems to facilitate sympathetic inhibition and vagal stimulation by its 5-HT_{1A} receptor in brain stem areas. However, when 5-HT_{1A} agonists are applied to forebrain nuclei, sympathoexcitation and hypertension is induced. These areas get there serotonergic input from the more rostral raphe nuclei B5–B9. To further complicate the issue, recent studies have shown that also the 5-HT_7 receptor may be involved in the effects of 5-HT in the brain stem, in particular in the NTS where it is mainly expressed [102]. Accordingly, the local application of antagonists for this receptor also blunts cardiovascular reflexes employing the autonomic nervous system. Moreover, 5-HT_2 and 5-HT_3 receptors have been shown to participate in central serotonergic regulation of cardiovascular physiology. The relative importance of each receptor and brain nucleus in this regulation needs to be clarified in genetically altered animal models being affected only in a single component of the serotonin system. Unfortunately, there are very few published studies in such animals assessing changes in autonomic cardiovascular control.

Serotonin has been implicated in the pathophysiology of migraine, and agonists at the 5-HT_{1B}, 5-HT_{1D} and 5-HT_{1F} receptor, the triptans, are the most effective drugs for this disease in the moment. Originally, it was assumed that the migraine headache is induced by a inadequate vasorelaxation in the brain, which could be counteracted by the activation of 5-HT_{1B} receptors on vascular smooth muscle cells. The fact that this issue is discussed in the brain section indicates that there is now evidence that it is not a vascular etiology, but the activation of a neuronal pathway, the trigeminovascular system, that is considered to be crucial for the initiation of a migraine attack [103]. The effect of triptans on neuronal 5-HT_{1F} receptor in this system may be clinically relevant as indicated by a drug only activating this receptor and not 5-HT_{1B} and lacking vasorelaxing capacity, which was anyhow effective in migraine therapy [104]. Although the exact pathomechanisms involved in migraine and the corresponding role of serotonin and its receptors need to be clarified, 5-HT_1 receptors remain a valuable target for therapeutic drugs.

9.4
Conclusions

With 13 different receptors expressed in nearly all cell types (Table 9.1), the serotonin system is maybe the most complicated hormone system in mammals. Moreover, it is divided in two independent subsystems initiated by two distinct TPH isoforms in the periphery and in the brain (Figure 9.1). Serotonin is involved in most physiological mechanisms, and has a particularly pivotal role in the CNS and peripheral nervous system as a major neurotransmitter. Cardiovascular regulation may not be the main duty of this system; nevertheless, it is of functional relevance in all cardiovascular tissues. In the meantime, nearly all components of the 5-HT system have been ablated in gene-targeted rodent models. However, only a few of these models have already been employed to study the cardiovascular functions of serotonin. Moreover, most likely the existing mouse models will not suffice and we will even need new tissue-specifically targeted models to clarify the actions of 5-HT. However, this effort is worthwhile since we may be recompensed by novel therapeutic options for cardiovascular diseases.

Acknowledgment

I would like to thank Natalia Alenina for the critical reading of the manuscript.

References

1 Janeway, T.C., Richardson, H.B. and Park, E.A. (1918) Experiments on the vasoconstrictor action of blood serum. *Archives of Internal Medicine*, **21**, 565–603.

2 Rapport, M.M., Green, A.A. and Page, I.H. (1948) Partial purification of the vasoconstrictor in beef serum. *Journal of Biological Chemistry*, **174**, 735–41.

3 Rapport, M.M. (1949) Serum vasoconstrictor (serotonin) the presence of creatinine in the complex; a proposed structure of the vasoconstrictor principle. *Journal of Biological Chemistry*, **180**, 961–9.

4 Vialli, M. and Erspamer, V. (1937) Ricerche sul secreto delle cellule enterocromaffini. IX. Intorno alla natura chimica della sostanza specifica. *Societa Medico-Chirurgica di Pisa*, **51**, 1111–18.

5 Erspamer, V. and Asero, B. (1952) Identification of enteramine, the specific hormone of the enterochromaffin cell system, as 5-hydroxytryptamine. *Nature*, **169**, 800–1.

6 Twarog, B.M. and Page, I.H. (1953) Serotonin content of some mammalian tissues and urine and a method for its determination. *American Journal of Physiology*, **175**, 157–61.

7 Dahlström, A. and Fuxe, K. (1964) Evidence for the existence of monoamine-containing neurons in the central nervous system. I. Demonstration of monoamines in the cell bodies of brain stem neurons. *Acta Physiologica Scandinavica* **62** (Suppl. 232), 1–55.

8 Azmitia, E.C. (2007) Serotonin and brain: evolution, neuroplasticity, and homeostasis. *International Review of Neurobiology*, **77**, 31–56.

9 Sze, J.Y., Victor, M., Loer, C., Shi, Y. and Ruvkun, G. (2000) Food and metabolic signalling defects in a *Caenorhabditis*

elegans serotonin-synthesis mutant. *Nature*, **403**, 560–4.

10 Petrascheck, M., Ye, X. and Buck, L.B. (2007) An antidepressant that extends lifespan in adult *Caenorhabditis elegans*. *Nature*, **450**, 553–6.

11 Johnson, E., Ringo, J. and Dowse, H. (1997) Modulation of *Drosophila* heartbeat by neurotransmitters. *Journal of Comparative Physiology B Biochemical, Systemic, and Environmental Physiology*, **167**, 89–97.

12 Parks, C.L., Robinson, P.S., Sibille, E., Shenk, T. and Toth, M. (1998) Increased anxiety of mice lacking the serotonin1A receptor. *Proceedings of the National Academy of Sciences of the United States of America*, **95**, 10734–9.

13 Ramboz, S., Oosting, R., Amara, D.A., Kung, H.F., Blier, P., Mendelsohn, M., Mann, J.J., Brunner, D. and Hen, R. (1998) Serotonin receptor 1A knockout: an animal model of anxiety-related disorder. *Proceedings of the National Academy of Sciences of the United States of America*, **95**, 14476–81.

14 Heisler, L.K., Chu, H.M., Brennan, T.J., Danao, J.A., Bajwa, P., Parsons, L.H. and Tecott, L.H. (1998) Elevated anxiety and antidepressant-like responses in serotonin 5-HT$_{1A}$ receptor mutant mice. *Proceedings of the National Academy of Sciences of the United States of America*, **95**, 15049–54.

15 Saudou, F., Amara, D.A., Dierich, A., LeMeur, M., Ramboz, S., Segu, L., Buhot, M.C. and Hen, R. (1994) Enhanced aggressive behavior in mice lacking 5-HT$_{1B}$ receptor. *Science*, **265**, 1875–8.

16 Weisstaub, N.V., Zhou, M., Lira, A., Lambe, E., Gonzalez-Maeso, J., Hornung, J.P., Sibille, E., Underwood, M., Itohara, S., Dauer, W.T., Ansorge, M.S., Morelli, E., Mann, J.J., Toth, M., Aghajanian, G., Sealfon, S.C., Hen, R. and Gingrich, J.A. (2006) Cortical 5-HT$_{2A}$ receptor signaling modulates anxiety-like behaviors in mice. *Science*, **313**, 536–40.

17 Nebigil, C.G., Choi, D.S., Dierich, A., Hickel, P., Le Meur, M., Messaddeq, N., Launay, J.M. and Maroteaux, L. (2000) Serotonin 2B receptor is required for heart development. *Proceedings of the National Academy of Sciences of the United States of America*, **97**, 9508–13.

18 Tecott, L.H., Sun, L.M., Akana, S.F., Strack, A.M., Lowenstein, D.H., Dallman, M.F. and Julius, D. (1995) Eating disorder and epilepsy in mice lacking 5-HT$_{2C}$ serotonin receptors. *Nature*, **374**, 542–6.

19 Zeitz, K.P., Guy, N., Malmberg, A.B., Dirajlal, S., Martin, W.J., Sun, L., Bonhaus, D.W., Stucky, C.L., Julius, D. and Basbaum, A.I. (2002) The 5-HT$_3$ subtype of serotonin receptor contributes to nociceptive processing via a novel subset of myelinated and unmyelinated nociceptors. *Journal of Neuroscience*, **22**, 1010–19.

20 Compan, V., Zhou, M., Grailhe, R., Gazzara, R.A., Martin, R., Gingrich, J., Dumuis, A., Brunner, D., Bockaert, J. and Hen, R. (2004) Attenuated response to stress and novelty and hypersensitivity to seizures in 5-HT$_4$ receptor knock-out mice. *Journal of Neuroscience*, **24**, 412–9.

21 Grailhe, R., Waeber, C., Dulawa, S.C., Hornung, J.P., Zhuang, X., Brunner, D., Geyer, M.A. and Hen, R. (1999) Increased exploratory activity and altered response to LSD in mice lacking the 5-HT$_{5A}$ receptor. *Neuron*, **22**, 581–91.

22 Bonasera, S.J., Chu, H.M., Brennan, T.J. and Tecott, L.H. (2006) A null mutation of the serotonin 6 receptor alters acute responses to ethanol. *Neuropsychopharmacology*, **31**, 1801–13.

23 Hedlund, P.B., Danielson, P.E., Thomas, E.A., Slanina, K., Carson, M.J. and Sutcliffe, J.G. (2003) No hypothermic response to serotonin in 5-HT$_7$ receptor knockout mice. *Proceedings of the National Academy of Sciences of the United States of America*, **100**, 1375–80.

24 Barnes, N.M. and Sharp, T. (1999) A review of central 5-HT receptors and their function. *Neuropharmacology*, **38**, 1083–152.

25 Kroeze, W.K., Kristiansen, K. and Roth, B.L. (2002) Molecular biology of serotonin receptors structure and function at the molecular level. *Current Topics in Medicinal Chemistry*, **2**, 507–28.

26 Kaumann, A.J. and Levy, F.O. (2006) 5-hydroxytryptamine receptors in the human cardiovascular system.

Pharmacology and Therapeutics, **111**, 674–706.
27 Nagatomo, T., Rashid, M., Abul, M.H. and Komiyama, T. (2004) Functions of 5-HT$_{2A}$ receptor and its antagonists in the cardiovascular system. *Pharmacology and Therapeutics*, **104**, 59–81.
28 Frishman, W.H., Huberfeld, S., Okin, S., Wang, Y.H., Kumar, A. and Shareef, B. (1995) Serotonin and serotonin antagonism in cardiovascular and non-cardiovascular disease. *Journal of Clinical Pharmacology*, **35**, 541–72.
29 McCall, R.B. and Clement, M.E. (1994) Role of serotonin1A and serotonin2 receptors in the central regulation of the cardiovascular system. *Pharmacological Reviews*, **46**, 231–43.
30 Villalon, C.M. and Centurion, D. (2007) Cardiovascular responses produced by 5-hydroxytryptamine:a pharmacological update on the receptors/mechanisms involved and therapeutic implications. *Naunyn-Schmiedeberg's Archives of Pharmacology*, **376**, 45–63.
31 Vanhoutte, P.M. (1991) Platelet-derived serotonin, the endothelium, and cardiovascular disease. *Journal of Cardiovascular Pharmacology*, **17** (Suppl. 5), S6–12.
32 Ni, W. and Watts, S.W. (2006) 5-hydroxytryptamine in the cardiovascular system: focus on the serotonin transporter (SERT). *Clinical and Experimental Pharmacology and Physiology*, **33**, 575–83.
33 Fitzpatrick, P.F. (1999) Tetrahydropterin-dependent amino acid hydroxylases. *Annual Review of Biochemistry*, **68**, 355–81.
34 Walther, D.J., Peter, J.U., Bashammakh, S., Hörtnagl, H., Voits, M., Fink, H. and Bader, M. (2003) Synthesis of serotonin by a second tryptophan hydroxylase isoform. *Science*, **299**, 76.
35 Walther, D.J. and Bader, M. (2003) A unique central tryptophan hydroxylase isoform. *Biochemical Pharmacology*, **66**, 1673–80.
36 Cote, F., Thevenot, E., Fligny, C., Fromes, Y., Darmon, M., Ripoche, M. A., Bayard, E., Hanoun, N., Saurini, F., Lechat, P., Dandolo, L., Hamon, M., Mallet, J. and Vodjdani, G. (2003) Disruption of the nonneuronal tph1 gene demonstrates the importance of peripheral serotonin in cardiac function. *Proceedings of the National Academy of Sciences of the United States of America*, 100, 13525–30.
37 Walther, D.J., Peter, J.U., Winter, S., Höltje, M., Paulmann, N., Grohmann, M., Vowinckel, J., Alamo-Bethencourt, V., Wilhelm, C.S., Ahnert-Hilger, G. and Bader, M. (2003) Serotonylation of small GTPases is a signal transduction pathway that triggers platelet α-granule release. *Cell*, **115**, 851–62.
38 Matsuda, M., Imaoka, T., Vomachka, A.J., Gudelsky, G.A., Hou, Z., Mistry, M., Bailey, J.P., Nieport, K.M., Walther, D.J., Bader, M. and Horseman, N.D. (2004) Serotonin regulates mammary gland development via a novel autocrine–paracrine loop. *Developmental Cell*, **6**, 193–203.
39 Lesurtel, M., Graf, R., Aleil, B., Walther, D.J., Tian, Y., Jochum, W., Gachet, C., Bader, M. and Clavien, P-A. (2006) Platelet-derived serotonin mediates liver regeneration. *Science*, **312**, 104–7.
40 Morecroft, I., Dempsie, Y., Bader, M., Walther, D.J., Kotnik, K., Loughlin, L., Nilsen, M. and MacLean, M.R. (2007) Effect of tryptophan hydroxylase 1 deficiency on the development of hypoxia-induced pulmonary hypertension. *Hypertension*, **49**, 232–6.
41 Cote, F., Fligny, C., Bayard, E., Launay, J.M., Gershon, M.D., Mallet, J. and Vodjdani, G. (2007) Maternal serotonin is crucial for murine embryonic development. *Proceedings of the National Academy of Sciences of the United States of America*, 104, 329–34.
42 Cases, O., Seif, I., Grimsby, J., Gaspar, P., Chen, K., Pournin, S., Muller, U., Aguet, M., Babinet, C., Shih, J.C. and De Maeyer, E. (1995) Aggressive behavior and altered amounts of brain serotonin and norepinephrine in mice lacking MAOA. *Science*, **268**, 1763–6.
43 Grimsby, J., Toth, M., Chen, K., Kumazawa, T., Klaidman, L., Adams, J.D., Karoum, F., Gal, J. and Shih, J.C. (1997) Increased stress response and beta-phenylethylamine in MAOB-

deficient mice. *Nature Genetics*, **17**, 206–10.
44 Chen, K., Holschneider, D.P., Wu, W., Rebrin, I. and Shih, J.C. (2004) A spontaneous point mutation produces monoamine oxidase A/B knock-out mice with greatly elevated monoamines and anxiety-like behavior. *Journal of Biological Chemistry*, **279**, 39645–52.
45 Pan, Y., Gembom, E., Peng, W., Lesch, K.P., Mossner, R. and Simantov, R. (2001) Plasticity in serotonin uptake in primary neuronal cultures of serotonin transporter knockout mice. *Brain Research. Developmental Brain Research*, **126**, 125–9.
46 Mossner, R., Simantov, R., Marx, A., Lesch, K.P. and Seif, I. (2006) Aberrant accumulation of serotonin in dopaminergic neurons. *Neuroscience Letters*, **401**, 49–54.
47 Homberg, J.R., Olivier, J.D., Smits, B.M., Mul, J.D., Mudde, J., Verheul, M., Nieuwenhuizen, O.F., Cools, A.R., Ronken, E., Cremers, T., Schoffelmeer, A.N., Ellenbroek, B.A. and Cuppen, E. (2007) Characterization of the serotonin transporter knockout rat: a selective change in the functioning of the serotonergic system. *Neuroscience*, **146**, 1662–76.
48 Zhang, Y.W. and Rudnick, G. (2006) The cytoplasmic substrate permeation pathway of serotonin transporter. *Journal of Biological Chemistry*, **281**, 36213–20.
49 Maurer-Spurej, E. (2005) Serotonin reuptake inhibitors and cardiovascular diseases: a platelet connection. *Cellular and Molecular Life Sciences*, **62**, 159–70.
50 Weinrieb, R.M., Auriacombe, M., Lynch, K.G. and Lewis, J.D. (2005) Selective serotonin re-uptake inhibitors and the risk of bleeding. *Expert Opinion on Drug Safety*, **4**, 337–44.
51 Loke, Y.K., Trivedi, A.N. and Singh, S. (2008) Meta-analysis: gastrointestinal bleeding due to interaction between selective serotonin uptake inhibitors and non-steroidal anti-inflammatory drugs. *Alimentary Pharmacology and Therapeutics*, **27**, 31–40.
52 Rothman, R.B. and Baumann, M.H. (2002) Therapeutic and adverse actions of serotonin transporter substrates. *Pharmacology and Therapeutics*, **95**, 73–88.
53 Bengel, D., Murphy, D.L., Andrews, A.M., Wichems, C.H., Feltner, D., Heils, A., Mossner, R., Westphal, H. and Lesch, K.P. (1998) Altered brain serotonin homeostasis and locomotor insensitivity to 3,4-methylenedioxymethamphetamine ("Ecstasy") in serotonin transporter-deficient mice. *Molecular Pharmacology*, **53**, 649–55.
54 Eddahibi, S., Hanoun, N., Lanfumey, L., Lesch, K.P., Raffestin, B., Hamon, M. and Adnot, S. (2000) Attenuated hypoxic pulmonary hypertension in mice lacking the 5-hydroxytryptamine transporter gene. *Journal of Clinical Investigation*, **105**, 1555–62.
55 Guignabert, C., Izikki, M., Tu, L.I., Li, Z., Zadigue, P., Barlier-Mur, A.M., Hanoun, N., Rodman, D., Hamon, M., Adnot, S. and Eddahibi, S. (2006) Transgenic mice overexpressing the 5-hydroxytryptamine transporter gene in smooth muscle develop pulmonary hypertension. *Circulation Research*, **98**, 1323–30.
56 Mekontso-Dessap, A., Brouri, F., Pascal, O., Lechat, P., Hanoun, N., Lanfumey, L., Seif, I., Benhaiem-Sigaux, N., Kirsch, M., Hamon, M., Adnot, S. and Eddahibi, S. (2006) Deficiency of the 5-hydroxytryptamine transporter gene leads to cardiac fibrosis and valvulopathy in mice. *Circulation*, **113**, 81–9.
57 Eiden, L.E., Schafer, M.K., Weihe, E. and Schutz, B. (2004) The vesicular amine transporter family (SLC18): amine/proton antiporters required for vesicular accumulation and regulated exocytotic secretion of monoamines and acetylcholine. *Pflugers Archiv: European Journal of Physiology*, **447**, 636–40.
58 Höltje, M., Winter, S., Walther, D., Pahner, I., Hörtnagl, H., Ottersen, O.P., Bader, M. and Ahnert-Hilger, G. (2003) The vesicular monoamine content regulates VMAT2 activity through $G\alpha_q$ in mouse platelets: Evidence for autoregulation of vesicular transmitter uptake. *Journal of Biological Chemistry*, **278**, 15850–8.
59 Wang, Y.M., Gainetdinov, R.R., Fumagalli, F., Xu, F., Jones, S.R., Bock,

C.B., Miller, G.W., Wightman, R.M. and Caron, M.G. (1997) Knockout of the vesicular monoamine transporter 2 gene results in neonatal death and supersensitivity to cocaine and amphetamine. *Neuron*, **19**, 1285–96.

60. Fon, E.A., Pothos, E.N., Sun, B.C., Killeen, N., Sulzer, D. and Edwards, R.H. (1997) Vesicular transport regulates monoamine storage and release but is not essential for amphetamine action. *Neuron*, **19**, 1271–83.

61. Takahashi, N., Miner, L.L., Sora, I., Ujike, H., Revay, R.S., Kostic, V., Jackson-Lewis, V., Przedborski, S. and Uhl, G.R. (1997) VMAT2 knockout mice: heterozygotes display reduced amphetamine-conditioned reward, enhanced amphetamine locomotion, and enhanced MPTP toxicity. *Proceedings of the National Academy of Sciences of the United States of America*, **94**, 9938–43.

62. Albert, P.R. and Tiberi, M. (2001) Receptor signaling and structure: insights from serotonin-1 receptors. *Trends in Endocrinology and Metabolism*, **12**, 453–60.

63. Berg, K.A., Harvey, J.A., Spampinato, U. and Clarke, W.P. (2005) Physiological relevance of constitutive activity of $5-HT_{2A}$ and $5-HT_{2C}$ receptors. *Trends in Pharmacological Sciences*, **26**, 625–30.

64. Kohen, R., Fashingbauer, L.A., Heidmann, D.E., Guthrie, C.R. and Hamblin, M.W. (2001) Cloning of the mouse $5-HT_6$ serotonin receptor and mutagenesis studies of the third cytoplasmic loop. *Brain Research. Molecular Brain Research*, **90**, 110–7.

65. Claeysen, S., Sebben, M., Becamel, C., Bockaert, J. and Dumuis, A. (1999) Novel brain-specific $5-HT_4$ receptor splice variants show marked constitutive activity: role of the C-terminal intracellular domain. *Molecular Pharmacology*, **55**, 910–20.

66. Locker, M., Bitard, J., Collet, C., Poliard, A., Mutel, V., Launay, J.M. and Kellermann, O. (2006) Stepwise control of osteogenic differentiation by $5-HT_{2B}$ receptor signaling: nitric oxide production and phospholipase A2 activation. *Cell Signalling*, **18**, 628–39.

67. Tohda, M., Nomura, M. and Nomura, Y. (2006) Molecular pathopharmacology of $5-HT_{2C}$ receptors and the RNA editing in the brain. *Journal of Pharmacological Sciences*, **100**, 427–32.

68. Niesler, B., Walstab, J., Combrink, S., Moller, D., Kapeller, J., Rietdorf, J., Bonisch, H., Gothert, M., Rappold, G. and Bruss, M. (2007) Characterization of the novel human serotonin receptor subunits $5-HT_{3C}$, $5-HT_{3D}$, and $5-HT_{3E}$. *Molecular Pharmacology*, **72**, 8–17.

69. Bender, E., Pindon, A., Van, O.I., Zhang, Y.B., Gommeren, W., Verhasselt, P., Jurzak, M., Leysen, J. and Luyten, W. (2000) Structure of the human serotonin $5-HT_4$ receptor gene and cloning of a novel $5-HT_4$ splice variant. *Journal of Neurochemistry*, **74**, 478–89.

70. Noda, M., Yasuda, S., Okada, M., Higashida, H., Shimada, A., Iwata, N., Ozaki, N., Nishikawa, K., Shirasawa, S., Uchida, M., Aoki, S. and Wada, K. (2003) Recombinant human serotonin 5A receptors stably expressed in C6 glioma cells couple to multiple signal transduction pathways. *Journal of Neurochemistry*, **84**, 222–32.

71. Beck, O., Wallen, N.H., Broijersen, A., Larsson, P.T. and Hjemdahl, P. (1993) On the accurate determination of serotonin in human plasma. *Biochemical and Biophysical Research Communications*, **196**, 260–6.

72. Lesurtel, M., Graf, R., Aleil, B., Walther, D.J., Tian, Y., Jochum, W., Gachet, C., Bader, M. and Clavien, P-A. (2006) Platelet-derived serotonin mediates liver regeneration. *Science*, **312**, 104–7.

73. Dale, G.L., Friese, P., Batar, P., Hamilton, S.F., Reed, G.L., Jackson, K.W., Clemetson, K.J. and Alberio, L. (2002) Stimulated platelets use serotonin to enhance their retention of procoagulant proteins on the cell surface. *Nature*, **415**, 175–9.

74. Walther, D.J., Peter, J.U., Winter, S., Höltje, M., Paulmann, N., Grohmann, M., Vowinckel, J., Alamo-Bethencourt, V., Wilhelm, C.S., Ahnert-Hilger, G. and Bader, M. (2003) Serotonylation of small GTPases is a signal transduction pathway

that triggers platelet α-granule release. *Cell*, **115**, 851–62.

75 Guilluy, C., Rolli-Derkinderen, M., Tharaux, P.L., Melino, G., Pacaud, P. and Loirand, G. (2007) Transglutaminase-dependent RhoA activation and depletion by serotonin in vascular smooth muscle cells. *Journal of Biological Chemistry*, **282**, 2918–28.

76 Sauer, W.H., Berlin, J.A. and Kimmel, S.E. (2001) Selective serotonin reuptake inhibitors and myocardial infarction. *Circulation*, **104**, 1894–8.

77 Serebruany, V.L., O'Connor, C.M. and Gurbel, P.A. (2001) Effect of selective serotonin reuptake inhibitors on platelets in patients with coronary artery disease. *American Journal of Cardiology*, **87**, 1398–400.

78 Ullmer, C., Schmuck, K., Kalkman, H.O. and Lubbert, H. (1995) Expression of serotonin receptor mRNAs in blood vessels. *FEBS Letters*, **370**, 215–21.

79 Centurion, D., Glusa, E., Sanchez-Lopez, A., Valdivia, L.F., Saxena, P.R. and Villalon, C.M. (2004) 5-HT$_7$, but not 5-HT$_{2B}$, receptors mediate hypotension in vagosympathectomized rats. *European Journal of Pharmacology*, **502**, 239–42.

80 Borgdorff, P., Fekkes, D. and Tangelder, G.J. (2002) Hypotension caused by extracorporeal circulation: serotonin from pump-activated platelets triggers nitric oxide release. *Circulation*, **106**, 2588–93.

81 Maurer-Spurej, E. (2005) Circulating serotonin in vertebrates. *Cellular and Molecular Life Sciences*, **62**, 1881–9.

82 MacLean, M.R., Deuchar, G.A., Hicks, M.N., Morecroft, I., Shen, S., Sheward, J., Colston, J., Loughlin, L., Nilsen, M., Dempsie, Y. and Harmar, A. (2004) Overexpression of the 5-hydroxytryptamine transporter gene: effect on pulmonary hemodynamics and hypoxia-induced pulmonary hypertension. *Circulation*, **109**, 2150–5.

83 Marcos, E., Adnot, S., Pham, M.H., Nosjean, A., Raffestin, B., Hamon, M. and Eddahibi, S. (2003) Serotonin transporter inhibitors protect against hypoxic pulmonary hypertension. *American Journal of Respiratory and Critical Care Medicine*, **168**, 487–93.

84 Li, M., Liu, Y., Dutt, P., Fanburg, B.L. and Toksoz, D. (2007) Inhibition of serotonin-induced mitogenesis, migration, and ERK MAPK nuclear translocation in vascular smooth muscle cells by atorvastatin. *American Journal of Physiology. Lung Cellular and Molecular Physiology*, **293**, L463–71.

85 Sullivan, C.C., Du, L., Chu, D., Cho, A.J., Kido, M., Wolf, P.L., Jamieson, S.W. and Thistlethwaite, P.A. (2003) Induction of pulmonary hypertension by an angiopoietin 1/TIE2/serotonin pathway. *Proceedings of the National Academy of Sciences of the United States of America*, **100**, 12331–6.

86 Eddahibi, S., Guignabert, C., Barlier-Mur, A.M., Dewachter, L., Fadel, E., Dartevelle, P., Humbert, M., Simonneau, G., Hanoun, N., Saurini, F., Hamon, M. and Adnot, S. (2006) Cross talk between endothelial and smooth muscle cells in pulmonary hypertension: critical role for serotonin-induced smooth muscle hyperplasia. *Circulation*, **113**, 1857–64.

87 Keegan, A., Morecroft, I., Smillie, D., Hicks, M.N. and MacLean, M.R. (2001) Contribution of the 5-HT$_{1B}$ receptor to hypoxia-induced pulmonary hypertension: converging evidence using 5-HT$_{1B}$-receptor knockout mice and the 5-HT$_{1B/1D}$-receptor antagonist GR127935. *Circulation Research*, **89**, 1231–9.

88 Launay, J.M., Herve, P., Peoc'h, K., Tournois, C., Callebert, J., Nebigil, C.G., Etienne, N., Drouet, L., Humbert, M., Simonneau, G. and Maroteaux, L. (2002) Function of the serotonin 5-hydroxytryptamine 2B receptor in pulmonary hypertension. *Nature Medicine*, **8**, 1129–35.

89 Lawrie, A., Spiekerkoetter, E., Martinez, E.C., Ambartsumian, N., Sheward, W.J., MacLean, M.R., Harmar, A.J., Schmidt, A.M., Lukanidin, E. and Rabinovitch, M. (2005) Interdependent serotonin transporter and receptor pathways regulate S100A4/Mts1, a gene associated with pulmonary vascular disease. *Circulation Research*, **97**, 227–35.

90 Nebigil, C.G. and Maroteaux, L. (2003) Functional consequence of serotonin/

5-HT$_{2B}$ receptor signaling in heart: role of mitochondria in transition between hypertrophy and heart failure? *Circulation*, **108**, 902–8.

91 Choi, D.S., Ward, S.J., Messaddeq, N., Launay, J.M. and Maroteaux, L. (1997) 5-HT$_{2B}$ receptor-mediated serotonin morphogenetic functions in mouse cranial neural crest and myocardiac cells. *Development*, **124**, 1745–55.

92 Nebigil, C.G., Jaffre, F., Messaddeq, N., Hickel, P., Monassier, L., Launay, J.M. and Maroteaux, L. (2003) Overexpression of the serotonin 5-HT$_{2B}$ receptor in heart leads to abnormal mitochondrial function and cardiac hypertrophy. *Circulation*, **107**, 3223–9.

93 Jaffre, F., Callebert, J., Sarre, A., Etienne, N., Nebigil, C.G., Launay, J.M., Maroteaux, L. and Monassier, L. (2004) Involvement of the serotonin 5-HT$_{2B}$ receptor in cardiac hypertrophy linked to sympathetic stimulation: control of interleukin-6, interleukin-1beta, and tumor necrosis factor-alpha cytokine production by ventricular fibroblasts. *Circulation*, **110**, 969–74.

94 Nebigil, C.G., Etienne, N., Messaddeq, N. and Maroteaux, L. (2003) Serotonin is a novel survival factor of cardiomyocytes: mitochondria as a target of 5-HT$_{2B}$ receptor signaling. *FASEB Journal*, **17**, 1373–5.

95 Gustafsson, B.I., Tommeras, K., Nordrum, I., Loennechen, J.P., Brunsvik, A., Solligard, E., Fossmark, R., Bakke, I., Syversen, U. and Waldum, H. (2005) Long-term serotonin administration induces heart valve disease in rats. *Circulation*, **111**, 1517–22.

96 von Bezold, A.V. and Hirt, L. (1867) Über die physiologischen Wirkungen des essigsauren Veratrins. *Untersuchungen aus dem Physiologischen Laboratorium Würzburg*, **1**, 75–156.

97 Jarisch, A. and Richter, H. (1939) Die Kreislaufwirkung des Veratrins. *Naunyn-Schmiedeberg's Archives of Pharmacology*, **193**, 347–54.

98 Mark, A.L. (1983) The Bezold-Jarisch reflex revisited: clinical implications of inhibitory reflexes originating in the heart. *Journal of the American College of Cardiology*, **1**, 90–102.

99 Ramage, A.G. (2001) Central cardiovascular regulation and 5-hydroxytryptamine receptors. *Brain Research Bulletin*, **56**, 425–39.

100 Jordan, D. (2005) Vagal control of the heart: central serotonergic (5-HT) mechanisms. *Experimental Physiology*, **90**, 175–81.

101 Kellett, D.O., Stanford, S.C., Machado, B.H., Jordan, D. and Ramage, A.G. (2005) Effect of 5-HT depletion on cardiovascular vagal reflex sensitivity in awake and anesthetized rats. *Brain Research*, **1054**, 61–72.

102 Damaso, E.L., Bonagamba, L.G., Kellett, D.O., Jordan, D., Ramage, A.G. and Machado, B.H. (2007) Involvement of central 5-HT$_7$ receptors in modulation of cardiovascular reflexes in awake rats. *Brain Research*, **1144**, 82–90.

103 Pietrobon, D. and Striessnig, J. (2003) Neurobiology of migraine. *Nature Reviews Neuroscience*, **4**, 386–98.

104 Ramadan, N.M., Skljarevski, V., Phebus, L.A. and Johnson, K.W. (2003) 5-HT$_{1F}$ receptor agonists in acute migraine treatment: a hypothesis. *Cephalalgia*, **23**, 776–85.

10
Adrenaline and Noradrenaline
Nadine Beetz and Lutz Hein

10.1
Introduction

Noradrenaline and adrenaline are prominent representatives of catecholamines that are involved in the regulation of diverse physiological functions of the sympathetic nervous system, including control of neurotransmitter release, endocrine function, cardiovascular and vegetative regulation, metabolic features, and energy homeostasis [1]. In the peripheral nervous system, noradrenaline is released from sympathetic nerves, and to a lesser degree also from the adrenal medulla. While noradrenaline acts primarily at its site of secretion as a classical neurotransmitter, adrenaline is released from the adrenal gland into the bloodstream eliciting its physiological actions as a hormone [1]. Noradrenaline and adrenaline are also synthesized by neurons in the central nervous system.

In addition to its role in the regulation of cardiovascular physiology, the adrenergic system is important for pathophysiology and pharmacology of cardiovascular disease states, including hypertension, coronary artery disease, chronic heart failure and angiogenesis. The peripheral adrenergic system may be divided into three functional parts. (1) A cascade of enzymes and transport systems is required for biosynthesis and vesicular storage of noradrenaline and adrenaline (Figure 10.1). (2) Catecholamines released from these vesicles mediate their biological effects via activation of adrenergic receptors, which belong to the class of G-protein-coupled receptors (Figure 10.2). (3) Membrane transporters and enzymes terminate the actions of the catecholamines and initiate biotransformation into inactive compounds. According to this classification, this chapter will first introduce the molecular components involved in biosynthesis, storage, release and metabolism of noradrenaline and adrenaline, and in the second part the functions of adrenergic receptors will be discussed. The main focus will be on cardiovascular functions of the catecholamines, but the physiological and pathophysiological significance of the molecular components will also be mentioned in brief.

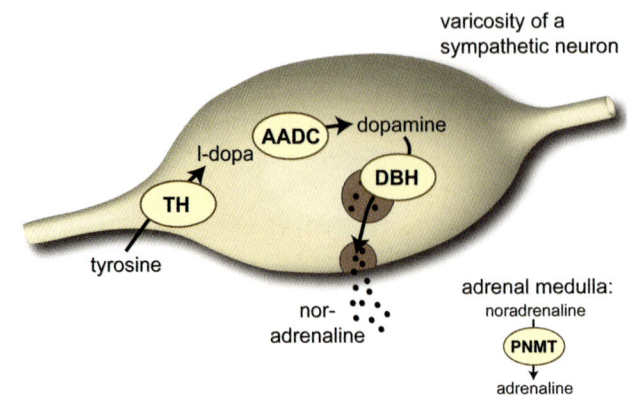

Figure 10.1 Overview of noradrenaline and adrenaline synthesis in sympathetic nerves and in the adrenal medulla. TH, tyrosine hydroxylase; AADC, aromatic L-amino acid decarboxylase; DBH, dopamine β-hydroxylase; PNMT, phenylethanolamine N-methyltransferase.

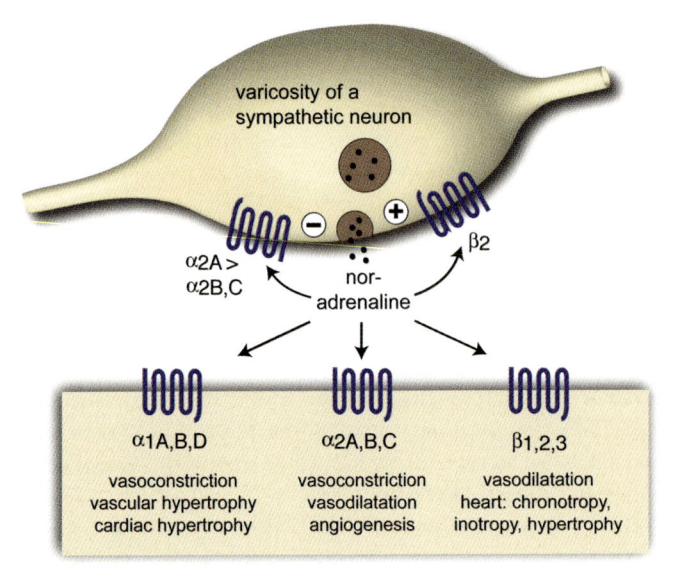

Figure 10.2 Adrenergic receptor subtypes and their pre- and postsynaptic localization and cardiovascular functions. For references, see text.

10.2
Biosynthesis and Degradation of Noradrenaline and Adrenaline

10.2.1
Biosynthesis

The endogenous catecholamines noradrenaline and adrenaline are derived from the amino acid precursor tyrosine (Figure 10.1). In a first step, tyrosine hydroxylase

(TH) converts tyrosine to L-3,4-dihydroxyphenylalanine (L-DOPA), which undergoes decarboxylation to dopamine catalyzed by the aromatic L-amino acid decarboxylase (AADC). A vesicular monoamine transporter guides dopamine into the interior of presynaptic storage vesicles where dopamine β-hydroxylase (DBH) catalyzes the conversion to noradrenaline. In adrenal chromaffin cells, phenylethanolamine N-methyltransferase (PNMT) methylates noradrenaline at the β-hydroxyl group to generate adrenaline.

10.2.1.1 Tyrosine Hydroxylase (TH)

10.2.1.1.1 Mouse Models In the biosynthesis of catecholamines the monooxygenase tyrosine hydroxylase catalyzes the rate-limiting step – the initial conversion of tyrosine to L-DOPA. The *Th* gene locus in mice was first disrupted by Zhou et al. [2], resulting in a deficiency in dopamine, noradrenaline and adrenaline. The majority of *Th*-deficient mice died during embryonic development and perinatally due to cardiovascular failure [2, 3]. Upon examination of $Th^{-/-}$ embryos cardiac alterations became apparent, ranging from dilated atria and thinning of the atrial wall, heterogenous appearance and limited organization of ventricular cardiomyocytes to bradycardia of embryonic hearts when compared to wild-type embryos [2]. Administration of L-DOPA to pregnant female mice completely rescued *Th*-deficient mice *in utero* [2]. In contrast, mice with selective dopamine deficiency were born at the expected Mendelian ratio [4]. In addition, $Pnmt^{-/-}$ mice that lack the enzyme catalyzing the final conversion of noradrenaline to adrenaline did not show any defects in embryonic development [5]. In conclusion, noradrenaline is essential for mouse fetal development. Reduction of gene dosage to 50% resulted in apparently normal and fertile heterozygous mice. Interestingly, tyrosinase, which is involved in the synthesis of melanin pigment, may assist to generate catecholamines in mice deficient in tyrosine hydroxylase [6].

10.2.1.1.2 Human Genetics and Function Loss of long tyrosine hydroxylase gene alleles was associated with longevity in male centenarians [7]. The authors described a 698 A to G transition (missense mutation Arg → His) in three independent members of Dutch families. The homozygous carriers presented with hypokinetic rigidity and severe delay in psychomotoric development and lower levels of catecholamine metabolites. In addition, a point mutation in exon 11 of the tyrosine hydroxylase gene was associated with the autosomal recessive form of Segawa syndrome – a DOPA-responsive dystonia caused by deficits in the biosynthesis of dopamine [8]. The kinetic disorder could symptomatically be managed by administration of L-DOPA in combination with a decarboxylase inhibitor [8]. Cardiovascular consequences of these genetic variants have not been reported in humans.

10.2.1.2 Aromatic L-Amino Acid Decarboxylase (AADC)

10.2.1.2.1 Mouse Models At present, no genetic mouse models deficient in AADC have been generated.

10.2.1.2.2 Human Genetics and Function Mutations in the human AADC gene may lead to severe enzyme deficiency and diverse symptoms from the neonatal period onwards, the most important of which are developmental delay, oculogyric crises and autonomic dysfunction with hypotension [9]. Clinical management of AADC deficiency consists of vitamin B6 supplementation, administration of dopamine agonists and monoamine oxidase (MAO) inhibitors [10].

10.2.1.3 Dopamine β-Hydroxylase (DBH)

10.2.1.3.1 Mouse Models Disruption of the *Dbh* locus resulted in animals deficient in the catecholamines noradrenaline and adrenaline [11]. Phenotypic characterization of this mouse model suggested diverse functions for adrenergic catecholamines in a wide variety of physiological fields, including embryonic development, cardiovascular function, metabolic regulation, thermogenesis, learning and memory, behavior, and immunology.

Deletion of the murine *Dbh* gene led to embryonic lethality and a histological phenotype of the heart which was similar to mice lacking functional tyrosine hydroxylase, indicating that cardiovascular failure may be the cause of embryonic death [11]. Transplacental noradrenaline transfer from the mother to the embryo may at least partially compensate for the loss of Dbh activity in the embryo proper, as 5% of *Dbh*-deficient embryos were born from heterozygous $Dbh^{+/-}$ mice and reached adulthood but no embryos from $Dbh^{-/-}$ mothers survived until birth [11].

Surviving $Dbh^{-/-}$ mice were bradycardic and hypotensive under basal conditions and showed blunted circadian blood pressure rhythms [12]. In addition, $Dbh^{-/-}$ mice exhibited enhanced cardiac contractility upon administration of the β_1-adrenoceptor agonist dobutamine [13]. Contractile function of adult myocytes isolated from $Dbh^{-/-}$ mice was significantly increased under basal conditions. Simultaneously, βARK1, the β-adrenergic receptor kinase that accounts at least partly for the uncoupling of β-receptors in human end-stage dilated cardiomyopathy, was found to be downregulated on the protein and enzyme activity levels in *Dbh*-deficient hearts [13]. After chronic left ventricular pressure overload, cardiac hypertrophy was significantly attenuated but contractile function was preserved in $Dbh^{-/-}$ mice [14]. Thus, cardiac hypertrophy and normalization of wall stress as induced by catecholamines are not necessary to maintain cardiac function after chronic pressure overload [14].

Lack of endogenous catecholamines resulted in higher food intake of $Dbh^{-/-}$ mice, which may be due to increased demand as a result of enhanced metabolic rate [15]. No alterations in plasma leptin levels and no signs of obesity occurred in these animals despite increased food intake and metabolic hypoactivity of brown adipose tissue [15]. In addition, an impaired cold resistance was found in $Dbh^{-/-}$ mice which could be attributed to several defects in thermoregulatory mechanisms [15]. Induction of uncoupling protein-1-mediated thermogenesis was reduced, and piloerectile dysfunction and an attenuated smooth muscle vasoconstrictor response in the vessel periphery was identified in $Dbh^{-/-}$ mice [15]. In a mouse model of hindlimb ischemia, $Dbh^{-/-}$ mice showed reduced angiogenesis and collateral growth as compared with wild-type mice, indicating that catecholamines can also contribute to vascular remodeling during ischemia [16]. In addition to these peripheral and cardiovascular phenotypes, several central nervous system and immune system functions were affected in mice lacking functional Dbh. $Dbh^{-/-}$ mice showed reduced swimming speed and attenuated coping strategies on rapidly rotating rotarods. While no alterations were detectable in a passive-avoidance task to test learning and memory, $Dbh^{-/-}$ mice were impaired in mastering an active-avoidance task and showed antecedent extinction [17–19].

10.2.1.3.2 Human Genetics and Function Humans with mutations in the *DBH* gene leading to deficiency in noradrenaline have been identified [20, 21]. These patients suffer from orthostatic hypotension, exercise intolerance and frequently develop syncope [22]. Zabetian *et al.* identified a C/T polymorphism in the promoter region of the *DBH* gene that is linked to plasma DBH activity levels [23, 24]. In particular, individuals carrying one or two copies of the T allele turned out to have reduced plasma DBH activity. This polymorphism accounted for more than 50% of total variability in plasma DBH activity.

10.2.1.4 Phenylethanolamine N-Methyltransferase (PNMT)

10.2.1.4.1 Mouse Models *Pnmt* has long been thought to be restricted in its expression to chromaffin cells of the adrenal gland, retinal horizontal cells and the brain stem [1]. There is increasing evidence in rat and mouse, however, that adrenergic cells of other than neurogenic origin might transiently arise in further tissues during embryogenesis, accounting for catecholamine synthesis before the adrenal gland and the sympathetic nervous system mature [5, 25].

Ebert *et al.* generated *Pnmt*-deficient mice that express β-galactosidase in adrenergic cells by replacing the *Pnmt* coding region with Cre recombinase and crossing these mice with the Rosa26 reporter mouse strain [5]. As the *Pnmt* gene encodes for the enzyme catalyzing the last step in catecholamine biosynthesis cascade – the conversion of noradrenaline to adrenaline – mice without functional Pnmt were exclusively deficient in adrenaline. In contrast to $Dbh^{-/-}$ animals that lack both noradrenaline and adrenaline, mice deficient in only the latter had no apparent developmental phenotype [5]. Homozygous $Pnmt^{-/-}$ mice were fertile, survived to adulthood and were indistinguishable from wild-type controls. Their capability to cope with stress situations, however, still remains elusive. By means of the *Pnmt*-Cre/R26R mice Ebert *et al.* found X-Gal staining in the heart as early as embryonic day 8.5 (intrinsic cardiac adrenergic cells) [5]. When these cells were followed up during the neonatal and adult period, they turned out to make up parts of the sinoatrial node and the atrioventricular node (i.e. the cardiac pacemaker and conductance system) [5].

Pnmt overexpression, on the contrary, lead to sustained elevation of adrenaline and subsequently to prolonged influence on metabolic and endocrine systems. Transgenic mice exhibited suppressed circulating leptin levels, augmented body fat and enlarged adipocytes due to enhanced lipid storage. They showed, however, normal body weight and no alterations in carbohydrate metabolism [26].

10.2.1.4.2 **Human Genetics and Function**   Ji et al. fully sequenced the human *PNMT* gene of African-American and Caucasian-American individuals to search for single nucleotide polymorphisms (SNPs) in the coding region as well as in the 5′-untranslated region. By means of luciferase reporter gene constructs the authors demonstrated graduated capacity of the constructs to drive transcription accounting for inherited variability in human PNMT expression [27]. Cui et al. found an association between two polymorphisms in the *PNMT* gene promoter region and the risk to develop essential hypertension in one study population of African-American origin [28]. Such an association could not be found in Caucasians of American and Greek origin. In conclusion, genetic variants within the *PNMT* promoter may increase the risk to develop hypertension [28].

10.2.2
Metabolism

The major part of noradrenaline released into the sympathetic cleft is recycled by reuptake into the synthesizing neurons via the noradrenaline transporter (NET) and subsequent transport across the membrane of storage vesicles by the vesicular monoamine transporter (VMAT) (Figure 10.3) [29]. Only a small proportion of cytoplasmic neurotransmitter coming either from outside the cell after degranulation or resulting from leakage from vesicular stores is not a substrate to VMAT, but is submitted to degradation by oxidative deamination through mitochondrial

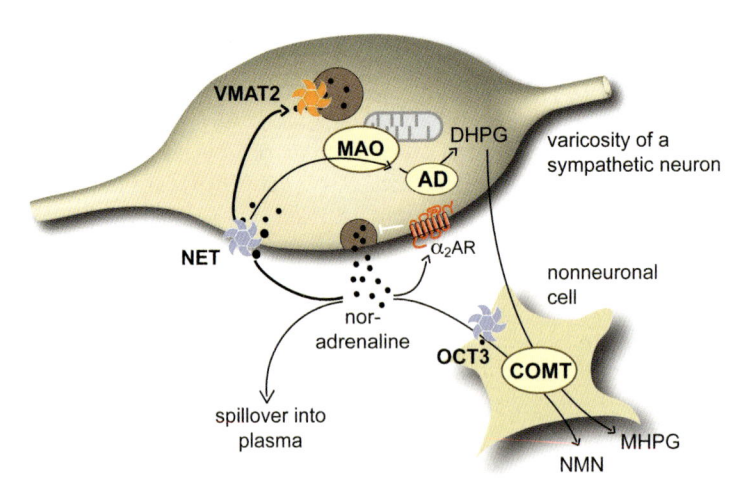

Figure 10.3 Enzymes and transporters involved in storage and metabolism of noradrenaline. AD, aldehyde reductase; NMN, normetanephrine; NET, noradrenaline transporter; VMAT2, vesicular monoamine transporter; MAO, monoamine oxidase; DHPG, dihydroxyphenylglycol; MHPG, methoxy-hydroxyphenyl-glycol; OCT3, organic cation transporter 3; α_2AR, α_2-adrenoceptor.

MAO-A. It is noteworthy, however, that under basal conditions most of the catecholamine metabolites occurring in the blood or urine come from molecules that leaked from storage vesicles into the cytoplasm and underwent subsequent deamination [29]. While toxic catecholaldehydes are formed during this first metabolizing step mediated by MAO-A (3,4-dihydroxyphenylglycolaldehyde from noradrenaline and adrenaline), these reactive substances are rapidly removed by conversion to non-toxic alcohols [dihydroxyphenylglycol (DHPG)] through enzymatic action of aldehyde or aldose reductase [29]. Due to the presence of the β-hydroxyl group on the catecholamine side-chain the reduction to the alcohol is favored over the oxidative reaction to the acid. Only recently, a possible role of those toxic intermediates in the development of neurodegenerative disorders was discussed [30]. In humans, the end-product of noradrenaline metabolism is vanillylmandelic acid formed by hepatocytes after extraction of precursors (methoxy-hydroxyphenylglycol and DHPG) from the bloodstream.

The uptake of synaptic noradrenaline molecules via organic cation transporters [(OCTs); extraneuronal membrane transporter (EMTs)] into extraneuronal cells leads to methylated products. However, only a small proportion of released neurotransmitter escapes the reuptake into the very same neuron via the NET. Non-neuronal cells play an essential role in further processing the already deaminated catecholamine derivatives from MAO-A catalysis (DHPG) to less hydrophilic compounds by catechol-*O*-methyltransferase (COMT)-mediated *O*-methylation (Figure 10.3). Due to lack of COMT in sympathetic neurons this reaction can only occur in extraneuronal tissues or chromaffin cells [29]. COMT facilitates the transfer of the methyl group of *S*-adenosyl-L-methionine to one of the phenolic groups of the catechol substrate in the presence of Mg^{2+} [29, 31].

10.2.2.1 Noradrenaline Transporter (NET)

NET is a high-affinity transporter located in the outer membrane of presynaptic sympathetic nerve varicosities which mediates neuronal noradrenaline reuptake. This Na^+- and Cl^--dependent transport is quantitatively superior to synaptic catecholamine clearance mediated by organic cation transporters, and accounts for about 90% of total catecholamine removal capacity. NET is claimed to be involved in the regulation of diverse physiological processes including learning and memory, mood, arousal and attention, blood flow, and metabolic function [32].

10.2.2.1.1 Mouse Models
Net-deficient mice displayed higher extracellular noradrenaline levels [33]. At rest, arterial pressure and heart rate were slightly higher in $Net^{-/-}$ mice as compared with wild-type control mice [34]. However, differences between genotypes were amplified during waking and activity phases [34]. Xu et al. additionally showed that deletion of *Net* lead to altered midbrain dopaminergic function. Mice deficient in *Net* were more susceptible to dopamine receptor D_2 and D_3 stimulation and to psychostimulants like amphetamine [35].

10.2.2.1.2 Human Genetics and Function
Similar to genetic defects in the *DBH* gene, mutations in the human *NET* gene result in behavioral and cardiovascular

dysfunction including severe orthostatic intolerance and postural tachycardia [36, 37]. Nonen et al. recently identified a polymorphism in the gene encoding the noradrenaline transporter in humans accounting at least in part for the heterogeneity in β-blocker responsiveness among patients with chronic heart failure [38]. A *NET* T 182C substitution together with two SNPs in the *ADRA1D* gene encoding the $α_{1D}$-adrenoceptor was linked to β-blocker-mediated improvement of left ventricular fractional shortening and may serve as predictive genetic markers for effectiveness of β-blocker therapy. Ksiazek et al. identified a nonsynonymous nucleotide exchange in exon 9 of the *NET* gene to be associated with enhanced susceptibility of type 2 diabetes patients to develop hypertension [39].

10.2.2.2 Organic Cation Transporter 3 (OCT3)

OCTs belong to the SLC22 family of solute carriers currently consisting of 12 members (human, rat), among which are the three organic cation transporters OCT1–3. OCT3 (also called EMT) was identified in 1998 [40, 41] and mediates low-affinity, corticosteroid-sensitive removal of catecholamines from the extracellular space into extraneuronal tissues [29].

Oct3-deficient mice were viable and fertile without any apparent neural or physiological dysfunction and with no signs of altered noradrenaline or dopamine levels [42]. *Oct3* was found to be expressed in a wide variety of tissues with highest expression levels seen in skeletal muscle, heart and uterus [42].

10.2.2.3 Catechol O-Methyltransferase (COMT)

10.2.2.3.1 Mouse Models
Homozygous deletion of *Comt* resulted in brain-region-specific changes in dopamine levels in male, but not in female mice [43]. Due to effective elimination of noradrenaline from the synaptic cleft via NET into presynaptic nerve terminals and metabolism via MAO-A, uptake into extraneuronal tissues that leads to degradation by COMT is only a minor pathway. This may explain why COMT inhibitors do not significantly alter catecholamine levels [44].

10.2.2.3.2 Human Genetics and Function
Variations of COMT enzyme activity have been described to be associated with a functional Val108Met (Val158Met) polymorphism. SNPs within the gene encoding for COMT that are associated with increased risk to develop cardiac diseases were mainly identified in accordance with the role of COMT in homocysteine metabolism. This was examined in a part of the Kuopio Ischemic Heart Disease Risk Factor Study where the COMT Val108Met polymorphism was associated with augmented risk of acute coronary events [45].

10.2.2.4 Monoamine Oxidase A (MAO-A)

10.2.2.4.1 Mouse Models
MAO-A is located in the outer mitochondrial membrane and catalyzes the oxidative deamination of catecholamines. *Maoa*-deficient

mice were first generated by random integration of an interferon-β transgene [46]. *Maoa*-deficient mice showed no MAO-A enzymatic activity, causing strongly elevated serotonin levels in young animals and also higher noradrenaline levels in older animals [46]. In addition to dysfunctional presynaptic transporters, presynaptic control of transmitter release by autoreceptors was found to be attenuated in this mouse strain [47]. This might present a molecular mechanism for the observed increased brain serotonin and noradrenaline levels.

10.2.2.4.2 **Human Genetics and Function** In a large Dutch kindred Brunner *et al.* described an X-linked mental retardation syndrome with impulsive aggressive and violent behavior [48]. By applying linkage analysis the defect could be located to the *MAO-A* gene locus on the X chromosome. Signs of disturbed monoamine metabolism were apparent in the results of 24-h urine analysis [48].

10.2.2.5 **Vesicular Monoamine Transporter 2 (VMAT2)**

10.2.2.5.1 **Mouse Models** Homozygous deletion of the vesicular monoamine transporter VMAT2 resulted in perinatal lethality [49]. Homozygotes survived only a few days after birth, and presented with severely impaired monoamine storage and release. Heterozygous knockout mice that appeared normal and survived to adulthood were affected by sudden lethal cardiac arrhythmias [50, 51].

10.2.2.5.2 **Human Genetics and Function** In humans, VMAT2 has been associated with various disease phenotypes (e.g. drug addiction and sensitivity to dopaminergic toxins) (for review, see [22]).

10.3
Adrenergic Receptors

Target structures of noradrenaline and adrenaline are G-protein-coupled α- and β-adrenergic receptors (ARs) that can be further subdivided into different groups and subtypes (Figure 10.3): three α_1-receptors (α_{1A}, α_{1B}, α_{1D}), α_2-receptors (α_{2A}, α_{2B}, α_{2C}) and β-receptors (β_{1-3}) [52]. α_1ARs primarily mediate $G_{q/11}$-dependent downstream signaling while α_2-receptors bind inhibitory G-proteins that slow down cAMP production, inhibit voltage-gated Ca^{2+} channels or activate K^+ channels upon activation. βARs mainly interact with G_s-proteins, but have been described to switch to inhibitory G-proteins as well. α_1ARs primarily mediate vasoconstriction of peripheral vessels [53], while presynaptically located α_2ARs in the sympathetic nervous system were identified as essential players in the feedback regulation of sympathetic neurotransmitter release [54, 55]. βARs on the postsynaptic side are crucial in cardiac signaling controlling heart rate and contractility, but are also part of control mechanisms for peripheral vessel resistance. All adrenoceptor genes have been deleted in the murine genome and the resulting phenotypes have been reviewed in great detail (for reviews, see [54, 56–59]). Here, only the cardiovascular phenotypes and their relevance for human disease are discussed.

10.3.1
α_1-Adrenoceptors

10.3.1.1 Mouse Models

All three α_1AR subtype genes have been targeted for deletion in the mouse [60–62]. Cardiovascular effects of these gene deletions have been reviewed previously [63–65]. α_{1A}-, α_{1B}- and α_{1D}-receptors contribute to agonist-induced vasoconstriction. However, resting blood pressure was only reduced in mice lacking α_{1A}AR or α_{1D}AR [60–62]. Differential distribution of α_1AR subtypes in the vasculature may be important to redirect organ blood flow [57]. Experimental evidence suggests that α_{1D}ARs affect compliance of large arteries, whereas α_{1A}ARs mediate constriction of smaller distributing arteries. In addition to regulation of vascular tone, α_1ARs are involved in vascular remodeling. Chronic infusion of noradrenaline caused remodeling of arteries with smaller vessel lumen in the absence of smooth muscle hypertrophy or hyperplasia in wild-type but not in α_{1B}-deficient mice [66]. In contrast, α_{1D}ARs were important for development of hypertension after salt loading and nephrectomy in mice [62, 67, 68].

In addition to their effects in the vasculature, α_1ARs were also implicated in the modulation of cardiac structure and function. Transgenic overexpression of α_{1B}ARs resulted in variable degrees of cardiac hypertrophy [69–73]. While deletion of single α_1AR genes did not affect cardiac function, double-receptor-deficient mice showed significant cardiac alterations. Mice lacking functional α_{1A}AR and α_{1B}AR presented with reduced cardiac growth and functional alterations which partly resembled characteristics observed in chronic heart failure (for review, see [74]). After induction of chronic pressure overload by aortic constriction, α_{1AB}-deficient mice showed reduced survival, lower ejection fraction and increased end-diastolic dimensions as compared with wild-type mice [75]. Isolated cardiomyocytes from α_{1AB}-deficient mice were more susceptible to apoptosis *in vitro*, suggesting that cardiac α_{1A}AR-mediated activation of extracellular signal-regulated kinase signaling is protective during pressure-induced remodeling [75, 76].

10.3.1.2 Human Genetics and Function

For long time, α_1-adrenoceptor antagonists have been among the first-line antihypertensive drugs. However, in the Antihypertensive and Lipid-Lowering Treatment to Prevent Heart Attack Trial (ALLHAT) clinical trial hypertensive patients receiving the α_1-antagonist doxazosin were more likely to show signs of chronic heart failure including peripheral edema than patients treated with other antihypertensives [77]. The above-mentioned studies in gene-targeted mice which demonstrated that inhibition of cardiac α_1AR function may precipitate cardiac failure provide a plausible explanation for the negative outcome of the doxazosin part of the ALLHAT trial.

10.3.2
α$_2$-Adrenoceptors

10.3.2.1 Mouse Models

Deletions of the genes encoding for α$_{2A}$AR, α$_{2B}$AR or α$_{2C}$AR in the mouse have been described previously [78–80]. In addition, a point mutation (D79N) was introduced into the α$_{2A}$AR gene by homologous recombination in embryonic stem cells [81]. All three α$_2$AR subtypes were identified as presynaptic feedback inhibitors of neurotransmitter release [78, 82, 83] (Figure 10.3). However, in vivo α$_{2A}$AR predominated to control noradrenaline secretion from sympathetic nerves, whereas α$_{2C}$AR inhibited adrenaline exocytosis from chromaffin cells of the adrenal medulla [84–86]. Intact feedback control of catecholamine release from sympathetic neurons as well as from the adrenal medulla is essential to limit the detrimental effects of catecholamines in the development of cardiac failure [85, 87]. The acute effects of α$_2$-agonists on blood pressure were mediated by distinct subtypes: whereas activation of α$_{2B}$AR elicited transient hypertensive effects, stimulation of α$_{2A}$AR resulted in long-lasting sympathoinhibition, bradycardia and hypotension [79, 81]. α$_{2C}$AR may also contribute to regulation of vascular tone by mediating cold-induced vasoconstriction [88, 89]. Endothelial α$_{2A}$ARs were shown to elicit vasodilatation in the mouse aorta [90]. Furthermore, α$_{2B}$ARs were essential for placenta vascular development as mice deficient in all three α$_2$AR subtypes died during mid-gestation from a severe vasculogenesis defect [91].

10.3.2.2 Human Genetics and Function

Thorough sequencing analysis revealed that all human adrenoceptor genes contain a number of SNPs and small deletions that may affect physiological function or pharmacological responses of these receptors [92]. To name two examples, the relevance of deletion variants in the genes encoding for α$_{2B}$AR and α$_{2C}$AR will be discussed briefly. A variant allele of the α$_{2B}$AR encodes a receptor lacking three out of 18 glutamate residues in the third intracellular receptor domain [92]. In a prospective population-based study from Finland homozygous carriers of this α$_{2B}$AR deletion had a 2.2 times higher risk of developing acute coronary syndromes than humans with other α$_{2B}$AR genotypes [93]. In addition, a deletion variant in the third intracellular loop of the α$_{2C}$AR (α$_{2C}$-Del322–325) has attracted particular attention. When expressed at low levels in Chinese hamster ovary cells, α$_{2C}$-Del receptors exhibited defective receptor–G-protein coupling [94]. Genetic association studies showed an increased prevalence of this polymorphism in patients with chronic heart failure and significantly reduced cardiac function [85, 95]. However, patients with dilated cardiomyopathy showed a longer event-free survival if they had at least one allele of the α$_{2C}$-Del variant [96]. Further studies are required to determine the precise role of this polymorphism for the progression of chronic heart failure.

10.3.3
β-Adrenoceptors

10.3.3.1 Mouse Models
Distinct physiological functions were identified in mice lacking β_1AR, β_2AR or β_3AR [58, 97, 98]. All three βAR subtypes were identified as vasorelaxing receptors in different mouse blood vessels [99]. In the heart, positive chronotropic and inotropic effects of isoproterenol were completely absent in β_1AR-deficient mice [100] and β_2AR-deficient mice showed normal cardiac responses to catecholamines [97]. Surprisingly, deletion of both β_1AR and β_2AR resulted in remarkably small alteration in basal cardiac function and metabolic rate, and both receptors were not essential for the development of cardiac hypertrophy after aortic constriction [58, 101]. Distinct signaling pathways for β_1AR and β_2AR were identified in cardiomyocytes derived from βAR-deficient mice (for review, see [102]).

10.3.3.2 Human Genetics and Function
In human chronic heart failure, cardiac βAR is downregulated by several molecular mechanisms (for a recent review, see [103]). Treatment of enhanced sympathetic tone by βAR antagonists can greatly improve survival of patients with chronic heart failure. However, the optimal way to interfere with sympathetic activity in heart failure has not yet been identified. In addition to the polymorphisms that have been identified within the α_2AR genes, βAR also exhibit significant genetic variation [92, 104]. In recombinant cell systems and in isolated cardiomyocytes Arg389 β_1ARs have demonstrated 3- to 4-fold better coupling to adenylyl cyclase or contractility than Gly389 β_1ARs [103, 105]. While this β_1AR polymorphism did not alter clinical outcome in patients with chronic heart failure, response to β-blocker therapy was significantly affected [105]. Arg389 β_1AR heart failure patients treated with bucindolol showed significant reduction in mortality and hospitalization, while Gly389 β_1AR carriers did not benefit from this treatment [105]. Using a fluorescence energy transfer system to monitor β_1AR activation *in vitro*, only carvedilol but not metoprolol or bisoprolol showed significant inverse agonism at Arg398 β_1AR [106]. Clinical studies will have to determine whether Arg389 β_1AR patients with chronic heart failure benefit most from carvedilol treatment.

10.4
Conclusions

Genetic mouse models have been instrumental in identifying the physiological functions and pathophysiological relevance of the molecular components of the adrenergic system. In order to better understand the significance of human polymorphisms in adrenergic genes, novel mouse models expressing wild-type and variant versions of the human genes are required. These models may then be used to test *in vivo* which drugs may be optimal for sympathetic blockade. The example of differential effects of selected β-blockers at Arg389Gly β$_1$ARs has demonstrated the power of this approach to delineate novel therapeutic strategies for human heart failure. The results of these trials will rapidly show whether personalized therapy based on adrenergic polymorphisms may yield additional benefits for patients with cardiovascular disease or whether the search for novel, potentially causative target genes would be more promising in the future treatment of hypertension, coronary artery disease and chronic heart failure.

References

1 Westfall, T.C. and Westfall, D.P. (2006) Neurotransmission: the autonomic and somatic motor nervous systems, in *Goodman and Gilman's the Pharmacological Basis of Therapeutics*, 11th edn (ed. L.L. Brunton), McGraw-Hill, New York, pp. 137–82.

2 Zhou, Q.Y., Quaife, C.J. and Palmiter, R.D. (1995) Targeted disruption of the tyrosine hydroxylase gene reveals that catecholamines are required for mouse fetal development. *Nature*, 374, 640–3.

3 Kobayashi, K., Morita, S., Sawada, H. et al. (1995) Targeted disruption of the tyrosine hydroxylase locus results in severe catecholamine depletion and perinatal lethality in mice. *Journal of Biological Chemistry*, 270, 27235–43.

4 Zhou, Q.Y. and Palmiter, R.D. (1995) Dopamine-deficient mice are severely hypoactive, adipsic, and aphagic. *Cell*, 83, 1197–209.

5 Ebert, S.N., Rong, Q., Boe, S., Thompson, R.P., Grinberg, A. and Pfeifer, K. (2004) Targeted insertion of the Cre-recombinase gene at the phenylethanolamine n-methyltransferase locus: a new model for studying the developmental distribution of adrenergic cells. *Developmental Dynamics*, 231, 849–58.

6 Rios, M., Habecker, B., Sasaoka, T. et al. (1999) Catecholamine synthesis is mediated by tyrosinase in the absence of tyrosine hydroxylase. *Journal of Neuroscience*, 19, 3519–26.

7 van den Heuvel, L.P., Luiten, B., Smeitink, J.A. et al. (1998) A common point mutation in the tyrosine hydroxylase gene in autosomal recessive L-DOPA-responsive dystonia in the Dutch population. *Human Genetics*, 102, 644–6.

8 Ludecke, B., Dworniczak, B. and Bartholome, K. (1995) A point mutation in the tyrosine hydroxylase gene associated with Segawa's syndrome. *Human Genetics*, 95, 123–5.

9 Hyland, K. and Clayton, P.T. (1990) Aromatic amino acid decarboxylase deficiency in twins. *Journal of Inherited Metabolic Disease*, 13, 301–4.

10 Pons, R., Ford, B., Chiriboga, C.A. et al. (2004) Aromatic L-amino acid decarboxylase deficiency: clinical features, treatment, and prognosis. *Neurology*, 62, 1058–65.

11 Thomas, S.A., Matsumoto, A.M. and Palmiter, R.D. (1995) Noradrenaline is essential for mouse fetal development. *Nature*, 374, 643–6.

12 Swoap, S.J., Weinshenker, D., Palmiter, R.D. and Garber, G. (2004) Dbh$^{-/-}$ mice are hypotensive, have altered circadian rhythms, and have abnormal responses to dieting and stress. *American Journal of Physiology*, 286, R108–113.

13 Cho, M.C., Rao, M., Koch, W.J., Thomas, S.A., Palmiter, R.D. and Rockman, H.A. (1999) Enhanced contractility and decreased β-adrenergic receptor kinase-1 in mice lacking endogenous norepinephrine and epinephrine. *Circulation*, 99, 2702–7.

14 Esposito, G., Rapacciuolo, A., Naga Prasad, S.V. et al. (2002) Genetic alterations that inhibit *in vivo* pressure-overload hypertrophy prevent cardiac dysfunction despite increased wall stress. *Circulation*, 105, 85–92.

15 Thomas, S.A. and Palmiter, R.D. (1997) Thermoregulatory and metabolic

phenotypes of mice lacking noradrenaline and adrenaline. *Nature*, **387**, 94–7.

16 Chalothorn, D., Zhang, H., Clayton, J.A., Thomas, S.A. and Faber, J.E. (2005) Catecholamines augment collateral vessel growth and angiogenesis in hindlimb ischemia. *American Journal of Physiology, Heart and Circulatory Physiology*, **289**, H947-959.

17 Thomas, S.A. and Palmiter, R.D. (1997) Disruption of the dopamine β-hydroxylase gene in mice suggests roles for norepinephrine in motor function, learning, and memory. *Behavioral Neuroscience*, **111**, 579–89.

18 Thomas, S.A. and Palmiter, R.D. (1997) Impaired maternal behavior in mice lacking norepinephrine and epinephrine. *Cell*, **91**, 583–92.

19 Alaniz, R.C., Thomas, S.A., Perez-Melgosa, M. *et al.* (1999) Dopamine β-hydroxylase deficiency impairs cellular immunity. *Proceedings of the National Academy of Sciences of the United States of America*, **96**, 2274–8.

20 Man in 't Veld, A.J., Boomsma, F., Moleman, P. and Schalekamp, M.A. (1987) Congenital dopamine-β-hydroxylase deficiency. A novel orthostatic syndrome. *Lancet*, **1**, 183–8.

21 Robertson, D., Goldberg, M.R., Onrot, J. *et al.* (1986) Isolated failure of autonomic noradrenergic neurotransmission. Evidence for impaired β-hydroxylation of dopamine. *New England Journal of Medicine*, **314**, 1494–7.

22 Usera, P.C., Vincent, S. and Robertson, D. (2004) Human phenotypes and animal knockout models of genetic autonomic disorders. *Journal of Biomedical Science*, **11**, 4–10.

23 Zabetian, C.P., Anderson, G.M., Buxbaum, S.G. *et al.* (2001) A quantitative-trait analysis of human plasma-DOPAmine β-hydroxylase activity: evidence for a major functional polymorphism at the DBH locus. *American Journal of Human Genetics*, **68**, 515–22.

24 Zabetian, C.P., Romero, R., Robertson, D. *et al.* (2003) A revised allele frequency estimate and haplotype analysis of the DBH deficiency mutation IVS1 + 2T → C in African- and European-Americans. *American Journal of Medical Genetics. Part A*, **123**, 190–2.

25 Ziegler, M.G., Bao, X., Kennedy, B.P., Joyner, A. and Enns, R. (2002) Location. development, control, and function of extraadrenal phenylethanolamine N-methyltransferase. *Annals of the New York Academy of Sciences*, **971**, 76–82.

26 Bottner, A., Haidan, A., Eisenhofer, G. *et al.* (2000) Increased body fat mass and suppression of circulating leptin levels in response to hypersecretion of epinephrine in phenylethanolamine-N-methyltransferase (PNMT)-overexpressing mice. *Endocrinology*, **141**, 4239–46.

27 Ji, Y., Salavaggione, O.E., Wang, L. *et al.* (2005) Human phenylethanolamine N-methyltransferase pharmacogenomics: gene re-sequencing and functional genomics. *Journal of Neurochemistry*, **95**, 1766–76.

28 Cui, J., Zhou, X., Chazaro, I. *et al.* (2003) Association of polymorphisms in the promoter region of the PNMT gene with essential hypertension in African Americans but not in whites. *American Journal of Hypertension*, **16**, 859–63.

29 Eisenhofer, G., Kopin, I.J. and Goldstein, D.S. (2004) Catecholamine metabolism: a contemporary view with implications for physiology and medicine. *Pharmacological Reviews*, **56**, 331–49.

30 Marchitti, S.A., Deitrich, R.A. and Vasiliou, V. (2007) Neurotoxicity and metabolism of the catecholamine-derived 3,4-dihydroxyphenylacetaldehyde and 3,4-dihydroxyphenylglycolaldehyde: the role of aldehyde dehydrogenase. *Pharmacological Reviews*, **59**, 125–50.

31 Huotari, M., Gogos, J.A., Karayiorgou, M. *et al.* (2002) Brain catecholamine metabolism in catechol-O-methyltransferase (COMT)-deficient mice. *European Journal of Neuroscience*, **15**, 246–56.

32 Bonisch, H. and Bruss, M. (2006) The norepinephrine transporter in physiology and disease. *Handbook of Experimental Pharmacology*, **175**, 485–524.

33 Wang, Y.M., Xu, F., Gainetdinov, R.R. and Caron, M.G. (1999) Genetic

approaches to studying norepinephrine function: knockout of the mouse norepinephrine transporter gene. *Biological Psychiatry*, **46**, 1124–30.
34 Keller, N.R., Diedrich, A., Appalsamy, M. *et al.* (2004) Norepinephrine transporter-deficient mice exhibit excessive tachycardia and elevated blood pressure with wakefulness and activity. *Circulation*, **110**, 1191–6.
35 Xu, F., Gainetdinov, R.R., Wetsel, W.C. *et al.* (2000) Mice lacking the norepinephrine transporter are supersensitive to psychostimulants. *Nature Neuroscience*, **3**, 465–71.
36 Robertson, D., Flattem, N., Tellioglu, T. *et al.* (2001) Familial orthostatic tachycardia due to norepinephrine transporter deficiency. *Annals of the New York Academy of Sciences*, **940**, 527–43.
37 Shannon, J.R., Flattem, N.L., Jordan, J. *et al.* (2000) Orthostatic intolerance and tachycardia associated with norepinephrine-transporter deficiency. *New England Journal of Medicine*, **342**, 541–9.
38 Nonen, S., Okamoto, H., Fujio, Y. *et al.* (2008) Polymorphisms of norepinephrine transporter and adrenergic receptor α_{1D} are associated with the response to β-blockers in dilated cardiomyopathy. *Pharmacogenomics Journal*, **8**, 78–84.
39 Ksiazek, P., Buraczynska, K. and Buraczynska, M. (2006) Norepinephrine transporter gene (NET) polymorphism in patients with type 2 diabetes. *Kidney and Blood Pressure Research*, **29**, 338–43.
40 Grundemann, D., Schechinger, B., Rappold, G.A. and Schomig, E. (1998) Molecular identification of the corticosterone-sensitive extraneuronal catecholamine transporter. *Nature Neuroscience*, **1**, 349–51.
41 Kekuda, R., Prasad, P.D., Wu, X. *et al.* (1998) Cloning and functional characterization of a potential-sensitive, polyspecific organic cation transporter (OCT3) most abundantly expressed in placenta. *Journal of Biological Chemistry*, **273**, 15971–9.
42 Zwart, R., Verhaagh, S., Buitelaar, M., Popp-Snijders, C. and Barlow, D.P. (2001) Impaired activity of the extraneuronal monoamine transporter system known as uptake-2 in Orct3/Slc22a3-deficient mice. *Molecular and Cellular Biology*, **21**, 4188–96.
43 Gogos, J.A., Morgan, M., Luine, V. *et al.* (1998) Catechol-*O*-methyltransferase-deficient mice exhibit sexually dimorphic changes in catecholamine levels and behavior. *Proceedings of the National Academy of Sciences of the United States of America*, **95**, 9991–6.
44 Mannisto, P.T. and Kaakkola, S. (1999) Catechol-*O*-methyltransferase (COMT): biochemistry, molecular biology, pharmacology, and clinical efficacy of the new selective COMT inhibitors. *Pharmacological Reviews*, **51**, 593–628.
45 Voutilainen, S., Tuomainen, T.P., Korhonen, M. *et al.* (2007) Functional COMT Val158Met polymorphism, risk of acute coronary events and serum homocysteine: the kuopio ischaemic heart disease risk factor study. *PLoS ONE*, **2**, e181.
46 Cases, O., Seif, I., Grimsby, J. *et al.* (1995) Aggressive behavior and altered amounts of brain serotonin and norepinephrine in mice lacking MAOA. *Science*, **268**, 1763–6.
47 Owesson, C.A., Hopwood, S.E., Callado, L.F., Seif, I., McLaughlin, D.P. and Stamford, J.A. (2002) Altered presynaptic function in monoaminergic neurons of monoamine oxidase-A knockout mice. *European Journal of Neuroscience*, **15**, 1516–22.
48 Brunner, H.G., Nelen, M.R., van Zandvoort, P. *et al.* (1993) X-linked borderline mental retardation with prominent behavioral disturbance: phenotype, genetic localization, and evidence for disturbed monoamine metabolism. *American Journal of Human Genetics*, **52**, 1032–9.
49 Wang, Y.M., Gainetdinov, R.R., Fumagalli, F. *et al.* (1997) Knockout of the vesicular monoamine transporter 2 gene results in neonatal death and supersensitivity to cocaine and amphetamine. *Neuron*, **19**, 1285–96.
50 Itokawa, K., Sora, I., Schindler, C.W., Itokawa, M., Takahashi, N. and Uhl, G.R. (1999) Heterozygous VMAT2 knockout mice display prolonged QT intervals:

possible contributions to sudden death. *Brain Research*, **71**, 354–7.
51 Uhl, G.R., Li, S., Takahashi, N. et al. (2000) The VMAT2 gene in mice and humans: amphetamine responses, locomotion, cardiac arrhythmias, aging, and vulnerability to dopaminergic toxins. *FASEB Journal*, **14**, 2459–65.
52 Bylund, D.B., Eikenberg, D.C., Hieble, J.P. et al. (1994) International Union of Pharmacology nomenclature of adrenoceptors. *Pharmacological Reviews*, **46**, 121–36.
53 Guimaraes, S. and Moura, D. (2001) Vascular adrenoceptors: an update. *Pharmacological Reviews*, **53**, 319–56.
54 Hein, L. (2001) Transgenic models of α_2-adrenergic receptor subtype function. *Reviews of Physiology, Biochemistry and Pharmacology*, **142**, 161–85.
55 Starke, K. (2001) Presynaptic autoreceptors in the third decade: focus on α_2-adrenoceptors. *Journal of neurochemistry*, **78**, 685–93.
56 Brede, M., Philipp, M., Knaus, A., Muthig, V. and Hein, L. (2004) α_2-adrenergic receptor subtypes – novel functions uncovered in gene-targeted mouse models. *Biology of the Cell*, **96**, 343–8.
57 Philipp, M. and Hein, L. (2004) Adrenergic receptor knockout mice: distinct functions of 9 receptor subtypes. *Pharmacology and Therapeutics*, **101**, 65–74.
58 Rohrer, D.K., Chruscinski, A., Schauble, E.H., Bernstein, D. and Kobilka, B.K. (1999) Cardiovascular and metabolic alterations in mice lacking both β_1- and β_2-adrenergic receptors. *Journal of Biological Chemistry*, **274**, 16701–8.
59 Knaus, A.E., Muthig, V., Schickinger, S. et al. (2007) α_2-Adrenoceptor subtypes – unexpected functions for receptors and ligands derived from gene-targeted mouse models. *Neurochemistry International*, **51**, 277–81.
60 Cavalli, A., Lattion, A.L., Hummler, E. et al. (1997) Decreased blood pressure response in mice deficient of the α_{1B}-adrenergic receptor. *Proceedings of the National Academy of Sciences of the United States of America*, **94**, 11589–94.

61 Rokosh, D.G. and Simpson, P.C. (2002) Knockout of the $\alpha_{1A/C}$-adrenergic receptor subtype: the $\alpha_{1A/C}$ is expressed in resistance arteries and is required to maintain arterial blood pressure. *Proceedings of the National Academy of Sciences of the United States of America*, **99**, 9474–9.
62 Tanoue, A., Nasa, Y., Koshimizu, T. et al. (2002) The α_{1D}-adrenergic receptor directly regulates arterial blood pressure via vasoconstriction. *Journal of Clinical Investigation*, **109**, 765–75.
63 Chen, Z.J. and Minneman, K.P. (2005) Recent progress in α_1-adrenergic receptor research. *Acta Pharmacologica Sinica*, **26**, 1281–7.
64 Koshimizu, T.A., Tanoue, A. and Tsujimoto, G. (2007) Clinical implications from studies of α_1 adrenergic receptor knockout mice. *Biochemical Pharmacology*, **73**, 1107–12.
65 Piascik, M.T. and Perez, D.M. (2001) α_1-adrenergic receptors: new insights and directions. *Journal of Pharmacology and Experimental Therapeutics*, **298**, 403–10.
66 Vecchione, C., Fratta, L., Rizzoni, D. et al. (2002) Cardiovascular influences of α_{1B}-adrenergic receptor defect in mice. *Circulation*, **105**, 1700–7.
67 Tanoue, A., Koba, M., Miyawaki, S. et al. (2002) Role of the α_{1D}-adrenergic receptor in the development of salt-induced hypertension. *Hypertension*, **40**, 101–6.
68 Tanoue, A., Koshimizu, T.A. and Tsujimoto, G. (2002) Transgenic studies of α_1-adrenergic receptor subtype function. *Life Sciences*, **71**, 2207–15.
69 Akhter, S.A., Milano, C.A., Shotwell, K.F. et al. (1997) Transgenic mice with cardiac overexpression of α_{1B}-adrenergic receptors. In vivo α_1-adrenergic receptor-mediated regulation of β-adrenergic signaling. *Journal of Biological Chemistry*, **272**, 21253–9.
70 Grupp, I.L., Lorenz, J.N., Walsh, R.A., Boivin, G.P. and Rindt, H. (1998) Overexpression of α_{1B}-adrenergic receptor induces left ventricular dysfunction in the absence of hypertrophy. *American Journal of Physiology*, **275**, H1338–1350.
71 Lemire, I., Allen, B.G., Rindt, H. and Hebert, T.E. (1998) Cardiac-specific overexpression of α_{1B}AR regulates βAR

activity via molecular crosstalk. *Journal of Molecular and Cellular Cardiology*, **30**, 1827–39.

72 Lemire, I., Ducharme, A., Tardif, J.C. et al. (2001) Cardiac-directed overexpression of wild-type α_{1B}-adrenergic receptor induces dilated cardiomyopathy. *American Journal of Physiology, Heart and Circulatory Physiology*, **281**, H931–938.

73 Zuscik, M.J., Sands, S., Ross, S.A. et al. (2000) Overexpression of the α_{1B}-adrenergic receptor causes apoptotic neurodegeneration: multiple system atrophy. *Nature Medicine*, **6**, 1388–94.

74 Woodcock, E.A. (2007) Roles of α_{1A}- and α_{1B}-adrenoceptors in heart: Insights from studies of genetically modified mice. *Clinical and Experimental Pharmacology and Physiology*, **34**, 884–8.

75 O'Connell, T.D., Swigart, P.M., Rodrigo, M.C. et al. (2006) α_1-adrenergic receptors prevent a maladaptive cardiac response to pressure overload. *Journal of Clinical Investigation*, **116**, 1005–15.

76 Huang, Y., Wright, C.D., Merkwan, C.L. et al. (2007) An α_{1A}-adrenergic-extracellular signal-regulated kinase survival signaling pathway in cardiac myocytes. *Circulation*, **115**, 763–72.

77 ALLHAT Collaborative Research Group (2000) Major cardiovascular events in hypertensive patients randomized to doxazosin vs chlorthalidone: the Antihypertensive and Lipid-Lowering Treatment to Prevent Heart Attack Trial (ALLHAT). *Journal of the American Medical Association*, **283**, 1967–75.

78 Altman, J.D., Trendelenburg, A.U., MacMillan, L. et al. (1999) Abnormal regulation of the sympathetic nervous system in α_{2A}-adrenergic receptor knockout mice. *Molecular Pharmacology*, **56**, 154–61.

79 Link, R.E., Desai, K., Hein, L. et al. (1996) Cardiovascular regulation in mice lacking α_2-adrenergic receptor subtypes b and c. *Science*, **273**, 803–5.

80 Link, R.E., Stevens, M.S., Kulatunga, M., Scheinin, M., Barsh, G.S. and Kobilka, B.K. (1995) Targeted inactivation of the gene encoding the mouse α_{2C}-adrenoceptor homolog. *Molecular Pharmacology*, **48**, 48–55.

81 MacMillan, L.B., Hein, L., Smith, M.S., Piascik, M.T. and Limbird, L.E. (1996) Central hypotensive effects of the α_{2A}-adrenergic receptor subtype. *Science*, **273**, 801–3.

82 Hein, L., Altman, J.D. and Kobilka, B.K. (1999) Two functionally distinct α_2-adrenergic receptors regulate sympathetic neurotransmission. *Nature*, **402**, 181–4.

83 Trendelenburg, A.U., Philipp, M., Meyer, A., Klebroff, W., Hein, L. and Starke, K. (2003) All three α_2-adrenoceptor types serve as autoreceptors in postganglionic sympathetic neurons. *Naunyn-Schmiedeberg's Archives of Pharmacology*, **368**, 504–12.

84 Brede, M., Nagy, G., Philipp, M., Sorensen, J.B., Lohse, M.J. and Hein, L. (2003) Differential control of adrenal and sympathetic catecholamine release by α_2-adrenoceptor subtypes. *Molecular Endocrinology*, **17**, 1640–6.

85 Brede, M., Wiesmann, F., Jahns, R. et al. (2002) Feedback inhibition of catecholamine release by two different α_2-adrenoceptor subtypes prevents progression of heart failure. *Circulation*, **106**, 2491–6.

86 Gilsbach, R., Brede, M., Beetz, N. et al. (2007) Heterozygous α_{2C}-adrenoceptor-deficient mice develop heart failure after transverse aortic constriction. *Cardiovascular Research*, **75**, 728–37.

87 Lymperopoulos, A., Rengo, G., Funakoshi, H., Eckhart, A.D. and Koch, W.J. (2007) Adrenal GRK2 upregulation mediates sympathetic overdrive in heart failure. *Nature Medicine*, **13**, 315–23.

88 Chotani, M.A., Flavahan, S., Mitra, S., Daunt, D. and Flavahan, N.A. (2000) Silent α_{2C}-adrenergic receptors enable cold-induced vasoconstriction in cutaneous arteries. *American Journal of Physiology, Heart and Circulatory Physiology*, **278**, H1075–1083.

89 Jeyaraj, S.C., Chotani, M.A., Mitra, S., Gregg, H.E., Flavahan, N.A. and Morrison, K.J. (2001) Cooling evokes redistribution of α_{2C}-adrenoceptors from Golgi to plasma membrane in transfected human embryonic kidney 293 cells. *Molecular Pharmacology*, **60**, 1195–200.

90 Shafaroudi, M.M., McBride, M., Deighan, C. et al. (2005) Two "knockout" mouse models demonstrate that aortic vasodilatation is mediated via α_{2A}-adrenoceptors located on the endothelium. *Journal of Pharmacology and Experimental Therapeutics*, **314**, 804–10.

91 Philipp, M., Brede, M., Hadamek, K., Gessler, M., Lohse, M.J. and Hein, L. (2002) Placental α_2-adrenoceptors control vascular development at the interface between mother and embryo. *Nature Genetics*, **31**, 311–15.

92 Small, K.M., McGraw, D.W. and Liggett, S.B. (2003) Pharmacology and physiology of human adrenergic receptor polymorphisms. *Annual Review of Pharmacology and Toxicology*, **43**, 381–411.

93 Snapir, A., Heinonen, P., Tuomainen, T.P. et al. (2001) An insertion/deletion polymorphism in the α_{2B}-adrenergic receptor gene is a novel genetic risk factor for acute coronary events. *Journal of the American College of Cardiology*, **37**, 1516–22.

94 Small, K.M., Forbes, S.L., Rahman, F.F., Bridges, K.M. and Liggett, S.B. (2000) A four amino acid deletion polymorphism in the third intracellular loop of the human α_{2C}-adrenergic receptor confers impaired coupling to multiple effectors. *Journal of Biological Chemistry*, **275**, 23059–64.

95 Small, K.M., Wagoner, L.E., Levin, A.M., Kardia, S.L. and Liggett, S.B. (2002) Synergistic polymorphisms of β_1- and α_{2C}-adrenergic receptors and the risk of congestive heart failure. *New England Journal of Medicine*, **347**, 1135–42.

96 Regitz-Zagrosek, V., Hocher, B., Bettmann, M. et al. (2006) α_{2C}-adrenoceptor polymorphism is associated with improved event-free survival in patients with dilated cardiomyopathy. *European Heart Journal*, **27**, 454–9.

97 Chruscinski, A.J., Rohrer, D.K., Schauble, E., Desai, K.H., Bernstein, D. and Kobilka, B.K. (1999) Targeted disruption of the β_2 adrenergic receptor gene. *Journal of Biological Chemistry*, **274**, 16694–700.

98 Susulic, V.S., Frederich, R.C., Lawitts, J. et al. (1995) Targeted disruption of the β_3-adrenergic receptor gene. *Journal of Biological Chemistry*, **270**, 29483–92.

99 Chruscinski, A., Brede, M.E., Meinel, L., Lohse, M.J., Kobilka, B.K. and Hein, L. (2001) Differential distribution of β-adrenergic receptor subtypes in blood vessels of knockout mice lacking β_1- or β_2-adrenergic receptors. *Molecular Pharmacology*, **60**, 955–62.

100 Rohrer, D.K., Desai, K.H., Jasper, J.R. et al. (1996) Targeted disruption of the mouse β_1-adrenergic receptor gene: developmental and cardiovascular effects. *Proceedings of the National Academy of Sciences of the United States of America*, **93**, 7375–80.

101 Palazzesi, S., Musumeci, M., Catalano, L. et al. (2006) Pressure overload causes cardiac hypertrophy in β_1-adrenergic and β_2-adrenergic receptor double knockout mice. *Journal of Hypertension*, **24**, 563–71.

102 Xiang, Y. and Kobilka, B.K. (2003) Myocyte adrenoceptor signaling pathways. *Science*, **300**, 1530–2.

103 Brodde, O.E. (2007) β-adrenoceptor blocker treatment and the cardiac β-adrenoceptor-G-protein(s)-adenylyl cyclase system in chronic heart failure. *Naunyn-Schmiedeberg's Archives of Pharmacology*, **374**, 361–72.

104 Leineweber, K., Büscher, R., Bruck, H. and Brodde, O.E. (2003) β-adrenoceptor polymorphisms. *Naunyn-Schmiedeberg's Archives of Pharmacology*, **369**, 1–22.

105 Liggett, S.B., Mialet-Perez, J., Thaneemit-Chen, S. et al. (2006) A polymorphism within a conserved β_1-adrenergic receptor motif alters cardiac function and β-blocker response in human heart failure. *Proceedings of the National Academy of Sciences of the United States of America*, **103**, 11288–93.

106 Rochais, F., Vilardaga, J.P., Nikolaev, V.O., Bünemann, M., Lohse, M.J. and Engelhardt, S. (2007) Real-time optical recording of β_1-adrenergic receptor activation reveals supersensitivity of the Agr389 variant to carvedilol. *Journal of Clinical Investigation*, **117**, 229–35.

11
Dopamine
Pedro Gomes and Patrício Soares-da-Silva

11.1
Introduction

Dopamine plays a dual role in the body by occurring in neurons and non-neuronal cells. Dopamine is a neurotransmitter in brain dopaminergic neurons, and its central role in the regulation of motor function and behavior is well established. In the peripheral nervous system, dopamine is the first catecholamine in the biosynthesis of neurotransmitters such as norepinephrine and epinephrine. The role of dopamine in the sympathetic nervous system is not only to provide a means for the synthesis of the neurotransmitter norepinephrine [1–5], as the amine can also work as cotransmitter in some circumstances [6–8]. Non-neuronal cells capable of producing dopamine are distributed to the kidney and intestine [9, 10]. These renal tubular and intestinal epithelial cells are endowed with a high capacity to synthesize dopamine from its immediate precursor, L-3,4-dihydroxyphenylalanine (L-DOPA), where the amine plays a pivotal role as a local hormone with autocrine and paracrine function, namely regulating water and electrolyte homeostasis [11]. This chapter will review evidence for autocrine and paracrine actions of dopamine formed within the kidney and intestine, their cellular signaling pathways, and discuss the potential role of dopamine in the pathophysiology of hypertension, heart failure, renal parenchymal disorders and diabetes.

11.2
Dopamine Synthesis

11.2.1
L-DOPA Uptake and Decarboxylation

The renal and intestinal dopaminergic systems are local non-neuronal systems constituted by epithelial cells of proximal convoluted renal tubules and the intestinal epithelial layer rich in aromatic L-amino acid decarboxylase (AADC) activity and using circulating or luminal L-DOPA as a source for dopamine [9, 12, 13]. Dopamine, produced in proximal convoluted renal tubules and the intestinal epithelium, is in close proximity to proximal tubular epithelial (PTE) or epithelial cells of the intestinal mucosa that contain receptors for the amine and acts as a paracrine or autocrine substance [14]. At the kidney level, the role of dopamine in the excretion of sodium after increased sodium intake has led to the hypothesis that an aberrant renal dopaminergic system is important in the pathogenesis of some forms of genetic hypertension [15–17]. Although the kidney and the intestinal mucosa are endowed with some of the highest levels of AADC in the body, and plasma levels of L-DOPA are in the nanomole per milliliter range [18, 19], the rate-limiting step for the synthesis of dopamine in renal tissues and the intestinal mucosa is still a matter of debate. In fact, large amounts of L-DOPA taken up in the kidney are not converted to dopamine [9, 13], suggesting that uptake of L-DOPA rather than its conversion to dopamine may rate-limit the formation of renal dopamine.

11.2.2
Amino Acid Transporters

Recent studies have shown that L-DOPA uptake in renal epithelial cells may be promoted through the L-type amino acid transporter [20, 21], as has been found in intestinal epithelial cells [22, 23] and at the level of brain capillary endothelium [24–29]. In this respect, it is interesting to underline the observation that the apical membrane in renal LLC-PK$_1$ cells is endowed with different transporters for the handling of L-DOPA [21, 30–32]. L-DOPA is transported from the extracellular fluid to the intracellular space across the apical membrane via the L-type amino acid transporter, whereas L-DOPA is transported from the intracellular space to the extracellular fluid across the membrane via a 4,4′-diisothiocyanatostilbene 2,2′-disulphonic acid-sensitive mechanism [21]. Candidate transport systems for apical-to-basal L-DOPA transfer may include the Na$^+$-dependent systems B, B$^{0,+}$ and y$^+$L, and the Na$^+$-independent systems L (LAT1 and LAT2) and b$^{0,+}$. Both b$^{0,+}$ and LAT1 were found to transport L-DOPA – the former in *Xenopus laevis* oocytes injected with poly(A)$^+$ RNA prepared from rabbit intestinal epithelium [33] and the latter in mouse brain capillary endothelial cells [27]. L-DOPA in OK$_{LC}$ and OK$_{HC}$ cells is transported quite efficiently across the apical cell border and several findings indicate that L-DOPA uses at least two major transporters (i.e. systems LAT2 and b$^{0,+}$) [34]. The transport of L-DOPA by LAT2 corresponds to a Na$^+$-independent transporter with a broad specificity for small and large neutral amino acids, stimulated by acid pH and inhibited by 2-aminobicyclo-(2,2,1)-heptane-2-carboxylic acid. The transport of L-DOPA by system b$^{0,+}$ is a Na$^+$-independent transporter for neutral and basic amino acids that also recognizes the di-amino acid cysteine. Transporters involved in Na$^+$-independent uptake of L-DOPA (systems LAT2 and b$^{0,+}$) also function as tightly coupled exchangers [34]. In line with this view is recent data from our laboratory showing for that overexpression of LAT2 in the spontaneous hypertensive rat (SHR) kidney is organ-specific and precedes the onset of hypertension that is accompanied by enhanced ability to take up L-DOPA [35]. This suggests that overexpression of renal LAT2 may constitute the basis for the enhanced renal production of dopamine in the SHR in an attempt overcome the deficient dopamine-mediated natriuresis generally observed in this genetic model of hypertension [17]. This adaptive mechanism may be limited to renal tissues, since at the intestinal level where defective transduction of the D_1 receptor signal also occurs [36], but is not accompanied by increases in either dopamine tissue levels or intestinal LAT2 expression [35].

In an attempt to understand better the differences in the handling of L-amino acids in hypertension, the function and expression of ASCT2 were also examined [37]. The inward transport of [^{14}C]L-alanine, an ASCT2 preferential substrate, was evaluated in monolayers of immortalized renal proximal tubular cells from SHR and normotensive control [Wistar-Kyoto (WKY)] rat strains. The quantification of ASCT2 mRNA and ASCT2 protein was performed in immortalized renal proximal tubular cells and kidney cortices from WKY and SHR strains. Immortalized SHR and WKY renal proximal tubular cells take up L-alanine mainly through a high-affinity, Na^+-dependent amino acid transporter, with functional features of ASCT2 transport. The activity and expression of ASCT2 transporter were considerably lower in the SHR cells. As a compensatory mechanism, in SHR cells, L-alanine is also transported by other amino acid transport systems, namely LAT1 and $b^{0,+}$, that account for approximately 10% of total [^{14}C]L-alanine uptake. Finally, findings obtained in immortalized cells match those *in vivo*: in the SHR model ASCT2 is underexpressed, at the kidney cortex level [37].

The abundance of LAT1 and 4F2hc was greater in SHR than in WKY renal proximal tubular cells, but this inversely correlated with the ability of the former cells to handle [^{14}C]L-leucine [38]. In fact, despite the increase in LAT1 and 4F2hc expression in SHR PTE cells, the spontaneous [^{14}C]L-leucine efflux was higher in WKY than in SHR PTE cells and the potency of L-leucine to stimulate [^{14}C]L-leucine efflux in WKY was 4.5-fold that in SHR PTE cells. Gene silencing with a LAT1 small interfering RNA and a 4F2hc small interfering RNA markedly reduced LAT1 and 4F2hc expression, which was accompanied by a marked reduction in Na^+-independent [^{14}C]L-leucine uptake in both SHR and WKY PTE cells. Altogether, it is suggested that SHR cells overexpress LAT1 units endowed with low affinity for L-leucine transport [38].

Recently, changes in urinary dopamine and dihydroxyphenylacetic acid (DOPAC; the deaminated metabolite) in WKY and SHR on normal salt (NS) and high salt (HS) intake were found to correlate well with differences in L-DOPA uptake in isolated renal tubules [39]. HS intake increased the tubular uptake of L-DOPA, in 4- and 12-week-old SHR, but not in the WKY, which parallels the changes observed in ASCT2 and B^0AT1 transcript abundance [39]. A more detailed analysis of the rates of L-DOPA uptake in renal tubules in 4- and 12-week-old WKY on NS intake, which were identical, reveals that these do not correlate with the marked differences in LAT1 and LAT2 mRNA levels, which in 4-week-old WKY were twice those in 12 week-old WKY. However, the 2-fold increase in LAT1 and LAT2 mRNA levels, to some extent, is compensated by decreases in ASCT2 and B⁰AT1 mRNA levels, which attain reduction levels of 58 and 74%, respectively. This constitutes circumstantial evidence for the involvement of LAT1, LAT2, B⁰AT1 and ASCT2 in L-DOPA uptake in renal proximal tubules. Other findings that fit well the observation that Na⁺-dependent L-DOPA transporters play an important role in the regulation of renal dopamine formation during HS intake are those concerning the renal delivery of L-DOPA and AADC activity. In fact, the renal delivery of L-DOPA during NS and HS intake in SHR was similar to that in WKY, both at 4 and 12 weeks of age. AADC activity in 4-week-old SHR and WKY was also unaffected by HS intake or was even decreased in 12-week-old SHR [39].

System B⁰ has a major role in the uptake of luminal neutral amino acids. It has been recently assigned to a specific gene in mouse kidney (i.e. *slc6a19* [40]), which belongs to the Na⁺- and Cl⁻-dependent neurotransmitter transporter family (SLC6). The SLC6 family members are structurally related; still, they display a variety of substrate specificity and different transport mechanisms. Distinct from other members of the SLC6 family, B⁰AT1-mediated transport is not chloride-dependent. As demonstrated by flux measurements and electrophysiological recordings in *Xenopus laevis* oocytes expressing mB⁰AT, this protein actively transports most neutral amino acids, but not anionic or cationic amino acids, in a Na⁺-dependent and Cl⁻ independent manner [40]. Recently, the expression in kidney and intestine of B⁰AT1 during development of high blood pressure (from weeks 4 to 12 after birth) in SHR and WKY was also evaluated. The hypothesis under evaluation also concerned whether the expression of renal B⁰AT1 correlates with changes in the expression of Na⁺ transporters NHE3 and Na⁺,K⁺-ATPase known to occur in the SHR model. The effect of HS intake on the expression of the renal and intestinal B⁰AT1 transcript abundance was also evaluated. It was found that underexpression of B⁰AT1 in the SHR kidney is organ-specific, precedes the onset of hypertension and correlates negatively with the renal tubular transport of Na⁺ [41]. The regulation of B⁰AT1 gene transcription appears to be under the influence of Na⁺ delivery, this also being organ specific [41].

11.2.3
Mechanisms Regulating Dopamine Availability

The physiological importance of the renal and intestinal actions of dopamine mainly depends on the sources of the amine, and on the availability of this dopamine to activate the amine-specific receptors. The proximal tubules, but not distal segments of the nephron, and the mucosa of jejunum, ileum and colon are endowed with a high AADC activity and epithelial cells have been demonstrated to synthesize dopamine from circulating or luminal L-DOPA [42–44]. This non-neuronal renal and intestinal dopaminergic system appears to be highly dynamic, and the basic mechanisms for the regulation of this system are thought to depend mainly on the availability of L-DOPA, its fast decarboxylation into dopamine and in precise and accurate cell outward amine transfer mechanisms [9, 10]. As a result of high monoamine oxidase (MAO) and catechol-*O*-methyltransferase (COMT) activities in the kidney and intestinal mucosa, dopamine has been found to undergo extensive degradation to DOPAC and to homovanillic acid (the deaminated and O-methylated metabolite, respectively) [45–54]. Although the metabolism of renal dopamine may limit the availability of dopamine to activate its specific receptors, factors affecting the renal synthesis of the amine, such as the amounts of L-DOPA and sodium delivered to kidney, have been suggested to represent a major role in determining the availability of dopamine. In fact, sodium has been found to constitute an important stimulus for the production of dopamine by renal tubular and intestinal epithelial cells [43, 55–57], resulting in increases in the urinary excretion of dopamine and dopamine metabolites DOPAC and homovanillic acid [39, 58–62].

11.3
Dopamine Receptors

11.3.1
Classification

Dopamine receptors are members of the superfamily of G-protein-coupled receptors, which possess a typical seven-transmembrane spanning domain architecture. The diverse physiological actions of dopamine are exerted through the activation of two families of cell surface receptors known as D_1-like (D_1 and D_5, also known as D_{1A} and D_{1B}, respectively, in rodents) and D_2-like (D_2, D_3 and D_4) dopamine receptors (reviewed in [63–65]). These receptor subclasses can be differentiated genetically, pharmacologically, physiologically and by their second messenger coupling properties (Table 11.1).

Dopamine binds to its receptors with nanomolar or submicromolar affinity constants. At higher concentrations, α- and β-adrenergic and serotonin receptors are occupied. Although circulating concentrations of dopamine are not sufficiently high to activate dopamine receptors, high nanomolar concentrations can be attained in dopamine-producing tissues (e.g. renal proximal tubule, jejunum).

The first indication that the five receptors could be divided into D_1-like and D_2-like subfamilies came from the comparison of their primary sequences. In their putative transmembrane core region, the D_1 and D_5 receptors share 79% sequence homology but only 40–45% with the D_2, D_3 and D_4 receptors. Conversely, the D_2, D_3 and D_4 receptors are between 75 and 51% identical to each other. The genomic organization of the dopamine receptors further supports the notion that they derive from the divergence of two gene subfamilies – the D_1-like and D_2-like receptor genes. The D_1 and D_5 receptors are encoded by homologous intronless genes [66, 67]. In contrast, the D_2, D_3 and D_4 genes have introns in their coding region, and exist in various isoforms by alternative splicing [66, 67]. Indeed, the D_2 receptor gene encodes two alternatively spliced variants – the short form (D_{2S}) and the long form (D_{2L}) [68–70]. Furthermore, most of the introns in the D_2-like receptor genes are located in similar positions. Drugs allowing the pharmacological

Table 11.1 Members of the D_1-like and D_2-like dopamine receptor subfamilies, their selective agonists/antagonists, and predominant cell signaling mechanisms.

Receptor family	D_1-like		D_2-like		
Receptor subtype	D_1[a]	D_5[b]	D_2 (D_{2S}/D_{2L})	D_3	D_4
Amino acids					
Rat	446	475	415/444	446	387
Human	446	477	414/443	400	387–515[c]
G-protein coupling	$G\alpha_s$	$G\alpha_s$	$G\alpha_i$	$G\alpha_i$	$G\alpha_i$
	$G\alpha_{olf}$	$G\alpha_{olf}$	$G\alpha_o$	$G\alpha_o$	$G\alpha_o$
	$G\alpha_1$	$G\alpha_z$			
	$G\alpha_2$	$G\alpha_{12/13}$			
	$G\alpha_3$				
	$G\alpha_{15/16}$				
	$G\alpha_o$				
	$G\alpha_q$				
Selective agonist	fenoldopam; SKF 38393	fenoldopam; SKF 38393	(+)-PHNO [(+)-4-propyl-9-hydroxynaphthoxazine]	7-hydroxy-N,N-di-n-propyl-2-aminotetralin [7-OH-DPAT]; PD 128907	PD 168077
Selective antagonist	SCH 23390; SKF 83566	SCH 23390; SKF 83566	raclopride	S-sulpiride	RBI 127
Effects	↑ adenylyl cyclase; ↑ Phospholipase C	↑ adenylyl cyclase	↓ adenylyl cyclase; ↓ Ca^{2+} channel; ↑ K^+ channel	↓ adenylyl cyclase; ↓ Ca^{2+} channel; ↑ K^+ channel	↓ adenylyl cyclase; ↓ Ca^{2+} channel; ↑ K^+ channel

a Also known as D1A in rodents.
b Also known as D1B in rodents.
c Number of amino acids in the human D4 receptor depends on the number of repeats in the third intracellular loop.

distinction of D_1-like from D_2-like receptors have been developed (Table 11.1). While there are currently no D_1 or D_5 subtype-selective agonists or antagonists, drugs that can distinguish among the D_2-like receptors have been synthesized. Both cloned members of the D_1-like receptor group are coupled to the stimulatory G (G_s)-protein and stimulate adenylyl cyclase. All three of the cloned D_2-like receptors are associated with the inhibitory G (G_i/G_o)-proteins, inhibit adenylyl cyclase and modulate ion channel activity. This topic will be discussed in more detail in Section 11.4.1.

11.3.2
Tissue Distribution

The actions of dopamine were originally ascribed to the activation of two central nervous system (CNS) dopamine receptor subtypes classified as D_1 and D_2 receptors. Dopamine receptors were subsequently found to be localized in peripheral tissues outside the CNS, such as various vascular beds, the heart, adrenal gland, parathyroid gland, carotid body, the gastrointestinal tract and the kidney [63], by radioligand binding, autoradiography and immunohistochemistry techniques.

The vast majority of studies analyzing the distribution of peripheral dopamine receptor subtypes have been conducted in the rat kidney. The kidney expresses the five different subtypes of dopamine receptors, displaying a heterogeneous vascular and tubular localization. D_1-like (D_1 and D_5) and D_3 receptors are located mainly within smooth muscle cells of the tunica media of renal arteries [71–73], whereas D_2-like receptors are expressed mainly in the intima and adventitia [74]. To date, no dopamine receptors have been reported in the endothelium of renal arteries. Glomeruli mainly express D_3 and D_4 receptor subtypes [72, 73, 75]. The proximal tubule and the collecting ducts (cortical and medullary) express D_1, D_5, D_3 and D_4 receptors [72, 73, 75–78]. The medullary thick ascending limb (mTAL) but not the cortical thick ascending limb of the loop of Henle expresses D_1, D_5 and D_3 receptors [72, 73, 77]. The macula densa and juxtaglomerular cell express D_1 and D_3 receptors [79, 80].

Dopamine receptors have been detected throughout the gastrointestinal tract. In particular, the D_1 receptor is expressed in the esophagus, gastroesophageal junction, stomach, pylorus, small intestine and colon. The receptor is distributed in epithelial and muscle layers, blood vessels and lamina propria cells in different gastrointestinal regions [81, 82].

11.4
Signaling Machinery and Effectors Downstream Dopamine Receptor Activation

11.4.1
Ion Transporters and Channels

One of the most prominent effects of dopamine is the regulation of sodium transport in renal and intestinal epithelia. In the kidney, dopamine has been shown to inhibit sodium reabsorption at multiple sites along the nephron – the overall effect being a marked increase in urinary sodium and water excretion (reviewed in [15, 63, 83, 84]). In the intestine, the action of dopamine on sodium metabolism seems to be restricted to the jejunum [85, 86]. The sodium transporters that have been identified as final effector proteins for the action of dopamine include sodium/hydrogen exchanger isoforms NHE1 and NHE3, Na^+/P_i cotransporter, Na^+/HCO_3^- cotransporter, $Na^+–K^+–2Cl^-$ cotransporter and Na^+,K^+-ATPase [11, 87–99]. More recently, the apical membrane Cl^-/HCO_3^- exchanger (SLC26A6), which works in parallel with the Na^+/H^+ exchanger and is responsible for the bulk of Na^+ and Cl^- reabsorption in the proximal tubule, has also been reported as a target for the action of dopamine [99, 100]. Among these transporters, the apical NHE3 and the basolateral Na^+,K^+-ATPase have been the focus of intensive research. In general, D_1-like receptors inhibit [101–103], whereas D_2-like receptors stimulate, Na^+,K^+-ATPase activity [104–106]. However, there are reports showing that D_2-like receptors can act synergistically with D_1-like receptors to inhibit Na^+,K^+-ATPase [107, 108]. Dopamine receptors have been reported to influence the activity of K^+ channels. This is well documented in the case of D_2-like receptors, whereas evidence to support a role for D_1-like receptors is still inconclusive. *In vitro* studies with renal epithelial cells demonstrate that D_2-like receptors increase outward K^+ currents, leading to membrane hyperpolarization that ultimately decreases Na^+,K^+-ATPase activity [109, 110].

11.4.2
G-Proteins

The coupling of dopamine receptors to second messenger pathways has been a subject of intense interest ever since their existence was recognized. As mentioned before, dopamine interacts with five distinct receptor subtypes, which transduce their signals by coupling to heterotrimeric G-proteins, comprising α, β and γ subunits [15, 63, 65, 83, 111]. The classical view was that dopamine receptors could only stimulate or inhibit adenylyl cyclase by coupling to either G_s or G_i/G_o, respectively. It is now widely accepted that dopamine receptor subtypes can couple to multiple and diverse G-proteins, eliciting functionally distinct physiological effects.

D_1-like receptors are coupled to the stimulatory G-proteins $G\alpha_s$ and $G\alpha_{olf}$ [112–114]. There is also evidence that both D_1 and D_5 receptor subtypes can interact with $G\alpha_{15/16}$, but not with $G\alpha_{14}$ [115]. The coupling of D_1 receptor to any of the three isoforms of $G\alpha_i$ ($G\alpha_{i1}$, $G\alpha_{i2}$ and $G\alpha_{i3}$) has been demonstrated under conditions of pertussis toxin treatment [116, 117]. The G-protein $\gamma 7$ subunit was identified as a component of the G-protein that couples the D_1 receptor to downstream effectors [118]. The D_1 receptor can couple to $G\alpha_o$ [113] and also to $G\alpha_q$, and stimulates phospholipase Cβ [119–123], in the presence of the adaptor protein, calcyon [124]. The D_5 receptor subtype can couple to $G\alpha_z$ and to $G\alpha_{12/13}$ [125–127]. The D_2-like receptors are coupled to the inhibitory G-proteins $G\alpha_i$ and $G\alpha_o$ [63, 65].

The ability of dopamine receptors to couple multiple G-proteins with opposing effects on signaling cascades can serve as a mechanism for signal termination and to simultaneously propagate that signal through the activation of additional effectors.

11.4.3
Adenylyl Cyclase/Protein Kinase A

The molecular mechanism by which dopamine regulates cAMP levels is well established. $G\alpha_s$ couples D_1-like receptors to adenylyl cyclase, the enzyme responsible for the synthesis of cAMP, such that the enzyme is stimulated upon receptor activation. In a variety of preparations, it was shown that the D_1-like receptors robustly stimulated cAMP accumulation. In the kidney proximal tubular cells and the mTAL of the loop of Henle, dopamine-mediated increase in cAMP production stimulates protein kinase A (PKA), which phosphorylates a variety of effector proteins, including the sodium/hydrogen exchanger and the Na^+,K^+-ATPase, and the subsequent inhibition of their activities [95–97, 128, 129]. DARPP-32 is a dopamine- and cAMP-regulated phosphoprotein of 32 kDa highly abundant in the mTAL, that has also been reported to play a role in dopamine-mediated inhibition of Na^+,K^+-ATPase activity in this nephron region [130]. The stimulation of D_2-like receptors causes a decrease in cAMP via the $G\alpha_i/G\alpha_o$-proteins [63, 65]. It is now becoming generally accepted that a decrease in cAMP is the first biochemical signal that leads to a cascade of events ultimately resulting in the stimulation of Na^+,K^+-ATPase.

11.4.4
Phospholipase C/Protein Kinase C

Another well-known mechanism through which dopamine modulates water and electrolyte transport in the kidney involves the activation of the phospholipase C/protein kinase C (PLC/PKC) signaling cascade. Indeed, early reports demonstrate that the renal D_1 receptor-mediated effects are exerted through activation of the $G\alpha_q$-protein and its downstream effector, PLCβ [119, 120, 131]. Dopamine and D_1-like receptor agonists have been reported to stimulate PLC activity in the kidney cortex by regulating the expression and activity of PLCβ1 and PLCγ1 isoforms [122]. Other studies have shown that D_1-like agonists stimulate PKC activity in proximal tubules [132, 133]. Furthermore, PLC and PKC inhibitors have been shown to prevent the inhibitory effect of D_1-like receptor activation on the phosphorylation and activity of Na^+,K^+-ATPase [88, 102, 134]. The cellular events set in motion following stimulation of the D_1 receptor–$G\alpha_q$–PLCβ pathway include the formation of inositol 1,4,5-triphosphate (and increase in cytosolic calcium and diacylglycerol) causing PKC activation, which in turn phosphorylates and inhibits Na^+,K^+-ATPase activity. It was also demonstrated that PKC activation can stimulate phosphatidylinositol 3-kinase (PI3K) and cause the internalization of Na^+,K^+-ATPase α subunits, and inhibition of its activity [135]. Other reports link D_2 and

D_3 receptor activation to PI3K stimulation [136, 137]. PKC can also mediate D_1-like receptor inhibition of NHE3 activity in the opossum but not rat proximal tubule cells [95].

11.4.5
Other Pathways

There is experimental evidence for the involvement of other signaling elements in dopamine-mediated inhibition of sodium transport in the kidney. It has been reported that D_1 receptor-mediated activation of PKC stimulates phospholipase A_2 (PLA_2), which in turn releases arachidonic acid from membrane phospholipids [103, 138]. In the rat kidney, arachidonic acid is mainly metabolized to 20-hydroxyeicosatetraenoic acid (20-HETE) by the cytochrome P450 pathway. 20-HETE modulates ion transport in the proximal tubules and mTAL by affecting the activities of Na^+,K^+-ATPase [139] and the $Na^+-K^+-2Cl^-$ cotransporter [140], respectively. In mTAL and cortical collecting duct (CCD), the PLA_2 pathway interacts with PKA to inhibit Na^+,K^+-ATPase activity [101].

Finally, it is becoming increasingly evident that D_2-like receptors can also activate tyrosine kinase and mitogen-activated protein kinase (MAPK) pathways. In renal proximal tubules, D_2-like receptor-mediated stimulation of Na^+,K^+-ATPase activity has been shown to be blocked by tyrosine kinase as well as MAPK inhibitors [141]. Also, a D_2-like agonist increased phosphorylation of p44/42 MAPK in proximal tubules, suggesting that D_2-like receptor activation causes stimulation of Na^+,K^+-ATPase via a tyrosine kinase and p44/42 MAPK pathway [141, 142]. Whereas D_2-like receptors activate the p44/42 MAPK pathways and promote mitogenesis, D_1-like receptors activate the p38 MAPK pathway, which plays a role in apoptosis [143].

During the last two decades, much effort has being put into the study of dopamine receptors and their signal transduction mechanisms. Many intracellular second messengers and effectors for these receptors have been identified. However, many conflicting reports are found in the literature about the modulation of various messengers or the mechanism by which an effector is modulated by the dopamine receptors. Some of the discrepancies probably arise from the use of different tissues or cell lines. It is now known that many of the components of signal transduction pathways have multiple isoforms, including receptors, G-proteins and effectors, and that these have differing patterns of expression and regulatory properties. Defining which of these specific signal transduction events is involved in the various physiological actions of dopamine may require the development of specific pharmacological agents or genetic animal models.

11.5
Peripheral Effects of Dopamine

11.5.1
Renal Function

11.5.1.1 Renal Blood Flow

Low concentrations of dopamine increase renal blood flow and decrease renal vascular resistance [144]. This effect occurs even when α- and β-adrenergic receptors are blocked, suggesting that postsynaptic receptors mediate dopamine-induced vasodilation [145]. Under these conditions, dopamine most likely activates D_1-like receptors, as dopamine's effect is mimicked by D_1-like agonists and blocked by the D_1-like antagonist SCH23390 [145]. However, the effect of D_1-like receptors on vascular tone may vary depending on the arterial segment. For example, a D_1-like agonist causes vasoconstriction in the rat tail artery [146], whereas in the renal artery D_1-like receptors display vasodilatory action [147, 148]. Furthermore, the vasorelaxant effect of the D_1 receptor is enhanced by the calcium channel blocker nifedipine, indicating that the Ca^{2+} channel is involved in the vasodilatory effect of D_1 receptors [149, 150].

The renal vascular response to D_2-like receptor stimulation can result in either vasodilation or vasoconstriction. The variable effect of D_2-like receptors on the renal vasculature is probably dependent on the state of renal nerve activity. Dopamine can vasodilate the renal vasculature via presynaptic D_2-like receptors [151–153]. This effect is mainly evident when renal nerve activity is increased, a situation that is seen during anesthesia and sodium-depleted states [154, 155]. Stimulation of postsynaptic D_2-like receptors (D_3 and/or D_4 subtypes) can result in either vasodilation or vasoconstriction. For example, under conditions of reduced renal nerve activity following volume expansion, postsynaptic D_2-like receptors (presumably the D_3 subtype) produce vasoconstriction [156]. However, when the renal vascular resistance is increased, the D_2-like effect at postsynaptic sites is that of vasodilation [150], because D_2-like receptors inhibit Ca^{2+} channels and stimulate K^+ channels. The effect of dopamine on vascular tone may differ greatly according to vascular bed (conduit versus resistance vessel).

11.5.1.2 Glomerular Filtration Rate

As stated above, dopamine at low doses increases renal blood flow. However, this effect is not consistently associated with an increase in glomerular filtration rate (GFR). For example, dopamine can improve the reduction in GFR caused by amphotericin B and in hypovolemic states [157, 158]. This action could be a direct glomerular effect, because dopamine has been shown to attenuate the contractile response to angiotensin (Ang) II in isolated glomeruli [159]. Nevertheless, the precise mechanism by which dopamine exerts its direct effect on GFR is not fully understood. An increase in glomerular cAMP levels is unlikely because only D_2-like receptors are expressed in glomeruli [73, 160, 161]. In isolated glomeruli, dopamine decreases adenylyl cyclase activity, in keeping with the presence of D_2-

like receptors [161]. *In vivo*, D_2-like receptors can decrease or increase GFR, depending upon the state of renal vascular D_2-like receptor activation. When the interaction between D_1-like and D_2-like receptors results in a greater vasodilatory effect on afferent than efferent arterioles, GFR can increase [162]. D_2-like receptors are thought to be involved in the increase in GFR associated with the infusion of certain amino acids [163, 164]. Renal nerves apparently mediate this action because it is absent after renal denervation [164]. When D_2-like receptors decrease renal blood flow, there is greater afferent than efferent constriction – a result producing a greater decrease in GFR than renal blood flow [156, 165].

11.5.1.3 Tubular Effect

Previous studies have demonstrated that dopamine and D_1-like agonists induce an increase in urine flow, and sodium, phosphate, calcium, magnesium and potassium excretion [166–170]. However, the natriuretic and diuretic effect of dopamine and D_1-like agonists can be dissociated from their effects on renal blood flow. When the renal vasodilation induced by a D_1-like agonist is abolished, natriuresis and diuresis are still observed, although to a less extent [171]. As the natriuretic effect of dopamine far exceeds its phosphaturic, calciuretic or kaliuretic action, it is suggested that inhibition of sodium transport also occurs in nephron segments beyond the proximal tubule (mTAL and CCD). Under conditions of moderate sodium excess, more than 50% of renal sodium excretion is regulated by dopamine receptors [14, 172–175]. In fact, the natriuretic effect of dopamine and D_1-like agonists is influenced by sodium balance, in contrast to the neutral effect of sodium balance on dopamine-mediated renal vasodilation. The D_1-like receptor involved in the natriuretic effect of dopamine is thought to be the D_1 subtype, as the D_1 receptor increases cAMP levels to a greater extent than the D_5 receptor in renal proximal tubules. However, the exact contribution of each receptor subtype to D_1-like receptor-mediated diuretic and natriuretic effects were not fully resolved due to the lack of subtype-specific ligands. Hence, alternative approaches able to discriminate between D_1 and D_5 receptors had to be employed. One such approach included the use of antisense oligonucleotides to selectively silence the expression of either the D_1 or D_5 receptor subtype. It was concluded that the increase in cAMP levels following D_1-like receptor stimulation is due mainly to the D_1 receptor relative to the D_5 receptor [176]. The D_5 receptor null mutant mice offered an alternative approach. These animals are hypertensive and a HS diet further increases blood pressure, suggesting that the D_5 receptor in the kidney plays a role in the control of blood pressure, at least in part by regulating sodium transport, which can be partially ascribed to an interaction between D_5 and Ang II type 1 receptors [177, 178]. Nevertheless, the major D_1-like receptor regulating sodium transport is the D_1 receptor. The effect of renal D_2-like receptors on ion and water transport remains unclear. The responses to D_2-like receptor stimulation on sodium transport and natriuresis can range from no effect [145, 179], to an attenuation of sodium excretion (antinatriuresis) [156, 180] and even to an increase in sodium excretion [165]. These discrepant effects can be related to the use of drugs with poor selectivity for a particular D_2-like receptor subtype. There are also reports of a synergistic action between D_1- and D_2-like receptors to increase renal sodium excretion, but the dopamine receptor subtypes involved in this action are not identified [108, 181].

11.5.2
Gastrointestinal Effect

There is substantial evidence that dopamine is involved in the regulation of bicarbonate, chloride and sodium transport in the gastrointestinal tract. This effect is modulated by sodium intake, as in the kidney [62, 182–184]. In young rats, jejunal Na^+,K^+-ATPase activity is reduced during a HS diet. This inhibition is abolished by a blocker of dopamine synthesis, suggesting that this is an effect mediated by endogenously formed dopamine [86]. The dopamine receptor subtype involved is probably a D_1-like receptor, because D_2-like receptors increase transport in the ileum [185]. The inhibitory effect of dopamine on sodium transport at the intestinal level is impaired in obesity, hypertension and aging [36, 61, 85, 86].

11.6
Dopamine and Pathophysiology

11.6.1
Hypertension

Hypertension has an estimated worldwide prevalence of approximately 25% in the adult population [186]. Essential hypertension results from complex interactions between internal derangement (primarily in the kidney) and the external environment. Sodium has long been considered the pivotal environmental factor in the disorder. Numerous studies suggest that dopamine plays a role in the pathophysiology of hypertension (reviewed in [15, 63, 81, 83, 84]). In human essential hypertension and rodent models of genetic hypertension, two renal dopaminergic abnormalities have been described: defective renal dopamine production and/or dopamine receptor function. Both defects may result in sodium retention and hypertension.

A reduction in renal dopamine synthesis has been reported in several forms of human hypertension. In some salt-sensitive subjects with or without hypertension, an acute NaCl or protein loading is not accompanied by an increase in renal dopamine production [187–191]. Suppressed dopaminergic activity has also been reported in the prehypertensive stage of essential hypertension [192, 193]. However, in many cases, a decreased renal production of dopamine does not provide a compelling explanation for the impaired function of endogenous dopamine in essential hypertension. As an example, urinary excretion of dopamine and its main metabolites are actually increased in young patients with essential hypertension [194–196]. The precise mechanisms underlying the renal dopaminergic deficiency in human primary hypertension are not completely elucidated. However, a defect in AADC has been shown in subjects with a family history of hypertension [187, 193, 197]. Others have reported a decrease in both the tubular uptake of L-DOPA and its conversion to dopamine in a subgroup of salt-sensitive hypertensive patients [189]. In contrast, recent studies have shown the overexpression of Na^+-independent and pH-sensitive LAT2 in the SHR kidney, which might contribute to enhanced L-DOPA uptake in the proximal tubule and increased dopamine production [35, 198].

The defective dopamine receptor function at the kidney level in hypertension is genetic and mainly involves an abnormal transduction of the dopamine receptor signal in renal proximal tubules. In both rats and humans, the impaired ability of dopamine and D_1-like receptor agonists to inhibit sodium transporters is related to a defective dopaminergic stimulation of second messenger production by adenylyl cyclase, PLC and PLA_2 [138, 199–202]. The aberrant dopamine receptor function does not result from abnormalities in G-proteins, second messenger-producing enzymes or sodium transporters [83, 98, 99, 117, 133, 138, 199, 201–209]. Rather, it results from the uncoupling of the D_1-like receptor from its G-protein/effector complex [83, 98, 99, 112, 117, 133, 138, 199, 201–204, 207, 210–214]. The uncoupling defect is specific to the D_1-like receptor because the responses to parathyroid hormone and β-adrenergic receptor agonist remain intact; it is organ-selective, as the defect is present in the kidney and intestine but not in the brain; it is nephron segment-specific (confined to the proximal tubule and thick ascending limb of Henle but not the CCD); and it precedes the onset of hypertension. From a mechanistic point of view, the defective D_1-like receptor-G-protein coupling in hypertension is similar to that seen in the homologous desensitization of the D_1 receptor (following agonist stimulation). However, in genetic hypertension, the D_1 receptor in renal proximal tubules is desensitized even in the absence of its ligand. This process is related to phosphorylation of the receptor by one of the G-protein-coupled receptor kinases (GRKs) resulting in their desensitization. In fact, it has been reported that the D_1 receptor is hyper-serine-phosphorylated [204, 214, 215]. In humans with essential hypertension, this is the result of a constitutively activated *GRK4* gene variant (*R65L, A142V, A486V*) that phosphorylates and desensitizes the receptor [204]. GRK4 activity is also increased in the kidney of SHR strains. Inhibiting renal GRK4 expression by antisense oligonucleotide infusion attenuates the age-related increase in blood pressure in SHR strains [216]. Furthermore, expressing GRK4γ A142V produces hypertension and impaired diuretic and natriuretic effects of D_1-like agonist stimulation [204].

Recent evidence suggests that reactive oxygen species (ROS) contribute to impaired dopamine receptor function in hypertension, diabetes, and aging. ROS production is increased in humans with essential hypertension and in experimental and genetic animal models of hypertension [217–220]. The converse is also true: mouse models deficient in ROS-producing enzymes have lower blood pressure than their wild-type controls [221]. However, the mechanisms by which oxidative stress participates in the pathogenesis of hypertension remain to be elucidated. In one study, Sprague-Dawley rats were treated with L-buthionine sulfoximine, an oxidant, which caused renal dopamine D_1 receptor dysfunction and hypertension by mechanisms that involve increased nuclear translocation of NF-κB and augmented PKC activity [222]. Nevertheless, additional studies are required to define the exact mechanisms of the oxidative stress-mediated impaired D_1 receptor signaling and hypertension.

The targeted deletion of the individual dopamine receptor genes in mice has provided valuable information regarding the role of dopamine receptors in hypertension. All of the five dopamine receptor genes and some of their regulators are in loci linked to hypertension in humans and in rodents. The disruption of any of the dopamine receptor genes in mice results in hypertension, by mechanisms specific to the subtype. Disruption of the D_1 receptor in mice leads to the development of hypertension as a result of decreased ability to excrete sodium in response to D_1-like agonist stimulation [199]. However, a mutation in the coding region of the D_1 receptor has not been found in human essential hypertension or in genetically hypertensive rats. Hypertension in D_2 null mice is related to increased noradrenergic discharge and oxidative stress, and is not related to impairment of sodium excretion [223, 224]. Disruption of the dopamine D_3 receptor gene produces renin-dependent hypertension and these mice also cannot excrete a sodium load [225]. D_4 receptor-deficient mice display increased blood pressure, in part, by increased Ang II type 1 receptor expression [226]. D_5 null mice are hypertensive due to increased sympathetic tone, apparently due to activation of oxytocin, V_1 vasopressin, and non-N-methyl-L-aspartate receptors in the central nervous system [227]. These studies underscore the potential role of dopamine receptor dysfunction (or their regulation) in the pathogenesis of essential hypertension.

11.6.2
Renal Failure

Abnormalities in the dopamine regulation of tubular sodium handling can occur and might, alone or in combination, contribute to the changes in sodium homeostasis observed in diverse cardiovascular and renal disorders. A number of studies have addressed the status of the renal dopaminergic system in patients with different renal disorders, based on the urinary excretion of L-DOPA, dopamine and its metabolites. Patients with chronic renal failure have a reduction in the urinary excretion of dopamine [228], the extent of which appears to correlate with the degree of deterioration of renal function [229]. In some studies, urinary dopamine excretion was measured after an acute saline loading, the result being a marked attenuation of the normal increase in urinary dopamine during salt loading [228]. This may be a consequence of the reduced number of functional tubular units endowed with the capacity to synthesize dopamine. However, these residual tubular units in patients with compromised renal function apparently maintain an intact ability to take up and decarboxylate L-DOPA, and deaminate newly formed dopamine to DOPAC. Therefore, the evaluation in patients of delivery of L-DOPA to sites of uptake – an important aspect in the process of renal dopamine production – is crucial for a better understanding of the role of the renal dopaminergic system in the pathogenesis of renal failure.

11.6.3
Heart Failure

Renal sodium and water handling becomes abnormal in the early stages of heart failure. This tendency to sodium and water retention results from a complex interplay between hemodynamic and neurohumoral factors activated in the course of heart failure. Patients with chronic heart failure have a decreased urinary excretion of L-DOPA and dopamine that appears to be related to the reduced delivery of L-DOPA to the renal proximal tubules, probably as a consequence of a decreased renal blood flow. However, these patients are endowed with an enhanced ability to take up (or decarboxylate) filtered L-DOPA, which might counteract the decreased delivery of L-DOPA to the kidney, contributing to a relative preservation of dopamine synthesis [230, 231]. Increasing evidence suggests that the renal dopaminergic system may be considered as a natriuretic and diuretic counter-regulatory system activated in patients with heart failure by stimuli leading to sodium and water reabsorption [230–232].

11.6.4
Diabetes Mellitus

Diabetes mellitus is a chronic metabolic disorder resulting from defects in insulin secretion, insulin action or both and is characterized by sustained hyperglycemia. Diabetes is also associated with sodium retention. Among many other mechanisms, the renal dopaminergic system is one of the candidates possibly involved in the disturbed renal sodium handling in diabetes. In both type 1 diabetes (T1D; insulin-dependent) and type 2 diabetes (T2D; noninsulin-dependent) there is a reduced ability to excrete sodium, and this correlates with decreased urinary dopamine excretion and an impaired response to renal dopamine D_1 receptor activation [233–235].

However, the use of experimental animal models of diabetes and cell lines has contributed to a better understanding of the dysfunction of the dopaminergic system in diabetes. Both streptozotocin-induced T1D rats and the obese Zucker rats, a model of hyperinsulinemia and T2D, confirm and further extend the findings in human patients. In both rat strains, there is a reduced renal dopamine D_1 receptor function, evidenced by receptor downregulation in proximal tubules, defective receptor–G-protein coupling, and the subsequent reduced inhibition of Na^+,K^+-ATPase and NHE3 activities, which culminate on a diminished natriuretic and diuretic response to dopamine D_1 receptor stimulation [236–238]. There is increasing evidence in the literature that recognizes oxidative stress as a common factor mediating the clustering of hypertension, diabetes and obesity. Antioxidant (tempol) treatment of obese Zucker rats reduces oxidative stress, improves insulin sensitivity, decreases renal dopamine D_1 receptor hyperphosphorylation, and restores D_1 receptor–G-protein coupling and function [239]. Furthermore, at the cellular level, tempol decreases PKC activity, which could at least in part be responsible for normalization of D_1 receptor serine phosphorylation and subsequent D_1 receptor–G-protein coupling [239]. These phenomena could account for restoration of dopamine-induced inhibition of Na^+,K^+-ATPase activity and the ability of dopamine to promote sodium excretion. Similar findings were reported in streptozotocin-induced diabetic rats [240].

Insulin has been shown to affect the functional response to a number of hormones by altering the receptor number or ligand affinity. It is possible that in T2D, hyperinsulinemia might downregulate the D_1 receptor and its function, which subsequently leads to a diminished natriuretic response to exogenously infused dopamine. In one recent study employing a renal cell line, it was shown that insulin, in a PI3K-dependent manner, increases PKC activity causing GRK2 membranous translocation; this in turn increases D_1 receptor serine phosphorylation which leads to D_1 receptor downregulation and uncoupling from G-proteins, and results in the failure of SKF38393 to stimulate G-proteins and inhibit Na^+,K^+-ATPase activity [241].

There is also evidence for a decrease in the number of tubular L-DOPA transport sites as a cause for renal dopaminergic deficiency in T1D [242].

11.6.5
Aging

Aging is characterized by a gradual deterioration of all organ systems, which reduces the ability to maintain homeostasis under conditions of stress. The incidence of pathologies such as cancer, diabetes, cardiovascular and renal diseases, among others, rises sharply with age. During aging, the kidney undergoes significant anatomical, biochemical and physiological alterations that result in a decline of some renal functions such as renal blood flow and GFR. Age-related changes in renal hemodynamics are accompanied by significant alterations of renal hormones and of renal sodium handling. There is ample evidence suggesting that old rats may present particular defects in the renal handling of L-DOPA, its subsequent conversion to dopamine and at the level of receptor number or coupling to G-proteins. Studies in old Fischer 344 rats, which have been extensively used for aging research, reported an impairment in dopamine receptor-mediated inhibition of renal Na^+,K^+-ATPase [61]. This deficient response to dopamine is accompanied by a decreased number of dopamine receptors and a marked attenuation of D_1-receptor-mediated activation of G-proteins [243, 244]. In another study, it was observed that the renal dopaminergic tonus in old rats is higher than in adult rats, but fails to respond to HS intake as observed in adult rats [61]. This may be due to the failure of dopamine to inhibit Na^+,K^+-ATPase activity and a marked decrease in the number of dopamine receptors in old rats. The increased renal dopaminergic tonus may be related to an attempt of the renal epithelial cells from old Fischer 344 rats to induce a compensatory response, and overcome the reduced expression of dopamine receptors and the functional consequences of this phenomenon.

11.7
Clinical Applications of Dopamine

11.7.1
Heart Failure

The complex pathophysiology of heart failure is characterized by the activation of different counter-regulatory systems with antagonistic actions. Some of these systems produce vasodilatation, diuresis and antiproliferative effects (nitric oxide, natriuretic peptides and circulating dopamine), whereas others produce vasoconstriction, antidiuresis and stimulate cell growth (angiotensin, aldosterone, norepinephrine and vasopressin). The infusion of dopamine in heart failure patients is accompanied by increases in effective renal plasma flow and GFR with a pronounced natriuresis, this being greater in heart failure patients than in controls [13]. This could be due to an increased hemodynamic effect of dopamine or to an increased effect on the renal tubules, as a result of receptor upregulation due to decreased availability of the endogenous agonist.

Low-dose dopamine functions as both an inotropic agent and a diuretic. Both of these actions would seem to be beneficial in patients with congestive heart failure. The initial studies by Goldberg et al. [245, 246] would support the use of this agent in the setting of congestive heart failure. Despite these early studies, two more recent investigations do not lend such support. Robinson et al. [247] showed in elderly patients with congestive heart failure that low-dose dopamine did not increase GFR, effective plasma flow or urine output. In another study, Leier et al. [248] studied patients with cardiomyopathic heart failure treated with dopamine at 4µg/kg/min; there was no improvement in creatinine clearance, renal blood flow or urine output. Cardiac index was improved; however, the number of premature ventricular contractions also increased.

Fenoldopam, a D_1 dopamine receptor agonist, improved hemodynamics in heart failure patients, but failed to increase natriuresis [249]. This apparent discrepancy may be related to the well-known activation of counter-regulatory mechanisms initiated by increases in renin secretion during dopamine-induced renal vasodilatation [250, 251]. In contrast, renal dopamine (originating in tubular cells) fails to directly affect the vasculature and is expected only to produce natriuresis.

As a result of decreased renal blood flow in heart failure it is expected that L-DOPA availability to the kidney might be reduced, and consequently its conversion to dopamine. Heart failure patients are endowed with reduced renal dopaminergic activity, which appears to result from reduced delivery of L-DOPA to the kidney and enhanced deamination of dopamine [230, 231, 252–255]. The increased ability to take up/decarboxylate filtered L-DOPA is a guarantee that its supplementation may result in an enhanced formation of dopamine. Supplementation with small doses of L-DOPA [256] or administration of inhibitors of COMT, by increasing renal dopamine, may contribute to increase the distal delivery of sodium and overcome the resistance to natriuretic peptides that occurs in heart failure. The increased delivery of sodium to distal segments of the nephron may contribute to negate the actions of aldosterone and improve renal sodium handling. Simultaneously, increased levels of dopamine may effectively antagonize the tubular actions of Ang II. It has been shown that high-dose L-DOPA treatment (1.5 ± 2.0 g/day) induces hemodynamic improvement in patients with heart failure [257, 258]. However, much lower doses (300 mg/day) significantly (5-fold) increased urinary dopamine, producing increased natriuresis and diuresis [256].

11.7.2
Renal Failure

Infusion of dopamine at a dose of between 2 and 5 µg/min (low-dose or renal dopamine) into experimental animals or healthy humans causes intrarenal vasodilatation with a dose-dependent increase in renal blood flow [246, 259–262]. This effect is mediated predominantly by the stimulation of the D_1 dopamine receptors located in the renal vasculature. In addition, dopamine binds to D_1 dopamine receptors in the proximal convoluted tubule, mTAL and CCD, inhibiting a variety of Na^+ transporters that ultimately leads to natriuresis [103, 132]. On the basis of these findings low-dose dopamine has been used in patients at risk of developing an acute deterioration in renal function or in patients who have suffered an acute renal insult. The clinical evidence to support this practice has, however, been questioned in editorials and reviews [263–271]. A large randomized, placebo-controlled trial [272] and two large case-controlled studies [273, 274] were unable to demonstrate a beneficial effect from the use of low-dose dopamine in patients with an acute deterioration in renal function due to sepsis, or an ischemic or toxic insult. The Australian and New Zealand Intensive Care Society (ANZICS) Clinical Trials Group, the largest randomized clinical trial published to date, randomized 324 patients with sepsis with early renal dysfunction to low-dose dopamine or placebo [272]. The primary endpoint of therapy in the ANZICS study was the absolute change in serum creatinine, which showed not to be significantly improved by low-dose dopamine. Durations of intensive care unit (ICU) stay and of hospital stay were also similar between the two groups. There were 69 deaths in the dopamine group and 66 in the placebo group.

Kellum and Decker published a meta-analysis on the use of dopamine in acute renal failure [275]. The primary endpoints in their meta-analysis were mortality, acute renal failure and the need for hemodialysis. The authors found that low-dose dopamine had no effect on these endpoints. It is, however, possible that the endpoints used by Kellum and Decker were insensitive to the renal effects of low-dose dopamine (i.e. low-dose dopamine could improve renal function without affecting mortality or the need for dialysis.) In addition, it is plausible that dopamine may have a renoprotective effects in certain circumstances/disease states and not in others. Subsequently, Marik published another meta-analysis that evaluated all relevant randomized trials that compared low-dose dopamine with placebo in adult patients at risk of developing an acute decline in renal function or in patients with acute renal insufficiency [276]. This work analyzed 15 studies containing 970 subjects. The main outcome measure was the absolute change in serum creatinine. In addition, the number of patients who developed an acute decline in renal function was recorded. The meta-analysis demonstrated no significant difference between the absolute change in serum creatinine and the incidence of acute renal dysfunction between those patients receiving low-dose dopamine and the control group. In addition, no subgroup of patients showed improved renal function with low-dose dopamine. The study by Kapoor et al. was the only study included in this meta-analysis that showed a beneficial effect of low-dose dopamine [277]. However, the four other studies included in this meta-analysis that studied the renoprotective effects of low-dose dopamine in patients receiving contrast media did not replicate these findings [278–281]. Indeed, by subgroup analysis, low-dose dopamine had no significant effect on renal function in patients receiving radiocontrast media (-11.8; 95% confidence interval -34.4, 10.6; $P = 0.3$). It is possible that the patients who received low-dose dopamine in the study by Kapoor et al. were better prehydrated than the control group.

A major problem in the assessment of benefits and risks associated with the use of low-dose dopamine infusion may be related to lack of information on the exposure to the amine, since it undergoes extensive metabolism and binds to plasma protein [282]. In fact, dopamine is partially protein-bound and usually reaches a steady plasma concentration within 5–30 min, with the unbound form responsible for its effects [283]. Dopamine is rapidly conjugated to sulfates and glucuronides, with an elimination half-life of 1–2 min [282]. Irreversible dopamine metabolism takes place in the liver, kidney and pulmonary endothelium by MAO and COMT [45, 48, 52]. Approximately 25% of infused dopamine is converted to norepinephrine. Therefore, considerable changes in exposure to dopamine may take place considering these factors. It should be noted, however, that although the dose of infusion is reported in clinical studies, actual plasma levels are not [282]. Weight-based dosing of dopamine in normal men has been shown to result in profound variation in plasma levels [284]. In the critically ill population, there is extreme variability in plasma dopamine concentrations, even at steady-state infusion rates [285, 286]. Given these findings, it is unclear whether patients are truly receiving "low-dose" dopamine. It makes extrapolation of data difficult and understanding its underlying physiologic mechanisms even more so. Thus, a low-dose dopamine infusion that is expected to be selective for the activation of D_1 dopamine receptors may originate circulating concentrations of the amine that also activate other receptors, such as α- and β-adrenoceptors, originating complex cardiovascular effects, including episodes of tachycardia, cardiac arrhythmias, and myocardial ischemia and infarction [287].

To circumvent these problems stable selective D_1 dopamine receptor agonists have been developed, fenoldopam being one of the most studied drugs. Fenoldopam mesylate is a selective D_1 dopamine agonist, with no effect on D_2 dopamine and $α_1$-adrenoceptors, producing a selective renal vasodilation. This may favor the kidney oxygen supply/demand ratio and prevent acute renal failure. Intravenous fenoldopam at a dose of 0.2 μg/kg/min improved renal hemodynamics and increased Na^+ and K^+ excretion in patients requiring mechanical ventilation of their lungs and positive end-expiratory pressure. These effects were probably caused by an increased kidney perfusion secondary to renal artery vasodilatation [288]. Fenoldopam increases renal blood flow in a dose-dependent manner compared with placebo and, at the lowest dose, significantly increases renal blood flow occurred without changes in systemic blood pressure or heart rate [289]. These findings suggest a role for fenoldopam in preventing or treating renal failure in patients who are not hypertensive. Impairment of renal and splanchnic perfusion during and after cardiopulmonary bypass may be responsible for acute renal failure and endotoxin-mediated systemic inflammation, respectively. It was hypothesized that fenoldopam, a selective dopamine receptor agonist, would preserve renal function after cardiopulmonary bypass through its selective renal vasodilatory and natriuretic effects, and increase gastrointestinal mucosal perfusion by selective splanchnic vasodilation. The findings obtained are consistent with the hypothesis that fenoldopam possesses a renoprotective effect in patients undergoing cardiopulmonary bypass [290]. In a prospective, multiple-center, randomized, controlled trial, involving 100 adult critically ill patients with early renal dysfunction (ICU stay less than 1 week, hemodynamic stability and urine output 0.5 ml/kg or less over a 6-h period and/or serum creatinine concentration 1.5–3.5 mg/dl), fenoldopam (0.1 µg/kg/min) produced a more significant reduction in creatinine values compared with dopamine (2 µg/kg/min) after 2, 3 and 4 days of infusion. The maximum decrease in creatinine compared with baseline was significantly greater in patients receiving fenoldopam than in patents treated with a low-dose dopamine. Moreover, 66% of patients in patients receiving fenoldopam had a creatinine decrease greater than 10% of the baseline value at the end of infusion, compared with only 46% in patients receiving dopamine. The conclusion from this study was that in critically ill patients, a continuous infusion of fenoldopam at doses that not cause any clinically significant hemodynamic impairment and improves renal function compared with "renal-dose" dopamine. Therefore, in the setting of acute early renal dysfunction, before severe renal failure has occurred, the attempt to reverse renal hypoperfusion with fenoldopam appears to be more effective than with low-dose dopamine [291].

11.7.3
Surgery and Transplantation

"Renal-dose" dopamine (1–3 µg/kg/min) is administered to patients after cardiac surgery to preserve or improve renal function. Many of these patients develop new-onset postoperative atrial fibrillation or atrial flutter that could be related to "renal-dose" dopamine administration. Myles *et al.* [292] saw no benefit with "renal-dose" dopamine on urine output, creatinine clearance or the development of renal insufficiency despite improvement in cardiac index. Davis *et al.* [293] studied patients who underwent coronary artery bypass grafting who were treated with dopamine at 200 µg/min. These investigators found that there was a significant increase in urine output, creatinine clearance, sodium excretion and cardiac index. However, heart rate, the main determinant of myocardial oxygen consumption, was also increased. When comparing dopamine and dobutamine in postoperative cardiac surgical patients, both agents increased renal plasma flow and GFR. However, diuresis, natriuresis and kaliuresis were higher in the dopamine group [294]. More recently, Argalious et al. [295] studied 1731 patients undergoing coronary artery bypass grafting with cardiopulmonary bypass; of these, 15.0% developed postoperative atrial fibrillation. The incidence of postoperative atrial fibrillation was 23.3% (41/176) among patients who received "renal-dose" dopamine (1–3 µg/kg/min) and 14.1% (219/1555) among those who did not receive "renal-dose" dopamine. In conclusion, "renal-dose" dopamine was associated with a 1.74 odds ratio of postoperative atrial fibrillation developing after coronary artery bypass grafting with cardiopulmonary bypass. This fits well with the views of Lassing et al. [296] and Tang et al. [297], which showed that dopamine given at "renal-dose" appears to offer no renal protection in patients with normal heart and kidney functions undergoing elective coronary surgery. On the contrary, it exacerbates the severity of renal tubular injury during the early postoperative period. Based on these findings we do not recommend the use of dopamine for routine renal prophylaxis in this group of patients.

In a randomized study [298] of 37 patients undergoing abdominal aortic aneurysm repair, placebo or dopamine was given for the first 24 h postoperatively. Urine output was higher in the dopamine group. However, there was no difference in plasma creatinine level, creatinine clearance, incidence of renal failure or mortality between the two groups. Of note, there were three myocardial infarctions in the dopamine group and only one in the placebo group. There were also two deaths in the dopamine group, whereas there were none in the placebo group. In another study of elective abdominal aortic aneurysm repair, Paul et al. [299] randomized patients to saline or dopamine plus mannitol, before aortic cross-clamp. Hemodynamic measures including pulmonary artery pressure were maintained at normal levels in both groups. In this study, the combination of dopamine and mannitol had no effect on urine output or fractional excretion of sodium and only produced a slight delay in the onset of postclamp decrease in GFR.

Patients undergoing orthotopic liver transplantation frequently receive dopamine infusions to preserve renal function. To test the benefit of such infusions on renal function, 48 nonanuric patients presenting for orthotopic liver transplantation were entered into a randomized double-blind protocol. Twenty-two patients received dopamine at a rate of 3 µg/kg/min during surgery and the first postoperative 48 h, and a control group of 25 patients received saline. During the hepatic vascular anastomoses, the donor liver was flushed with cold saline. In seven patients, the flush contained mannitol (50 g) as part of a surgical protocol. Initially, it appeared that there was an increase in urine output during the neohepatic phase in those patients receiving dopamine versus controls (4.20 ± 3.3 versus 2.10 ± 1.3 ml/kg/h, respectively). After excluding all patients that received flush containing mannitol, there was no significant difference in urine output during the neohepatic phase between the dopamine group and controls. The GFRs at 1 month after surgery were similar and decreased approximately 40% in each group. Although a beneficial effect of dopamine in all situations cannot be ruled out, the authors [300] conclude that routine perioperative use of dopamine is of little value in nonanuric patients presenting for orthotopic liver transplantation.

Recently, Zhang et al. [301] studied dopamine or norepinephrine combined with dopamine infused during orthotopic liver transplantation (15 patients in each group) to observe and compare their effects on hemodynamics, oxygenation and renal function during different stages of the operation. Vasopressors were infused after induction of anesthesia. Data of hemodynamics, oxygenation and renal function were collected after induction, 1 h in preanhepatic, anhepatic, neohepatic phase and at the end of operation. The findings in this study suggest that during orthotopic liver transplantation, both dopamine and norepinephrine combined with dopamine can maintain hemodynamics stable, whereas the latter may create better condition. Norepinephrine has positive effects on maintenance of renal function.

In a study aimed to determine the risk factors of postoperative acute renal failure in orthotopic liver transplantation, 184 consecutive cases were reviewed [302]. Postoperative acute renal failure was defined as a persistent rise of 50% increase or more of the serum creatinine. The patients were classified as early postoperative acute renal failure (E-ARF) (first week) and late postoperative acute renal failure (L-ARF) (second to fourth week). The development of E-ARF was influenced by preoperative factors such as acute renal failure and hypoalbuminemia, as well as postoperative factors such as liver dysfunction and prolonged treatment with dopamine. The predicting factors of L-ARF differ from E-ARF and correspond to postoperative causes such as bacterial infection and surgical re-operation.

To test the relative effects on serum creatinine, blood urea nitrogen and urine output of low-dose dopamine and fenoldopam in patients undergoing liver transplantation, 43 patients were randomized to one of two continuous infusions over 48 h, starting with anesthesia induction: fenoldopam (0.1 µg/kg/min) or dopamine (2 µg/kg/min). At postoperative day 3, the median creatinine increase was 0.2 mg/dl with fenoldopam and 0.5 mg/dl in the dopamine group. The median increase in blood urea nitrogen was 2 mg/dl versus 8.5 mg/dl, respectively, with fenoldopam versus dopamine. Urine output was similar; however, significantly fewer fenoldopam patients required furosemide compared with dopamine patients. The hemodynamic effects of dopamine and fenoldopam were similar. Compared with dopamine, in the setting of liver transplantation, fenoldopam is associated with better serum creatinine and blood urea nitrogen values [303].

With the aim of assessing whether fenoldopam can help to preserve renal function after liver transplantation, Biancofiore et al. [304] randomized 140 consecutive recipients with comparable preoperative renal function to receive fenoldopam (0.1 µg/kg/min) (46 patients), dopamine (3 µg/kg/min) (48 patients) or placebo (46 patients) from the time of anesthesia induction to 96 h postoperatively. There were no differences between the groups in intraoperative urinary output or furosemide administration. Daily recordings made during the first 4 postoperative days revealed no significant differences in urinary output, serum creatinine, incidence of renal insufficiency, or need for loop diuretics or vasoactive drugs. In comparison with preoperative levels, creatinine clearance at the end of the study in the patients receiving fenoldopam remained substantially unchanged, whereas it decreased by 39 and 12.3%, respectively, in the patients receiving placebo or dopamine; blood cyclosporine A levels were similar in the three groups. These results confirm the inefficacy of dopamine in preventing or limiting early renal dysfunction after liver transplantation, and suggest that fenoldopam may preserve creatinine clearance by counterbalancing the renal vasoconstrictive effect of cyclosporine A.

11.7.4
Sepsis and Inflammation

Insufficient blood flow to the splanchnic organs is believed to be an important contributory factor for the development of organ failure after septic shock. It has been suggested that increasing systemic flow also may improve splanchnic blood flow in septic patients. In a porcine endotoxin model, dopamine showed no beneficial effect on splanchnic or renal blood flow compared with norepinephrine or placebo [305]. In an ovine model of sepsis, Bersten et al. [306] found that dopamine increased renal blood flow as measured by an electromagnetic flow probe in healthy controls but not in septic animals. Dopamine failed to increase creatinine clearance or alter renal arteriovenous oxygen extraction in either group. The study of Hiltebrand et al. [307] compared the effects of dopamine (5 and 10 µg/kg/min), dobutamine (5 and 10 µg/kg/min) and dopexamine (1 and 2 µg/kg/min), on systemic (cardiac index), regional (superior mesenteric artery) and local (microcirculatory) blood flow during septic shock in pigs. All three drugs markedly increased cardiac output in this sepsis model. However, increased systemic flow did not reach the microcirculation in the gastrointestinal tract. This may in part explain why some of the clinical trials, in which systemic oxygen delivery was deliberately increased by administration of inotropic drugs, have failed to improve survival in critically ill patients. Infusion of dopamine in septic mice increased splenocyte apoptosis and decreased splenocyte proliferation and interleukin-2 release of septic mice. Furthermore, an inhibitory effect of dopamine infusion on splenocyte proliferation and the release of the T helper type 1 cytokines interleukin-2 and interferon-γ was observed in sham-operated control mice. These effects were paralleled by a decreased survival of dopamine-treated septic animals (47 versus 67%) [308].

A recent observational study investigated whether dopamine administration influences the outcome from shock [309]. The ICU mortality rate for shock was 38.3 and for septic shock was 47.4%. Of patients in shock, 375 (35.4%) received dopamine and 683 (64.6%) never received dopamine. The dopamine group had higher ICU (42.9 versus 35.7%, $P = 0.02$) and hospital (49.9 versus 41.7%, $P = 0.01$) mortality rates. A Kaplan–Meier survival curve showed diminished 30-day survival in the dopamine group. In a multivariate analysis with ICU outcome as the dependent factor, age, cancer, medical admissions, higher mean Sequential Organ Failure Assessment score, higher mean fluid balance and dopamine administration were independent risk factors for ICU mortality in patients with shock.

Morelli et al. [310] recently reported on the use of fenoldopam. Three hundred septic patients with baseline serum creatinine concentrations below 150 µmol/l were randomized to a continuous infusion of either fenoldopam (0.09 µg/kg/min) ($n = 150$) or placebo ($n = 150$) while in the ICU. The incidence of acute renal failure was significantly lower in the fenoldopam group compared with the control group (29 versus 51 patients; $P = 0.006$). The odds ratio of developing acute renal failure for patients treated with fenoldopam was estimated to be 0.47 ($P = 0.005$). The difference in the incidence of severe acute renal failure (creatinine above 300 µmol/l), however, failed to achieve statistical significance (10 versus 21; $P = 0.056$). The length of ICU stay in surviving patients was significantly lower in the fenoldopam group compared with the control group. A direct effect of treatment on the probability of death, beyond its effect on acute renal failure, was not significant.

Dopamine has important immunological and endocrine effects. Both T and B lymphocytes are endowed with dopamine receptors [311]. Dopamine in low concentrations inhibits stimulated lymphocyte proliferation, immunoglobulin synthesis and cytokine production, and promotes lymphocyte apoptosis [312–315]. Dopamine has been found to decrease serum prolactin [316], which is an important immunostimulatory hormone through the activation of prolactin receptors present on T and B lymphocytes [317]. Dopamine also decreases growth hormone secretion and thyrotropin release [316]. Growth hormone deficiency may contribute to a negative nitrogen balance in critical illness. Dopamine has also been demonstrated to increase expression of human immunodeficiency virus in cells of the immune system [318].

References

1 Caramona, M.M. and Soares-da-Silva, P. (1985) The effects of chemical sympathectomy on dopamine, noradrenaline and adrenaline content in some peripheral tissues. *British Journal of Pharmacology*, **86**, 351–6.

2 Soares-da-Silva, P. (1986) Evidence for a non-precursor dopamine pool in noradrenergic neurones of the dog mesenteric artery. *Naunyn-Schmiedebergs Archives of Pharmacology*, **333**, 219–23.

3 Soares-da-Silva, P. (1988) Further evidence for a noradrenaline-independent storage of dopamine in the dog kidney. *Journal of Autonomic Pharmacology*, **8**, 127–33.

4 Soares-da-Silva, P. (1992) A study of the neuronal and non-neuronal stores of dopamine in rat and rabbit kidney. *Pharmacological Research*, **26**, 161–71.

5 Soares-da-Silva, P. and Davidson, R. (1985) Effects of 6-hydroxydopamine on dopamine and noradrenaline content in dog blood vessels and heart. Evidence for a noradrenaline-independent dopamine pool. *Naunyn-Schmiedebergs Archives of Pharmacology*, **329**, 253–7.

6 Soares-da-Silva, P. (1987) A comparison between the pattern of dopamine and noradrenaline release from sympathetic neurones of the dog mesenteric artery. *British Journal of Pharmacology*, **90**, 91–8.

7 Soares-da-Silva, P. (1987) Dopamine released from nerve terminals activates prejunctional dopamine receptors in dog mesenteric arterial vessels. *British Journal of Pharmacology*, **91**, 591–9.

8 Soares-da-Silva, P. (1988) Evidence for dopaminergic co-transmission in dog mesenteric arterial vessels. *British Journal of Pharmacology*, **95**, 218–24.

9 Soares-da-Silva, P. (1994) Source and handling of renal dopamine: its physiological importance. *News in Physiological Sciences*, **9**, 128–34.
10 Vieira-Coelho, M.A., Gomes, P., Serrão, M.P. and Soares-da-Silva, P. (1997) Renal and intestinal autocrine monoaminergic systems: dopamine versus 5-hydroxytryptamine. *Clinical and Experimental Hypertension*, **19**, 43–58.
11 Pinto-do, O.P., Chibalin, A.V., Katz, A.I., Soares-da-Silva, P. and Bertorello, A.M. (1997) Short-term vs. sustained inhibition of proximal tubule Na,K-ATPase activity by dopamine: cellular mechanisms. *Clinical and Experimental Hypertension*, **19**, 73–86.
12 Aperia, A.C. (2000) Intrarenal dopamine: a key signal in the interactive regulation of sodium metabolism. *Annual Review of Physiology*, **62**, 621–47.
13 Lee, M.R. (1993) Dopamine and the kidney. Ten years on. *Clinical Science*, **84**, 357–75.
14 Siragy, H.M., Felder, R.A., Howell, N.L., Chevalier, R.L., Peach, M.J. and Carey, R.M. (1989) Evidence that intrarenal dopamine acts as a paracrine substance at the renal tubule. *American Journal of Physiology*, **257**, F469–77.
15 Hussain, T. and Lokhandwala, M.F. (2003) Renal dopamine receptors and hypertension. *Experimental Biology and Medicine of Maywood*, **228**, 134–42.
16 Jose, P.A., Eisner, G.M. and Felder, R.A. (2003) Dopamine and the kidney: a role in hypertension? *Current Opinion in Nephrology and Hypertension*, **12**, 189–94.
17 Jose, P.A., Eisner, G.M. and Felder, R.A. (2002) Role of dopamine receptors in the kidney in the regulation of blood pressure. *Current Opinion in Nephrology and Hypertension*, **11**, 87–92.
18 Grossman, E., Goldstein, D.S., Hoffman, A., Wacks, I.R. and Epstein, M. (1992) Effects of water immersion on sympathoadrenal and dopa–dopamine systems in humans. *American Journal of Physiology*, **262**, R993–9.
19 Soares-da-Silva, P., Pestana, M., Vieira-Coelho, M.A., Fernandes, M.H. and Albino-Teixeira, A. (1995) Assessment of renal dopaminergic system activity in the nitric oxide-deprived hypertensive rat model. *British Journal of Pharmacology*, **114**, 1403–13.
20 Soares-da-Silva, P. and Serrão, M.P. (2004) High- and low-affinity transport of L-leucine and L-DOPA by the hetero amino acid exchangers LAT1 and LAT2 in LLC-PK$_1$ renal cells. *American Journal of Physiology – Renal Physiology*, **287**, F252–61.
21 Soares-da-Silva, P. and Serrão, M.P. (2000) Molecular modulation of inward and outward apical transporters of L-dopa in LLC-PK$_1$ cells. *American Journal of Physiology – Renal Physiology*, **279**, F736–46.
22 Fraga, S., Serrão, M.P. and Soares-da-Silva, P. (2002) The L-3,4-dihydroxyphenylalanine transporter in human and rat epithelial intestinal cells is a type 2 hetero amino acid exchanger. *European Journal of Pharmacology*, **441**, 127–35.
23 Fraga, S., Serrão, M.P. and Soares-da-Silva, P. (2002) L-Type amino acid transporters in two intestinal epithelial cell lines function as exchangers with neutral amino acids. *Journal of Nutrition*, **132**, 733–8.
24 Audus, K.L. and Borchardt, R.T. (1986) Characteristics of the large neutral amino acid transport system of bovine brain microvessel endothelial cell monolayers. *Journal of Neurochemistry*, **47**, 484–8.
25 Gomes, P. and Soares-da-Silva, P. (1999) Interaction between L-DOPA and 3-O-methyl-L-DOPA for transport in immortalised rat capillary cerebral endothelial cells. *Neuropharmacology*, **38**, 1371–80.
26 Gomes, P. and Soares-da-Silva, P. (1999) L-DOPA transport properties in an immortalised cell line of rat capillary cerebral endothelial cells, RBE 4. *Brain Research*, **829**, 143–50.
27 Kageyama, T., Nakamura, M., Matsuo, A., Yamasaki, Y., Takakura, Y., Hashida, M., Kanai, Y., Naito, M., Tsuruo, T., Minato, N. and Shimohama, S. (2000) The 4F2hc/LAT1 complex transports L-DOPA across

the blood–brain barrier. *Brain Research*, **879**, 115–21.
28. Sampaio-Maia, B., Serrão, M.P. and Soares-da-Silva, P. (2001) Regulatory pathways and uptake of L-DOPA by capillary cerebral endothelial cells, astrocytes, and neuronal cells. *American Journal of Physiology – Cell Physiology*, **280**, C333–42.
29. Wade, L.A. and Katzman, R. (1975) Synthetic amino acids and the nature of L-DOPA transport at the blood–brain barrier. *Journal of Neurochemistry*, **25**, 837–42.
30. Soares-da-Silva, P. and Serrão, M.P. (2000) Outward transfer of dopamine precursor L-3,4-dihydroxyphenylalanine (L-dopa) by native and human P-glycoprotein in LLC-PK$_1$ and LLC-GA5 col300 renal cells. *Journal of Pharmacology and Experimental Therapeutics*, **293**, 697–704.
31. Soares-da-Silva, P., Serrão, M.P. and Vieira-Coelho, M.A. (1998) Apical and basolateral uptake and intracellular fate of dopamine precursor l-dopa in LLC-PK$_1$ cells. *American Journal of Physiology*, **274**, F243–51.
32. Soares-da-Silva, P., Serrão, M.P., Vieira-Coelho, M.A. and Pestana, M. (1998) Evidence for the involvement of P-glycoprotein on the extrusion of taken up L-DOPA in cyclosporine A treated LLC-PK$_1$ cells. *British Journal of Pharmacology*, **123**, 13–22.
33. Ishii, H., Sasaki, Y., Goshima, Y., Kanai, Y., Endou, H., Ayusawa, D., Ono, H., Miyamae, T. and Misu, Y. (2000) Involvement of rBAT in Na$^+$-dependent and -independent transport of the neurotransmitter candidate L-DOPA in *Xenopus laevis* oocytes injected with rabbit small intestinal epithelium poly A$^+$ RNA. *Biochimica et Biophysica Acta*, **1466**, 61–70.
34. Gomes, P. and Soares-da-Silva, P. (2002) Na$^+$-independent transporters, LAT-2 and b$^{0,+}$, exchange L-DOPA with neutral and basic amino acids in two clonal renal cell lines. *Journal of Membrane Biology*, **186**, 63–80.
35. Pinho, M.J., Gomes, P., Serrão, M.P., Bonifacio, M.J. and Soares-da-Silva, P. (2003) Organ-specific overexpression of renal LAT2 and enhanced tubular L-DOPA uptake precede the onset of hypertension. *Hypertension*, **42**, 613–18.
36. Lucas-Teixeira, V.A., Vieira-Coelho, M.A., Serrão, P., Pestana, M. and Soares-da-Silva, P. (2000) Salt intake and sensitivity of intestinal and renal Na$^+$-K$^+$ atpase to inhibition by dopamine in spontaneous hypertensive and Wistar-Kyoto rats. *Clinical and Experimental Hypertension*, **22**, 455–69.
37. Pinho, M.J., Pinto, V., Serrão, M.P., Jose, P.A. and Soares-da-Silva, P. (2007) Underexpression of the Na$^+$-dependent neutral amino acid transporter ASCT2 in the spontaneously hypertensive rat kidney. *American Journal of Physiology – Regulatory Integrative and Comparative Physiology*, **293**, R538–47.
38. Pinho, M.J., Serrão, M.P., Jose, P.A. and Soares-da-Silva, P. (2007) Overexpression of non-functional LAT1/4F2hc in renal proximal tubular epithelial cells from the spontaneous hypertensive rat. *Cellular Physiology and Biochemistry*, **20**, 535–48.
39. Pinho, M.J., Serrão, M.P. and Soares-da-Silva, P. (2007) High-salt intake and the renal expression of amino acid transporters in spontaneously hypertensive rats. *American Journal of Physiology – Renal Physiology*, **292**, F1452–63.
40. Broer, A., Klingel, K., Kowalczuk, S., Rasko, J.E., Cavanaugh, J. and Broer, S. (2004) Molecular cloning of mouse amino acid transport system B0, a neutral amino acid transporter related to Hartnup disorder. *Journal of Biological Chemistry*, **279**, 24467–76.
41. Pinho, M.J., Serrão, M.P., Jose, P.A. and Soares-da-Silva, P. (2007) Organ specific underexpression renal of Na$^+$-dependent B^{0}AT1 in the SHR correlates positively with overexpression of NHE3 and salt intake. *Molecular and Cellular Biochemistry*, **306**, 9–18.
42. Hayashi, M., Yamaji, Y., Kitajima, W. and Saruta, T. (1990) Aromatic L-amino acid decarboxylase activity along the rat nephron. *American Journal of Physiology*, **258**, F28–33.
43. Soares-da-Silva, P. and Fernandes, M.H. (1990) Regulation of dopamine synthesis

in the rat kidney. *Journal of Autonomic Pharmacology*, **10**, s25–30.
44 Vieira-Coelho, M.A. and Soares-da-Silva, P. (1993) Dopamine formation, from its immediate precursor 3,4-dihydroxyphenylalanine, along the rat digestive tract. *Fundamental and Clinical Pharmacology*, **7**, 235–43.
45 Fernandes, M.H., Pestana, M. and Soares-da-Silva, P. (1991) Deamination of newly-formed dopamine in rat renal tissues. *British Journal of Pharmacology*, **102**, 778–82.
46 Fernandes, M.H. and Soares-da-Silva, P. (1990) Effects of MAO-A and MAO-B selective inhibitors Ro 41-1049 and Ro 19-6327 on the deamination of newly formed dopamine in the rat kidney. *Journal of Pharmacology and Experimental Therapeutics*, **255**, 1309–13.
47 Fernandes, M.H. and Soares-da-Silva, P. (1990) Role of monoamine oxidase A and B in the deamination of newly-formed dopamine in the rat kidney. *Journal of Neural Transmission Supplement*, **32**, 155–9.
48 Fernandes, M.H. and Soares-da-Silva, P. (1994) Role of monoamine oxidase and catechol-*O*-methyltransferase in the metabolism of renal dopamine. *Journal of Neural Transmission Supplement*, **41**, 101–5.
49 Fernandes, M.H. and Soares-da-Silva, P. (1992) Type A and B monoamine oxidase activities in the human and rat kidney. *Acta Physiologica Scandinavica*, **145**, 363–7.
50 Guimaraes, J.T. and Soares-da-Silva, P. (1998) The activity of MAO A and B in rat renal cells and tubules. *Life Sciences*, **62**, 727–37.
51 Guimaraes, J.T., Vieira-Coelho, M.A., Serrão, M.P. and Soares-da-Silva, P. (1997) Opossum kidney (OK) cells in culture synthesize and degrade the natriuretic hormone dopamine: a comparison with rat renal tubular cells. *International Journal of Biochemistry and Cell Biology*, **29**, 681–8.
52 Pestana, M. and Soares-da-Silva, P. (1994) Effect of type A and B monoamine oxidase selective inhibition by Ro 41-1049 and Ro 19-6327 on dopamine outflow in rat kidney slices. *British Journal of Pharmacology*, **113**, 1269–74.
53 Pestana, M. and Soares-Da-Silva, P. (1994) Outflow of dopamine and noradrenaline originating from L-dopa and L-threo-DOPS in rat renal tissues. *General Pharmacol*, **25**, 879–85.
54 Pestana, M. and Soares-da-Silva, P. (1994) The renal handling of dopamine originating from L-dopa and gamma-glutamyl-L-dopa. *British Journal of Pharmacology*, **112**, 417–22.
55 Soares-da-Silva, P. and Fernandes, M.H. (1992) Sodium-dependence and ouabain-sensitivity of the synthesis of dopamine in renal tissues of the rat. *British Journal of Pharmacology*, **105**, 811–16.
56 Soares-da-Silva, P. and Fernandes, M.H. (1990) Synthesis and metabolism of dopamine in the kidney. Effects of sodium chloride, monoamine oxidase inhibitors and alpha-human atrial natriuretic peptide. *American Journal of Hypertension*, **3**, 7S–10S.
57 Soares-da-Silva, P., Fernandes, M.H. and Pestana, M. (1993) Studies on the role of sodium on the synthesis of dopamine in the rat kidney. *Journal of Pharmacology and Experimental Therapeutics*, **264**, 406–14.
58 Ball, S.G., Oats, N.S. and Lee, M.R. (1978) Urinary dopamine in man and rat: effects of inorganic salts on dopamine excretion. *Clinical Science and Molecular Medicine*, **55**, 167–73.
59 Goldstein, D.S., Stull, R., Eisenhofer, G. and Gill, J.R.J. (1989) Urinary excretion of dihydroxyphenylalanine and dopamine during alterations of dietary salt intake in humans. *Clinical Science*, **76**, 517–22.
60 Grossman, E., Hoffman, A., Tamrat, M., Armando, I., Keiser, H.R. and Goldstein, D.S. (1991) Endogenous dopa and dopamine responses to dietary salt loading in salt-sensitive rats. *Journal of Hypertension*, **9**, 259–63.
61 Vieira-Coelho, M.A., Hussain, T., Kansra, V., Serrão, M.P., Guimaraes, J.T., Pestana, M., Soares-da-Silva, P. and Lokhandwala, M.F. (1999) Aging, high salt intake, and renal dopaminergic activity in Fischer 344 rats. *Hypertension*, **34**, 666–72.

62 Vieira-Coelho, M.A., Serrão, P., Hussain, T., Lokhandwala, M.F. and Soares-da-Silva, P. (2001) Salt intake and intestinal dopaminergic activity in adult and old Fischer 344 rats. *Life Sciences*, **69**, 1957–68.

63 Jose, P.A., Eisner, G.M. and Felder, R.A. (1998) Renal dopamine receptors in health and hypertension. *Pharmacology and Therapeutics*, **80**, 149–82.

64 Missale, C., Nash, S.R., Robinson, S.W., Jaber, M. and Caron, M.G. (1998) Dopamine receptors: from structure to function. *Physiological Reviews*, **78**, 189–225.

65 Sibley, D.R. and Monsma F.J. Jr (1992) Molecular biology of dopamine receptors. *Trends in Pharmacological Sciences*, **13**, 61–9.

66 Civelli, O., Bunzow, J.R. and Grandy, D.K. (1993) Molecular diversity of the dopamine receptors. *Annual Review of Pharmacology and Toxicology*, **33**, 281–307.

67 O'Dowd, B.F. (1993) Structures of dopamine receptors. *Journal of Neurochemistry*, **60**, 804–16.

68 Dal Toso, R., Sommer, B., Ewert, M., Herb, A., Pritchett, D.B., Bach, A., Shivers, B.D. and Seeburg, P.H. (1989) The dopamine D_2 receptor: two molecular forms generated by alternative splicing. *EMBO Journal*, **8**, 4025–34.

69 Giros, B., Sokoloff, P., Martres, M.P., Riou, J.F., Emorine, L.J. and Schwartz, J.C. (1989) Alternative splicing directs the expression of two D_2 dopamine receptor isoforms. *Nature*, **342**, 923–6.

70 Monsma, F.J. Jr, McVittie, L.D., Gerfen, C.R., Mahan, L.C. and Sibley, D.R. (1989) Multiple D_2 dopamine receptors produced by alternative RNA splicing. *Nature*, **342**, 926–9.

71 Amenta, F., Barili, P., Bronzetti, E., Felici, L., Mignini, F. and Ricci, A. (2000) Localization of dopamine receptor subtypes in systemic arteries. *Clinical and Experimental Hypertension*, **22**, 277–88.

72 Amenta, F., Barili, P., Bronzetti, E. and Ricci, A. (1999) Dopamine D_1-like receptor subtypes in the rat kidney: a microanatomical study. *Clinical and Experimental Hypertension*, **21**, 17–23.

73 O'Connell, D.P., Vaughan, C.J., Aherne, A.M., Botkin, S.J., Wang, Z.Q., Felder, R.A. and Carey, R.M. (1998) Expression of the dopamine D_3 receptor protein in the rat kidney. *Hypertension*, **32**, 886–95.

74 Amenta, F. (1990) Density and distribution of dopamine receptors in the cardiovascular system and in the kidney. *Journal of Autonomic Pharmacology*, **10** (Suppl. 1), S11–18.

75 Ricci, A., Marchal-Victorion, S., Bronzetti, E., Parini, A., Amenta, F. and Tayebati, S.K. (2002) Dopamine D_4 receptor expression in rat kidney: evidence for pre- and postjunctional localization. *Journal of Histochemistry and Cytochemistry*, **50**, 1091–6.

76 Ozono, R., O'Connell, D.P., Wang, Z.Q., Moore, A.F., Sanada, H., Felder, R.A. and Carey, R.M. (1997) Localization of the dopamine D_1 receptor protein in the human heart and kidney. *Hypertension*, **30**, 725–9.

77 Sun, D., Wilborn, T.W. and Schafer, J.A. (1998) Dopamine D_4 receptor isoform mRNA and protein are expressed in the rat cortical collecting duct. *American Journal of Physiology*, **275**, F742–51.

78 Yamaguchi, I., Jose, P.A., Mouradian, M.M., Canessa, L.M., Monsma, F.J. Jr, Sibley, D.R., Takeyasu, K. and Felder, R.A. (1993) Expression of dopamine D_{1A} receptor gene in proximal tubule of rat kidneys. *American Journal of Physiology*, **264**, F280–5.

79 Sanada, H., Yao, L., Jose, P.A., Carey, R.M. and Felder, R.A. (1997) Dopamine D_3 receptors in rat juxtaglomerular cells. *Clinical and Experimental Hypertension*, **19**, 93–105.

80 Yamaguchi, I., Yao, L., Sanada, H., Ozono, R., Mouradian, M.M., Jose, P.A., Carey, R.M. and Felder, RA. (1997) Dopamine D_{1A} receptors and renin release in rat juxtaglomerular cells. *Hypertension*, **29**, 962–8.

81 Amenta, F., Ricci, A., Rossodivita, I., Avola, R. and Tayebati, S.K. (2001) The dopaminergic system in hypertension. *Clinical and Experimental Hypertension*, **23**, 15–24.

82 Vaughan, C.J., Aherne, A.M., Lane, E., Power, O., Carey, R.M. and O'Connell, D.P. (2000) Identification and regional distribution of the dopamine D_{1A} receptor in the gastrointestinal tract. *American Journal of Physiology – Regulatory Integrative and Comparative Physiology*, **279**, R599–609.

83 Carey, R.M. (2001) Theodore Cooper Lecture: renal dopamine system: paracrine regulator of sodium homeostasis and blood pressure. *Hypertension*, **38**, 297–302.

84 Jose, P.A., Eisner, G.M. and Felder, R.A. (2000) Renal dopamine and sodium homeostasis. *Current Hypertension Reports*, **2**, 174–83.

85 Finkel, Y., Eklof, A.C., Granquist, L., Soares-da-Silva, P. and Bertorello, A.M. (1994) Endogenous dopamine modulates jejunal sodium absorption during high-salt diet in young but not in adult rats. *Gastroenterology*, **107**, 675–9.

86 Vieira-Coelho, M.A., Teixeira, V.A., Finkel, Y., Soares-da-Silva, P. and Bertorello, A.M. (1998) Dopamine-dependent inhibition of jejunal Na^+–K^+-ATPase during high-salt diet in young but not in adult rats. *American Journal of Physiology*, **275**, G1317–23.

87 Aoki, Y., Albrecht, F.E., Bergman, K.R. and Jose, PA. (1996) Stimulation of Na^+–K^+–$2Cl^-$ cotransport in rat medullary thick ascending limb by dopamine. *American Journal of Physiology*, **271**, R1561–7.

88 Asghar, M., Hussain, T. and Lokhandwala, M.F. (2001) Activation of dopamine D_1-like receptor causes phosphorylation of alpha$_1$-subunit of Na^+,K^+-ATPase in rat renal proximal tubules. *European Journal of Pharmacology*, **411**, 61–6.

89 Bacic, D., Kaissling, B., McLeroy, P., Zou, L., Baum, M. and Moe, O.W. (2003) Dopamine acutely decreases apical membrane Na/H exchanger NHE3 protein in mouse renal proximal tubule. *Kidney International*, **64**, 2133–41.

90 Baines, A.D. and Drangova, R. (1998) Does dopamine use several signal pathways to inhibit Na-P$_i$ transport in OK cells? *Journal of the American Society of Nephrology*, **9**, 1604–12.

91 Chibalin, A.V., Zierath, J.R., Katz, A.I., Berggren, P.O. and Bertorello, A.M. (1998) Phosphatidylinositol 3-kinase-mediated endocytosis of renal Na^+, K^+-ATPase alpha subunit in response to dopamine. *Molecular Biology of the Cell*, **9**, 1209–20.

92 Efendiev, R., Bertorello, A.M., Zandomeni, R., Cinelli, A.R. and Pedemonte, C.H. (2002) Agonist-dependent regulation of renal Na^+,K^+-ATPase activity is modulated by intracellular sodium concentration. *Journal of Biological Chemistry*, **277**, 11489–96.

93 Felder, C.C., Albrecht, F.E., Campbell, T., Eisner, G.M. and Jose, P.A. (1993) cAMP-independent, G protein-linked inhibition of Na^+/H^+ exchange in renal brush border by D_1 dopamine agonists. *American Journal of Physiology*, **264**, F1032–7.

94 Glahn, R.P., Onsgard, M.J., Tyce, G.M., Chinnow, S.L., Knox, F.G. and Dousa, T.P. (1993) Autocrine/paracrine regulation of renal Na^+-phosphate cotransport by dopamine. *American Journal of Physiology*, **264**, F618–22.

95 Gomes, P. and Soares-da-Silva, P. (2004) Dopamine acutely decreases type 3 Na^+/H^+ exchanger activity in renal OK cells through the activation of protein kinases A and C signalling cascades. *European Journal of Pharmacology*, **488**, 51–9.

96 Gomes, P. and Soares-da-Silva, P. (2002) Role of cAMP–PKA–PLC signaling cascade on dopamine-induced PKC-mediated inhibition of renal Na^+–K^+-ATPase activity. *American Journal of Physiology – Renal Physiology*, **282**, F1084–96.

97 Hu, M.C., Fan, L., Crowder, L.A., Karim-Jimenez, Z., Murer, H. and Moe, O.W. (2001) Dopamine acutely stimulates Na^+/H^+ exchanger (NHE3) endocytosis via clathrin-coated vesicles: dependence on protein kinase A-mediated NHE3 phosphorylation. *Journal of Biological Chemistry*, **276**, 26906–15.

98 Kunimi, M., Seki, G., Hara, C., Taniguchi, S., Uwatoko, S., Goto, A., Kimura, S. and Fujita, T. (2000)

Dopamine inhibits renal $Na^+:HCO_3^-$ cotransporter in rabbits and normotensive rats but not in spontaneously hypertensive rats. *Kidney International*, **57**, 534–43.

99 Pedrosa, R., Jose, P.A. and Soares-da-Silva, P. (2004) Defective D_1-like receptor-mediated inhibition of the $Cl^-/HCO3^-$ exchanger in immortalized SHR proximal tubular epithelial cells. *American Journal of Physiology – Renal Physiology*, **286**, F1120–6.

100 Fraga, S., Luo, Y., Jose, P., Zandi-Nejad, K., Mount, D.B. and Soares-da-Silva, P. (2006) Dopamine D_1-like receptor-mediated inhibition of Cl/HCO_3^- exchanger activity in rat intestinal epithelial IEC-6 cells is regulated by G protein-coupled receptor kinase 6 (GRK 6). *Cellular Physiology and Biochemistry*, **18**, 347–60.

101 Bertorello, A.M. and Katz, A.I. (1993) Short-term regulation of renal Na–K-ATPase activity: physiological relevance and cellular mechanisms. *American Journal of Physiology*, **265**, F743–55.

102 Chen, C. and Lokhandwala, M.F. (1993) Inhibition of Na^+, K^+-ATPase in rat renal proximal tubules by dopamine involved DA-1 receptor activation. *Naunyn-Schmiedebergs Archives of Pharmacology*, **347**, 289–95.

103 Satoh, T., Cohen, H.T. and Katz, A.I. (1993) Different mechanisms of renal Na–K-ATPase regulation by protein kinases in proximal and distal nephron. *American Journal of Physiology*, **265**, F399–405.

104 Hussain, T., Abdul-Wahab, R. and Lokhandwala, M.F. (1997) Bromocriptine stimulates Na^+, K^+-ATPase in renal proximal tubules via the cAMP pathway. *European Journal of Pharmacology*, **321**, 259–63.

105 Narkar, V.A., Hussain, T. and Lokhandwala, M.F. (2002) Activation of D_2-like receptors causes recruitment of tyrosine-phosphorylated NKA alpha 1-subunits in kidney. *American Journal of Physiology – Renal Physiology*, **283**, F1290–5.

106 Yamaguchi, I., Walk, S.F., Jose, P.A. and Felder, R.A. (1996) Dopamine D_{2L} receptors stimulate Na^+/K^+-ATPase activity in murine LTK- cells. *Molecular Pharmacology*, **49**, 373–8.

107 Bertorello, A. and Aperia, A. (1990) Inhibition of proximal tubule Na^+–K^+-ATPase activity requires simultaneous activation of DA1 and DA2 receptors. *American Journal of Physiology*, **259**, F924–8.

108 Jose, P.A., Asico, L.D., Eisner, G.M., Pocchiari, F., Semeraro, C. and Felder, R.A. (1998) Effects of costimulation of dopamine D_1- and D_2-like receptors on renal function. *American Journal of Physiology*, **275**, R986–94.

109 Gomes, P. and Soares-da-Silva, P. (2002) D_2-like receptor-mediated inhibition of Na^+–K^+-ATPase activity is dependent on the opening of K^+ channels. *American Journal of Physiology – Renal Physiology*, **283**, F114–23.

110 Gomes, P. and Soares-da-Silva, P. (2003) Dopamine D_2-like receptor-mediated opening of K^+ channels in opossum kidney cells. *British Journal of Pharmacology*, **138**, 968–76.

111 Zeng, C., Sanada, H., Watanabe, H., Eisner, G.M., Felder, R.A. and Jose, P.A. (2004) Functional genomics of the dopaminergic system in hypertension. *Physiol Genomics*, **19**, 233–46.

112 Hussain, T. and Lokhandwala, M.F. (1997) Renal dopamine DA1 receptor coupling with G_s and $G_{q/11}$ proteins in spontaneously hypertensive rats. *American Journal of Physiology*, **272**, F339–46.

113 Kimura, K., White, B.H. and Sidhu, A. (1995) Coupling of human D-1 dopamine receptors to different guanine nucleotide binding proteins. Evidence that D-1 dopamine receptors can couple to both G_s and G_o. *Journal of Biological Chemistry*, **270**, 14672–8.

114 Zhuang, X., Belluscio, L. and Hen, R. (2000) G_{olf}alpha mediates dopamine D_1 receptor signaling. *Journal of Neuroscience*, **20**, RC91.

115 Ho, M.K., Yung, L.Y., Chan, J.S., Chan, J.H., Wong, C.S. and Wong, Y.H. (2001) Galpha$_{14}$ links a variety of G_i- and G_s-coupled receptors to the stimulation of phospholipase C. *British Journal of Pharmacology*, **132**, 1431–40.

116 Albrecht, F.E., Xu, J., Moe, O.W., Hopfer, U., Simonds, W.F., Orlowski, J. and Jose, P.A. (2000) Regulation of NHE3 activity by G protein subunits in renal brush-border membranes. *American Journal of Physiology – Regulatory Integrative and Comparative Physiology*, **278**, R1064–73.

117 Uh, M., White, B.H. and Sidhu, A. (1998) Alteration of association of agonist-activated renal D_{1A} dopamine receptors with G proteins in proximal tubules of the spontaneously hypertensive rat. *Journal of Hypertension*, **16**, 1307–13.

118 Wang, Q., Jolly, J.P., Surmeier, J.D., Mullah, B.M., Lidow, M.S., Bergson, C.M. and Robishaw, J.D. (2001) Differential dependence of the D_1 and D_5 dopamine receptors on the G protein gamma 7 subunit for activation of adenylylcyclase. *Journal of Biological Chemistry*, **276**, 39386–93.

119 Felder, C.C., Blecher, M. and Jose, P.A. (1989) Dopamine-1-mediated stimulation of phospholipase C activity in rat renal cortical membranes. *Journal of Biological Chemistry*, **264**, 8739–45.

120 Felder, C.C., Jose, P.A. and Axelrod, J. (1989) The dopamine-1 agonist, SKF 82526, stimulates phospholipase-C activity independent of adenylate cyclase. *Journal of Pharmacology and Experimental Therapeutics*, **248**, 171–5.

121 Friedman, E., Jin, L.Q., Cai, G.P., Hollon, T.R., Drago, J., Sibley, D.R. and Wang, H.Y. (1997) D_1-like dopaminergic activation of phosphoinositide hydrolysis is independent of D_{1A} dopamine receptors: evidence from D_{1A} knockout mice. *Molecular Pharmacology*, **51**, 6–11.

122 Yu, P.Y., Asico, L.D., Eisner, G.M. and Jose, P.A. (1995) Differential regulation of renal phospholipase C isoforms by catecholamines. *Journal of Clinical Investigation*, **95**, 304–8.

123 Yu, P.Y., Eisner, G.M., Yamaguchi, I., Mouradian, M.M., Felder, R.A. and Jose, PA. (1996) Dopamine D_{1A} receptor regulation of phospholipase C isoform. *Journal of Biological Chemistry*, **271**, 19503–8.

124 Lezcano, N., Mrzljak, L., Eubanks, S., Levenson, R., Goldman-Rakic, P. and Bergson, C. (2000) Dual signaling regulated by calcyon, a D_1 dopamine receptor interacting protein. *Science*, **287**, 1660–4.

125 Mao, J., Yuan, H., Xie, W., Simon, M.I. and Wu, D. (1998) Specific involvement of G proteins in regulation of serum response factor-mediated gene transcription by different receptors. *Journal of Biological Chemistry*, **273**, 27118–23.

126 Sidhu, A., Kimura, K., Uh, M., White, B.H. and Patel, S. (1998) Multiple coupling of human D_5 dopamine receptors to guanine nucleotide binding proteins G_s and G_z. *Journal of Neurochemistry*, **70**, 2459–67.

127 Zheng, S., Yu, P., Zeng, C., Wang, Z., Yang, Z., Andrews, P.M., Felder, R.A. and Jose, P.A. (2003) Galpha12- and Galpha13-protein subunit linkage of D_5 dopamine receptors in the nephron. *Hypertension*, **41**, 604–10.

128 Felder, C.C., Campbell, T., Albrecht, F. and Jose, P.A. (1990) Dopamine inhibits Na^+-H^+ exchanger activity in renal BBMV by stimulation of adenylate cyclase. *American Journal of Physiology*, **259**, F297–303.

129 Horiuchi, A., Takeyasu, K., Mouradian, M.M., Jose, P.A. and Felder, R.A. (1993) D_{1A} dopamine receptor stimulation inhibits Na^+/K^+-ATPase activity through protein kinase A. *Molecular Pharmacology*, **43**, 281–5.

130 Meister, B., Fryckstedt, J., Schalling, M., Cortes, R., Hokfelt, T., Aperia, A., Hemmings, H.C. Jr, Nairn, A.C., Ehrlich, M. and Greengard, P. (1989) Dopamine- and cAMP-regulated phosphoprotein (DARPP-32) and dopamine DA1 agonist-sensitive Na^+,K^+-ATPase in renal tubule cells. *Proceedings of the National Academy of Sciences of the United States of America*, **86**, 8068–72.

131 Vyas, S.J., Eichberg, J. and Lokhandwala, M.F. (1992) Characterization of receptors involved in dopamine-induced activation of phospholipase-C in rat renal cortex. *Journal of Pharmacology and Experimental Therapeutics*, **260**, 134–9.

132 Kansra, V., Chen, C. and Lokhandwala, M.F. (1995) Dopamine causes stimulation of protein kinase C in rat renal proximal tubules by activating dopamine D_1 receptors. *European Journal of Pharmacology*, **289**, 391–4.

133 Yao, L.P., Li, X.X., Yu, P., Xu, J., Asico, L.D. and Jose, P.A. (1998) Dopamine D_1 receptor and protein kinase C isoforms in spontaneously hypertensive rats. *Hypertension*, **32**, 1049–53.

134 Efendiev, R., Bertorello, A.M. and Pedemonte, C.H. (1999) PKC-beta and PKC-zeta mediate opposing effects on proximal tubule Na^+,K^+-ATPase activity. *FEBS Letters*, **456**, 45–8.

135 Chibalin, A.V., Ogimoto, G., Pedemonte, C.H., Pressley, T.A., Katz, A.I., Feraille, E., Berggren, P.O. and Bertorello, A.M. (1999) Dopamine-induced endocytosis of Na^+,K^+-ATPase is initiated by phosphorylation of Ser-18 in the rat alpha subunit and is responsible for the decreased activity in epithelial cells. *Journal of Biological Chemistry*, **274**, 1920–7.

136 Cussac, D., Newman-Tancredi, A., Pasteau, V. and Millan, M.J. (1999) Human dopamine D_3 receptors mediate mitogen-activated protein kinase activation via a phosphatidylinositol 3-kinase and an atypical protein kinase C-dependent mechanism. *Molecular Pharmacology*, **56**, 1025–30.

137 Welsh, G.I., Hall, D.A., Warnes, A., Strange, P.G. and Proud, C.G. (1998) Activation of microtubule-associated protein kinase (Erk) and p70, S6 kinase by D_2 dopamine receptors. *Journal of Neurochemistry*, **70**, 2139–46.

138 Hussain, T. and Lokhandwala, M.F. (1996) Altered arachidonic acid metabolism contributes to the failure of dopamine to inhibit Na^+,K^+-ATPase in kidney of spontaneously hypertensive rats. *Clinical and Experimental Hypertension*, **18**, 963–74.

139 Nowicki, S., Chen, S.L., Aizman, O., Cheng, X.J., Li, D., Nowicki, C., Nairn, A., Greengard, P. and Aperia, A. (1997) 20-Hydroxyeicosa-tetraenoic acid (20 HETE) activates protein kinase C. Role in regulation of rat renal Na^+,K^+-ATPase. *Journal of Clinical Investigation*, **99**, 1224–30.

140 Amlal, H., Legoff, C., Vernimmen, C., Paillard, M. and Bichara, M. (1996) Na^+-$K^+(NH4^+)$-2Cl– cotransport in medullary thick ascending limb. control by PKA, PKC, and 20-HETE. *American Journal of Physiology*, **271**, C455–63.

141 Narkar, V., Hussain, T. and Lokhandwala, M. (2002) Role of tyrosine kinase and p44/42 MAPK in D_2-like receptor-mediated stimulation of Na^+, K^+-ATPase in kidney. *American Journal of Physiology–Renal Physiology*, **282**, F697–702.

142 Narkar, V.A., Hussain, T., Pedemonte, C. and Lokhandwala, M.F. (2001) Dopamine D_2 receptor activation causes mitogenesis via p44/42 mitogen-activated protein kinase in opossum kidney cells. *Journal of the American Society of Nephrology*, **12**, 1844–52.

143 Zhen, X., Uryu, K., Wang, H.Y. and Friedman, E. (1998) D_1 dopamine receptor agonists mediate activation of p38 mitogen-activated protein kinase and c-Jun amino-terminal kinase by a protein kinase A-dependent mechanism in SK-N-MC human neuroblastoma cells. *Molecular Pharmacology*, **54**, 453–8.

144 Goldberg, L.I. (1972) Cardiovascular and renal actions of dopamine: potential clinical applications. *Pharmacological Reviews*, **24**, 1–29.

145 Frederickson, E.D., Bradley, T. and Goldberg, L.I. (1985) Blockade of renal effects of dopamine in the dog by the DA1 antagonist SCH 23390. *American Journal of Physiology*, **249**, F236–40.

146 Rashed, S.M. and Songu-Mize, E. (1995) Regulation of Na^+-pump activity by dopamine in rat tail arteries. *European Journal of Pharmacology*, **284**, 289–97.

147 O'Connell, D.P., Ragsdale, N.V., Boyd, D.G., Felder, R.A. and Carey, R.M. (1997) Differential human renal tubular responses to dopamine type 1 receptor stimulation are determined by blood pressure status. *Hypertension*, **29**, 115–22.

148 Ventura, H.O., Messerli, F.H., Frohlich, E.D., Kobrin, I., Oigman, W., Dunn, F.G. and Carey, R.M. (1984) Immediate hemodynamic effects of a dopamine-receptor agonist (fenoldopam) in patients

with essential hypertension. *Circulation*, **69**, 1142–5.
149 Zeng, C., Wang, D., Asico, L.D., Welch, W.J., Wilcox, C.S., Hopfer, U., Eisner, G.M., Felder, R.A. and Jose, P.A. (2004) Aberrant D_1 and D_3 dopamine receptor transregulation in hypertension. *Hypertension*, **43**, 654–60.
150 Zeng, C., Wang, D., Yang, Z., Wang, Z., Asico, L.D., Wilcox, C.S., Eisner, G.M., Welch, W.J., Felder, R.A. and Jose, P.A. (2004) Dopamine D_1 receptor augmentation of D_3 receptor action in rat aortic or mesenteric vascular smooth muscles. *Hypertension*, **43**, 673–9.
151 Rump, L.C., Schwertfeger, E., Schuster, M.J., Schaible, U., Frankenschmidt, A. and Schollmeyer, P.J. (1993) Dopamine DA2-receptor activation inhibits noradrenaline release in human kidney slices. *Kidney International*, **43**, 197–204.
152 Seri, I., Eklof, A.C. and Aperia, A. (1987) Role of dopamine2-receptors in mediating renal vascular response to low dose dopamine infusion in the rat. *Acta Physiologica Scandinavica*, **130**, 563–9.
153 Szabo, B., Crass, D. and Starke, K. (1992) Effect of the dopamine D_2 receptor agonist quinpirole on renal sympathetic nerve activity and renal norepinephrine spillover in anesthetized rabbits. *Journal of Pharmacology and Experimental Therapeutics*, **263**, 806–15.
154 Bass, A.S. and Robie, N.W. (1984) Stereoselectivity of *S*- and *R*-sulpiride for pre- and postsynaptic dopamine receptors in the canine kidney. *Journal of Pharmacology and Experimental Therapeutics*, **229**, 67–71.
155 Lokhandwala, M.F., Tadepalli, A.S. and Jandhyala, B.S. (1979) Cardiovascular actions of bromocriptine: evidence for a neurogenic mechanism. *Journal of Pharmacology and Experimental Therapeutics*, **211**, 620–5.
156 Siragy, H.M., Felder, R.A., Peach, M.J. and Carey, R.M. (1992) Intrarenal DA2 dopamine receptor stimulation in the conscious dog. *American Journal of Physiology*, **262**, F932–8.
157 Reiner, N.E. and Thompson, W.L. (1979) Dopamine and saralasin antagonism of renal vasoconstriction and oliguria caused by amphotericin B in dogs. *Journal of Infectious Diseases*, **140**, 564–75.
158 ter Wee, P.M. and Donker, A.J. (1994) Pharmacologic manipulation of glomerular function. *Kidney International*, **45**, 417–24.
159 Barnett, R., Singhal, P.C., Scharschmidt, L.A. and Schlondorff, D. (1986) Dopamine attenuates the contractile response to angiotensin II in isolated rat glomeruli and cultured mesangial cells. *Circulation Research*, **59**, 529–33.
160 Amenta, F. (1997) Light microscope autoradiography of peripheral dopamine receptor subtypes. *Clinical and Experimental Hypertension*, **19**, 27–41.
161 Felder, R.A., Blecher, M., Eisner, G.M. and Jose, P.A. (1984) Cortical tubular and glomerular dopamine receptors in the rat kidney. *American Journal of Physiology*, **246**, F557–68.
162 Seri, I. and Aperia, A. (1988) Contribution of dopamine 2 receptors to dopamine-induced increase in glomerular filtration rate. *American Journal of Physiology*, **254**, F196–201.
163 Mendez, R.E., Lopez, R., Lopez, G., Marti, M.S. and Martinez-Maldonado, M. (1991) Effects of dopamine-receptor antagonists and renal denervation on amino acid-induced hyperfiltration. *American Journal of Physiology*, **261**, F70–5.
164 Muhlbauer, B., Spohr, F., Schmidt, R. and Osswald, H. (1997) Role of renal nerves and endogenous dopamine in amino acid-induced glomerular hyperfiltration. *American Journal of Physiology*, **273**, F144–9.
165 Siragy, H.M., Felder, R.A., Howell, N.L., Chevalier, R.L., Peach, M.J. and Carey, R.M. (1990) Evidence that dopamine-2 mechanisms control renal function. *American Journal of Physiology*, **259**, F793–800.
166 Bhat, S., Churchill, M., Churchill, P. and McDonald, F. (1986) Renal effects of SK & F 82526 in anesthetized rats. *Life Sciences*, **38**, 1565–71.
167 Felder, R.A., Felder, C.C., Eisner, G.M. and Jose, P.A. (1989) The dopamine

receptor in adult and maturing kidney. *American Journal of Physiology*, 257, F315–27.

168 Massry, S.G. and Kleeman, C.R. (1972) Calcium and magnesium excretion during acute rise in glomerular filtration rate. *Journal of Laboratory and Clinical Medicine*, 80, 654–64.

169 McGrath, B., Bode, K., Luxford, A., Howden, B. and Jablonski, P. (1985) Effects of dopamine on renal function in the rat isolated perfused kidney. *Clinical and Experimental Pharmacology and Physiology*, 12, 343–52.

170 Olsen, N.V., Hansen, J.M., Ladefoged, S.D., Fogh-Andersen, N. and Leyssac, P.P. (1990) Renal tubular reabsorption of sodium and water during infusion of low-dose dopamine in normal man. *Clinical Science*, 78, 503–7.

171 Jose, P.A., Eisner, G.M. and Robillard, J.E. (1987) Renal hemodynamics and natriuresis induced by the dopamine-1 agonist, SKF 82526. *American Journal of the Medical Sciences*, 294, 181–6.

172 Chen, C.J. and Lokhandwala, M.F. (1992) An impairment of renal tubular DA-1 receptor function as the causative factor for diminished natriuresis to volume expansion in spontaneously hypertensive rats. *Clinical and Experimental Hypertension Part A – Theory and Practice*, 14, 615–28.

173 Felder, R.A., Seikaly, M.G., Cody, P., Eisner, G.M. and Jose, P.A. (1990) Attenuated renal response to dopaminergic drugs in spontaneously hypertensive rats. *Hypertension*, 15, 560–9.

174 Hansell, P. and Fasching, A. (1991) The effect of dopamine receptor blockade on natriuresis is dependent on the degree of hypervolemia. *Kidney International*, 39, 253–8.

175 Wang, Z.Q., Siragy, H.M., Felder, R.A. and Carey, R.M. (1997) Intrarenal dopamine production and distribution in the rat. Physiological control of sodium excretion. *Hypertension*, 29, 228–34.

176 Sanada, H., Xu, J. and Watanabe, H. (2000) Differential expression and regulation of dopamine-1 (D-1) and dopamine-5 (D-5) receptor function in human kidney. *American Journal of Hypertension*, 13, 156A.

177 Yang, Z., Sibley, D.R. and Jose, P.A. (2004) D_5 dopamine receptor knockout mice and hypertension. *Journal of Receptor and Signal Transduction Research*, 24, 149–64.

178 Zeng, C., Yang, Z., Wang, Z., Jones, J., Wang, X., Altea, J., Mangrum, A.J., Hopfer, U., Sibley, D.R., Eisner, G.M., Felder, R.A. and Jose, P.A. (2005) Interaction of angiotensin II type 1 and D_5 dopamine receptors in renal proximal tubule cells. *Hypertension*, 45, 804–10.

179 Stier, C.T. Jr, Cowden, E.A. and Allison, M.E. (1982) Effects of bromocriptine on single nephron and whole-kidney function in rats. *Journal of Pharmacology and Experimental Therapeutics*, 220, 366–70.

180 Bennett, E.D., Tighe, D. and Wegg, W. (1982) Abolition, by dopamine blockade, of the natriuretic response produced by lower-body positive pressure. *Clinical Science*, 63, 361–6.

181 Eklof, A.C. (1997) The natriuretic response to a dopamine DA1 agonist requires endogenous activation of dopamine DA2 receptors. *Acta Physiologica Scandinavica*, 160, 311–14.

182 Fraga, S., Jose, P.A. and Soares-da-Silva, P. (2004) Involvement of G protein-coupled receptor kinase 4 and 6 in rapid desensitization of dopamine D_1 receptor in rat IEC-6 intestinal epithelial cells. *American Journal of Physiology – Regulatory Integrative and Comparative Physiology*, 287, R772–9.

183 Safsten, B. (1993) Duodenal bicarbonate secretion and mucosal protection. Neurohumoral influence and transport mechanisms. *Acta Physiologica Scandinavica Supplement*, 613, 1–43.

184 Vieira-Coelho, M.A. and Soares-da-Silva, P. (2001) Comparative study on sodium transport and Na^+,K^+-ATPase activity in Caco-2 and rat jejunal epithelial cells: effects of dopamine. *Life Sciences*, 69, 1969–81.

185 Donowitz, M., Elta, G., Battisti, L., Fogel, R. and Label-Schwartz, E. (1983) Effect of dopamine and bromocriptine on rat ileal and colonic transport. Stimulation of absorption and reversal of cholera toxin-

induced secretion. *Gastroenterology*, **84**, 516–23.
186 Kearney, P.M., Whelton, M., Reynolds, K., Muntner, P., Whelton, P.K. and He, J. (2005) Global burden of hypertension: analysis of worldwide data. *Lancet*, **365**, 217–23.
187 Clark, B.A., Rosa, R.M., Epstein, F.H., Young, J.B. and Landsberg, L. (1992) Altered dopaminergic responses in hypertension. *Hypertension*, **19**, 589–94.
188 Damasceno, A., Santos, A., Serrão, P., Caupers, P., Soares-da-Silva, P. and Polonia, J. (1999) Deficiency of renal dopaminergic-dependent natriuretic response to acute sodium load in black salt-sensitive subjects in contrast to salt-resistant subjects. *Journal of Hypertension*, **17**, 1995–2001.
189 Gill, J.R. Jr, Grossman, E. and Goldstein, D.S. (1991) High urinary dopa and low urinary dopamine-to-dopa ratio in salt-sensitive hypertension. *Hypertension*, **18**, 614–21.
190 Gill, J.R. Jr, Gullner, G., Lake, C.R., Lakatua, D.J. and Lan, G. (1988) Plasma and urinary catecholamines in salt-sensitive idiopathic hypertension. *Hypertension*, **11**, 312–19.
191 Lee, M.R. (1987) Dopamine, the kidney and essential hypertension studies with gludopa. *Clinical and Experimental Hypertension Part A – Theory and Practice*, **9**, 977–86.
192 Iimura, O. and Shimamoto, K. (1990) Suppressed dopaminergic activity and water-sodium handling in the kidneys at the prehypertensive stage of essential hypertension. *Journal of Autonomic Pharmacology*, **10** (Suppl. 1), S73–7.
193 Iimura, O., Shimamoto, K. and Ura, N. (1990) Dopaminergic activity and water-sodium handling in the kidneys of essential hypertensive subjects: is renal dopaminergic activity suppressed at the prehypertensive stage? *Journal of Cardiovascular Pharmacology*, **16** (Suppl. 7), S56–8.
194 Kuchel, O.G. and Kuchel, G.A. (1991) Peripheral dopamine in pathophysiology of hypertension. Interaction with aging and lifestyle. *Hypertension*, **18**, 709–21.
195 Saito, I., Itsuji, S., Takeshita, E., Kawabe, H., Nishino, M., Wainai, H., Hasegawa, C., Saruta, T., Nagano, S. and Sekihara, T. (1994) Increased urinary dopamine excretion in young patients with essential hypertension. *Clinical and Experimental Hypertension*, **16**, 29–39.
196 Saito, I., Takeshita, E., Saruta, T., Nagano, S. and Sekihara, T. (1986) Urinary dopamine excretion in normotensive subjects with or without family history of hypertension. *Journal of Hypertension*, **4**, 57–60.
197 Kuchel, O. and Shigetomi, S. (1992) Defective dopamine generation from dihydroxyphenylalanine in stable essential hypertensive patients. *Hypertension*, **19**, 634–8.
198 Pinho, M.J., Serrão, M.P., Gomes, P., Hopfer, U., Jose, P.A. and Soares-da-Silva, P. (2004) Over-expression of renal LAT1 and LAT2 and enhanced L-DOPA uptake in SHR immortalized renal proximal tubular cells. *Kidney International*, **66**, 216–26.
199 Albrecht, F.E., Drago, J., Felder, R.A., Printz, M.P., Eisner, G.M., Robillard, J.E., Sibley, D.R., Westphal, H.J. and Jose, P.A. (1996) Role of the D_{1A} dopamine receptor in the pathogenesis of genetic hypertension. *Journal of Clinical Investigation*, **97**, 2283–8.
200 Chen, C.J., Vyas, S.J., Eichberg, J. and Lokhandwala, M.F. (1992) Diminished phospholipase C activation by dopamine in spontaneously hypertensive rats. *Hypertension*, **19**, 102–8.
201 Kinoshita, S., Sidhu, A. and Felder, R.A. (1989) Defective dopamine-1 receptor adenylate cyclase coupling in the proximal convoluted tubule from the spontaneously hypertensive rat. *Journal of Clinical Investigation*, **84**, 1849–56.
202 Nishi, A., Eklof, A.C., Bertorello, A.M. and Aperia, A. (1993) Dopamine regulation of renal Na^+,K^+-ATPase activity is lacking in Dahl salt-sensitive rats. *Hypertension*, **21**, 767–71.
203 Chen, C., Beach, R.E. and Lokhandwala, M.F. (1993) Dopamine fails to inhibit renal tubular sodium pump in hypertensive rats. *Hypertension*, **21**, 364–72.

204 Felder, R.A., Sanada, H., Xu, J., Yu, P., Wang, Z., Watanabe, H., Asico, L.D., Wang, W., Zheng, S., Yamaguchi, I., Williams, S.M., Gainer, J., Brown, N.J., Hazen-Martin, D., Wong, L.J., Robillard, J.E., Carey, R.M., Eisner, G.M. and Jose, P.A. (2002) G protein-coupled receptor kinase 4 gene variants in human essential hypertension. *Proceedings of the National Academy of Sciences of the United States of America*, **99**, 3872–7.

205 Hussain, T. and Lokhandwala, M.F. (1997) Dopamine-1 receptor G-protein coupling and the involvement of phospholipase A2 in dopamine-1 receptor mediated cellular signaling mechanisms in the proximal tubules of SHR. *Clinical and Experimental Hypertension*, **19**, 131–40.

206 Li, X.X., Xu, J., Zheng, S., Albrecht, F.E., Robillard, J.E., Eisner, G.M. and Jose, P.A. (2001) D_1 dopamine receptor regulation of NHE3 during development in spontaneously hypertensive rats. *American Journal of Physiology – Regulatory Integrative and Comparative Physiology*, **280**, R1650–6.

207 Pedrosa, R., Gomes, P., Zeng, C., Hopfer, U., Jose, P.A. and Soares-da-Silva, P. (2004) Dopamine D_3 receptor-mediated inhibition of Na^+/H^+ exchanger activity in normotensive and spontaneously hypertensive rat proximal tubular epithelial cells. *British Journal of Pharmacology*, **142**, 1343–53.

208 Xu, J., Li, X.X., Albrecht, F.E., Hopfer, U., Carey, R.M. and Jose, P.A. (2000) Dopamine$_1$ receptor, G_{salpha}, and $Na^+–H^+$ exchanger interactions in the kidney in hypertension. *Hypertension*, **36**, 395–9.

209 Yu, P., Asico, L.D., Eisner, G.M., Hopfer, U., Felder, R.A. and Jose, P.A. (2000) Renal protein phosphatase 2A activity and spontaneous hypertension in rats. *Hypertension*, **36**, 1053–8.

210 Felder, R.A., Kinoshita, S., Ohbu, K., Mouradian, M.M., Sibley, D.R., Monsma, F.J. Jr, Minowa, T., Minowa, M.T., Canessa, L.M. and Jose, P.A. (1993) Organ specificity of the dopamine1 receptor/adenylyl cyclase coupling defect in spontaneously hypertensive rats. *American Journal of Physiology*, **264**, R726–32.

211 Jose, P.A., Eisner, G.M., Drago, J., Carey, R.M. and Felder, R.A. (1996) Dopamine receptor signaling defects in spontaneous hypertension. *American Journal of Hypertension*, **9**, 400–5.

212 Ohbu, K., Kaskel, F.J., Kinoshita, S. and Felder, R.A. (1995) Dopamine-1 receptors in the proximal convoluted tubule of Dahl rats: defective coupling to adenylate cyclase. *American Journal of Physiology*, **268**, R231–5.

213 Sela, S., White, B.H., Uh, M., Kimura, K., Patel, S. and Sidhu, A. (1997) Dysfunctional D_{1A} receptor-G-protein coupling in proximal tubules of spontaneously hypertensive rats is not due to abnormal G-proteins. *Journal of Hypertension*, **15**, 259–67.

214 Yu, P., Asico, L.D., Luo, Y., Andrews, P., Eisner, G.M., Hopfer, U., Felder, R.A. and Jose, P.A. (2006) D_1 dopamine receptor hyperphosphorylation in renal proximal tubules in hypertension. *Kidney International*, **70**, 1072–9.

215 Sanada, H., Jose, P.A., Hazen-Martin, D., Yu, P., Xu, J., Bruns, D.E., Phipps, J., Carey, R.M. and Felder, R.A. (1999) Dopamine-1 receptor coupling defect in renal proximal tubule cells in hypertension. *Hypertension*, **33**, 1036–42.

216 Sanada, H., Yatabe, J., Midorikawa, S., Katoh, T., Hashimoto, S., Watanabe, T., Xu, J., Luo, Y., Wang, X., Zeng, C., Armando, I., Felder, R.A. and Jose, P.A. (2006) Amelioration of genetic hypertension by suppression of renal G protein-coupled receptor kinase type 4 expression. *Hypertension*, **47**, 1131–9.

217 Beswick, R.A., Dorrance, A.M., Leite, R. and Webb, R.C. (2001) NADH/NADPH oxidase and enhanced superoxide production in the mineralocorticoid hypertensive rat. *Hypertension*, **38**, 1107–11.

218 White, B.H. and Sidhu, A. (1998) Increased oxidative stress in renal proximal tubules of the spontaneously hypertensive rat: a mechanism for defective dopamine D_{1A} receptor/G-protein coupling. *Journal of Hypertension*, **16**, 1659–65.

219 Wilcox, C.S. (2005) Oxidative stress and nitric oxide deficiency in the kidney: a critical link to hypertension? *American Journal of Physiology – Regulatory Integrative and Comparative Physiology*, **289**, R913–35.

220 Yang, Z., Asico, L.D., Yu, P., Wang, Z., Jones, J.E., Escano, C.S., Wang, X., Quinn, M.T., Sibley, D.R., Romero, G.G., Felder, R.A. and Jose, P.A. (2006) D_5 dopamine receptor regulation of reactive oxygen species production, NADPH oxidase, and blood pressure. *American Journal of Physiology – Regulatory Integrative and Comparative Physiology*, **290**, R96–104.

221 Landmesser, U., Cai, H., Dikalov, S., McCann, L., Hwang, J., Jo, H., Holland, S.M. and Harrison, D.G. (2002) Role of $p47^{phox}$ in vascular oxidative stress and hypertension caused by angiotensin II. *Hypertension*, **40**, 511–15.

222 Banday, A.A., Fazili, F.R. and Lokhandwala, M.F. (2007) Oxidative stress causes renal dopamine D_1 receptor dysfunction and hypertension via mechanisms that involve nuclear factor-kappaB and protein kinase C. *Journal of the American Society of Nephrology*, **18**, 1446–57.

223 Li, X.X., Bek, M., Asico, L.D., Yang, Z., Grandy, D.K., Goldstein, D.S., Rubinstein, M., Eisner, G.M. and Jose, P.A. (2001) Adrenergic and endothelin B receptor-dependent hypertension in dopamine receptor type-2 knockout mice. *Hypertension*, **38**, 303–8.

224 Ueda, A., Ozono, R., Oshima, T., Yano, A., Kambe, M., Teranishi, Y., Katsuki, M. and Chayama, K. (2003) Disruption of the type 2 dopamine receptor gene causes a sodium-dependent increase in blood pressure in mice. *American Journal of Hypertension*, **16**, 853–8.

225 Asico, L.D., Ladines, C., Fuchs, S., Accili, D., Carey, R.M., Semeraro, C., Pocchiari, F., Felder, R.A., Eisner, G.M. and Jose, P.A. (1998) Disruption of the dopamine D_3 receptor gene produces renin-dependent hypertension. *Journal of Clinical Investigation*, **102**, 493–8.

226 Bek, M.J., Wang, X., Asico, L.D., Jones, J.E., Zheng, S., Li, X., Eisner, G.M., Grandy, D.K., Carey, R.M., Soares-da-Silva, P. and Jose, P.A. (2006) Angiotensin-II type 1 receptor-mediated hypertension in D_4 dopamine receptor-deficient mice. *Hypertension*, **47**, 288–95.

227 Hollon, T.R., Bek, M.J., Lachowicz, J.E., Ariano, M.A., Mezey, E., Ramachandran, R., Wersinger, S.R., Soares-da-Silva, P., Liu, Z.F., Grinberg, A., Drago, J., Young, W.S. 3rd , Westphal, H., Jose, P.A. and Sibley, D.R. (2002) Mice lacking D_5 dopamine receptors have increased sympathetic tone and are hypertensive. *Journal of Neuroscience*, **22**, 10801–10.

228 Casson, I.F., Lee, M.R., Brownjohn, A.M., Parsons, F.M., Davison, A.M., Will, E.J. and Clayden, A.D. (1983) Failure of renal dopamine response to salt loading in chronic renal disease. *British Medical Journal (Clinical Research Edition)*, **286**, 503–6.

229 Itskovitz, H.D. and Wynn, N.C. (1985) Renal functional status and patterns of catecholamine excretion. *Journal of Clinical Hypertension*, **1**, 223–7.

230 Ferreira, A., Bettencourt, P., Dias, P., Pestana, M., Serrão, P., Soares-da-Silva, P. and Cerqueira-Gomes, M. (2001) Neurohormonal activation, the renal dopaminergic system and sodium handling in patients with severe heart failure under vasodilator therapy. *Clinical Science*, **100**, 557–66.

231 Ferreira, A., Bettencourt, P., Pimenta, J., Frioes, F., Pestana, M., Soares-da-Silva, P. and Cerqueira-Gomes, M. (2002) The renal dopaminergic system, neurohumoral activation, and sodium handling in heart failure. *American Heart Journal*, **143**, 391–7.

232 Dzau, V.J. (1987) Renal and circulatory mechanisms in congestive heart failure. *Kidney International*, **31**, 1402–15.

233 O'Hare, J.P., Roland, J.M., Walters, G. and Corrall, R.J. (1986) Impaired sodium excretion in response to Volume expansion induced by water immersion in insulin-dependent diabetes mellitus. *Clinical Science*, **71**, 403–9.

234 Segers, O., Anckaert, E., Gerlo, E., Dupont, A.G. and Somers, G. (1996) Dopamine-sodium relationship in type 2 diabetic patients. *Diabetes Research and Clinical Practice*, **34**, 89–98.

235 Stenvinkel, P., Saggar-Malik, A.K., Wahrenberg, H., Diczfalusy, U., Bolinder, J. and Alvestrand, A. (1991) Impaired intrarenal dopamine production following intravenous sodium chloride infusion in type 1 (insulin-dependent) diabetes mellitus. *Diabetologia*, **34**, 114–18.

236 Hussain, T., Becker, M., Beheray, S. and Lokhandwala, M.F. (2001) Dopamine fails to inhibit Na,H-exchanger in proximal tubules of obese Zucker rats. *Clinical and Experimental Hypertension*, **23**, 591–601.

237 Hussain, T., Beheray, S.A. and Lokhandwala, M.F. (1999) Defective dopamine receptor function in proximal tubules of obese zucker rats. *Hypertension*, **34**, 1091–6.

238 Marwaha, A., Banday, A.A. and Lokhandwala, M.F. (2004) Reduced renal dopamine D_1 receptor function in streptozotocin-induced diabetic rats. *American Journal of Physiology – Renal Physiology*, **286**, F451–7.

239 Banday, A.A., Marwaha, A., Tallam, L.S. and Lokhandwala, M.F. (2005) Tempol reduces oxidative stress, improves insulin sensitivity, decreases renal dopamine D_1 receptor hyperphosphorylation, and restores D_1 receptor-G-protein coupling and function in obese Zucker rats. *Diabetes*, **54**, 2219–26.

240 Marwaha, A. and Lokhandwala, M.F. (2006) Tempol reduces oxidative stress and restores renal dopamine D_1-like receptor–G protein coupling and function in hyperglycemic rats. *American Journal of Physiology – Renal Physiology*, **291**, F58–66.

241 Banday, A.A., Fazili, F.R. and Lokhandwala, M.F. (2007) Insulin causes renal dopamine D_1 receptor desensitization via GRK2-mediated receptor phosphorylation involving phosphatidylinositol 3-kinase and protein kinase C. *American Journal of Physiology – Renal Physiology*, **293**, F877–84.

242 Carranza, A., Karabatas, L., Barontini, M. and Armando, I. (2001) Decreased tubular uptake of L-3,4-dihydroxyphenylalanine in streptozotocin-induced diabetic rats. *Hormone Research*, **55**, 282–7.

243 Beheray, S., Kansra, V., Hussain, T. and Lokhandwala, M.F. (2000) Diminished natriuretic response to dopamine in old rats is due to an impaired D_1-like receptor-signaling pathway. *Kidney International*, **58**, 712–20.

244 Kansra, V., Hussain, T. and Lokhandwala, M.F. (1997) Alterations in dopamine DA1 receptor and G proteins in renal proximal tubules of old rats. *American Journal of Physiology*, **273**, F53–9.

245 Goldberg, L.I., McDonald, R.H. and Zimmerman, A.M. (1963) Sodium diuresis produced by dopamine in patients with congestive heart failure. *New England Journal of Medicine*, **269**, 1060–3.

246 McDonald, R.H., Goldberg, L.I., McNay, J.L. and Tuttle, E. (1964) Effects of dopamine in man. Augmentation of sodium excretion, glomerular filtration rate, and renal plasma flow. *Journal of Clinical Investigation*, **43**, 1116–24.

247 Robinson, T., Gariballa, S., Fancourt, G., Potter, J. and Castleden, M. (1994) The acute effects of a single dopamine infusion in elderly patients with congestive cardiac failure. *British Journal of Clinical Pharmacology*, **37**, 261–3.

248 Leier, C.V., Heban, P.T., Huss, P., Bush, C.A. and Lewis, R.P. (1978) Comparative systemic and regional hemodynamic effects of dopamine and dobutamine in patients with cardiomyopathic heart failure. *Circulation*, **58**, 466–75.

249 Patel, J.J., Mitha, A.S., Sareli, P. and de Vaal, J.B. (1993) Intravenous fenoldopam infusion in severe heart failure. *Cardiovascular Drugs and Therapy*, **7**, 97–101.

250 Clarck, K.L., Hilditch, A., Robertson, M.J. and Drew, G.M. (1991) Effects of dopamine DA1-receptor blockade and angiotensin converting enzyme inhibition on the renal actions of fenoldopam in the anaesthetized dog. *Journal of Hypertension*, **9**, 1143–50.

251 van Veldhuisen, D.J., Girbes, A.R., de Graeff, P.A. and Lie, K.I. (1992) Effects of dopaminergic agents on cardiac and renal function in normal man and in patients

with congestive heart failure. *International Journal of Cardiology*, **37**, 293–300.

252 Alvelos, M., Ferreira, A., Bettencourt, P., Pimenta, J., Azevedo, A., Serrão, P., Rocha-Goncalves, F. and Soares-da-Silva, P. (2005) Effect of saline load and metoclopramide on the renal dopaminergic system in patients with heart failure and healthy controls. *Journal of Cardiovascular Pharmacology*, **45**, 197–203.

253 Alvelos, M., Ferreira, A., Bettencourt, P., Serrão, P., Pestana, M., Cerqueira-Gomes, M. and Soares-Da-Silva, P. (2004) The effect of dietary sodium restriction on neurohumoral activity and renal dopaminergic response in patients with heart failure. *European Journal of Heart Failure*, **6**, 593–9.

254 Ferreira, A., Bettencourt, P., Pestana, M., Correia, F., Serrão, P., Martins, L., Cerqueira-Gomes, M. and Soares-da-Silva, P. (2001) Heart failure, aging, and renal synthesis of dopamine. *American Journal of Kidney Diseases*, **38**, 502–9.

255 Ferreira, A., Bettencourt, P., Pestana, M., Oliveira, N., Serrão, P., Maciel, M.J., Cerqueira-Gomes, M. and Soares-da-Silva, P. (2000) Renal synthesis of dopamine in asymptomatic post-infarction left ventricular systolic dysfunction. *Clinical Science*, **99**, 195–200.

256 Grossman, E., Shenkar, A., Peleg, E., Thaler, M. and Goldstein, D.S. (1999) Renal effects of L-DOPA in heart failure. *Journal of Cardiovascular Pharmacology*, **33**, 922–8.

257 Rafjer, S.I., Anton, A.H., Rossen, J.D. and Goldberg, L.I. (1984) Beneficial hemodynamic effects of oral levodopa in heart failure. *New England Journal of Medicine*, **310**, 1357–62.

258 Rafjer, S.I., Rossem, J.D., Nemanich, J.W., Douglas, F.L., Davis, F. and Osinski, J. (1987) Sustained hemodynamic improvement during long-term therapy with levodopa in heart failure: role of plasma catecholamines. *Journal of the American College of Cardiology*, **10**, 1286–93.

259 Carey, R.M., Siragy, H.M., Ragsdale, N.V., Howell, N.L., Felder, R.A., Peach, M.J. and Chevalier, R.L. (1990) Dopamine-1 and dopamine-2 mechanisms in the control of renal function. *American Journal of Hypertension*, **3**, 59S–63S.

260 Kapusta, D.R. and Robie, N.W. (1988) Plasma dopamine in regulation of canine renal blod flow. *American Journal of Physiology*, **255**, R379–87.

261 McNay, J.L., McDonald, R.H. and Goldberg, L.I. (1965) Direct vasodilation produced by dopamine in the dog. *Circulation Research*, **16**, 510–17.

262 Olsen, N.V., Lund, J., Jensen, P.F., Espersen, K., Kanstrup, I.L., Plum, I. and Leyssac, P.P. (1993) Dopamine, dobutamine, and dopexamine. A comparison of renal effects in unanesthetized human volunteers. *Anesthesiol*, **79**, 685–94.

263 Cottee, D.B. and Saul, W.P. (1996) Is renal dose dopamine protective or therapeutic? *Critical Care Clinics*, **12**, 687–95.

264 Denton, M.D., Chertow, G.M. and Brady, H.R. (1998) "Renal-dose" dopamine for the treatment of acute renal failure: scientific rationale, experimental studies, and clinical trials. *Kidney International*, **49**, 4–14.

265 Denton, R. and Slater, R. (1997) Just how benign is renal dopamine? *European Journal of Anaesthesiology*, **14**, 347–9.

266 Dugger, B. (1997) Peripheral dopamine infusions: are they worth the risk of infiltration? *Journal of Intravenous Nursing*, **20**, 95–9.

267 Galley, H.F. (2000) Renal-dose dopamine: will the message now get through? *Lancet*, **356**, 2112–13.

268 Girbes, A.R. and Smit, A.J. (1997) Use of dopamine in the ICU. Hope, hype, belief and facts. *Clinical and Experimental Hypertension*, **19**, 191–9.

269 Szerlip, H.M. (1991) Renal-dose dopamine: fact and fiction. *Annals of Internal Medicine*, **115**, 153–4.

270 Thompson, B.T. and Cockrill, B.A. (1994) Renal-dose dopamine: a siren song? *Lancet*, **344**, 7–8.

271 Vincent, J.L. (1994) Renal effects of dopamine: can our dream ever come true? *Critical Care Medicine*, **22**, 5–6.

272 Bellomo, R., Chapman, M., Finfer, S., Hickling, K. and Myburgh, J. (2000) Low-dose dopamine in patients with early renal dysfunction: a placebo-controlled randomised trial. Australian and New Zealand Intensive Care Society (ANZICS) Clinical Trials Group. *Lancet*, **356**, 2139–43.

273 Chertow, G., Sayegh, M. and Allgren, R.L. (1996) Is the administration of dopamine associated with adverse or favourable outcomes in acute renal failure? *American Journal of Medicine*, **101**, 49–53.

274 Marik, P.E. and Iglesias, J. (1999) Low-dose dopamine does not prevent acute renal failure in patients with septic shock and oliguria. NORASEPT II Study Investigators. *American Journal of Medicine*, **107**, 387–90.

275 Kellum, J.A. and Decker, J.M. (2001) Use of dopamine in acute renal failure: a meta-analysis. *Critical Care Medicine*, **29**, 1526–31.

276 Marik, P.E. (2002) Low-dose dopamine: a systematic review. *Intensive Care Medicine*, **28**, 877–83.

277 Kapoor, A., Sinha, N., Sharma, R.K., Shrivastava, S., Radhakrishnan, S., Goel, P.K. and Bajaj, R. (1996) Use of dopamine in prevention of contrast induced acute renal failure – a randomised study. *International Journal of Cardiology*, **53**, 233–6.

278 Abizaid, A.S., Clark, C.E., Mintz, G.S., Dosa, S., Popma, J.J., Pichard, A.D., Satler, L.F., Harvey, M., Kent, K.M. and Leon, M.B. (1999) Effects of dopamine and aminophylline on contrast-induced acute renal failure after coronary angioplasty in patients with preexisting renal insufficiency. *American Journal of Cardiology*, **83**, 260–3.

279 Gare, M., Haviv, Y.S., Ben Yehuda, A., Rubinger, D., Bdolah-Abram, T., Fuchs, S., Gat, O., Popovtzer, M.M., Gotsman, M.S. and Mosseri, M. (1999) The renal effect of low-dose dopamine in high-risk patients undergoing coronary angiography. *Journal of the American College of Cardiology*, **34**, 1682–8.

280 Hans, B., Hans, S.S., Mittal, V.K., Khan, T.A., Patel, N. and Dahn, M.S. (1990) Renal functional response to dopamine during and after arteriography in patients with chronic renal insufficiency. *Radiology*, **176**, 651–4.

281 Weisberg, L.S., Kurnik, P.B. and Kurnik, B.R. (1993) Dopamine and renal blood flow in radiocontrast-induced nephropathy in humans. *Renal Failure*, **15**, 61–8.

282 Yoshizumi, M., Kitagawa, T., Ozawa, Y., Tano, K., Tsuchiya, K., Houchi, H., Minakuchi, K. and Tamaki, T. (1998) Changes in plasma free and sulfoconjugated catecholamines during the perioperative period of cardiac surgery: effect of continuous infusion of dopamine. *Biological and Pharmaceutical Bulletin*, **21**, 787–91.

283 Olsen, N.V. (1998) Effects of dopamine on renal haemodynamics, tubular function, and sodium excretion in normal humans. *Danish Medical Bulletin*, **45**, 282–97.

284 MacGregor, D.A. and Smith, A.J. (2000) Pharmacokinetics of dopamine in healthy male subjects. *Anesthesiology*, **92**, 338–46.

285 Allen, E., Pettigrew, A. and Frank, D. (1998) Alterations in dopamine clearance and catechol-*O*-methyltransferase activity in children. *Critical Care Medicine*, **25**, 181–9.

286 Juste, R.N., Moran, L., Hooper, J. and Soni, N. (1998) Dopamine in critically ill patients. *Intensive Care Medicine*, **24**, 1217–20.

287 Schenarts, P.J., Sagraves, S.G., Bard, M.R., Toschlog, E.A., Goettler, C.E., Newell, M.A. and Rotondo, M.F. (2006) Low-dose dopamine: a physiologically based review. *Current Surgery*, **63**, 219–25.

288 Poinsot, O., Romand, J.A., Favre, H. and Suter, P.M. (1993) Fenoldopam improves renal hemodynamics impaired by positive end-expiratory pressure. *Anesthesiology*, **79**, 680–4.

289 Mathur, V.S., Swan, S.K., Lambrecht, L.J., Anjum, S., Fellmann, J., McGuire, D., Epstein, M. and Luther, R.R. (1999) The effects of fenoldopam, a selective dopamine receptor agonist, on systemic and renal hemodynamics in normotensive subjects. *Critical Care Medicine*, **27**, 1832–7.

290 Halpenny, M., Lekshmi, S., O'Donnell, A., O'Callaghan-Enright, S., Shorten,

G.D. (2001) Fenoldopam. renal and splanchniceffects in patients undergoing coronary artery bypass grafting. *Anaesthesia*, **56**, 953–60.

291 Brienza, N., Malcangi, V., Dalfino, L., Trerotoli, P., Guagliardi, C., Bortone, D., Faconda, G., Ribezzi, M., Ancona, G., Bruno, F. and Fiore, T. (2006) A comparison between fenoldopam and low-dose dopamine in early renal dysfunction of critically ill patients. *Critical Care Medicine*, **34**, 707–14.

292 Myles, P.S., Buckland, M.R., Schenk, N.J., Cannon, G.B., Langley, M., Davis, B.B. and Weeks, A.M. (1993) Effect of "renal-dose" dopamine on renal function following cardiac surgery. *Anaesth Intensive Care*, **21**, 56–61.

293 Davis, R.F., Lappas, D.G., Kirklin, J.K., Buckley, M.J. and Lowenstein, E. (1982) Acute oliguria after cardiopulmonary bypass: renal functional improvement with low-dose dopamine infusion. *Critical Care Medicine*, **10**, 852–6.

294 Hilberman, M., Maseda, J., Stinson, E.B., Derby, G.C., Spencer, R.J., Miller, D.C., Oyer, P.E. and Myers, B.D. (1984) The diuretic properties of dopamine in patients after open-heart operation. *Anesthesiology*, **61**, 489–94.

295 Argalious, M., Motta, P., Khandwala, F., Samuel, S., Koch, C.G., Gillinov, A.M., Yared, J.P., Starr, N.J. and Bashour, C.A. (2005) "Renal dose" dopamine is associated with the risk of new-onset atrial fibrillation after cardiac surgery. *Critical Care Medicine*, **33**, 1327–32.

296 Lassnigg, A., Donner, E., Grubhofer, G., Presterl, E., Druml, W. and Hiesmayr, M. (2000) Lack of renoprotective effects of dopamine and furosemide during cardiac surgery. *Journal of the American Society of Nephrology*, **11**, 97–104.

297 Tang, A.T., El-Gamel, A., Keevil, B., Yonan, N. and Deiraniya, A.K. (1999) The effect of "renal-dose" dopamine on renal tubular function following cardiac surgery: assessed by measuring retinol binding protein (RBP). *European Journal of Cardio-Thoracic Surgery*, **15**, 717–21. discussion 712–21.

298 Baldwin, L., Henderson, A. and Hickman, P. (1994) Effect of postoperative low-dose dopamine on renal function after elective major vascular surgery. *Annals of Internal Medicine*, **120**, 744–7.

299 Paul, M.D., Mazer, C.D. and Byrick, R.J. (1986) Influence of mannitol and dopamine on renal function during elective infrarenal aortic cross clamping in man. *American Journal of Nephrology*, **6**, 427–34.

300 Swygert, T.H., Roberts, L.C., Valek, T.R., Brajtbord, D., Brown, M.R., Gunning, T.C., Paulsen, A.W. and Ramsay, M.A. (1991) Effect of intraoperative low-dose dopamine on renal function in liver transplant recipients. *Anesthesiology*, **75**, 571–6.

301 Zhang, L.P., Li, M. and Yang, L. (2005) Effects of different vasopressors on hemodynamics in patients undergoing orthotopic liver transplantation. *Chinese Medical Journal – English Edition*, **118**, 1952–58.

302 Cabezuelo, J.B., Ramirez, P., Rios, A., Acosta, F., Torres, D., Sansano, T., Pons, J.A., Bru, M., Montoya, M., Bueno, F.S., Robles, R. and Parrilla, P. (2006) Risk factors of acute renal failure after liver transplantation. *Kidney International*, **69**, 1073–80.

303 Della Rocca, G., Pompei, L., Costa, M.G., Coccia, C., Scudeller, L., Di Marco, P., Monaco, S. and Pietropaoli, P. (2004) Fenoldopam mesylate and renal function in patients undergoing liver transplantation: a randomized, controlled pilot trial. *Anesthesia and Analgesia*, **99**, 1604–9. Table of contents.

304 Biancofiore, G., Della Rocca, G., Bindi, L., Romanelli, A., Esposito, M., Meacci, L., Urbani, L., Filipponi, F. and Mosca, F. (2004) Use of fenoldopam to control renal dysfunction early after liver transplantation. *Liver Transplantation*, **10**, 986–92.

305 Breslow, M.J., Miller, C.F., Parker, S.D., Walman, A.T. and Traystman, R.J. (1987) Effect of vasopressors on organ blood flow during endotoxin shock in pigs. *American Journal of Physiology*, **252**, H291–300.

306 Bersten, A.D. and Rutten, A.J. (1995) Renovascular interaction of epinephrine,

dopamine, and intraperitoneal sepsis. *Critical Care Medicine*, **23**, 537–44.
307 Hiltebrand, L.B., Krejci, V. and Sigurdsson, G.H. (2004) Effects of dopamine, dobutamine, and dopexamine on microcirculatory blood flow in the gastrointestinal tract during sepsis and anesthesia. *Anesthesiology*, **100**, 1188–97.
308 Oberbeck, R., Schmitz, D., Wilsenack, K., Schuler, M., Husain, B., Schedlowski, M. and Exton, M.S. (2006) Dopamine affects cellular immune functions during polymicrobial sepsis. *Intensive Care Medicine*, **32**, 731–9.
309 Sakr, Y., Reinhart, K., Vincent, J.L., Sprung, C.L., Moreno, R., Ranieri, V. M., De Backer, D. and Payen, D. (2006) Does dopamine administration in shock influence outcome? Results of the Sepsis Occurrence in Acutely Ill Patients (SOAP) Study. *Critical Care Medicine*, **34**, 589–97.
310 Morelli, A., Ricci, Z., Bellomo, R., Ronco, C., Rocco, M., Conti, G., De Gaetano, A., Picchini, U., Orecchioni, A., Portieri, M., Coluzzi, F., Porzi, P., Serio, P., Bruno, A. and Pietropaoli, P. (2005) Prophylactic fenoldopam for renal protection in sepsis: a randomized, double-blind, placebo-controlled pilot trial. *Critical Care Medicine*, **33**, 2451–6.
311 Santambrogio, L., Lipartiti, M., Bruni, A. and Dal Toso, R. (1993) Dopamine receptors on human T- and B-lymphocytes. *Journal of Neuroimmunology*, **45**, 113–19.
312 Bergquist, J., Josefsson, E., Tarkowski, A., Ekman, R. and Ewing, A. (1997) Measurement of catecholamine-mediated apoptosis of immunocompetent cells by capillary electrophoresis. *Electrophoresis*, **18**, 1760–6.
313 Devins, S.S., Miller, A., Herndon, B.L., O'Toole, L. and Reisz, G. (1992) Effects of dopamine on T-lymphocyte proliferative responses and serum prolactin concentrations in critically ill patients. *Critical Care Medicine*, **20**, 1644–9.
314 Josefsson, E., Bergquist, J., Ekman, R. and Tarkowski, A. (1996) Catecholamines are synthesized by mouse lymphocytes and regulate function of these cells by induction of apoptosis. *Immunology*, **88**, 140–6.
315 Kouassi, E., Li, Y.S., Boukhris, W., Millet, I. and Revillard, J.P. (1988) Opposite effects of catecholamines dopamine and norepinephrine on murine polyclonal B-cell activation. *Immunopharmacology*, **16**, 125–37.
316 Van den Berghe, G. and de Zegher, F. (1996) Anterior pituitary function during critical illness and dopamine treatment. *Critical Care Medicine*, **24**, 1580–90.
317 Russell, D.H., Kibler, R., Matrisian, L., Larson, D.F., Poulos, B. and Magun, B.E. (1985) Prolactin receptors on human T and B lymphocytes: antagonism of prolactin binding by cyclosporine. *Journal of Immunology*, **134**, 3027–31.
318 Rohr, O., Sawaya, B.E., Lecestre, D., Aunis, D. and Schaeffer, E. (1999) Dopamine stimulates expression of the human immunodeficiency virus type1 via NFkappaB in cells of the immune system. *Nucleic Acids Research*, **27**, 3291–9.

12
Histamine

Izabela Rozenberg, Felix C. Tanner, and Thomas F. Lüscher

12.1
Introduction

Due to its numerous effects and possible therapeutic potential, histamine is one of the most intensely studied mediators in medicine. Histamine is an endogenous amine produced in a one-step reaction from L-histidine. It regulates vasoreactivity, coagulation as well as immunological responses. An increased production of histamine has been noted during allergic action, within atherosclerotic plaques, in patients with stable coronary artery disease and in patients with acute coronary syndromes [1–3]. Most of the histamine is released by activated mast cells and basophils, but also by macrophages, lymphocytes, platelets and other cells [3–8]. Mast cells store histamine in secretory vesicles and only release it upon stimulation with specific ligands (i.e. IgE) activating surface receptors. Degranulation of mast cells causes fast, but transient increases in histamine concentration, resulting in hypersensitivity, vascular permeability, vasodilation or vasoconstriction [9].

Recently, it has been demonstrated that the histamine-producing enzyme L-histidine decarboxylase (HDC) is expressed in activated macrophages from human atherosclerotic plaques. Since macrophages, unlike mast cells, do not contain secretory vesicles, macrophage-derived histamine is not stored, but rather produced *de novo* [10, 11]. Therefore, macrophage stimulation does not cause sudden increases in local histamine concentration, but rather provides persistently increased histamine tissue levels.

The diverse effects of histamine are due to differential expression and regulation of the four histamine receptors (see Section 12.3).

12.2
Biochemistry

12.2.1
Synthesis

Histamine (2-[4-imidazole]-ethylamine) is an endogenous amine, the product of L-histidine decarboxylation. Its synthesis is controlled primarily by the availability of L-histidine and the activity of HDC [12], the rate-limiting enzyme controlling histamine biosynthesis (Figure 12.1). Activity of HDC is regulated mainly at the transcriptional level, at least in cultured enterochromaffin-like cells [13, 14]. The enzyme has been detected in various cells of the immune system, gastrointestinal and bronchial endocrine cells, neuroendocrine cells, as well as some tumor cells [15]. Histamine may be released in response to various immunological (IgE, cytokines) and nonimmunological (compound 48/80, calcium ionophore, opioids) stimuli.

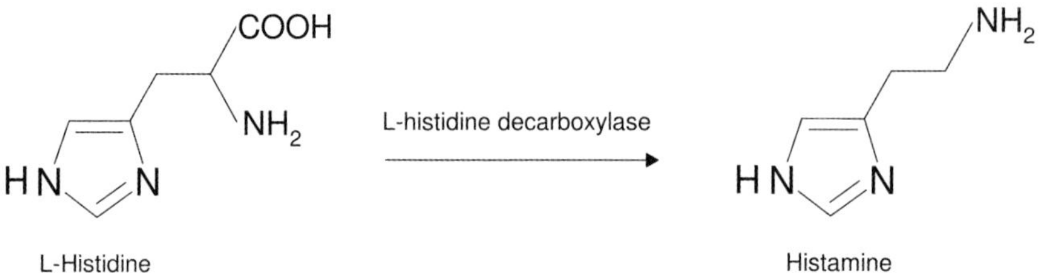

Figure 12.1 Biosynthesis of histamine from L-histidine.

12.2.2
Degradation

In order to avoid excessive inflammatory reactions, histamine's action has to be highly regulated by limited production but also degradation. Histamine inactivation occurs either through metabolization into inactive compounds and/or cellular uptake. Indeed, many cells express high-affinity receptors specific for monoamines like histamine. One recently characterized receptor is the organic cation transporter-2. On the other hand, histamine is inactivated is by two catabolic pathways: (i) methylation by histamine N-metyltransferase (HNMT) and (ii) oxidative deamination by diamine oxidase. Since histamine cannot easily enter cells and HNMT seems to be localized primarily in the cytoplasm, there are two hypotheses of histamine metabolization – the transporter hypothesis and the membrane hypothesis. The transporter hypothesis assumes that histamine is transported via special histamine transporters into the intracellular space, where it is metabolized by HNMT. According to the second hypothesis, HNMT upon stimulation is translocated to the plasma membrane, where it has easy access to histamine [16].

12.3
Receptors

The pleiotropic and tissue-specific effects of histamine are mediated via four G-coupled protein membrane receptors (histamine H_1–H_4 receptors; Table 12.1). Effects of histamine upon histamine receptor ligation are very complex and often contradictory; depending on the pattern of receptors it activates [17].

Histamine H_1 receptor is coupled to the $G_{q/11}$-protein and the phosphoinositol hydrolysis pathway resulting in enhancement of inflammatory processes. Its activation induces signs of allergic reactions, such as bronchoconstriction, increased vascular permeability and vasodilation due to nitric oxide (NO) release from vascular endothelial cells (ECs). Mice lacking the histamine H_1 receptor develop normally, but exhibit behavioral changes, such as reduced exploratory behavior and altered circadian rhythm as well as increased daily food intake [18]. These observations confirm the importance of histamine H_1 receptor signaling in the brain. Histamine H_1 receptor deficiency also attenuates airway allergic inflammation, which is associated with decreased levels of T helper (T_h) 2 cells, cytokine production and circulating IgE in bronchoalveolar lavage fluid [19].

The histamine H_2 receptor is coupled to the $G\alpha_s$-protein and cAMP signaling. Acting via the H_2 receptor, histamine regulates inflammatory responses, angiogenesis and gastric acid secretion. Homozygous mutant mice lacking histamine H_2 receptor are viable and, surprisingly, exhibit normal basal gastric pH. This observation is clearly distinct from the elevated gastric pH observed in mice treated with H_2 receptor antagonists [20].

Table 12.1 Expression and function of histamine receptors.

Type of receptor	Expression	Molecular pathway involved	Function
Histamine H_1 receptor	ECs, SMCs, leukocytes, nerve cells	phospholipase C via $G_{q/11}$ family of G-proteins	allergy and inflammation
Histamine H_2 receptor	ECs, VSMCs, leukocytes, gastric parietal cells, nerve cells	couple to adelylyl cyclase via GTP-binding protein G_s	gastric acid secretion, inflammation
Histamine H_3 receptor	neurons, leukocytes	inhibition of cAMP via $G_{i/o}$	neurotransmission
Histamine H_4 receptor	leukocytes, nerve cells	inhibition of cAMP via $G_{i/os}$	allergy and inflammation

The histamine H_3 receptor has been detected in the central as well as peripheral nervous system and negatively regulates the release of histamine and other neurotransmitters as well. Mice treated with histamine H_3 receptor antagonists exhibit behavioral changes, such as enhanced spatial learning and reduced anxiety. A study with H_3 receptor knockout animals as well as with H_3 receptor agonist-treated animals demonstrated a role of H_3 receptor in energy homeostasis, and suggested a therapeutic potential for H_3 receptor ligands in the treatment of obesity and diabetes mellitus [21]. The recently discovered histamine H_4 receptor involves $G_{i/os}$-protein signaling and enhances intracellular Ca^{2+} levels. Activation of histamine H_4 receptor promotes the accumulation of inflammatory cells, primarily eosinophils and mast cells, and triggers proinflammatory responses. H_4 receptor-deficient mice as well as mice treated with H_4 receptor antagonists exhibit diminished allergic pulmonary inflammation, decreased infiltration of lung eosinophils and lymphocytes, and decreased T_h2 responses [22].

12.4
Vasomotion

The role of histamine in the regulation of vasomotion is complex, and varies depending on its concentration and the receptor it activates. Both the contracting and the relaxing responses in bronchial arteries are significantly inhibited by mepyramine, suggesting that the H_1 receptor is involved [23]. In contrast, another study demonstrated that H_2 receptor activation is crucial for NO production and vascular relaxation in human arteries and veins [24]. Hence, the involvement of histamine receptors in vasomotion regulation is complex. Evidence indicates that the vasodilator response is related to the release of NO from ECs, and that the vasoconstrictor effect is resulting from a direct action of histamine on vascular smooth muscle cells (VSMCs). Indeed, histamine-dependent arterial dilation, but not constriction, is blunted in mice lacking endothelial NO synthase (eNOS, also classified as NOSIII) [25]. However, contraction of VSMCs triggered by histamine can be at least in part endothelium-dependent due to the action of endothelium-derived contracting factor (EDCF). It has been proposed that stimulation with low concentrations of histamine results in relaxation of human vessels, whereas higher concentrations exert an opposite effect [26]. Importantly, the action of histamine differs depending on the vascular bed involved. In mammary artery, it induces endothelium-dependent relaxation, whereas it promotes contraction in saphenous vein (Figure 12.2) [27].

12.4.1
EDRF

In the past, VSMCs were considered to be the sole regulator of vascular tone. VSMCs, the most abundant component of the vessel wall, are indeed known to contain large numbers of contractile proteins, determining the resistance of vessels

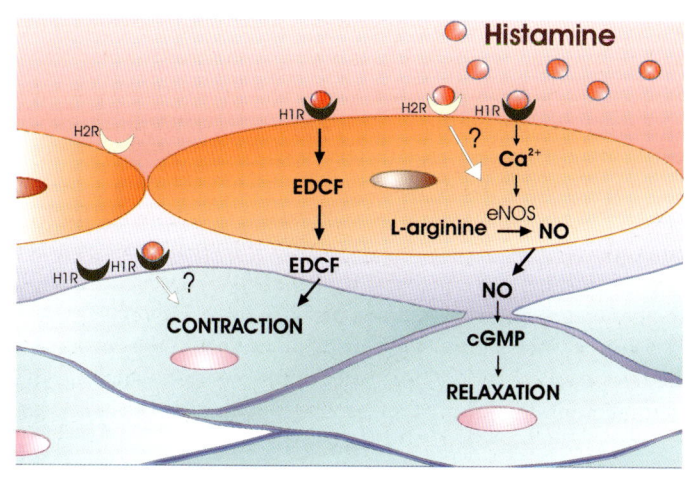

Figure 12.2 Dual role of histamine in regulation of vascular tone. Histamine acting via histamine H$_1$ receptor (H1R) increases eNOS activity and enhances NO production in ECs through increasing intracellular Ca^{2+} levels. Release of NO from EC causes relaxation of VSMCs. Concomitantly, histamine-triggered release of EDCF from ECs may cause contraction of VSMCs.

and therefore regulating blood flow. However, in 1980, Furchgott and Zawadzki observed that anatomical integrity of the EC layer is crucial for acetylcholine-mediated relaxation of VSMCs [28]. This observation proved the importance of the endothelium as a modulator of the vasoconstriction-vasodilatation balance. The compound discovered in 1980 was initially, based on its function, named endothelium-derived relaxing factor (EDRF) and later characterized as NO [29]. Since this discovery, various endothelium-derived vasodilators (prostacyclin) and vasoconstrictors (endothelin-1, angiotensin II, prostaglandins) have been identified as well. Together with NO, these compounds contribute to the regulation of vascular tone [30]. Histamine plays a role in endothelial NO release, thereby promoting vasodilation. NO dilates blood vessels by stimulating soluble guanylyl cyclase and increasing cGMP in VSMCs [31, 32]. NO is synthesized from L-arginine by at least three distinct NOS isoforms. Production of NO in ECs is regulated by the constitutively expressed eNOS. In response to stimuli such as shear stress, acetylcholine or histamine, eNOS activity and NO production in ECs increases.

In addition to its vasoactive properties, NO also prevents platelet aggregation and leukocyte adhesion as well as VSMC proliferation, which may be relevant in the development of cardiovascular diseases like atherosclerosis [33–35]. The importance of NO release by ECs in blood pressure regulation has been confirmed with eNOS-deficient animals, which develop hypertension [36]. It has also been reported, that NO production is diminished in patients with elevated blood pressure when compared to normotensive individuals [37].

12.4.1.1 Signaling Role of Ca^{2+}

Production of NO is regulated primarily by modulation of eNOS activity in a Ca^{2+}- and calmodulin-dependent manner. After synthesis eNOS is translocated to specialized signal-transducing plasmalemmal compartments called caveolae. Caveolin, the main structural component of caveolae, interacts with eNOS leading to enzyme inhibition. It has been demonstrated in human ECs that the inhibitory action of calveolin on eNOS activity is modulated by Ca^{2+}/calmodulin. Addition of calmodulin indeed disrupts the eNOS–calveolin complex in a Ca^{2+}-dependent manner [38].

H_1 receptor stimulation results in activation of phospholipase C and the production of inositol-1,4,5-triphosphate from phosphatidylinositol (4,5)-bisphosphate, which triggers the release of Ca^{2+}. Accordingly, histamine increases eNOS activity and enhances NO production at least in part through increase of intracellular Ca^{2+} [39].

12.4.1.2 Activation of eNOS

Phosphorylation represents a crucial post-translational regulatory mechanism of enzyme activation. In the case of eNOS, phosphorylation on serine or tyrosine residues is probably involved in intracellular enzyme translocation followed by its activation. Interestingly, histamine mediates eNOS phosphorylation at serine 1177, independent of Ca^{2+} signaling. It has been shown that inhibition of protein kinase A as well as 5′-AMP-activated protein kinase strongly inhibits phosphorylation and activation of eNOS triggered by histamine stimulation [40].

12.4.1.3 Transcriptional Regulation of eNOS

Regulation of gene expression is a very common way to modulate biological pathways. Recently, it has been shown that histamine does not only play a role in activation of eNOS, but also affects the expression level of the enzyme. Indeed, histamine upregulates gene expression of eNOS in cultured human ECs in a Ca^{2+}-dependent manner. The enhanced expression of eNOS could be prevented by mepyramine, a selective antagonist of histamine H_1 receptor, but not by H_2 receptor and H_3 receptor antagonists [41].

12.4.2
EDCF

ECs cannot only induce relaxation, but also contraction of underlying VSMCs. The molecule responsible for endothelium-dependent contraction was initially called EDCF. Indeed, various mediators produced by the endothelium such as endothelin-1 or thromboxane A_2 were later found to induce vascular constriction. It has been demonstrated that histamine may trigger either constriction or relaxation depending on its concentration and the animal model used. The histamine-mediated constriction of VSMCs was blocked by histamine H_1 receptor inhibitors, but increased after application of histamine H_2 receptor inhibitors in a canine model [42]. Similarly, histamine induced strong constriction in rabbit basilar artery [43].

These results suggest that histamine, activating the H_1 receptor, plays a role in endothelium-derived contraction, presumably via the generation of EDCF, which can be modulated by the concomitant release of endothelium-derived NO via activation of the H_2 receptor.

12.5
Thrombosis

Coronary atherosclerosis complicated by plaque rupture and thrombosis is primarily responsible for the development of acute coronary syndromes. Under normal conditions, however, the endothelial layer provides a barrier between the blood and the vessel wall; it maintains blood fluidity and is crucial for regulating the interaction between blood and solid tissue components. Hence, ECs are able to prevent procoagulant and thrombotic events. They synthesize negatively charged heparin sulfate proteoglycans (glycocalix), maintaining the vascular wall in an antithrombotic state by accelerating the inhibition of factor Xa and thrombin via antithrombin III. Endothelial cells are also the main source of tissue factor pathway inhibitor (TFPI). Binding of TFPI to activated factor Xa leads to the formation of an inactive tissue factor–VIIa–Xa complex. Furthermore, the endothelium expresses the receptor for thrombomodulin, which modifies thrombin specificity and accelerates thrombin mediated conversion of precursor protein C to activated protein C. Finally, the sequestration of two prothrombotic molecules – tissue factor as well as von Willebrand factor (vWF) – accounts for the antithrombotic properties of ECs. Under certain conditions like inflammatory stimulation, however, ECs participate in activation of coagulation. Mast cells, the main source of histamine, tend to redistribute and accumulate at sites of thrombosis. Histamine indeed promotes a series of prothrombotic events like activation of ECs, induction of tissue factor expression and stimulation of vWF release. Furthermore, histamine is released by platelets during thrombotic events, and exogenous histamine dose-dependently enhances platelet aggregation induced by various stimuli. Conversely, it has been demonstrated that mast cell-deficient mice are hyperresponsive to thrombogenic stimuli, probably due to lack of mast cell-derived heparin [44].

12.5.1
Tissue Factor Expression

Tissue factor is a 263-residue membrane-bound glycoprotein, a member of the type II cytokine receptor family, controlling initiation of the extrinsic coagulation cascade, since it is the cellular receptor for activated factor VII. The endothelial layer provides a barrier between the blood and the tissue factor-rich vessel wall, which prevents thrombotic events. However, upon stimulation with pro-inflammatory agents like tumor necrosis factor (TNF)-α, CD40 ligand, interleukin (IL)-1β or histamine, ECs express tissue factor, resulting in activation of the coagulation cascade. Tissue factor, by inducing coagulation, seems to be involved in the pathogenesis of atherosclerosis, and the initiation and propagation of acute coronary syndromes as well [45]. Elevated tissue factor level and activity have indeed been detected in patients with unstable angina. Moreover, tissue factor is highly expressed in atherosclerotic plaques in numerous cell types including ECs, smooth muscle cells and monocytes/macrophages [46, 47]. However, additional studies are required to elucidate the role of tissue factor in the development of atherosclerosis *in vivo*.

Recently, it has been observed that histamine induces tissue factor expression in a concentration-dependent manner on both human ECs and VSMCs [48]. Histamine-induced tissue factor expression is mediated by activation of the H_1 receptor, since H_1 receptor antagonists prevent this effect. Although mRNA stability, intracellular storage and localization of tissue factor may significantly modulate its activity, tissue factor induction in response to inflammatory stimuli is predominantly regulated at the transcriptional level. Functional studies indicate that Sp1 controls basal tissue factor gene expression, while a promoter (−227 to −172) containing two AP-1 sites and an NF-κB site mediates induction of tissue factor mRNA in monocytes and ECs [49]. Accordingly, tissue factor is activated by several signal transduction pathways including mitogen-activated protein kinases (MAPKs) and protein kinase C. Consistent with these data, p38, extracellular signal-regulated kinase and c-Jun N-terminal kinase inhibitors attenuate histamine-induced tissue factor expression, suggesting a crucial role of MAPK in this process (Figure 12.3) [48].

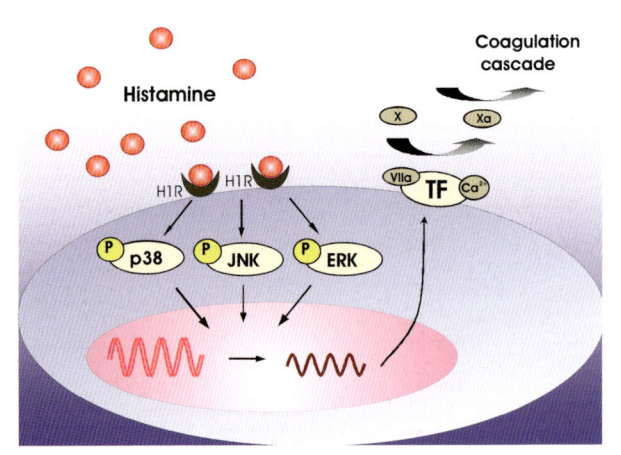

Figure 12.3 Histamine and thrombosis. Histamine induces tissue factor expression on both human ECs and VSMCs via activation of MAPKs. This histamine effect is mediated by histamine H_1 receptor (H1R). Tissue factor expression leads to coagulation cascade activation and eventually thrombus formation. ERK, extracellular signal-regulated kinase; JNK, c-Jun N-terminal kinase; TF, tissue factor.

12.5.2
Weibel–Palade Bodies

ECS contain rod-shaped storage vesicles called Weibel–Palade bodies (WPBs) harboring preformed vWF multimers and P-selectin. Stimulation of ECS with molecules raising intracellular Ca^{2+} levels like histamine or thrombin result in exocytosis of the WPB content to the EC surface (Figure 12.4). Histamine H_1 receptor, but not H_2 receptor antagonists abolish this effect [50].

Both vWF and P-selectin are known to play a crucial role in thrombus formation. vWF is a large glycoprotein, stabilizing coagulation factor VIII and participating in the binding of activated platelets to exposed collagen through receptors such as glycoprotein Ib on the platelet surface. Deficiency or mutation of vWF causes von Willebrand disease, which is characterized by an increased bleeding time [51]. In contrast, P-selectin is a member of the selectin adhesion molecule family, which mediates initial rolling and attachment of leukocytes to the activated endothelium, promoting the development of inflammation [52]. Recent studies, however, reveal additional possible role of P-selectin in thrombus formation. Binding of P-selectin to P-selectin glycoprotein ligand-1 indeed mediates the interaction of platelets with the endothelium as well as interaction of platelets with leukocytes and other platelets, thereby enhancing thrombotic events. What is more, P-selectin has been reported to induce the expression of tissue factor on cultured human monocytes [53]. Consistent with these observations, P-selectin-deficient mice exhibit an abnormally prolonged bleeding time and unusual thrombus geometry during *ex vivo* thrombus formation [54]. In contrast, transgenic animals having high soluble P-selectin level exhibit a procoagulant state [55]. In humans, elevated P-selectin

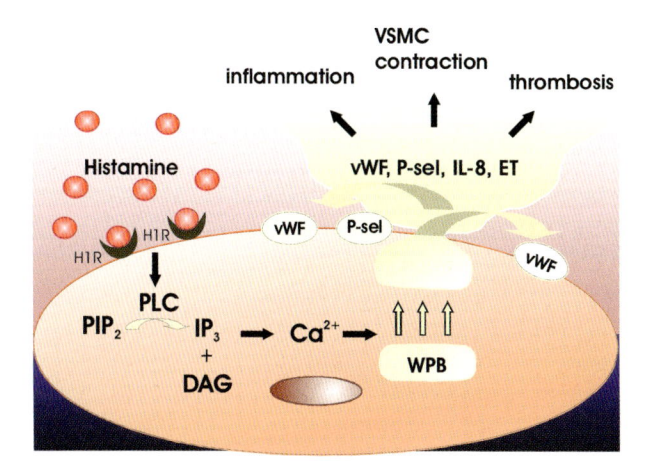

Figure 12.4 Histamine induces WPB release. Histamine induces exocytosis of endothelial WPBs in a Ca^{2+}-dependent manner. Released molecules like vWF, P-selectin (P-sel), IL-8 and endothelins (ET) modulate several important physiologic processes like inflammation, homeostasis and vasomotion. DAG, diacylglycerol; H1R, histamine H_1 receptor; IP_3, inositol-1,4,5-triphosphate; PIP_2, phosphatidylinositol (4,5)-bisphosphate.

level is associated with prothrombotic disorders such as stroke or peripheral artery disease [56].

Hence, histamine induces a dose-dependent secretion of both vWF and P-selectin from WPBs, by a Ca^{2+}-dependent mechanism, which can be blocked by mepyramine, a histamine H_1 receptor antagonist. These observations strongly suggest a crucial role of histamine in thrombus formation by stimulating the release of WPBs.

12.5.3
Platelet Aggregation

When vessel integrity is compromised, activated platelets bind to exposed collagen and vWF to form a hemostatic plug stabilized by fibrin. Similar to mast cells and other inflammatory cells, platelets contain intracellular granules containing numerous inflammatory and thrombogenic mediators like histamine, platelet factor 4, platelet-derived growth factor, ADP and reactive oxygen species. Exposure of platelets to thrombin indeed results in aggregation and histamine release. The presence of histamine receptors on the platelet membrane, high activity of HDC and the ability to take up preformed histamine suggest that histamine plays a role in platelet activation. Indeed, histamine enhances aggregation of isolated human platelets induced by a variety of agonists. The effect of histamine on platelet aggregation is abolished by H_1 receptor, but not H_2 receptor antagonists [57, 58]. On the other hand, histamine enhances eNOS activation resulting in endothelial NO production, which inhibits platelet activation and represents an important negative feedback mechanism [33, 35]. In conclusion, histamine plays a role not only in initiation of coagulation by stimulating tissue factor expression as well as WPB release, but also regulates platelet aggregation during thrombus formation.

12.6
Inflammation

Histamine is a well-known mediator of inflammation and allergic responses. Released from mast cells or macrophages during inflammatory processes, it can affect immune responses at several levels. Most of the proinflammatory actions of histamine are mediated via the H_1 receptor and activation of MAPKs as well as NF-κB [59]. In contrast, the immunoregulatory effects of histamine are primarily mediated via the H_2 receptor [60]. Histamine modulates progression of inflammation at many levels. It increases vascular permeability, enhances adhesion molecule expression, probably plays a role in activation of monocytes/macrophages, modulates the T_h1 versus T_h2 response, and the production of pro- as well as anti-inflammatory cytokines (Figure 12.5) [15].

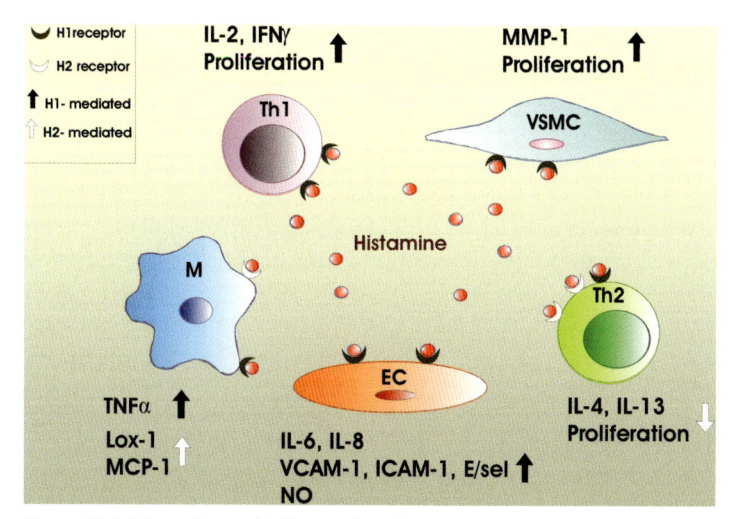

Figure 12.5 Histamine and inflammation. Histamine modulates expression of several proinflammatory molecules by different types of cells – monocyte/macrophages (M), T_h1, T_h2, ECs and VSMCs. E/sel, E-selectin; IFN, interferon; MMP, matrix metalloproteinase.

12.6.1
Vascular Permeability

An increased permeability leading to tissue edema and accumulation of leukocytes is a hallmark of inflammation. Similarly, early atherosclerosis is characterized by the accumulation of low-density lipoproteins in the arterial intima, occurring secondary to increased EC permeability. Histamine, acting via both H_1 receptor and H_2 receptor, is one of the most important regulators of this process. It has been observed that histamine causes transient phosphorylation of myosin light chains, actin cytoskeleton rearrangement and intracellular gap formation in ECs occurring in a Ca^{2+}-dependent manner, which leads to increased vascular permeability [61, 62]. Inhibition of the MAPK kinase 1/2 with a specific peptide inhibitor significantly attenuates the histamine-mediated changes in venular permeability [63].

12.6.2
Adhesion Molecule Expression

Under normal conditions, the endothelial layer resists prolonged contact with blood leukocytes. Upon activation, however, ECs start to express molecules, which facilitate firm adhesion and extravasation of leukocytes into sites of inflammation. This process is regulated by different families of adhesion molecules expressed on both, leukocytes and ECs. The types of adhesion molecules and their interaction regulates the specificity of leukocyte migration into tissues. The selectin family of adhesion molecules (E-, P-, and L-selectin) mediates the initial attachment of leukocytes to the endothelial layer before firm adhesion mediated via integrins, vascular cell adhesion molecule (VCAM-1)-1, intercellular adhesion molecule (ICAM) and platelet EC adhesion molecule-1 (CD31) occurs. Genetic deletion of either E-selectin, P-selectin or VCAM-1 significantly diminishes the formation of atherosclerotic lesions. Histamine, acting via H_1 receptor and NF-κB, enhances IL-4- and TNF-α-mediated expression of VCAM-1, TNF-α-mediated expression of E-selectin and ICAM-1, as well as surface translocation of P-selectin in human ECs *in vitro* [64, 65]. These observations suggest that histamine plays an important role in regulating adhesion of leukocytes to the activated endothelium. Indeed, in a rat model, histamine induces rolling of polymorphonuclear cells in a P-selectin dependent manner, which is almost completely abolished by H_1 receptor, but not H_2 receptor blockade [66, 67].

12.6.3
Leukocyte Accumulation

Leukocyte accumulation is controlled by chemoattractant molecules and represents a key process in atherosclerosis development, leading to plaque formation and disease progression. Once adherent to the endothelial layer, the cells enter the intima by diapedesis between ECs. Histamine enhances expression of numerous chemoattractant cytokines such as monocyte chemoattractant protein (MCP)-1 and IL-8 [68, 69]. *In vivo* studies prove the importance of these chemoattractants: mice lacking MCP-1 or its receptor chemokine (CC motif) receptor 2 exhibit a striking decrease in mononuclear phagocyte accumulation and atherosclerotic plaque development.

12.6.4
Regulation of T_h1/T_h2 Balance

Many inflammatory processes are associated with accumulation of CD4$^+$ T cells subsets. Stimulated by antigen, CD4$^+$ T cells can differentiate into two types of effector cells, namely T_h1 and T_h2. In general, T_h1 cells activate monocytes/macrophages and induce B cells to produce IgG antibodies; in contrast T_h2 cells initiate the humoral immune response by activating naive antigen-specific B cells to produce IgM antibodies. The immune responses enhancing T_h1 effector mechanisms are considered to be pro-atherogenic. T_h1 and T_h2 cells exhibit differential expression of histamine receptors – T_h1 cells preferentially express H_1 receptor, whereas T_h2 cells show predominant expression of H_2 receptor. Stimulation of T_h1 cells, but not T_h2 cells, with histamine results in increased intracellular calcium level, translocation of NF-AT and activation of IL-2. Indeed, histamine enhances CD3-induced proliferation of T_h1 cells and blocks proliferation of T_h2 cells. This effect of histamine is abolished by H_1 receptor or H_2 receptor blockers, respectively. T cells lacking the H_1 receptor exhibit significant suppression of interferon-γ secretion, enhancement of T_h2 cytokine production – IL-4 and IL-13 – and reduction of T cell proliferation. In contrast, T cells lacking H_2 receptor exhibit significant enhancement of both T_h1- and T_h2-specific cytokines [60]. These observations suggest that histamine can modulate T cell-mediated immune responses, and thereby the development of atherosclerosis and other diseases.

12.6.5
Macrophage Activation

Macrophage-derived foam cells are one of the most abundant components of atherosclerotic plaques. Foam cells do not only scavenge modified low-density lipoprotein particles rich in cholesterol, but also express many proinflammatory molecules. Monocytes switch their receptor expression profile from H_2 receptor to H_1 receptor upon stimulation, suggesting that histamine signaling may influence the transition from monocytes to activated macrophages [70]. Moreover, histamine, acting via the H_2 receptor, enhances expression of lectin-like oxidized low-density lipoprotein receptor (LOX)-1, which scavenges modified lipoprotein particles, and that of the chemoattractant MCP-1, thereby possibly enhancing both macrophage accumulation and foam cell formation [71]. Accordingly, histamine mediated upregulation of both LOX-1 and MCP-1 is abrogated in activated macrophages which express predominantly H_1 receptor.

12.6.6
Obesity

Obesity is an important risk factor for the development of cardiovascular diseases. Obesity is regulated by both, environmental and genetic factors. Leptin, the product of the *ob* gene, exerts wide-ranging effects including food intake and energy expenditure. Consistently, disruption of leptin protein or leptin receptor leads to the development of obesity. Interestingly, hypothalamic neuronal histamine contributes to the leptin signaling pathway via the H_1 receptor. Indeed, mice with targeted deletion of H_1 receptor exhibit a disrupted diurnal feeding rhythm, increased food consumption, and with time develop obesity [18]. In contrast, H_2 receptor blockers improve excess weight in patients with type 2 diabetes mellitus as well as in patients treated with antipsychotic drugs [72].

12.7
Atherosclerosis

Complications of atherosclerosis are the leading cause of death in Western societies. Atherosclerosis is a progressive inflammatory arterial disease characterized by accumulation of lipids and leukocytes, proliferation of VSMCs, and excessive production of extracellular matrix. The first stage of atherosclerosis development is characterized by activation of the endothelial layer lining the blood vessels, leading to functional activation, increased permeability and recruitment of leukocytes. The first cells found in the intima are macrophages, which scavenge oxidized low density lipoproteins, leading to lipid loaded foam cell formation. Foam cells express numerous cytokines, metalloproteinases and reactive oxygen species, which all promote development of inflammation and progression of atherosclerosis. Advanced plaques with a necrotic core composed of monocytes, foam cells and lipids as well as a fibrous cap consisting of connective tissue is a hallmark of advanced atherosclerosis leading to narrowing of the vessel and clinical manifestations [73].

Many cell types involved in the development and progression of atherosclerosis, such as ECs, macrophages, mast cells and VSMCs, are able to produce histamine and express histamine receptors. Indeed, HDC, the rate-limiting enzyme for histamine production, is highly upregulated in murine atherosclerotic lesions [3]. Moreover, numerous processes crucial for atherosclerosis development, such as vascular permeability, adhesion molecule expression, leukocyte activation and lipid metabolism, are known to be affected by histamine. These observations strongly suggest that histamine modulates the development and progression of atherosclerosis, and pharmacological or genetic blockade of histamine receptors may have beneficial effects in this context.

12.8
Autoimmune Diseases and Allergy

12.8.1
Autoimmune Diseases

The incidence of autoimmune diseases has increased significantly during recent years in developed countries. The reason for this phenomenon, however, is not clear. Under normal conditions, the immune system is able to distinguish between self components and external pathological antigens since self-tolerance is controlled by clonal selection of T and B cells during their maturation. Autoimmune disease occurs when this normal state of self-tolerance is altered, for example, when antigen is not abundant/organ specific, or not expressed in thymus, or when tolerance to an abundant antigen is broken. Autoimmune diseases result from a disrupted balance between anti- and proinflammatory signaling. Hence, degranulation of mast cells and the resulting increase in plasma histamine levels has been reported during the development of autoimmune diseases [74, 75]. Indeed, due to its modulating effect on immune responses, histamine receptors are interesting targets in therapy of autoimmune diseases.

12.8.2
Allergy

Histamine is an important mediator of allergic disease, as it is mainly stored in mast cells and basophils. Allergic reactions can be induced when allergens cross-link IgE bound to the FcεRI receptor on mast cells, but several other stimuli such as cytokines and complement components can induce activation of mast cells as well. Activated mast cells are involved in inflammation by secreting molecules stored in their granules such as histamine as well as by producing cytokines and leukotrienes. The release of inflammatory mediators, cytokines and chemokines at sites of allergic reactions results in the recruitment of eosinophils and basophils. Overall, leukocyte activation provokes a variety of inflammatory symptoms from sneezing and a running nose to life-threatening bronchoconstriction, hypotension and shock depending on the dose of allergen and the route of entry. The inflammatory reaction triggered by mast cell activation can be classified as an immediate response, occurring within minutes after mast cell activation, and a late-phase response lasting hours or days. Degranulation of mast cells leading to the release of histamine, prostaglandins and other molecules enhancing vascular permeability, and inducing inflammation is a main mediator of the immediate response. The late-phase response is generated via production of proinflammatory cytokines, chemokines and leukotrienes.

During experimental asthma, allergic rhinitis and in patients with urticaria, histamine level is significantly higher when compared with healthy subjects. Indeed, it has been demonstrated that histamine is affecting both immediate and late phase allergic responses [1, 76].

12.8.2.1 Immediate Response

Mast cells and basophils are the main sources of stored histamine. Activation of these cells results in degranulation and a sudden transient increase in histamine concentration. Hence, release of stored histamine from these cells accounts for histamine-mediated immediate allergic reactions. Acting via the H_1 receptor, histamine increases vascular permeability leading to edema, bronchoconstriction and mucus production. These alterations result in the development of typical allergic symptoms, such as bronchial obstruction, sneezing, rhinitis and itching. Consistent with the pathological role of the H_1 receptor, administration of histamine H_1 receptor blockers is known to have beneficial effects on such responses.

12.8.2.2 Long-Term Response

Histamine is also involved in the regulation and modulation of long-term inflammatory responses. Recent studies demonstrate that macrophages and neutrophils are the main source of inducible histamine. Many cell types involved in the development of allergic responses are known to express histamine receptors on their surface. Neutrophils and eosinophils express both H_1 receptor and H_2 receptor, whereas basophils express predominantly H_2 receptor. In contrast to the immediate response, where histamine promotes the allergic reaction, late allergic response mediated by basophils and neutrophils is rather attenuated by histamine. Indeed, histamine, acting via H_2 receptor, suppresses leukotriene synthesis [77]. This observation confirms the ambiguous role of histamine in the development of inflammation and allergy.

12.9
Conclusions

Histamine is an endogenous amine produced in a one-step reaction from L-histidine. Activity of HDC, a rate-limiting enzyme crucial for histamine formation, is detected in many different cell types including various cells of the immune system, gastrointestinal and bronchial endocrine cells, neuroendocrine cells, and some tumor cells.

Histamine mediates several processes like homeostasis, vasomotion and inflammation crucial for normal organism function. The action of histamine depends on several factors such as its concentration, the type of receptor it activates and the vascular bed it acts on.

Elevated plasma and/or tissue histamine levels are often associated with pathogenesis of cardiovascular, allergic and autoimmune diseases. Indeed, histamine is known to modulate numerous processes such as vascular permeability, adhesion molecule expression and leukocyte activation, which are crucial for the development of atherosclerosis, allergy or autoimmune disease. Therefore, pharmacological or genetic blockade of histamine receptors, for some of these disorders, represent a well established, for others a promising therapy.

Reference

1 Bachert, C. (2002) The role of histamine in allergic disease: re-appraisal of its inflammatory potential. *Allergy*, **57**, 287–96.

2 Clejan, S., Japa, S., Clemetson, C., Hasabnis, S.S., David, O. and Talano, J.V. (2002) Blood histamine is associated with coronary artery disease, cardiac events and severity of inflammation and atherosclerosis. *Journal of Cellular and Molecular Medicine*, **6**, 583–92.

3 Sasaguri, Y., Wang, K.Y., Tanimoto, A., Tsutsui, M., Ueno, H., Murata, Y., Kohno, Y., Yamada, S. and Ohtsu, H. (2005) Role of histamine produced by bone marrow-derived vascular cells in pathogenesis of atherosclerosis. *Circulation Research*, **96**, 974–81.

4 Brown, M.J., Ind, P.W. and Jenner, D.A. (1980) Platelet histamine. *New England Journal of Medicine*, **303**, 756.

5 Hollis, T.M., Gallik, S.G., Orlidge, A. and Yost, J.C. (1983) Aortic endothelial and smooth muscle histamine metabolism. Relationship to aortic ^{125}I-albumin accumulation in experimental diabetes. *Arteriosclerosis*, **3**, 599–606.

6 Delwaide, J., Vivario, M., Belaiche, J., Louis, E., Courtoy, R., Gast, P. and Boniver, J. (1991) Ultrastructural demonstration of histamine in human enterochromaffin like cell granules. *Gut*, **32**, 834.

7 Panula, P., Airaksinen, M.S., Pirvola, U. and Kotilainen, E. (1990) A histamine-containing neuronal system in human brain. *Neuroscience*, **34**, 127–32.

8 Reite, O.B. (1965) A phylogenetical approach to the functional significance of tissue mast cell histamine. *Nature*, **206**, 1334–6.

9 Macquin, I., Zerah, F., Harf, A., Sabatier, C. and Lhoste, F. (1984) Plasma histamine and catecholamines during carbachol-induced bronchoconstriction in normal subjects. *Journal of Allergy and Clinical Immunology*, **74** (3 Pt 1), 291–5.

10 Taguchi, Y., Tsuyama, K., Watanabe, T., Wada, H. and Kitamura, Y. (1982) Increase in histidine decarboxylase activity in skin of genetically mast-cell-deficient W/Wv

mice after application of phorbol 12-myristate 13-acetate: evidence for the presence of histamine-producing cells without basophilic granules. *Proceedings of the National Academy of Sciences of the United States of America*, **79**, 6837–41.
11 Takamatsu, S., Nakashima, I. and Nakano, K. (1996) Modulation of endotoxin-induced histamine synthesis by cytokines in mouse bone marrow-derived macrophages. *Journal of Immunology*, **156**, 778–85.
12 Zwadlo-Klarwasser, G., Vogts, M., Hamann, W., Belke, K., Baron, J. and Schmutzler, W. (1998) Generation and subcellular distribution of histamine in human blood monocytes and monocyte subsets. *Inflammation Research*, **47**, 434–9.
13 Nakagawa, S., Okaya, Y., Yatsunami, K., Tanaka, S., Ohtsu, H., Fukui, T., Watanabe, T. and Ichikawa, A. (1997) Identification of multiple regulatory elements of human L-histidine decarboxylase gene. *Journal of Biochemistry*, **121**, 935–40.
14 Zhang, Z., Hocker, M., Koh, T.J. and Wang, T.C. (1996) The human histidine decarboxylase promoter is regulated by gastrin and phorbol 12-myristate 13-acetate through a downstream *cis*-acting element. *Journal of Biological Chemistry*, **271**, 14188–97.
15 Sasaguri, Y. and Tanimoto, A. (2004) Role of macrophage-derived histamine in atherosclerosis – chronic participation in the inflammatory response. *Journal of Atherosclerosis and Thrombosis*, **11**, 122–30.
16 Ogasawara, M., Yamauchi, K., Satoh, Y., Yamaji, R., Inui, K., Jonker, J.W., Schinkel, A.H. and Maeyama, K. (2006) Recent advances in molecular pharmacology of the histamine systems: organic cation transporters as a histamine transporter and histamine metabolism. *Journal of Pharmacy and Pharmaceutical Sciences*, **101**, 24–30.
17 Akdis, C.A. and Simons, F.E. (2006) Histamine receptors are hot in immunopharmacology. *European Journal of Pharmacology*, **533**, 69–76.
18 Masaki, T., Chiba, S., Yasuda, T., Noguchi, H., Kakuma, T., Watanabe, T., Sakata, T. and Yoshimatsu, H. (2004) Involvement of hypothalamic histamine H_1 receptor in the regulation of feeding rhythm and obesity. *Diabetes*, **53**, 2250–60.
19 Miyamoto, K., Iwase, M., Nyui, M., Arata, S., Sakai, Y., Gabazza, E.C., Kimura, H. and Homma, I. (2006) Histamine type 1 receptor deficiency reduces airway inflammation in a murine asthma model. *International Archives of Allergy and Immunology*, **140**, 215–22.
20 Kobayashi, T., Tonai, S., Ishihara, Y., Koga, R., Okabe, S. and Watanabe, T. (2000) Abnormal functional and morphological regulation of the gastric mucosa in histamine H_2 receptor-deficient mice. *Journal of Clinical Investigation*, **105**, 1741–9.
21 Yoshimoto, R., Miyamoto, Y., Shimamura, K., Ishihara, A., Takahashi, K., Kotani, H., Chen, A.S., Chen, H.Y., Macneil, D.J., Kanatani, A. and Tokita, S. (2006) Therapeutic potential of histamine H_3 receptor agonist for the treatment of obesity and diabetes mellitus. *Proceedings of the National Academy of Sciences of the United States of America*, **103**, 13866–71.
22 Dunford, P.J., O'Donnell, N., Riley, J.P., Williams, K.N., Karlsson, L. and Thurmond, R.L. (2006) The histamine H_4 receptor mediates allergic airway inflammation by regulating the activation of CD4$^+$ T cells. *Journal of Immunology*, **176**, 7062–70.
23 Liu, S.F., Yacoub, M. and Barnes, P.J. (1990) Effect of histamine on human bronchial arteries *in vitro*. *Naunyn-Schmiedebergs Archives of Pharmacology*, **342**, 90–3.
24 Stahli, B.E., Greutert, H., Mei, S., Graf, P., Frischknecht, K., Stalder, M., Englberger, L., Kunzli, A., Scharer, L., Luscher, T.F., Carrel, T.P. and Tanner, F.C. (2006) Absence of histamine-induced nitric oxide release in the human radial artery: implications for vasospasm of coronary artery bypass vessels. *American Journal of Physiology – Heart and Circulatory Physiology*, **290**, H1182–9.
25 Payne, G.W., Madri, J.A., Sessa, W.C. and Segal, S.S. (2003) Abolition of arteriolar dilation but not constriction to histamine in cremaster muscle of eNOS$^{-/-}$ mice.

American Journal of Physiology – Heart and Circulatory Physiology, 285, H493–8.
26 Sessa, C., Morasch, M.D., Friedland, M. and Kline, R.A. (2001) Risk factors of atherosclerosis and saphenous vein endothelial function. International Angiology, 20, 152–63.
27 Yang, Z.H., Diederich, D., Schneider, K., Siebenmann, R., Stulz, P., von Segesser, L., Turina, M., Buhler, F.R. and Luscher, T.F. (1989) Endothelium-derived relaxing factor and protection against contractions induced by histamine and serotonin in the human internal mammary artery and in the saphenous vein. Circulation, 80, 1041–8.
28 Furchgott, R.F. and Zawadzki, J.V. (1980) The obligatory role of endothelial cells in the relaxation of arterial smooth muscle by acetylcholine. Nature, 288, 373–6.
29 Palmer, R.M., Ashton, D.S. and Moncada, S. (1988) Vascular endothelial cells synthesize nitric oxide from L-arginine. Nature, 333, 664–6.
30 Furchgott, R.F. and Vanhoutte, P.M. (1989) Endothelium-derived relaxing and contracting factors. FASEB Journal, 3, 2007–18.
31 Förstermann, U., Mülsch, A., Böhme, E. and Busse, R. (1986) Stimulation of soluble guanylate cyclase by an acetylcholine-induced endothelium-derived factor from rabbit and canine arteries. Circulation Research, 58, 531–8.
32 Rapoport, R.M. and Murad, F. (1983) Agonist-induced endothelium-dependent relaxation in rat thoracic aorta may be mediated through cGMP. Circulation Research, 52, 352–7.
33 Alheid, U., Frölich, J.C. and Förstermann, U. (1987) Endothelium-derived relaxing factor from cultured human endothelial cells inhibits aggregation of human platelets. Thrombosis Research, 47, 561–71.
34 Li, H. and Forstermann, U. (2000) Nitric oxide in the pathogenesis of vascular disease. Journal of Pathology, 190, 244–54.
35 Radomski, M.W., Palmer, R.M. and Moncada, S. (1987) The role of nitric oxide and cGMP in platelet adhesion to vascular endothelium. Biochemical and Biophysical Research Communications, 148, 1482–9.
36 Huang, P.L., Huang, Z., Mashimo, H., Bloch, K.D., Moskowitz, M.A., Bevan, J.A. and Fishman, M.C. (1995) Hypertension in mice lacking the gene for endothelial nitric oxide synthase. Nature, 377, 239–42.
37 Linder, L., Kiowski, W., Bühler, F.R. and Lüscher, T.F. (1990) Indirect evidence for release of endothelium-derived relaxing factor in human forearm circulation in vivo. Blunted response in essential hypertension. Circulation, 81, 1762–7.
38 Michel, J.B., Feron, O., Sase, K., Prabhakar, P. and Michel, T. (1997) Caveolin versus calmodulin. Counterbalancing allosteric modulators of endothelial nitric oxide synthase. Journal of Biological Chemistry, 272, 25907–12.
39 Lantoine, F., Iouzalen, L., Devynck, M.A., Millanvoye-Van Brussel, E. and David-Dufilho, M. (1998) Nitric oxide production in human endothelial cells stimulated by histamine requires Ca^{2+} influx. Biochemical Journal, 330, 695–9.
40 Thors, B., Halldorsson, H. and Thorgeirsson, G. (2004) Thrombin and histamine stimulate endothelial nitric-oxide synthase phosphorylation at Ser1177 via an AMPK mediated pathway independent of PI3K–Akt. FEBS Letters, 573, 175–80.
41 Li, H., Burkhardt, C., Heinrich, U.R., Brausch, I., Xia, N. and Forstermann, U. (2003) Histamine upregulates gene expression of endothelial nitric oxide synthase in human vascular endothelial cells. Circulation, 107, 2348–54.
42 Toshimitsu, Y., Uchida, K., Kojima, S. and Shimo, Y. (1984) Histamine responses mediated via H_1- and H_2-receptors in the isolated portal vein of the dog. Journal of Pharmacy and Pharmacology, 36, 404–5.
43 Ohnuki, A. and Ogawa, Y. (1997) Differential modulation by the endothelium of contractile responses to 5-hydroxytryptamine, noradrenaline, and histamine in the rabbit isolated basilar artery. General Pharmacology, 28, 681–7.
44 Hatanaka, K., Minamiyama, M., Tanaka, K., Taguchi, T., Tajima, S., Kitamura, Y. and Yamamoto, A. (1985) High susceptibility to ADP-induced thrombus

formation in mast cell-deficient W/Wv mice. *Thrombosis Research*, **40**, 453–64.

45 Steffel, J., Luscher, T.F. and Tanner, F.C. (2006) Tissue factor in cardiovascular diseases: molecular mechanisms and clinical implications. *Circulation*, **113**, 722–31.

46 Wilcox, J.N., Smith, K.M., Schwartz, S.M. and Gordon, D. (1989) Localization of tissue factor in the normal vessel wall and in the atherosclerotic plaque. *Proceedings of the National Academy of Sciences of the United States of America*, **86**, 2839–43.

47 Thiruvikraman, S.V., Guha, A., Roboz, J., Taubman, M.B., Nemerson, Y. and Fallon, J.T. (1996) *In situ* localization of tissue factor in human atherosclerotic plaques by binding of digoxigenin-labeled factors VIIa and X. *Laboratory Investigation*, **75**, 451–61.

48 Steffel, J., Akhmedov, A., Greutert, H., Luscher, T.F. and Tanner, F.C. (2005) Histamine induces tissue factor expression: implications for acute coronary syndromes. *Circulation*, **112**, 341–9.

49 Mackman, N. (1997) Regulation of the tissue factor gene. *Thrombosis and Haemostasis*, **78**, 747–54.

50 Hamilton, K.K. and Sims, P.J. (1987) Changes in cytosolic Ca^{2+} associated with von Willebrand factor release in human endothelial cells exposed to histamine. Study of microcarrier cell monolayers using the fluorescent probe indo-1. *Journal of Clinical Investigation*, **79**, 600–8.

51 Denis, C., Methia, N., Frenette, P.S., Rayburn, H., Ullman-Cullere, M., Hynes, R.O. and Wagner, D.D. (1998) A mouse model of severe von Willebrand disease: defects in hemostasis and thrombosis. *Proceedings of the National Academy of Sciences of the United States of America*, **95**, 9524–9.

52 Tedder, T.F., Steeber, D.A., Chen, A. and Engel, P. (1995) The selectins: vascular adhesion molecules. *FASEB Journal*, **9**, 866–73.

53 Celi, A., Pellegrini, G., Lorenzet, R., De Blasi, A., Ready, N., Furie, B.C. and Furie, B. (1994) P-selectin induces the expression of tissue factor on monocytes. *Proceedings of the National Academy of Sciences of the United States of America*, **91**, 8767–71.

54 Subramaniam, M., Frenette, P.S., Saffaripour, S., Johnson, R.C., Hynes, R.O. and Wagner, D.D. (1996) Defects in hemostasis in P-selectin-deficient mice. *Blood*, **87**, 1238–42.

55 Andre, P., Hartwell, D., Hrachovinova, I., Saffaripour, S. and Wagner, D.D. (2000) Pro-coagulant state resulting from high levels of soluble P-selectin in blood. *Proceedings of the National Academy of Sciences of the United States of America*, **97**, 13835–40.

56 Merten, M. and Thiagarajan, P. (2004) P-selectin in arterial thrombosis. *Zeitschrift Fur Kardiologie*, **93**, 855–63.

57 Mannaioni, P.F., Di Bello, M.G., Raspanti, S., Romano, V., Bani Sacchi, T., Cappugi, P. and Masini, E. (1995) Storage and release of histamine in human platelets. *Inflammation Research*, **44** (Suppl. 1), S16–17.

58 Masini, E., Di Bello, M.G., Raspanti, S., Fomusi Ndisang, J., Baronti, R., Cappugi, P. and Mannaioni, P.F. (1998) The role of histamine in platelet aggregation by physiological and immunological stimuli. *Inflammation Research*, **47**, 211–20.

59 Matsubara, M., Tamura, T., Ohmori, K. and Hasegawa, K. (2005) Histamine H_1 receptor antagonist blocks histamine-induced proinflammatory cytokine production through inhibition of Ca^{2+}-dependent protein kinase C, Raf/MEK/ERK and IKK/I kappa B/NF-kappa B signal cascades. *Biochemical Pharmacology*, **69**, 433–49.

60 Jutel, M., Watanabe, T., Klunker, S., Akdis, M., Thomet, O.A., Malolepszy, J., Zak-Nejmark, T., Koga, R., Kobayashi, T., Blaser, K. and Akdis, C.A. (2001) Histamine regulates T-cell and antibody responses by differential expression of H_1 and H_2 receptors. *Nature*, **413**, 420–5.

61 Moy, A.B., Shasby, S.S., Scott, B.D. and Shasby, D.M. (1993) The effect of histamine and cyclic adenosine monophosphate on myosin light chain phosphorylation in human umbilical vein endothelial cells. *Journal of Clinical Investigation*, **92**, 1198–206.

62 Moy, A.B., Van Engelenhoven, J., Bodmer, J., Kamath, J., Keese, C., Giaever, I., Shasby, S. and Shasby, D.M. (1996) Histamine and thrombin modulate endothelial focal adhesion through centripetal and centrifugal forces. *Journal of Clinical Investigation*, **97**, 1020–7.

63 Wu, M.H., Yuan, S.Y. and Granger, H.J. (2005) The protein kinase MEK1/2 mediate vascular endothelial growth factor- and histamine-induced hyperpermeability in porcine coronary venules. *Journal of Physiology*, **563**, 95–104.

64 Miki, I., Kusano, A., Ohta, S., Hanai, N., Otoshi, M., Masaki, S., Sato, S. and Ohmori, K. (1996) Histamine enhanced the TNF-alpha-induced expression of E-selectin and ICAM-1 on vascular endothelial cells. *Cellular Immunology*, **171**, 285–8.

65 Saito, H., Shimizu, H., Mita, H., Maeda, Y. and Akiyama, K. (1996) Histamine augments VCAM-1 expression on IL-4- and TNF-alpha-stimulated human umbilical vein endothelial cells. *International Archives of Allergy and Immunology*, **111**, 126–32.

66 Kubes, P. and Kanwar, S. (1994) Histamine induces leukocyte rolling in post-capillary venules. A P-selectin-mediated event. *Journal of Immunology*, **152**, 3570–7.

67 Yamaki, K., Thorlacius, H., Xie, X., Lindbom, L., Hedqvist, P. and Raud, J. (1998) Characteristics of histamine-induced leukocyte rolling in the undisturbed microcirculation of the rat mesentery. *British Journal of Pharmacology*, **123**, 390–9.

68 Jeannin, P., Delneste, Y., Gosset, P., Molet, S., Lassalle, P., Hamid, Q., Tsicopoulos, A. and Tonnel, A.B. (1994) Histamine induces interleukin-8 secretion by endothelial cells. *Blood*, **84**, 2229–33.

69 Kimura, S., Wang, K.Y., Tanimoto, A., Murata, Y., Nakashima, Y. and Sasaguri, Y. (2004) Acute inflammatory reactions caused by histamine via monocytes/macrophages chronically participate in the initiation and progression of atherosclerosis. *Pathology International*, **54**, 465–74.

70 Triggiani, M., Petraroli, A., Loffredo, S., Frattini, A., Granata, F., Morabito, P., Staiano, R.I., Secondo, A., Annunziato, L. and Marone, G. (2007) Differentiation of monocytes into macrophages induces the upregulation of histamine H_1 receptor. *Journal of Allergy and Clinical Immunology*, **119**, 472–81.

71 Tanimoto, A., Murata, Y., Nomaguchi, M., Kimura, S., Arima, N., Xu, H., Hamada, T. and Sasaguri, Y. (2001) Histamine increases the expression of LOX-1 via H_2 receptor in human monocytic THP-1 cells. *FEBS Letters*, **508**, 345–9.

72 Stoa-Birketvedt, G., Paus, P.N., Ganss, R., Ingebretsen, O.C. and Florholmen, J. (1998) Cimetidine reduces weight and improves metabolic control in overweight patients with type 2 diabetes. *International Journal of Obesity and Related Metabolic Disorders*, **22**, 1041–5.

73 Libby, P. (2002) Inflammation in atherosclerosis. *Nature*, **420**, 868–74.

74 Frewin, D.B., Cleland, L.G., Jonsson, J.R. and Robertson, P.W. (1986) Histamine levels in human synovial fluid. *Journal of Rheumatology*, **13**, 13–14.

75 Gill, D.S., Barradas, M.A., Fonseca, V.A. and Dandona, P. (1989) Plasma histamine concentrations are elevated in patients with diabetes mellitus and peripheral vascular disease. *Metabolism*, **38**, 243–7.

76 White, M.V. (1990) The role of histamine in allergic diseases. *Journal of Allergy and Clinical Immunology*, **86** (4 Pt 2), 599–605.

77 Flamand, N., Plante, H., Picard, S., Laviolette, M. and Borgeat, P. (2004) Histamine-induced inhibition of leukotriene biosynthesis in human neutrophils: involvement of the H_2 receptor and cAMP. *British Journal of Pharmacology*, **141**, 552–61.

13
Prostaglandins and Leukotrienes
Katharina Lötzer and Andreas J. R. Habenicht

Prostaglandins (PGs) and leukotrienes (LTs) – also referred to as eicosanoids – are potent lipid mediators derived from arachidonic acid (AA) via the cyclooxygenase (COX) and 5-lipoxygenase (5-LO) pathways, respectively. PGs and LTs are implicated in a wide range of physiological and pathophysiological reactions – the maintenance of tissue homeostasis, host defense, inflammation, pain, fever, immediate hypersensitivity, allergy, asthma, atherosclerosis and cancer. In this chapter we will focus on the potential role of eicosanoids in the cardiovascular system.

13.1
AA Metabolism by the COX and 5-LO Pathways

PGs and LTs share a common polyunsaturated fatty acid precursor substrate, AA, which is released from the *sn*-2 position of membrane glycerophospholipids following activation of cytosolic phospholipase A_2 in response to diverse physical, chemical, inflammatory and mitogenic stimuli. The subsequent formation of the intermediate metabolites, PGH_2 and LTA_4, is catalyzed by PGG/H synthase, widely known as COX, and 5-LO, respectively [1] (Figure 13.1). Two COX isoforms have been identified: the constitutively expressed COX-1 and the inducible COX-2. A splice variant of COX-1 has been described and has been referred to as COX-3, but this terminology has been disputed [2]. COXs possess bifunctional enzymatic activities: catalyzed by the COX activity two oxygen molecules are incorporated into AA to form the labile intermediate peroxide, PGG_2, which is reduced to the unstable endoperoxide PGH_2 by the enzyme's second activity hydroperoxidase. PGH_2 is metabolized by tissue-specific isomerases and synthases to form the biological active prostanoids prostacyclin (PGI_2), thromboxane (Tx) A_2, PGD_2, PGE_2 and PGF_2 [3].

LT biosynthesis is initiated by 5-LO, in concert with 5-LO activating protein (FLAP)[4], resulting in the sequential formation of 5(*S*)-hydroperoxy-6-*trans*-8,11,14-*cis*-eicosatetraenoic acid and the unstable epoxide, LTA_4, via dual dioxygenase and synthase activities of 5-LO [5]. FLAP is supposed to be an AA-binding and

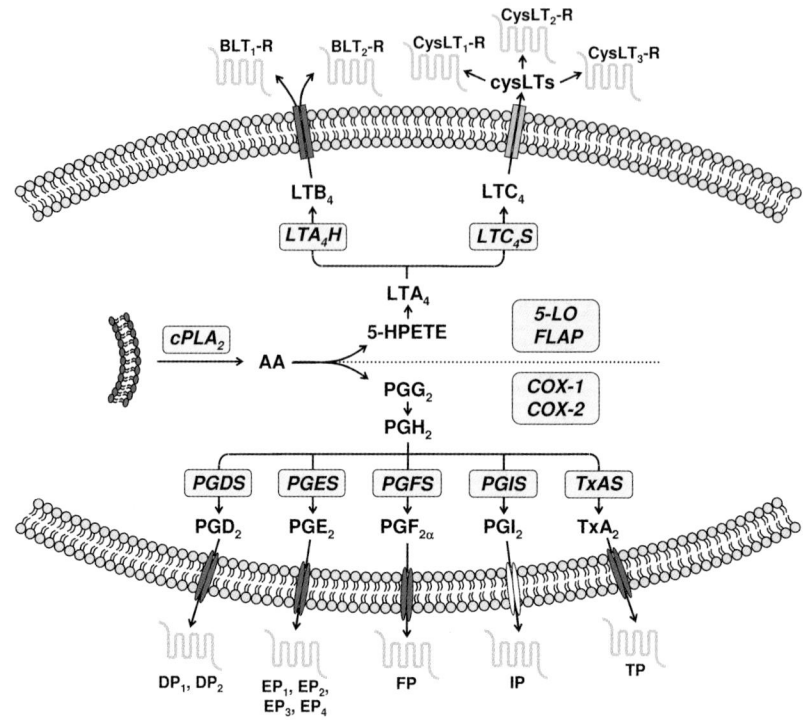

Figure 13.1 AA metabolism by the COX and 5-LO pathways. Upon cell activation AA is released by cytosolic phospholipase A_2 (cPLA$_2$) from membrane glycerophospholipids. Subsequent formation of PGs and LTs is catalyzed by the COX and 5-LO pathway, respectively. After transporter facilitated transfer out of the cell PGs and LTs activate specific G-protein-coupled receptors in an autocrine or paracrine manner. R, receptor.

transfer protein, but the precise mechanisms of 5-LO and FLAP interactions are not completely understood [6]. Although nonhematopoietic cells are also implicated in LT formation, 5-LO is prominently expressed by myeloid cells. However, LTs can be formed by non-5-LO-expressing cells via transcellular metabolism of LTA$_4$ [7–9]. LTA$_4$ is either hydrolyzed to LTB$_4$ or conjugated with glutathione to form LTC$_4$ and its metabolites, LTD$_4$ and LTE$_4$, by the downstream enzymes LTA$_4$ hydrolase (LTA$_4$H) and LTC$_4$ synthase (LTC$_4$S), respectively [5]. Due to their short half-lives PGs and LTs must act close to their sites of synthesis, by autocrine or paracrine activation of distinct G-protein-coupled receptors (Figure 13.1). Individual high-affinity eicosanoid receptors have been identified for PGI$_2$, TxA$_2$ and PGF$_2$ (IP, TP and FP, respectively), two for PGD$_2$ (DP$_1$ and DP$_2$) and four for PGE$_2$ (EP$_1$–EP$_4$) [10]. Two subtypes of BLT receptors are known, BLT$_1$ receptor, a high-affinity receptor for LTB$_4$, and BLT$_2$ receptor, that binds LTB$_4$ with low affinity [11].

CysLTs exert their biological effects by activation of at least two CysLT receptors, $CysLT_1$, and $CysLT_2$, with different agonist response profiles ($LTD_4 \gg LTC_4 > LTE_4$ and $LTC_4 = LTD_4 > LTE_4$, respectively)[12]. Moreover, the orphan receptor, GPR17, was recently identified as a new dual uracil nucleotide/CysLT receptor. Its biological role remains to be investigated [13].

13.2
PGs in Cardiovascular Physiology and Pathophysiology

13.2.1
Diversity of Prostanoid Effects in the Cardiovascular System

PGI_2, the major COX product in vascular endothelial cells (ECs), and TxA_2, mainly produced by platelets, have potent and contrasting effects on vascular tone and platelet function [14]. PGI_2 is a potent vasodilator, inhibits platelet aggregation, decreases leukocyte adhesion and inhibits vascular smooth muscle cell (SMC) proliferation [15]. It is believed to modulate cardiovascular homeostasis acting as a physiological antagonist of TxA_2, a strong platelet activator, vasoconstrictor and SMC mitogen. In mice lacking IP increased susceptibility to thrombosis was observed, whereas inflammatory and pain responses were reduced, indicating a role of PGI_2 as an antithrombotic agent, and as a mediator of acute inflammation *in vivo* [16]. Moreover, other mouse studies revealed enhanced proliferative response to catheter-induced carotid artery injury and platelet activation in IP-deficient mice, whereas both were depressed in TP-deficient mice or mice treated with a TP antagonist [17]. The role of PGE_2 is more complex, due to activation of four distinct EPs, which mediate diverse biological effects: PGE_2 triggers vasoconstriction or vasodilation depending on the EP involved, induces proinflammatory cytokine production, platelet activation and regulates matrix metalloproteinase (MMP)-2 and -9 expression in monocytes [18–20]. Studies with EP2 receptor-deficient mice revealed a role of this receptor in PGE_2-induced arterial dilatation [21].

13.2.2
PGs and Atherosclerosis

Expression of COX in human atherosclerotic arteries indicated a potential role of the enzyme in atherosclerotic lesion formation [22–24]. COX-1 is expressed abundantly in normal and diseased arteries, while COX-2 expression is restricted to atherosclerotic lesions. COX-2 localizes to macrophages, SMCs and ECs. Moreover, excretion of both PGI_2 and TxA_2 metabolites are increased in patients with severe atherosclerosis [23, 25]. To elucidate a role of COXs and their products atherosclerosis-prone mouse models were studied. Nonselective inhibition of COX by indomethacin reduced atherosclerotic plaque formation in low-density lipoprotein (LDL) receptor-deficient mice fed a cholesterol-containing diet, whereas the selective COX-2 inhibitor, nimesulfide, tended to retard atherosclerosis under these conditions. Indomethacin inhibits both TxA_2 and PGI_2 formation, whereas nimesulfide selectively inhibits PGI_2 formation, suggesting that COX-1-derived PGs are more relevant in atherogenesis [26]. These data are consistent with the observation that selective inhibition of COX-1 retarded atherogenesis in apolipoprotein (apo) E-deficient mice, while selective inhibition of COX-2 had no effect on lesion development [27]. In contrast, apoE/IP-double-deficient mice showed a significant acceleration in atherogenesis, indicating a protective role of PGI_2 in initiation and progression of atherogenesis [28]. Moreover, deletion of IP in female LDL receptor-deficient mice abolished the atheroprotective effect of estrogen receptor-α-mediated upregulation of COX-2-derived PGI_2 [29]. Deletion of TP in apoE-deficient mice [28] and antagonism of TP in apoE-deficient mice and Apobec-1/LDL receptor double-deficient mice retarded atherosclerosis [30, 31], but treatment with aspirin or indomethacin had no significant effect on atherogenesis in apoE-deficient mice and Apobec-1/LDL receptor double-deficient mice, respectively. The authors proposed that blockade of TP receptors inhibits atherosclerosis by a mechanism independent of TxA_2 and implicated other prostanoids instead. Alternatively, concomitant inhibition of PGI_2 synthesis may abolish the otherwise protective effect of COX-1 inhibition. In fact, in LDL receptor-deficient mice fed a high-fat diet and treated with low-dose aspirin, development of atherosclerotic lesions was significantly retarded [32]. Thus, the role of COX-2 in mouse atherogenesis is clearly controversial, since mouse studies examining the effects of COX-2 inhibitors in atherogenesis have yielded conflicting results. As mentioned above, two studies – one of nimesulfide in LDL receptor-deficient mice and the other of SC-236 in apoE-deficient mice – did not find significant effects in atherogenesis [26, 27]. Additional studies revealed that treatment with the COX-2 inhibitor rofecoxib significantly reduced the extent of atherosclerosis in cholesterol-fed LDL receptor-deficient mice and in apoE-deficient mice, indicating a role for COX-2 products in early atherosclerosis [33, 34]. In contrast to these findings, treatment with MF-tricyclic, a selective COX-2 inhibitor, increased lipid aortic accumulation in apoE-deficient mice [35]. In conclusion the contrasting results concerning pharmacological COX-2 inhibition in atherogenesis may be model-, therapy duration- and drug-dependent. Moreover, although PGI_2 is the main COX-2-dependent prostanoid in ECs, other cells of the atherosclerotic lesion are able to produce other types of prostanoids. For example, TxA_2 and PGE_2 are the main products of human blood monocyte COX-2 [36]. A role for macrophage COX-2 in atherogenesis was suggested in LDL receptor-deficient mice null for expression of the COX-2 gene in macrophages by reconstituting irradiated mice with $COX-2^{-/-}$ hemopoietic cells. LDL receptor-deficient mice were transplanted with $COX-2^{-/-}$ or $COX-2^{+/+}$ fetal liver cells and then fed a Western-type diet for 8 weeks to induce fatty streak lesions. Atherosclerotic lesion area in LDL receptor-deficient mice reconstituted with $COX-2^{-/-}$ fetal liver cells was significantly reduced compared with mice transplanted with $COX-2^{+/+}$ fetal liver cells [33]. COX-2, microsomal PGE synthase (mPGES)-1 and EP4 were detected in human atherosclerotic lesions of patients undergoing carotid endarterectomy afflicted with recent transient ischemic attack or stroke [18, 37]. Since symptomatic plaques contained active MMP-2 and -9, and COX-2-dependent PGE_2 synthesis regulated monocyte MMP-2 and -9 expression, a role for PGE_2 was suggested in plaque instability. A recent mouse study also indicated a role for mPGES-1 in mouse atherogenesis. Vascular mPGES-1 was upregulated during atherogenesis in LDL receptor-deficient mice on a high-fat diet, and deletion of mPGES-1 retarded atherosclerosis in LDL receptor-deficient mice. Moreover, systemic production of PGE_2 was significantly decreased in mPGES-1/LDL receptor double-deficient mice, whereas biosynthesis of PGI_2 but not TxA_2 was augmented [38].

13.2.3
COX-2 Inhibition and Cardiovascular Risk in Humans

Coxibs are a subclass of nonsteroidal antiinflammatory drugs (NSAIDs) that inhibit COX-2, while COX-1 appears to be less affected. Their development was based on the "COX-2 hypothesis" assuming that gastrointestinal adverse effects of traditional nonselective NSAIDs are related to inhibiting COX-1-dependent prostanoid formation, whereas selective inhibition of inducible COX-2 should lower the risk of gastric ulceration and/or bleeding [39]. However, both COX-1 and -2 are constitutively expressed in healthy human and animal gastric mucosa [40], and emerging evidence indicates an important role of COX-2 in gastrointestinal mucosal defense and repair [41]. Moreover, results of clinical trials have raised doubts about the cardiovascular safety of coxibs (see below). Two coxibs, rofecoxib and valdecoxib, were withdrawn from the market, when their cardiovascular risk was indicated in randomized controlled clinical trials (see below). Consequently, patients with elevated cardiovascular risk were subsequently excluded from treatment with the remaining coxibs [42, 43]. The first evidence of deleterious cardiovascular adverse effects emerged from the Vioxx Gastrointestinal Outcome Research (VIGOR) study [44]. VIGOR was conducted to assess whether rofecoxib, a selective inhibitor of COX-2, would be associated with a lower incidence of gastrointestinal adverse effects than the nonselective NSAID naproxen among patients with rheumatoid arthritis. Indeed, the incidence of serious gastrointestinal events was half among those receiving rofecoxib compared to the naproxen-treated patients. However, an increase by a factor of five in the incidence of myocardial infarction was observed. An antithrombotic, cardioprotective effect of naproxen was considered, but subsequent epidemiological studies did not reveal clear results [45]. About the same time the Celecoxib Long-term Arthritis Safety Study (CLASS) was published. The CLASS trial compared gastrointestinal toxicity of celecoxib versus the traditional NSAIDs ibuprofen or diclofenac in patients with osteoarthritis and rheumatoid arthritis [46]. Six months of treatment with celecoxib was associated with a lower incidence of symptomatic ulcers and ulcer complications compared to NSAIDs, but no difference in the incidence of cardiovascular events was observed between the two groups. However, only partial data were published and analysis of the full data set revealed no difference in the incidence of gastrointestinal side-effects, but signs of increased cardiovascular risk [42]. Moreover, a meta-analysis of randomized double blind trials of celecoxib indicated an increased risk of myocardial infarction, but not for composite cardiovascular events, deaths or stroke [47]. A third arthritis trial, the Therapeutic Arthritis Research and Gastrointestinal Event Trial, was conducted to assess gastrointestinal and cardiovascular safety of the second generation COX-2 inhibitor lumiracoxib compared with two traditional NSAID, naproxen and ibuprofen, in patients with osteoarthritis. Reduction in gastrointestinal adverse effects was observed in the lumiracoxib group without any significant increase in the rate of serious cardiovascular events [48, 49]. This trial is not without controversy regarding the risk assessment of cardiovascular events, since most of the patients with known and significant pre-existing coronary disease were excluded and the study was underpowered to detect significant differences in cardiovascular events [50]. Several smaller clinical trials and observational studies with opposing results in terms of cardiovascular hazard associated with coxibs had been published up to 2004 [51, 52]. The withdrawal of rofecoxib from the market in October 2004 was the consequence of the results of the Adenomatous Polyp Prevention on Vioxx (APPROVe) study. In this randomized, placebo-controlled trial the cardiovascular outcomes associated with the use of rofecoxib was investigated in patients with a history of colorectal adenomas. [53]. Daily use of 25 mg rofecoxib was associated with an increased cardiovascular risk that became apparent after 18 month of treatment. Coincident with the APPROVe study a dose-dependent increased risk of cardiovascular events was observed after treatment with 200 or 400 mg celecoxib twice daily compared with placebo in the Colorectal Adenoma Prevention Trial [54]. Given the emerging evidence of COX-2 inhibition and increased cardiovascular risk, and since several studies comparing COX-2-specific inhibitors and traditional NSAIDs failed to detect differences in cardiovascular risk, the cardiovascular safety of traditional NSAIDs is also under debate. However, placebo-controlled studies have to be performed to support this hypothesis. Although data from basic research, observational studies, and randomized trials strongly support an increased cardiovascular risk with COX-2 inhibition the question of the underlying mechanisms still remains to be fully understood. An imbalance between TxA_2 and PGI_2 production that results from selective COX-2 inhibition was supposed but this hypothesis has not been proven until now.

13.3
LTs in Cardiovascular Physiology and Pathophysiology

13.3.1
Activities of LTs in the Cardiovascular System

Cardiovascular actions of LTs were first described in the 1980s, when investigators noted that LTC_4 and LTD_4 induced a systemic arterial hypotensive effect [55], and promoted plasma leakage and vasoconstriction, whereas LTB_4 caused leukocyte adhesion to the endothelium in postcapillary venules [56]. CysLTs were shown to induce vasoconstriction in different species, including human, in a vascular bed-dependent manner [57–64]. Nonatherosclerotic coronary artery ring segments were unresponsive to LTC_4 and LTD_4, whereas both LTs induced concentration-dependent contractions in atherosclerotic coronary arteries, indicating a role for LTs in atherosclerosis pathophysiology [65]. Mice deficient in constituents of the 5-LO cascade and LT receptors were generated to elucidate biological and pathological functions. Increased vascular permeability in zymosan A-induced peritoneal inflammation, in IgE-dependent passive cutaneous anaphylaxis, and in bleomycin-induced chronic pulmonary inflammation and fibrosis was reduced in LTC_4 synthase-deficient mice [66, 67], and targeted disruption of $CysLT_1$ or $CysLT_2$ receptor revealed a CysLT receptor-specific role in this alteration of vascular permeability [68, 69]. These findings may also be relevant to cardiovascular responses. In another mouse study, the $CysLT_2$ receptor was overexpressed in ECs. The resulting phenotype also suggested a role of $CysLT_2$ receptor in blood vessel permeability and indicated that the receptor participates in blood pressure regulation [70]. These mouse models should now allow characterization of additional $CysLT_2$ receptor-mediated events in cardiovascular responses to evaluate the potential impact of CysLTs *in vivo*. Cardiovascular-related biological activities of LTs have also emerged from *in vitro* data at the cellular level. CysLTs induced P-selectin surface expression [71, 72], von Willebrand factor secretion [71] and platelet-activating factor synthesis [73] in cultured ECs, while LTB_4 increased EC adhesiveness for PMN [74], all important events in leukocyte activation and subsequent emigration to sites of inflammation. Moreover, identification of the gene expression profile induced by $CysLT_2$ receptor-dependent signaling in ECs revealed an early proinflammatory and prothrombotic EC phenotype, but also up-regulation of genes with anti-inflammatory properties [75]. Prominently LTD_4-triggered genes in ECs were macrophage-inflammatory protein (MIP)-2, interleukin-8, transcription factor early growth response 1 and several additional transcription factors. These chemokines and their receptors may exacerbate cardiovascular inflammation by recruiting leukocytes into the arterial wall. A recent study revealed that interferon-γ induces $CysLT_2$ receptor expression and enhances CysLT-induced inflammatory responses in ECs [76]. However, LTD_4 also promotes formation of the atherosclerosis-protective and antithrombotic eicosanoid PGI_2 by markedly upregulating EC COX-2 [77]. In human and mouse monocytic cell lines, LTD_4 induced MCP-1α, a T cell chemoattractant [78], and a study on the effects of LTB_4 in cultured monocytes demonstrated that BLT_1 receptor activation mediated induction of the proinflammatory chemokine MCP-1 [79].

13.3.2
5-LO Atherosclerosis Hypothesis

Although former studies revealed that atherosclerotic vessels are capable to produce LTB_4 [80], a role of 5-LO in atherogenesis remained uncertain, since an early study failed to detect the enzyme in macrophages/foam cells of the human vessel wall [81], while other investigators reported the expression of constituents of the 5-LO pathway in both, normal and diseased, coronary arteries [65]. We proposed a potential connection of the 5-LO pathway with atherogenesis in 2003 when we observed expression of 5-LO in a large number of specimens of carotid and coronary arteries and aortae of patients afflicted with various degrees of atherosclerosis including clinically overt disease [82]. 5-LO localized to macrophages, dendritic cells and foam cells, and the number of 5-LO expressing leukocytes in the atherosclerotic lesions markedly increased during disease progression. Moreover, all other constituents of the LT biosynthetic cascade (FLAP, LTA$_4$H, and LTC$_4$S, and the four LT receptors: BLT$_1$, BLT$_2$, CysLT$_1$, and CysLT$_2$) were also expressed in human diseased arteries. Further evidence for a link between 5-LO and atherogenesis came from two studies of symptomatic compared with asymptomatic human atherosclerotic plaques in patients undergoing carotid endarterectomy. Increased 5-LO and LTA$_4$H expression, and elevated plaque concentration of LTB$_4$ was observed in association with increased expression and activity of MMP-2 and -9, and decreased collagen content, indicating a possible role for 5-LO in plaque instability in humans [83, 84]. The cell type that expressed LT receptors in atherosclerotic lesions has not been identified, but *in vitro* data indicated preferential expression of CysLT$_1$ receptor in monocyte/macrophages and T cells, and of CysLT$_2$ receptor in ECs [85]. Moreover, CysLT$_1$ and CysLT$_2$ receptors are expressed in mast cells [86, 87], and SMCs [88, 89], whereas expression of both BLT$_1$ and BLT$_2$ receptors was reported in several cell types, including monocyte/macrophages, ECs, SMCs, and T cells [85, 90, 91]. Thus, the formation of 5-LO$^+$ cell infiltrates observed in clinical samples at predilection sites of atherosclerosis, that is coronary and carotid arteries and the aorta, are compatible with participation of LTs to EC dysfunction, intimal edema, leukocyte infiltration, aberrant contractility, SMC proliferation and immune reactivity. These data support a model of atherogenesis in which macrophages/foam cells or mast cells from within the lesions and/or from the adventitial connective tissue produce LTs which act on LT receptor-expressing cells in the atherosclerotic vessel in an autocrine and/or paracrine manner. These events may promote 5-LO cascade-dependent inflammatory circuits within the blood vessel wall during atherosclerotic lesion development [92]. However, much work is required to demonstrate the validity of the hypothesis.

13.3.3
5-LO Pathway in Mouse Models of Atherosclerosis

Support for a link between the 5-LO cascade and atherogenesis emerged from several mouse studies. A locus on mouse chromosome 6, where the 5-LO gene is located, was identified which conferred resistance of CAST/Ei mice towards atherosclerosis [93]. It was hypothesized that 5-LO may represent a proatherogenic susceptibility gene, and in a congenic mouse strain, containing the resistant chromosome 6 region derived from the CAST/Ei strain, 5-LO mRNA and protein levels were reduced, compared to the parent strain. Moreover, in 5-LO$^{+/-}$/LDL receptor-deficient mice a significant decrease in aortic lesion development was observed [94]. On the other hand, no or only a minor difference was observed in atherosclerotic lesion development between wild-type and 5-LO-deficient mice crossbred to apoE- or LDL receptor-deficient mice, whereas disruption of the 5-LO gene reduced both incidence and extent of aortic aneurysm development [78]. In addition, pharmacological studies supported a role of 5-LO in the formation of lipid-rich lesions of the aorta in hyperlipidemic mice. In both LDL receptor-deficient mice and apoE-deficient mice a BLT receptor antagonist, CP-105696, significantly reduced lipid accumulation, monocyte infiltration and lesion size. Moreover, expression of CD11b on circulating blood monocytes and in atherosclerotic lesions was reduced, indicating a role for LTB_4 in monocyte chemotaxis, adhesion and endothelial migration [95]. In BLT_1 receptor/apoE double-deficient mice lesion formation was significantly reduced during initial stages (4 and 8 weeks) of lesion formation compared to BLT_1 receptor$^{+/+}$/apoE$^{-/-}$, with lesser numbers of infiltrating monocytes, but no reduction in lesions in mice on an atherogenic diet for 19 weeks was observed, indicating a potential role for BLT_2 receptor in advanced disease [96]. Another study revealed that a decrease in early lesion formation in BLT_1 receptor/apoE double-deficient mice was accompanied by significant decreases in SMCs, macrophages and T cells [97]. When taken together, data from mouse models do not yield a consistent proatherogenic role of vascular LTs.

13.3.4
Population Genetic Studies Indicate a Role of the 5-LO Pathway in Cardiovascular Disease

Several population genetic studies indicated a link between the 5-LO cascade and cardiovascular disease (CVD). Specific variant genotypes of the 5-LO gene were associated with increased atherosclerosis as assessed by sonographic determination of the carotid intima/media thickness and increased plasma levels of a marker of inflammation, C-reactive protein [98]. Moreover, increased dietary intake of *n*-6 polyunsaturated fatty acids (AA and its precursor linoleic acid) enhanced the apparent atherogenic effect of the genotype, whereas increased dietary *n*-3 polyunsaturated fatty acids (eicosapentaenoic and docosahexaenoic acid) blunted the effect in the 5-LO gene variant carriers. In another report polymorphisms in the FLAP gene were associated with an increased risk of myocardial infarction and stroke [99]. A particular FLAP gene haplotype (HapA) was associated with a 2-fold greater risk of myocardial infarction and stroke in Iceland, while another cohort in the United Kingdom carrying another variant haplotype (HapB) correlated with an increased risk of myocardial infarction, but not stroke [99], and in a Scottish population HapA, but not HapB, was associated with ischemic stroke [100]. In a cohort in southern Germany sequence variants of FLAP distinct from the reported HapA variants showed a significant correlation with stroke, particularly in males, whereas no increased risk for stroke was observed in HapA carriers of the German population [101]. Moreover, a haplotype (HapK) spanning the LTA_4H gene encoding LTA_4H, conferred modest risk of myocardial infarction in an Icelandic cohort and also in three cohorts from the United States. A much stronger association was observed in African-Americans compared to European Americans [102]. In conclusion, these studies indicated but did not prove connections between genetic alterations of constituents of the 5-LO cascade and CVD. Following the hypothesis of a role for the 5-LO pathway in CVD a crossover trial of an inhibitor of FLAP (DG-031) was conducted with patients that had experienced a myocardial infarction and who carried at-risk variants in the FLAP gene or in the LTA_4H gene, and changes in levels of biomarkers associated with risk of myocardial infarction were measured. DG-031 significantly reduced LTB_4 formation and also reduced myeloperoxidase, whereas C-reactive protein was not significantly affected by the drug [103]. Decode Genetics announced initiation of a large phase III 2-year outcome study of 1500 patients with previous myocardial infarction to test whether FLAP inhibition will prevent recurrent heart attacks, but this trial has recently been abandoned because the drug formulation used did not allow for reproducible blood drug levels. The company, however, announced that the trial will continue once these problems have been overcome. If the upcoming trials to be conducted between 2007 and 2010 confirm the data of the phase IIa trial and indicate beneficial effects on clinical outcome, conditions for larger drug development programs to treat arterial inflammation using anti-LT drugs will be set. While all the evidence so far obtained for a link between 5-LO and CVD has been accumulating over the last several years, major questions remain. We have pointed out a particularly important aspect–the need to link genotype and phenotype, that is the necessity to demonstrate either loss-of-function or gain-of-function phenotypes and the need to relate clinical disease to functional mutations rather than mutations based solely on genotypes. This cannot be overstated as recent genetic evidence indicates that many genotypes develop randomly without affecting phenotypes (i.e. they are functionally irrelevant). It needs to be pointed out that none of the above-mentioned genetic trials–while raising interesting questions–has met these criteria and, therefore, we are still far away from a solid connection between 5-LO and CVD. Thus, since the total impact of LTs in the cardiovascular systems is only beginning to be understood, large-scale clinical trials and drug development programs should be limited to preclinical trials. Unfortunately, the current mouse models of atherosclerosis appear to be unsuitable animal models since 5-LO deficiency does not affect plaque formation [78].

References

1 Funk, C.D. (2001) Prostaglandins and leukotrienes: advances in eicosanoid biology. *Science*, **294**, 1871–5.
2 Chandrasekharan, N.V., Dai, H., Roos, K.L., Evanson, N.K., Tomsik, J., Elton, T.S. and Simmons, D.L. (2002) COX-3, a cyclooxygenase-1 variant inhibited by acetaminophen and other analgesic/antipyretic drugs: cloning, structure, and expression. *Proceedings of the National Academy of Sciences of the United States of America*, **99**, 13926–31.
3 Smith, W.L., DeWitt, D.L. and Garavito, R.M. (2000) Cyclooxygenases: structural, cellular, and molecular biology. *Annual Review of Biochemistry*, **69**, 145–82.

4 Dixon, R.A., Diehl, R.E., Opas, E., Rands, E., Vickers, P.J., Evans, J.F., Gillard, J.W. and Miller, D.K. (1990) Requirement of a 5-lipoxygenase-activating protein for leukotriene synthesis. *Nature*, **343**, 282–4.

5 Samuelsson, B. (1983) Leukotrienes: mediators of immediate hypersensitivity reactions and inflammation. *Science*, **220**, 568–75.

6 Peters-Golden, M. (1998) Cell biology of the 5-lipoxygenase pathway. *American Journal of Respiratory and Critical Care Medicine*, **157**, S227–32.

7 Claesson, H.E. and Haeggstrom, J. (1988) Human endothelial cells stimulate leukotriene synthesis and convert granulocyte released leukotriene A_4 into leukotrienes B_4, C_4, D_4 and E_4. *European Journal of Biochemistry*, **173**, 93–100.

8 Maclouf, J., Murphy, R.C. and Henson, P.M. (1989) Transcellular sulfidopeptide leukotriene biosynthetic capacity of vascular cells. *Blood*, **74**, 703–7.

9 McGee, J.E. and Fitzpatrick, F.A. (1986) Erythrocyte–neutrophil interactions: formation of leukotriene B_4 by transcellular biosynthesis. *Proceedings of the National Academy of Sciences of the United States of America*, **83**, 1349–53.

10 Narumiya, S. and FitzGerald, G.A. (2001) Genetic and pharmacological analysis of prostanoid receptor function. *Journal of Clinical Investigation*, **108**, 25–30.

11 Tager, A.M. and Luster, A.D. (2003) BLT_1 and BLT_2: the leukotriene B_4 receptors. *Prostaglandins Leukotrienes and Essential Fatty Acids*, **69**, 123–34.

12 Evans, J.F. (2003) The cysteinyl leukotriene receptors. *Prostaglandins Leukotrienes and Essential Fatty Acids*, **69**, 117–22.

13 Ciana, P., Fumagalli, M., Trincavelli, M.L., Verderio, C., Rosa, P., Lecca, D., Ferrario, S., Parravicini, C., Capra, V., Gelosa, P., Guerrini, U., Belcredito, S., Cimino, M., Sironi, L., Tremoli, E., Rovati, G.E., Martini, C. and Abbracchio, M.P. (2006) The orphan receptor GPR17 identified as a new dual uracil nucleotides/cysteinyl-leukotrienes receptor. *EMBO Journal*, **25**, 4615–27.

14 Bunting, S., Moncada, S. and Vane, J.R. (1983) The prostacyclin–thromboxane A_2 balance: pathophysiological and therapeutic implications. *British Medical Bulletin*, **39**, 271–6.

15 Vane, J.R. and Botting, R.M. (1995) Pharmacodynamic profile of prostacyclin. *American Journal of Cardiology*, **75**, 3A–10A.

16 Murata, T., Ushikubi, F., Matsuoka, T., Hirata, M., Yamasaki, A., Sugimoto, Y., Ichikawa, A., Aze, Y., Tanaka, T., Yoshida, N., Ueno, A., Oh-ishi, S. and Narumiya, S. (1997) Altered pain perception and inflammatory response in mice lacking prostacyclin receptor. *Nature*, **388**, 678–82.

17 Cheng, Y., Austin, S.C., Rocca, B., Koller, B.H., Coffman, T.M., Grosser, T., Lawson, J.A. and FitzGerald, G.A. (2002) Role of prostacyclin in the cardiovascular response to thromboxane A_2. *Science*, **296**, 539–41.

18 Cipollone, F., Prontera, C., Pini, B., Marini, M., Fazia, M., De Cesare, D., Iezzi, A., Ucchino, S., Boccoli, G., Saba, V., Chiarelli, F., Cuccurullo, F. and Mezzetti, A. (2001) Overexpression of functionally coupled cyclooxygenase-2 and prostaglandin E synthase in symptomatic atherosclerotic plaques as a basis of prostaglandin E_2-dependent plaque instability. *Circulation*, **104**, 921–7.

19 Hinson, R.M., Williams, J.A. and Shacter, E. (1996) Elevated interleukin 6 is induced by prostaglandin E_2 in a murine model of inflammation: possible role of cyclooxygenase-2. *Proceedings of the National Academy of Sciences of the United States of America*, **93**, 4885–90.

20 Lopez, J.A., Armstrong, M.L., Harrison, D.G., Piegors, D.J. and Heistad, D.D. (1989) Vascular responses to leukocyte products in atherosclerotic primates. *Circulation Research*, **65**, 1078–86.

21 Kennedy, C.R., Zhang, Y., Brandon, S., Guan, Y., Coffee, K., Funk, C.D., Magnuson, M.A., Oates, J.A., Breyer, M.D. and Breyer, R.M. (1999) Salt-sensitive hypertension and reduced fertility in mice lacking the prostaglandin EP2 receptor. *Nature Medicine*, **5**, 217–20.

22. Baker, C.S., Hall, R.J., Evans, T.J., Pomerance, A., Maclouf, J., Creminon, C., Yacoub, M.H. and Polak, J.M. (1999) Cyclooxygenase-2 is widely expressed in atherosclerotic lesions affecting native and transplanted human coronary arteries and colocalizes with inducible nitric oxide synthase and nitrotyrosine particularly in macrophages. *Arteriosclerosis Thrombosis and Vascular Biology*, **19**, 646–55.

23. Belton, O., Byrne, D., Kearney, D., Leahy, A. and Fitzgerald, D.J. (2000) Cyclooxygenase-1 and -2-dependent prostacyclin formation in patients with atherosclerosis. *Circulation*, **102**, 840–5.

24. Schonbeck, U., Sukhova, G.K., Graber, P., Coulter, S. and Libby, P. (1999) Augmented expression of cyclo-oxygenase-2 in human atherosclerotic lesions. *American Journal of Pathology*, **155**, 1281–91.

25. FitzGerald, G.A., Smith, B., Pedersen, A.K. and Brash, A.R. (1984) Increased prostacyclin biosynthesis in patients with severe atherosclerosis and platelet activation. *New England Journal of Medicine*, **310**, 1065–8.

26. Pratico, D., Tillmann, C., Zhang, Z.B., Li, H. and FitzGerald, G.A. (2001) Acceleration of atherogenesis by COX-1-dependent prostanoid formation in low density lipoprotein receptor knockout mice. *Proceedings of the National Academy of Sciences of the United States of America*, **98**, 3358–63.

27. Belton, O.A., Duffy, A., Toomey, S. and Fitzgerald, D.J. (2003) Cyclooxygenase isoforms and platelet vessel wall interactions in the apolipoprotein E knockout mouse model of athero-sclerosis. *Circulation*, **108**, 3017–23.

28. Kobayashi, T., Tahara, Y., Matsumoto, M., Iguchi, M., Sano, H., Murayama, T., Arai, H., Oida, H., Yurugi-Kobayashi, T., Yamashita, J.K., Katagiri, H., Majima, M., Yokode, M., Kita, T. and Narumiya, S. (2004) Roles of thromboxane A_2 and prostacyclin in the development of atherosclerosis in apoE-deficient mice. *Journal of Clinical Investigation*, **114**, 784–94.

29. Egan, K.M., Lawson, J.A., Fries, S., Koller, B., Rader, D.J., Smyth, E.M. and Fitzgerald, G.A. (2004) COX-2-derived prostacyclin confers atheroprotection on female mice. *Science*, **306**, 1954–7.

30. Cayatte, A.J., Du, Y., Oliver-Krasinski, J., Lavielle, G., Verbeuren, T.J. and Cohen, R.A. (2000) The thromboxane receptor antagonist S18886 but not aspirin inhibits atherogenesis in apo E-deficient mice: evidence that eicosanoids other than thromboxane contribute to atherosclerosis. *Arteriosclerosis Thrombosis and Vascular Biology*, **20**, 1724–8.

31. Egan, K.M., Wang, M., Fries, S., Lucitt, M.B., Zukas, A.M., Pure, E., Lawson, J.A. and FitzGerald, G.A. (2005) Cyclooxygenases, thromboxane, and atherosclerosis: plaque destabilization by cyclooxygenase-2 inhibition combined with thromboxane receptor antagonism. *Circulation*, **111**, 334–42.

32. Cyrus, T., Sung, S., Zhao, L., Funk, C.D., Tang, S. and Pratico, D. (2002) Effect of low-dose aspirin on vascular inflammation, plaque stability, and atherogenesis in low-density lipoprotein receptor-deficient mice. *Circulation*, **106**, 1282–7.

33. Burleigh, M.E., Babaev, V.R., Oates, J.A., Harris, R.C., Gautam, S., Riendeau, D., Marnett, L.J., Morrow, J.D., Fazio, S. and Linton, M.F. (2002) Cyclooxygenase-2 promotes early atherosclerotic lesion formation in LDL receptor-deficient mice. *Circulation*, **105**, 1816–23.

34. Burleigh, M.E., Babaev, V.R., Yancey, P.G., Major, A.S., McCaleb, J.L., Oates, J.A., Morrow, J.D., Fazio, S. and Linton, M.F. (2005) Cyclooxygenase-2 promotes early atherosclerotic lesion formation in ApoE-deficient and C57BL/6 mice. *Journal of Molecular and Cellular Cardiology*, **39**, 443–52.

35. Rott, D., Zhu, J., Burnett, M.S., Zhou, Y.F., Zalles-Ganley, A., Ogunmakinwa, J. and Epstein, S.E. (2003) Effects of MF-tricyclic, a selective cyclooxygenase-2 inhibitor, on atherosclerosis progression and susceptibility to cytomegalovirus replication in apolipoprotein-E knockout mice. *Journal of the American College of Cardiology*, **41**, 1812–19.

36 Fu, J.Y., Masferrer, J.L., Seibert, K., Raz, A. and Needleman, P. (1990) The induction and suppression of prostaglandin H_2 synthase (cyclooxygenase) in human monocytes. *Journal of Biological Chemistry*, **265**, 16737–40.

37 Cipollone, F., Fazia, M.L., Iezzi, A., Cuccurullo, C., De Cesare, D., Ucchino, S., Spigonardo, F., Marchetti, A., Buttitta, F., Paloscia, L., Mascellanti, M., Cuccurullo, F. and Mezzetti, A. (2005) Association between prostaglandin E receptor subtype EP_4 overexpression and unstable phenotype in atherosclerotic plaques in human. *Arteriosclerosis Thrombosis and Vascular Biology*, **25**, 1925–31.

38 Wang, M., Zukas, A.M., Hui, Y., Ricciotti, E., Pure, E. and FitzGerald, G.A. (2006) Deletion of microsomal prostaglandin E synthase-1 augments prostacyclin and retards atherogenesis. *Proceedings of the National Academy of Sciences of the United States of America*, **103**, 14507–12.

39 Mardini, I.A. and FitzGerald, G.A. (2001) Selective inhibitors of cyclooxygenase-2: a growing class of anti-inflammatory drugs. *Molecular Interventions*, **1**, 30–8.

40 Zimmermann, K.C., Sarbia, M., Schror, K. and Weber, A.A. (1998) Constitutive cyclooxygenase-2 expression in healthy human and rabbit gastric mucosa. *Molecular Pharmacology*, **54**, 536–40.

41 Wallace, J.L. and Devchand, P.R. (2005) Emerging roles for cyclooxygenase-2 in gastrointestinal mucosal defense. *British Journal of Pharmacology*, **145**, 275–82.

42 Fitzgerald, G.A. (2004) Coxibs and cardiovascular disease. *New England Journal of Medicine*, **351**, 1709–11.

43 Furberg, C.D. (2006) The COX-2 inhibitors – an update. *American Heart Journal*, **152**, 197–9.

44 Bombardier, C., Laine, L., Reicin, A., Shapiro, D., Burgos-Vargas, R., Davis, B., Day, R., Ferraz, M.B., Hawkey, C.J., Hochberg, M.C., Kvien, T.K. and Schnitzer, T.J. (2000) Comparison of upper gastrointestinal toxicity of rofecoxib and naproxen in patients with rheumatoid arthritis. VIGOR Study Group. *New England Journal of Medicine*, **343**, 1520–8.

45 FitzGerald, G.A. (2003) COX-2 and beyond: approaches to prostaglandin inhibition in human disease. *Nature Reviews Drug Discovery*, **2**, 879–90.

46 Silverstein, F.E., Faich, G., Goldstein, J.L., Simon, L.S., Pincus, T., Whelton, A., Makuch, R., Eisen, G., Agrawal, N.M., Stenson, W.F., Burr, A.M., Zhao, W.W., Kent, J.D., Lefkowith, J.B., Verburg, K.M. and Geis, G.S. (2000) Gastrointestinal toxicity with celecoxib vs nonsteroidal anti-inflammatory drugs for osteoarthritis and rheumatoid arthritis: the CLASS study: a randomized controlled trial. Celecoxib Long-term Arthritis Safety Study. *Journal of the American Medical Association*, **284**, 1247–55.

47 Caldwell, B., Aldington, S., Weatherall, M., Shirtcliffe, P. and Beasley, R. (2006) Risk of cardiovascular events and celecoxib: a systematic review and meta-analysis. *Journal of the Royal Society of Medicine*, **99**, 132–40.

48 Farkouh, M.E., Kirshner, H., Harrington, R.A., Ruland, S., Verheugt, F.W., Schnitzer, T.J., Burmester, G.R., Mysler, E., Hochberg, M.C., Doherty, M., Ehrsam, E., Gitton, X., Krammer, G., Mellein, B., Gimona, A., Matchaba, P., Hawkey, C.J. and Chesebro, J.H. (2004) Comparison of lumiracoxib with naproxen and ibuprofen in the Therapeutic Arthritis Research and Gastrointestinal Event Trial (TARGET), cardiovascular outcomes: randomised controlled trial. *Lancet*, **364**, 675–84.

49 Schnitzer, T.J., Burmester, G.R., Mysler, E., Hochberg, M.C., Doherty, M., Ehrsam, E., Gitton, X., Krammer, G., Mellein, B., Matchaba, P., Gimona, A. and Hawkey, C.J. (2004) Comparison of lumiracoxib with naproxen and ibuprofen in the Therapeutic Arthritis Research and Gastrointestinal Event Trial (TARGET), reduction in ulcer complications: randomised controlled trial. *Lancet*, **364**, 665–74.

50 Topol, E.J. and Falk, G.W. (2004) A coxib a day won't keep the doctor away. *Lancet*, **364**, 639–40.

51 Grosser, T., Fries, S. and FitzGerald, G.A. (2006) Biological basis for the

51 cardiovascular consequences of COX-2 inhibition: therapeutic challenges and opportunities. *Journal of Clinical Investigation*, **116**, 4–15.
52 Mitchell, J.A. and Warner, T.D. (2006) COX isoforms in the cardiovascular system: understanding the activities of non-steroidal anti-inflammatory drugs. *Nature Reviews Drug Discovery*, **5**, 75–86.
53 Bresalier, R.S., Sandler, R.S., Quan, H., Bolognese, J.A., Oxenius, B., Horgan, K., Lines, C., Riddell, R., Morton, D., Lanas, A., Konstam, M.A. and Baron, J.A. (2005) Cardiovascular events associated with rofecoxib in a colorectal adenoma chemoprevention trial. *New England Journal of Medicine*, **352**, 1092–102.
54 Solomon, S.D., McMurray, J.J., Pfeffer, M.A., Wittes, J., Fowler, R., Finn, P., Anderson, W.F., Zauber, A., Hawk, E. and Bertagnolli, M. (2005) Cardiovascular risk associated with celecoxib in a clinical trial for colorectal adenoma prevention. *New England Journal of Medicine*, **352**, 1071–80.
55 Drazen, J.M., Austen, K.F., Lewis, R.A., Clark, D.A., Goto, G., Marfat, A. and Corey, E.J. (1980) Comparative airway and vascular activities of leukotrienes C-1 and D *in vivo* and *in vitro*. *Proceedings of the National Academy of Sciences of the United States of America*, **77**, 4354–8.
56 Dahlen, S.E., Bjork, J., Hedqvist, P., Arfors, K.E., Hammarstrom, S., Lindgren, J.A. and Samuelsson, B. (1981) Leukotrienes promote plasma leakage and leukocyte adhesion in postcapillary venules: *in vivo* effects with relevance to the acute inflammatory response. *Proceedings of the National Academy of Sciences of the United States of America*, **78**, 3887–91.
57 Allen, S.P., Chester, A.H., Dashwood, M.R., Tadjkarimi, S., Piper, P.J. and Yacoub, M.H. (1994) Preferential vasoconstriction to cysteinyl leukotrienes in the human saphenous vein compared with the internal mammary artery. Implications for graft performance. *Circulation*, **90**, 515–24.
58 Allen, S.P., Chester, A.H., Piper, P.J., Sampson, A.P., Akl, E.S. and Yacoub, M.H. (1992) Effects of leukotrienes C4 and D4 on human isolated saphenous veins. *British Journal of Clinical Pharmacology*, **34**, 409–14.
59 Berkowitz, B.A., Zabko-Potapovich, B., Valocik, R. and Gleason, J.G. (1984) Effects of the leukotrienes on the vasculature and blood pressure of different species. *Journal of Pharmacology and Experimental Therapeutics*, **229**, 105–12.
60 Feuerstein, G. (1984) Leukotrienes and the cardiovascular system. *Prostaglandins*, **27**, 781–802.
61 Lawson, D.L., Smith, C., Mehta, J.L., Mehta, P. and Nichols, W.W. (1988) Leukotriene D4 potentiates the contractile effects of epinephrine and norepinephrine on rat aortic rings. *Journal of Pharmacology and Experimental Therapeutics*, **247**, 953–7.
62 Letts, L.G., Newman, D.L., Greenwald, S.E. and Piper, P.J. (1983) Effects of intracoronary administration of leukotriene D4 in the anaesthetized dog. *Prostaglandins*, **26**, 563–72.
63 Letts, L.G. and Piper, P.J. (1983) Cardiac actions of leukotrienes B_4, C_4, D_4, and E_4 in guinea pig and rat *in vitro*. *Advances in Prostaglandin Thromboxane and Leukotriene Research*, **11**, 391–5.
64 Michelassi, F., Landa, L., Hill, R.D., Lowenstein, E., Watkins, W.D., Petkau, A.J. and Zapol, W.M. (1982) Leukotriene D_4: a potent coronary artery vasoconstrictor associated with impaired ventricular contraction. *Science*, **217**, 841–3.
65 Allen, S., Dashwood, M., Morrison, K. and Yacoub, M. (1998) Differential leukotriene constrictor responses in human atherosclerotic coronary arteries. *Circulation*, **97**, 2406–13.
66 Beller, T.C., Friend, D.S., Maekawa, A., Lam, B.K., Austen, K.F. and Kanaoka, Y. (2004) Cysteinyl leukotriene 1 receptor controls the severity of chronic pulmonary inflammation and fibrosis. *Proceedings of the National Academy of Sciences of the United States of America*, **101**, 3047–52.
67 Kanaoka, Y., Maekawa, A., Penrose, J.F., Austen, K.F. and Lam, B.K. (2001) Attenuated zymosan-induced peritoneal vascular permeability and IgE-dependent

passive cutaneous anaphylaxis in mice lacking leukotriene C_4 synthase. *Journal of Biological Chemistry*, **276**, 22608–13.
68 Beller, T.C., Maekawa, A., Friend, D.S., Austen, K.F. and Kanaoka, Y. (2004) Targeted gene disruption reveals the role of the cysteinyl leukotriene 2 receptor in increased vascular permeability and in bleomycin-induced pulmonary fibrosis in mice. *Journal of Biological Chemistry*, **279**, 46129–34.
69 Maekawa, A., Austen, K.F. and Kanaoka, Y. (2002) Targeted gene disruption reveals the role of cysteinyl leukotriene 1 receptor in the enhanced vascular permeability of mice undergoing acute inflammatory responses. *Journal of Biological Chemistry*, **277**, 20820–4.
70 Hui, Y., Cheng, Y., Smalera, I., Jian, W., Goldhahn, L., Fitzgerald, G.A. and Funk, C.D. (2004) Directed vascular expression of human cysteinyl leukotriene 2 receptor modulates endothelial permeability and systemic blood pressure. *Circulation*, **110**, 3360–6.
71 Datta, Y.H., Romano, M., Jacobson, B.C., Golan, D.E., Serhan, C.N. and Ewenstein, B.M. (1995) Peptido-leukotrienes are potent agonists of von Willebrand factor secretion and P-selectin surface expression in human umbilical vein endothelial cells. *Circulation*, **92**, 3304–11.
72 Pedersen, K.E., Bochner, B.S. and Undem, B.J. (1997) Cysteinyl leukotrienes induce P-selectin expression in human endothelial cells via a non-CysLT1 receptor-mediated mechanism. *Journal of Pharmacology and Experimental Therapeutics*, **281**, 655–62.
73 McIntyre, T.M., Zimmerman, G.A. and Prescott, S.M. (1986) Leukotrienes C_4 and D_4 stimulate human endothelial cells to synthesize platelet-activating factor and bind neutrophils. *Proceedings of the National Academy of Sciences of the United States of America*, **83**, 2204–8.
74 Heimburger, M. and Palmblad, J.E. (1996) Effects of leukotriene C_4 and D_4, histamine and bradykinin on cytosolic calcium concentrations and adhesiveness of endothelial cells and neutrophils. *Clinical and Experimental Immunology*, **103**, 454–60.
75 Uzonyi, B., Lotzer, K., Jahn, S., Kramer, C., Hildner, M., Bretschneider, E., Radke, D., Beer, M., Vollandt, R., Evans, J.F., Funk, C.D. and Habenicht, A.J. (2006) Cysteinyl leukotriene 2 receptor and protease-activated receptor 1 activate strongly correlated early genes in human endothelial cells. *Proceedings of the National Academy of Sciences of the United States of America*, **103**, 6326–31.
76 Woszczek, G., Chen, L.Y., Nagineni, S., Alsaaty, S., Harry, A., Logun, C., Pawliczak, R. and Shelhamer, J.H. (2007) IFN-γ induces cysteinyl leukotriene receptor 2 expression and enhances the responsiveness of human endothelial cells to cysteinyl leukotrienes. *Journal of Immunology*, **178**, 5262–70.
77 Lotzer, K., Jahn, S., Kramer, C., Hildner, M., Nusing, R., Funk, C.D. and Habenicht, A.J.R. (2007) 5-Lipoxygenase/cyclooxygenase-2 cross-talk through cysteinyl leukotriene receptor 2 in endothelial cells. *Prostaglandins & Other Lipid Mediators*, **84**, 108–15.
78 Zhao, L., Moos, M.P., Grabner, R., Pedrono, F., Fan, J., Kaiser, B., John, N., Schmidt, S., Spanbroek, R., Lotzer, K., Huang, L., Cui, J., Rader, D.J., Evans, J.F., Habenicht, A.J. and Funk, C.D. (2004) The 5-lipoxygenase pathway promotes pathogenesis of hyperlipidemia-dependent aortic aneurysm. *Nature Medicine*, **10**, 966–73.
79 Huang, L., Zhao, A., Wong, F., Ayala, J.M., Struthers, M., Ujjainwalla, F., Wright, S.D., Springer, M.S., Evans, J. and Cui, J. (2004) Leukotriene B_4 strongly increases monocyte chemoattractant protein-1 in human monocytes. *Arteriosclerosis Thrombosis and Vascular Biology*, **24**, 1783–8.
80 De Caterina, R., Mazzone, A., Giannessi, D., Sicari, R., Pelosi, W., Lazzerini, G., Azzara, A., Forder, R., Carey, F., Caruso, D. (1988) Leukotriene B_4 production in human atherosclerotic plaques. *Biomedica Biochimica Acta*, **47**, S182–5.
81 Yla-Herttuala, S., Rosenfeld, M.E., Parthasarathy, S., Sigal, E., Sarkioja, T., Witztum, J.L. and Steinberg, D. (1991)

Gene expression in macrophage-rich human atherosclerotic lesions. 15-lipoxygenase and acetyl low density lipoprotein receptor messenger RNA colocalize with oxidation specific lipid-protein adducts. *Journal of Clinical Investigation*, **87**, 1146–52.

82 Spanbroek, R., Grabner, R., Lotzer, K., Hildner, M., Urbach, A., Ruhling, K., Moos, M.P., Kaiser, B., Cohnert, T.U., Wahlers, T., Zieske, A., Plenz, G., Robenek, H., Salbach, P., Kuhn, H., Radmark, O., Samuelsson, B. and Habenicht, A.J. (2003) Expanding expression of the 5-lipoxygenase pathway within the arterial wall during human atherogenesis. *Proceedings of the National Academy of Sciences of the United States of America*, **100**, 1238–43.

83 Cipollone, F., Mezzetti, A., Fazia, M.L., Cuccurullo, C., Iezzi, A., Ucchino, S., Spigonardo, F., Bucci, M., Cuccurullo, F., Prescott, S.M. and Stafforini, D.M. (2005) Association between 5-lipoxygenase expression and plaque instability in humans. *Arteriosclerosis Thrombosis and Vascular Biology*, **25**, 1665–70.

84 Qiu, H., Gabrielsen, A., Agardh, H.E., Wan, M., Wetterholm, A., Wong, C.H., Hedin, U., Swedenborg, J., Hansson, G.K., Samuelsson, B., Paulsson-Berne, G. and Haeggstrom, J.Z. (2006) Expression of 5-lipoxygenase and leukotriene A$_4$ hydrolase in human atherosclerotic lesions correlates with symptoms of plaque instability. *Proceedings of the National Academy of Sciences of the United States of America*, **103**, 8161–6.

85 Lotzer, K., Spanbroek, R., Hildner, M., Urbach, A., Heller, R., Bretschneider, E., Galczenski, H., Evans, J.F. and Habenicht, A.J. (2003) Differential leukotriene receptor expression and calcium responses in endothelial cells and macrophages indicate 5-lipoxygenase-dependent circuits of inflammation and atherogenesis. *Arteriosclerosis Thrombosis and Vascular Biology*, **23**, e32–6.

86 Mellor, E.A., Frank, N., Soler, D., Hodge, M.R., Lora, J.M., Austen, K.F. and Boyce, J.A. (2003) Expression of the type 2 receptor for cysteinyl leukotrienes (CysLT2R) by human mast cells: Functional distinction from CysLT1R. *Proceedings of the National Academy of Sciences of the United States of America*, **100**, 11589–93.

87 Mellor, E.A., Maekawa, A., Austen, K.F. and Boyce, J.A. (2001) Cysteinyl leukotriene receptor 1 is also a pyrimidinergic receptor and is expressed by human mast cells. *Proceedings of the National Academy of Sciences of the United States of America*, **98**, 7964–9.

88 Amrani, Y., Moore, P.E., Hoffman, R., Shore, S.A. and Panettieri, R.A. Jr (2001) Interferon-gamma modulates cysteinyl leukotriene receptor-1 expression and function in human airway myocytes. *American Journal of Respiratory and Critical Care Medicine*, **164**, 2098–101.

89 Kamohara, M., Takasaki, J., Matsumoto, M., Matsumoto, S., Saito, T., Soga, T., Matsushime, H. and Furuichi, K. (2001) Functional characterization of cysteinyl leukotriene CysLT$_2$ receptor on human coronary artery smooth muscle cells. *Biochemical and Biophysical Research Communications*, **287**, 1088–92.

90 Back, M., Bu, D.X., Branstrom, R., Sheikine, Y., Yan, Z.Q. and Hansson, G.K. (2005) Leukotriene B4 signaling through NF-kappaB-dependent BLT1 receptors on vascular smooth muscle cells in atherosclerosis and intimal hyperplasia. *Proceedings of the National Academy of Sciences of the United States of America*, **102**, 17501–6.

91 Schoenberger, S.P. (2003) BLT for speed. *Nature Immunology*, **4**, 937–9.

92 Lotzer, K., Funk, C.D. and Habenicht, A.J. (2005) The 5-lipoxygenase pathway in arterial wall biology and atherosclerosis. *Biochimica et Biophysica Acta*, **1736**, 30–7.

93 Mehrabian, M., Wong, J., Wang, X., Jiang, Z., Shi, W., Fogelman, A.M. and Lusis, A.J. (2001) Genetic locus in mice that blocks development of atherosclerosis despite extreme hyperlipidemia. *Circulation Research*, **89**, 125–30.

94 Mehrabian, M., Allayee, H., Wong, J., Shi, W., Wang, X.P., Shaposhnik, Z., Funk, C.D. and Lusis, A.J. (2002) Identification of 5-lipoxygenase as a

major gene contributing to atherosclerosis susceptibility in mice. *Circulation Research*, **91**, 120–6.
95. Aiello, R.J., Bourassa, P.A., Lindsey, S., Weng, W., Freeman, A. and Showell, H.J. (2002) Leukotriene B_4 receptor antagonism reduces monocytic foam cells in mice. *Arteriosclerosis Thrombosis and Vascular Biology*, **22**, 443–9.
96. Subbarao, K., Jala, V.R., Mathis, S., Suttles, J., Zacharias, W., Ahamed, J., Ali, H., Tseng, M.T. and Haribabu, B. (2004) Role of leukotriene B_4 receptors in the development of atherosclerosis: potential mechanisms. *Arteriosclerosis Thrombosis and Vascular Biology*, **24**, 369–75.
97. Heller, E.A., Liu, E., Tager, A.M., Sinha, S., Roberts, J.D., Koehn, S.L., Libby, P., Aikawa, E.R., Chen, J.Q., Huang, P., Freeman, M.W., Moore, K.J., Luster, A.D. and Gerszten, R.E. (2005) Inhibition of atherogenesis in BLT_1-deficient mice reveals a role for LTB_4 and BLT_1 in smooth muscle cell recruitment. *Circulation*, **112**, 578–86.
98. Dwyer, J.H., Allayee, H., Dwyer, K.M., Fan, J., Wu, H., Mar, R., Lusis, A.J. and Mehrabian, M. (2004) Arachidonate 5-lipoxygenase promoter genotype, dietary arachidonic acid, and atherosclerosis. *New England Journal of Medicine*, **350**, 29–37.
99. Helgadottir, A., Manolescu, A., Thorleifsson, G., Gretarsdottir, S., Jonsdottir, H., Thorsteinsdottir, U., Samani, N.J., Gudmundsson, G., Grant, S.F., Thorgeirsson, G., Sveinbjornsdottir, S., Valdimarsson, E.M., Matthiasson, S.E., Johannsson, H., Gudmundsdottir, O., Gurney, M.E., Sainz, J., Thorhallsdottir, M., Andresdottir, M., Frigge, M.L., Topol, E.J., Kong, A., Gudnason, V., Hakonarson, H., Gulcher, J.R. and Stefansson, K. (2004) The gene encoding 5-lipoxygenase activating protein confers risk of myocardial infarction and stroke. *Nature Genetics*, **36**, 233–9.
100. Helgadottir, A., Gretarsdottir, S., St Clair, D., Manolescu, A., Cheung, J., Thorleifsson, G., Pasdar, A., Grant, S.F., Whalley, L.J., Hakonarson, H., Thorsteinsdottir, U., Kong, A., Gulcher, J., Stefansson, K. and MacLeod, M.J. (2005) Association between the gene encoding 5-lipoxygenase-activating protein and stroke replicated in a Scottish population. *American Journal of Human Genetics*, **76**, 505–9.
101. Lohmussaar, E., Gschwendtner, A., Mueller, J.C., Org, T., Wichmann, E., Hamann, G., Meitinger, T. and Dichgans, M. (2005) ALOX5AP gene and the PDE4D gene in a central European population of stroke patients. *Stroke*, **36**, 731–6.
102. Helgadottir, A., Manolescu, A., Helgason, A., Thorleifsson, G., Thorsteinsdottir, U., Gudbjartsson, D.F., Gretarsdottir, S., Magnusson, K.P., Gudmundsson, G., Hicks, A., Jonsson, T., Grant, S.F., Sainz, J., O'Brien, S.J., Sveinbjornsdottir, S., Valdimarsson, E.M., Matthiasson, S.E., Levey, A.I., Abramson, J.L., Reilly, M.P., Vaccarino, V., Wolfe, M.L., Gudnason, V., Quyyumi, A.A., Topol, E.J., Rader, D.J., Thorgeirsson, G., Gulcher, J.R., Hakonarson, H., Kong, A. and Stefansson, K. (2006) A variant of the gene encoding leukotriene A_4 hydrolase confers ethnicity-specific risk of myocardial infarction. *Nature Genetics*, **38**, 68–74.
103. Hakonarson, H., Thorvaldsson, S., Helgadottir, A., Gudbjartsson, D., Zink, F., Andresdottir, M., Manolescu, A., Arnar, D.O., Andersen, K., Sigurdsson, A., Thorgeirsson, G., Jonsson, A., Agnarsson, U., Bjornsdottir, H., Gottskalksson, G., Einarsson, A., Gudmundsdottir, H., Adalsteinsdottir, A.E., Gudmundsson, K., Kristjansson, K., Hardarson, T., Kristinsson, A., Topol, E.J., Gulcher, J., Kong, A., Gurney, M., Thorgeirsson, G. and Stefansson, K. (2005) Effects of a 5-lipoxygenase-activating protein inhibitor on biomarkers associated with risk of myocardial infarction: a randomized trial. *Journal of the American Medical Association*, **293**, 2245–56.

14
Cytochrome P450-Dependent Eicosanoids
Wolf-Hagen Schunck and Cosima Schmidt

14.1
Introduction

This chapter has been written from two perspectives: (i) that of structure and function of a single cytochrome P450 (CYP) molecule and (ii) that of physiology and cardiovascular disease.

The CYP enzymes to be described are hemoproteins and reside in the endoplasmic reticulum (ER) of many cell types throughout the cardiovascular system. CYP enzymes require access to their substrate, arachidonic acid (AA) in our case. Moreover, CYP enzymes need molecular oxygen and electrons to catalyze product formation. As will be shown in detail below, these requirements at the molecular level already determine a deep integration of CYP function into a broad physiological and pathophysiological context.

The other perspective considers the CYP-dependent metabolites as second messengers of various hormones and growth factors regulating vascular, renal and cardiac function. Here, the CYP-dependent metabolites are components of signaling pathways that modulate the activity of ion channels, transcription factors and mitogen-activated protein kinases (MAPKs). Moreover, we will consider the cause–effect relationships associating alterations in CYP-dependent AA metabolism with the development of cardiovascular disease.

14.2
Prospects of the Research Field

AA is oxygenated by cyclooxygenases (COXs), lipoxygenases (LOXs) and CYP enzymes to different classes of biologically active metabolites, collectively termed eicosanoids (Figure 14.1). The name is derived from the Greek word for the number 20 (*eicosa*) and refers to the chain length of AA as the principal precursor.

14 Cytochrome P450-Dependent Eicosanoids

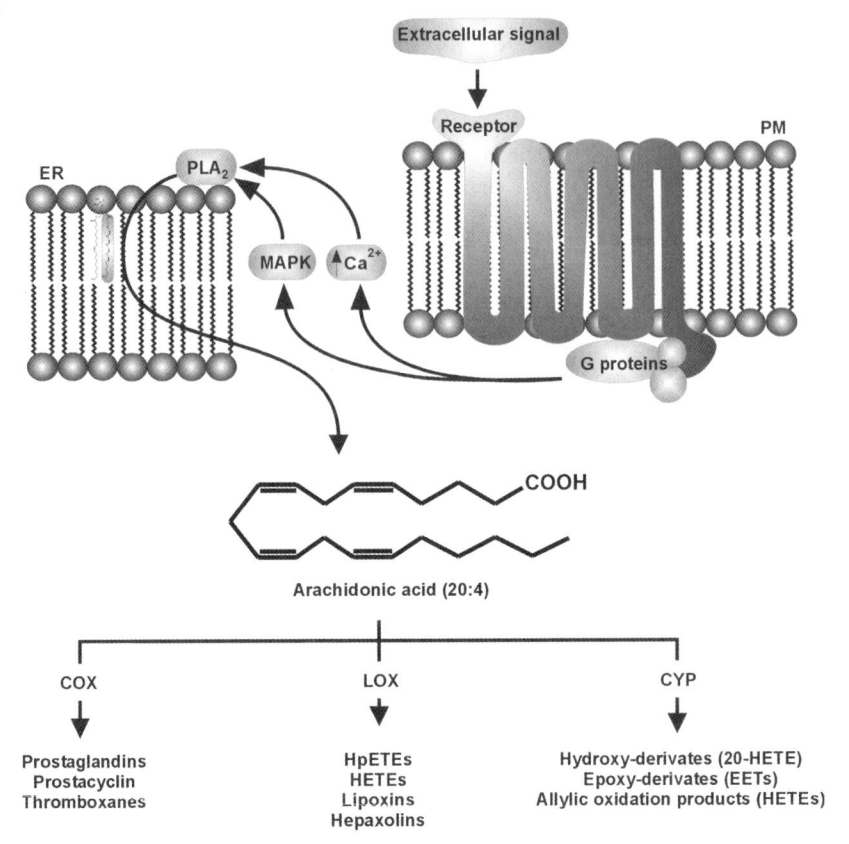

Figure 14.1 The three branches of the AA cascade. Extracellular signals induce AA release from the ER and the nuclear envelope after binding to GPCRs and subsequent activation of PLA$_2$ or other types of phospholipases. COXs, LOXs and CYP enzymes utilize free AA as substrate and initiate the formation of different classes of biologically active metabolites.

COX enzymes initiate the formation of prostaglandins and thromboxanes, and LOX enzymes that of leukotrienes and other lipid mediators [1]. The COX- and LOX-dependent pathways are important clinical targets for the treatment of inflammation, cardiovascular disease, asthma, fever and pain. Today, a wide variety of drugs is available that either affect prostaglandin and leukotriene formation (COX and LOX inhibitors) or block their action (receptor antagonists). Examples involve nonsteroidal antiinflammatory drugs, specific COX-2 inhibitors and leukotriene C$_4$ antagonists. The COX- and LOX-dependent pathways are the topic of a separate chapter of this volume (see Chapter 12) and the reader is referred there for further details.

The CYP-catalyzed hydroxylation and epoxidation of AA was established only recently as the so-called third branch of the AA cascade (see Section 14.3). Physiologically important metabolites include 20-hydroxyeicosatetraenoic acid (20-HETE) and a set of regio- and stereoisomeric epoxyeicosatrienoic acids (EETs).

Recent studies revealed crucial physiological and pathophysiological roles of this unique class of AA metabolites. These developments raise the possibility that the CYP branch of the AA cascade may become a clinical target of similar importance as the COX- and LOX-dependent pathways. Specific application fields are emerging for the treatment and prevention of cardiovascular disease, and of hypertension- and ischemia-associated injuries of the vasculature, heart, kidney and brain, in particular.

14.3
How CYP Enzymes Became Established Members of the AA Cascade

The first reports [2, 3] indicating an involvement of CYP enzymes in AA metabolism were published at the end of the 1970s. In 1981, microsomal and purified CYP enzymes were directly shown to catalyze AA oxygenation. These seminal studies were reported by Capdevila *et al.* [4, 5], Morrison and Pascoe [6] and Oliw *et al.* [7]. At that time, research on the biological relevance of the first two branches of the AA cascade was at least a decade ahead, and the CYP research field was shaped by discoveries in pharmacology and toxicology.

However, soon it became obvious that CYP-dependent eicosanoids are also produced under *in vivo* conditions and display potent biological activities. Early studies revealed effects on renal salt reabsorption and vascular tone. In 1989, Sacerdoti *et al.* showed that increased renal 20-HETE production contributes to the elevation of blood pressure in the spontaneously hypertensive rat (SHR) model [8]. McGiff proposed then that CYP-dependent AA metabolism is involved in the pathophysiology of hypertension [9]. Later on, Dahl salt-sensitive rats were shown to develop hypertension due to their inability to upregulate renal EET levels in response to salt loading [10]. Studies by Roman *et al.* revealed that there is also a deficiency in tubular 20-HETE production which significantly impairs salt excretion [11]. This was the starting point for noticing and elucidating the partially opposite and partially parallel roles of EETs and 20-HETE in blood pressure regulation.

In 1996, EETs were described as endothelium-derived hyperpolarizing factors (EDHFs) by Campbell *et al.* [12]. Three years later, Fisslthaler *et al.* identified a CYP2C enzyme as the EDHF synthase in coronary arteries [13]. However, the specific mechanisms of EET action in different vascular beds and the relative importance of EETs compared to other EDHF-like factors are still issues of intensive research and discussion [14–17].

Recent studies have indicated that the relevance of EETs and 20-HETE goes far beyond their role as mediators in the regulation of renal function and vascular tone. Node *et al.* reported in 1999 that the cytokine-induced inflammatory response of vascular endothelial cells (ECs) is repressed by EETs [18]. Later on, EETs were shown to have general cell- and organ-protective properties that ameliorate ischemia/reperfusion (I/R) injury. Noteworthy, 20-HETE appears to play a detrimental role under the same conditions. Moreover, Falck et al. synthesized stable analogs of EETs and 20-HETE that became important tools as agonists or antagonists.

A new field was also opened by reports showing an involvement of EETs and 20-HETE as mediators of mitogenic and angiogenic responses [19–25]. These responses are relevant to physiological adaptation and tissue repair. On the other hand, an imbalance in the proliferation and migration of EC and vascular smooth muscle cells (VSMCs) contributes to pathophysiological vascular remodeling and restenosis. Although beyond the scope of this chapter, it is important to note that EETs and 20-HETE were shown to mediate mitogenic responses also in several tumor cell lines. Thus, combined with their angiogenic potential, CYP-dependent eicosanoids may significantly contribute to cancer development. Recent studies by Jiang et al. demonstrated that EETs promote the neoplastic phenotype of carcinoma cells, tumor growth and tumor metastasis [26, 27].

Recently, soluble epoxide hydrolase (sEH) became an important target to increase cardiovascular EET levels [28, 29]. To mention only one example at this point, Hammock et al. demonstrated that pharmacological sEH inhibition protects against cardiac hypertrophy in a mouse model [30]. Currently, there are great expectations regarding the development of clinically applicable sEH inhibitors in this research field.

Nutrition and the potential role of CYP-dependent AA metabolism in mediating the development of the metabolic syndrome may become a further long-term objective of research. Nutrition determines the availability of essential polyunsaturated fatty acids (PUFAs) for CYP-dependent production of lipid mediators. Wang et al. showed that rats receiving a "Western-type" high-fat diet display reduced renal EET and 20-HETE generation, and suffer from hypertension [31]. Our laboratory and other groups showed that fish oil ω-3 PUFAs (*n*-3 PUFAs) are efficient alternative substrates for various AA-metabolizing CYP isoforms [32], and protect against end-organ damage and sudden cardiac death in a rat model of angiotensin (Ang) II-induced hypertension [33].

Moreover, after initial studies of Goldstein, Zeldin and collaborators [34, 35], the search for functionally relevant polymorphisms in CYP and sEH genes and their association with human cardiovascular disease became a highly active novel research field.

14.4
Structure and Function of CYP Enzymes and Their Role in AA Metabolism

14.4.1
Unique Spectral and Catalytic Features of CYP Enzymes

CYP proteins have a molecular weight of about 50 kDa and carry iron-protoporphyrin IX as prosthetic group. A cysteinyl-SH group functions as the fifth ligand to the heme iron. The heme-binding cysteine is located within a C-terminal region, which harbors a set of invariant amino positions characteristic for the whole CYP superfamily: **FXXGXBZCXG** (the heme-binding cysteine is underlined; B = a basic residue, Z = a hydrophobic or neutral residue).

This type of heme-binding is an important prerequisite for the activation of molecular oxygen. At the same time, it is responsible for the name-giving spectral feature of CYP (P450) proteins, which were discovered in the 1950s as "pigments" showing an absorption maximum at 450 nm after binding carbon monoxide as the sixth ligand to the reduced heme iron. [36]

CYP enzymes function as monooxygenases and thereby catalyze reactions according to the following general equation:

$$RH + O_2 + NAD(P)H + H^+ \rightarrow ROH + H_2O + NAD(P)^+$$

14.4.2
CYP Systems and Their Reaction Cycle in the ER

14.4.2.1 Membrane Integration and Substrate Access

All currently known AA-metabolizing CYP isoforms are integral membrane proteins anchored via N-terminal hydrophobic sequences to the ER (Figure 14.2A). The remainder of the CYP molecule forms a large cytosolic domain, which surrounds the heme prosthetic group. Some hydrophobic regions of the cytosolic domain are attached to the membrane surface. This may provide direct access to the substrate-binding channel for those compounds distributed in the phospholipid bilayer (Figure 14.2A).

14.4.2.2 Electron Transfer and Activation of Molecular Oxygen

Activation of molecular oxygen requires two electrons, which are donated by NADPH. To meet the requirements of the catalytic cycle (Figure 14.2B), the two-electron delivery from NADPH has to be translated into two one-electron transfer steps to the heme iron. ER-bond (microsomal) CYP proteins cooperate for this purpose with a NADPH-dependent reductase [CYP reductase (CPR)], which contains flavin adenine dinucleotide (FAD) and flavin mononucleotide (FMN) as prosthetic groups (Figure 14.2A). Transfer of the first electron results in a reduction of the heme iron. This allows binding of molecular oxygen, which becomes then further activated by the second electron transfer.

14.4.2.3 Product Formation and Specificity

In the last but one step of the reaction cycle, the reduced dioxygen structure at the heme iron is cleaved to release a water molecule. Finally, the remaining oxygen atom is introduced into the substrate and the product is released from the active site. Substrate orientation to this activated oxygen atom at the heme iron is the major determinant for regio- and stereoselective product formation.

The substrate binding sites of CYP enzymes evolved based on a selection for specific amino acid residues and their spatial arrangement. Consequently, individual CYP isoforms can differ in (i) the principal type of their substrates

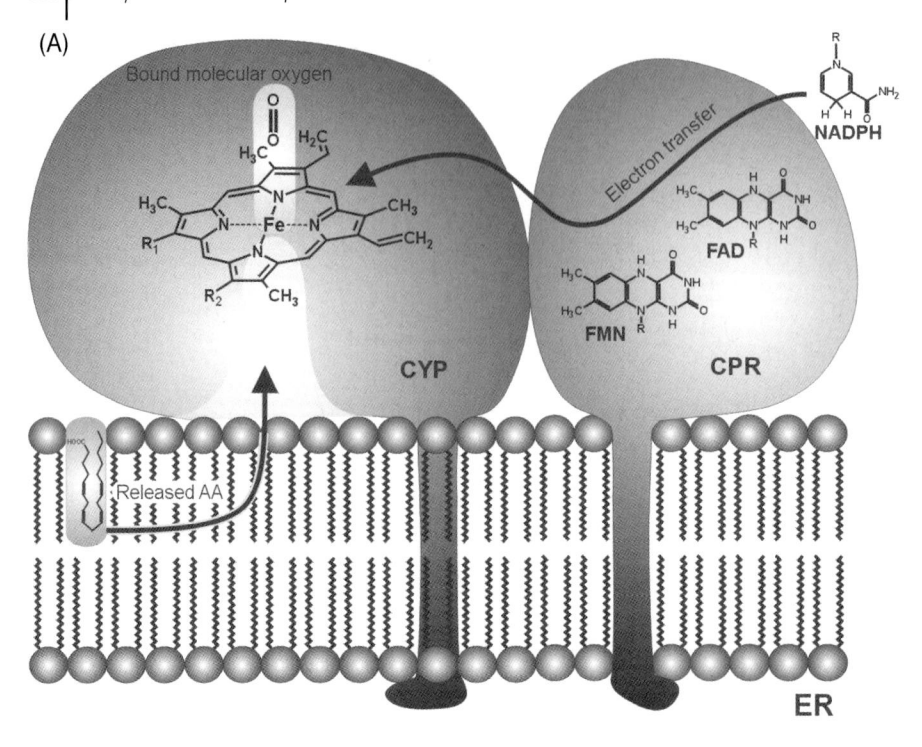

Figure 14.2 Structure and function of CYP enzymes in the ER. (a) Microsomal CYP systems consist of a CYP protein and the NADPH-CYP reductase (CPR). The CPR transfers electrons donated by NADPH via FAD and FMN to the heme of the CYP protein. The CYP component binds free AA, activates molecular oxygen at the heme iron and introduces one oxygen atom into the substrate. (b) Simplified scheme of the CYP-catalyzed reaction cycle. RH, substrate; Fe, heme iron. For further details see text.

("substrate specificity"), (ii) the atom of the same substrate, which becomes oxygenated ("regioselectivity"), and (iii) the sterical side of attack of activated oxygen towards the same atom and of the same substrate ("stereoselectivity").

14.4.3
Reaction Types and Primary Products of CYP-Dependent AA Metabolism

CYP enzymes produce biologically active AA metabolites by catalyzing hydroxylation and epoxidation reactions (Figure 14.3) [37, 38].

Hydroxylation may yield 16- through 20-HETE depending on the regioselectivity of the CYP isoform. 16- through 19-HETE can be formed either as R- or S-enantiomers.

Epoxidation results from an attack of the double bonds by activated oxygen. This reaction produces a family of metabolites comprising four regioisomers (5,6-, 8,9-, 11,12- and 14,15-EETs) and their stereoisomers (the R,S and S,R enantiomers).

Moreover, CYP enzymes catalyze allylic oxidation of AA, a reaction type resembling that of LOX enzymes. Potential metabolites are regioisomeric hydroxy-derivatives such as 7-, 8-, 9-, 10-, 11-, 12-, 13- and 15-HETE. The characteristic feature of these "subterminal" HETEs is the presence of a dienol functionality arising from a double-bond shift during the reaction.

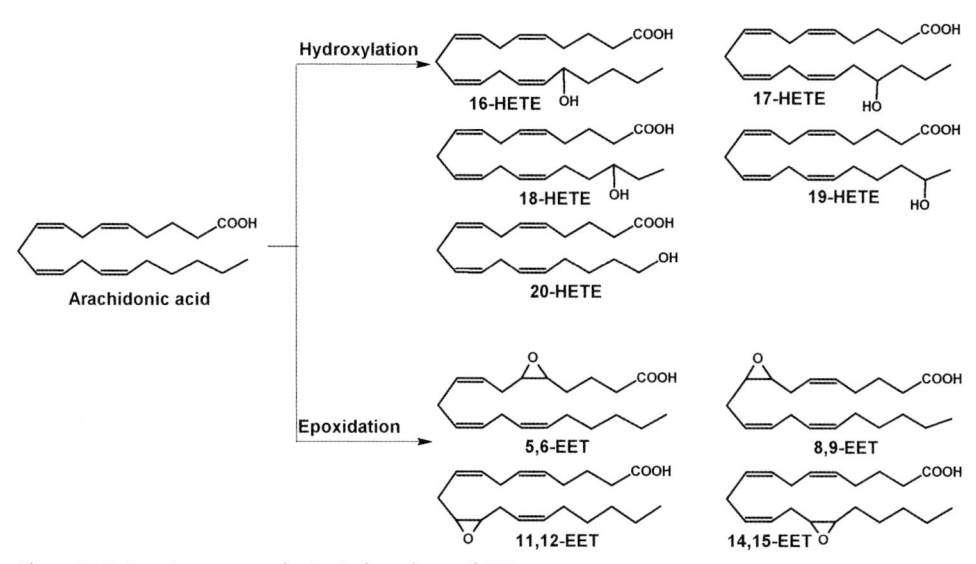

Figure 14.3 Reaction types and principal products of CYP-dependent AA metabolism. Hydroxylation yields HETEs and epoxidation a set of regio- and stereoisomeric EETs.

14.4.4
AA Metabolizing CYP Isoforms and Their Orthologs Among Rodents and Human

14.4.4.1 CYP Superfamily

CYP enzymes are the products of one of the largest gene superfamilies ever described in the biological kingdom (for further details, see http://drnelson. utmem.edu/Genome.list.htm.) The number of functional CYP genes in mammalian genomes ranges from 102 in mice and more than 60 in rats, to 57 in chimps and humans, and 54 in dogs.

CYP proteins sharing more than 40% sequence identity constitute a CYP family, which is designated by a number (e.g. CYP2 or CYP4). Subfamilies comprise members with more than 55% sequence identity and are designated by letters (e.g. CYP2C, CYP2J, CYP4A or CYP4F). Individual members of a subfamily are counted in the order of their discovery (e.g. CYP2C8, CYP2C9 or CYP4A11, CYP4A22) [39, 40].

14.4.4.2 Identity of AA-Metabolizing CYP Isoforms

Most of the mammalian AA-metabolizing CYP isoforms belong to the subfamilies 1A, 2C, 2E, 2J, 4A and 4F (Table 14.1). CYP2C and CYP2J enzymes function as

Table 14.1 AA-metabolizing CYP isoforms in mouse, rat and human.

	Mouse		Rat		Human		Function
	Principal candidates [number][a]	AA metabolizer [isoform][b]	Principal candidates [number][a]	AA metabolizer [isoform][b]	Principal candidates [number][a]	AA metabolizer [isoform][b]	
CYPs	102		>60		57		Functional CYP genes in the genome
CYP4A	4	12a	4	1, 2, 3, 8	2	11	20-HETE synthases
CYP4F	9	?	9	?	6	2, 3A	
CYP2C	15	44, . . . ?	7	11, 23	4	8, 9	EET synthases
CYP2J	8	?	2	3	1	2	
CYP1A	2	1, 2	2	1, 2	2	1, 2	Subterminal hydroxylases
CYP2E	1	1	1	1	1	1	

a   Gives the total number of CYP genes in the genome or the number of CYP genes comprising a specific subfamily.
b   Gives the name of a specific CYP subfamily member known to be involved in AA metabolism (e.g. 12a means the isoform Cyp4a12a in mouse).

AA epoxygenases, and are the most important source of EETs in the cardiovascular system. In contrast, CYP4A and CYP4F enzymes are AA hydroxylases, and produce 20-HETE in the kidney, vasculature and various other organs and tissues.

14.4.4.3 Problem of Overlapping Substrate Specificities

Several CYP isoforms show a rather broad substrate specificity. To mention only a few examples, CYP2C8 metabolizes AA, cerivastatin and paclitaxel, and CYP2C9 metabolizes AA, warfarin and estradiol. Moreover, some of the CYP4A and CYP4F enzymes do not only hydroxylate AA, but also a wide variety of related acyl compounds ranging from simple saturated fatty acids to leukotriene B_4.

Thus, drugs may interfere with CYP-dependent AA metabolism and vice versa. Moreover, despite the existence of candidates, the source of CYP-dependent eicosanoids has to be always unequivocally demonstrated for any new cell type or tissue as well as for any specific pathophysiological condition.

14.4.4.4 Problem of Orthologous Genes

This problem arises when trying to transfer the results from animal studies to human cardiovascular disease. Some of the candidate CYP subfamilies evolved differentially in rodents and human (Table 14.1) [41]. In general, the problem of orthologous CYP genes can only be solved by combined efforts of sequence and functional analysis and of the sex and tissue specificity of expression. Providing an example, among the four individual *Cyp4a* genes in mouse, there is one androgen-inducible isoform (*Cyp4a12a*) that is responsible for renal 20-HETE generation in male mice [42, 43]. The functional counterparts of the rodent 20-HETE synthases are CYP4A11 and CYP4F2 in human [44–46]. There are some indications for a sex-specific 20-HETE production also in human [47]. However, the induction of human 20-HETE synthases by androgens remains to be directly shown.

14.5 Physiological and Pathophysiological Context of CYP-Dependent Eicosanoid Formation and Action

14.5.1 Extracellular Signal-Induced AA Release

AA is mainly esterified to the *sn*-2 position of glycerophospholipids under basal conditions. Initiating eicosanoid production, free AA becomes readily accessible to the CYP enzymes after extracellular signal-induced activation of phospholipases as triggered by various G-protein coupled receptors (GPCRs). Cytosolic phospholipase A_2 (cPLA$_2$) translocates to the nuclear envelope and to the ER after binding Ca^{2+} ions. cPLA$_2$ is further activated by phosphorylation through cell- and context-specific kinases. Alternatively, AA can be released by the subsequent action of phospholipase C (PLC) and diacylglycerol lipase. Specific conditions exist in inflamed tissues due to the involvement of highly active secretory phospholipases.

14.5.2
Second Messenger Function

The requirement for phospholipase activation and subsequent AA release strictly couples the action of CYP enzymes to that of various hormones, cytokines and growth factors. Primary extracellular signals known to elicit EET and/or 20-HETE production are among vasodilatory hormones [bradykinin (BK), acetylcholine], and hormones that promote vasoconstriction and cell proliferation or modulate renal tubular ion transport (Ang II, endothelin-1, norepinephrine, dopamine) [48].

14.5.3
Cellular Context and the Multiplicity of Signaling

Each cell or tissue type contains unique sets of GPCRs, phospholipases and eicosanoid-generating enzymes and final effectors. Two examples may illustrate the importance of the cellular context. EETs are generated in ECs upon activation of cPLA$_2$ via a BK receptor. They act then as paracrine mediators and relax VSMCs by activating calcium-dependent potassium (BK) channels [17]. EETs are also produced in renal epithelial cells of the collecting duct. There, however, phospholipase activation is triggered after binding of adenosine to the (A1) adenosine receptor, EETs inhibit the epithelial sodium channel (ENaC) and thereby reduce salt reabsorption [49].

14.5.4
Physiological Context

The latter case provides also an example of how the interplay of the different signaling components can be regulated in adaptation to physiological needs. Under conditions of sodium deficiency, the ENaC is strongly upregulated and the expression of the EET-generating CYP isoform (CYP2C23 in rat) is suppressed [50]. Just the opposite occurs in response to high-salt diets.

14.5.5
Role of I/R

The role of I/R is described here since it appears as one of the most important examples of how EET and 20-HETE generation may become linked to the sudden occurrence of pathophysiological conditions. Our current understanding relies on the fact that CYP-dependent eicosanoids are partially incorporated into phospholipids after being synthesized from free AA [51, 52]. In this way, a membrane pool of "silent" eicosanoids is assembled that can be activated and utilized "on demand". This mechanism allows to bypass the necessity for *de novo* synthesis, which is only possible under conditions of sufficient oxygen supply since CYP enzymes are oxygenases. Therefore, under ischemic conditions, the membrane pool established during normoxia provides the only accessible source of EETs and 20-HETE. Ischemia and Ca^{2+} overload trigger phospholipase activation. Thus, the ability of EETs and 20-HETE to shuttle between an esterified and a free form may have a high physiological relevance.

14.6
Systemic and Tissue-Specific Metabolic Factors Modulating CYP-Dependent Eicosanoid Formation

As outlined above, CYP-dependent eicosanoids are produced as second messengers after extracellular signal-induced release of AA. However, there are additional and important modulators, which may come into play dependent on nutrition, cellular context and disease state (Figure 14.4).

14.6.1
Essential Fatty Acids Compete as Precursors for Oxygenated Metabolites

14.6.1.1 ω-6 Fatty Acids
AA (20:4 *n*-6) belongs to the group of *n*-6 PUFAs. Mammals do not possess the desaturases required to introduce the ω-6 double bond. Therefore, AA and its precursor (linoleic acid = 18:2 *n*-6) are essential fatty acids that are only available via the diet – linoleic acid mainly from plant sources and AA from meat.

14.6.1.2 ω-3 Fatty Acids
Another group of essential fatty acids is characterized by a double bond starting only three carbon atoms backwards from the ω-end. Physiologically important *n*-3 PUFAs are eicosapentaenoic acid (EPA; 20:5 *n*-3) and docosahexaenoic acid (DHA; 22:6 *n*-3). α-Linolenic acid (18:3 *n*-3) may serve as a precursor and is present among others in leafy green vegetables. However, this metabolic chain is rather inefficient making a direct supplementation with EPA and DHA superior. Oily fish and other sea food provides a rich source of EPA and DHA due to the marine food chain starting with *n*-3 PUFA-producing phytoplankton.

ω-6 and ω-3 fatty acids are not interconvertible due to a lack of an "ω-3 desaturase" in mammals. Therefore, nutrition determines in which ratio these essential fatty acids are available to the organism. Some organs developed specific systems for the uptake and conservation of *n*-3 PUFAs (DHA, in particular). This is important for the normal function of brain and retina, and perhaps also of the heart.

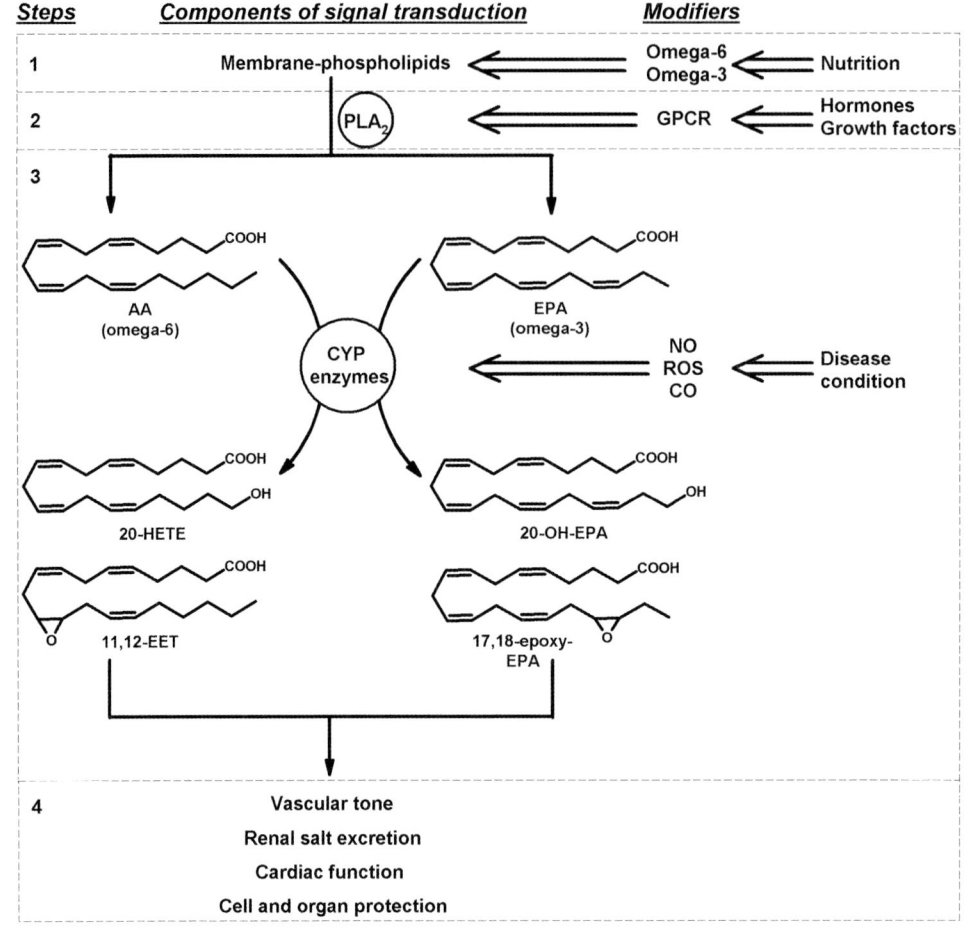

Figure 14.4 Generation and function of CYP-dependent eicosanoids in the cardiovascular system. (1) Nutritional uptake of essential fatty acids and their cell-type specific incorporation determines the acyl-chain composition of membrane phospholipids. (2) AA and after appropriate nutrition also ω-3 fatty acids such as EPA are released by PLA$_2$ which is activated by various hormones and growth factors via their GPCR. (3) Free AA and EPA are now accessible and compete for conversion by CYP enzymes. Specificity of metabolite production is provided by cell-type-specific expression of individual CYP4A/4F and CYP2C/2J family members which are distinguished by unique regio- and stereo-selectivities. CYP enzymes are inhibited by NO and CO. Reactive oxygen species (ROS) may react with NO resulting in NO deficiency and unmasking of CYP enzymes under various disease conditions. (4) CYP-dependent eicosanoids serve as second messengers in pathways regulating vascular, renal and cardiac function. Individual eicosanoids modulate (i) the activity of ion channels adjusting cell membrane potential or mediating salt reabsorption and (ii) the activation of proinflammatory transcription factors and MAPKs.

14.6.1.3 Health Benefits from ω-3 Fatty Acids

It has been frequently discussed that our ancestors evolved on a diet with a ratio of *n*-6 to *n*-3 PUFAs of about 1:1, whereas in recent Western diets the typical ratio is 15:1 [53]. The general hypothesis that this imbalance promotes cardiovascular and other chronic diseases has been supported by numerous epidemiological and clinical studies [54–56]. These studies revealed that the mortality from cardiovascular disease is inversely correlated to the relative amount of *n*-3 PUFAs in the diet. Most impressively, the risk of sudden cardiac death can be significantly reduced in patients with myocardial infarction treating them with EPA/DHA supplements [57].

14.6.1.4 ω-3 Fatty Acids Are the Precursors of Novel CYP-Dependent Eicosanoids

EPA and DHA can partially replace AA at the *sn*-2 position of glycerophospholipids. We and other authors have demonstrated that *n*-3 PUFAs are efficient alternative substrates for hepatic and renal microsomal CYP enzymes [33, 58, 59]. Actually, recent results show that each of the major EET and 20-HETE synthases in human, rat and mouse shares this capacity [32, 43, 51, 60–63].

Importantly, the CYP enzymes respond to the altered double bond structure and chain-length with changes in their regio- and stereoselectivity. Most of the AA epoxygenases show a clear preference for the ω-3 double bond when utilizing EPA as substrate. The product – 17,18-epoxy-EPA – is unique in having no homolog within the series of AA epoxides. AA hydroxylases show a different regio- and also reaction specificity when converting EPA instead of AA. The regioselectivity is shifted from almost exclusive ω- to moderate (ω-1)-hydroxylation. The reaction specificity changes from pure hydroxylation to hydroxylation as well as epoxidation. The novel capacity of the AA hydroxylases to catalyze an epoxidation reaction concerns exclusively the ω-3 double bond.

14.6.2 Role of Nitric Oxide

Nitric oxide (NO) binds to the heme iron and is a potent inhibitor of CYP enzymes (Figure 14.2B). Therefore, CYP activities may be repressed, if extracellular signals induce both AA release and NO production such as in ECs. Under these conditions, the EET (EDHF) component of the vasodilator response only becomes obvious after pharmacological inhibition of NO synthase [17]. However, EETs can come into play as an important "backup system" under conditions of NO deficiency as typical for endothelial dysfunction [64, 65]. The same conditions may also unmask 20-HETE production as indicated by increased urinary 20-HETE levels in patients with endothelial dysfunction [47, 66] and augmented myogenic constriction in endothelial NO synthase knockout mice [67].

Further demonstrating the relevance of CYP–NO interactions, NO-mediated inhibition of 20-HETE production contributes with almost 50% to the vasodilator response in cerebral arteries and renal microvessels [68, 69]. This is a novel mechanism compared to the generally predominant one that relies on NO-mediated activation of cGMP cyclase.

14.6.3
Carbon Monoxide and Heme Oxygenase

Carbon monoxide (CO) is the classical test ligand of the heme iron and a strong inhibitor of CYP enzymes. Heme degradation by heme oxygenases (HOs) is a major endogenous source of CO. HO-1 gene expression is induced under inflammatory conditions and provides multiple protective mechanisms [70]. Tin- or cobalt-mediated induction of HO-1 inhibited renal 20-HETE production and prevented the development of hypertension in SHRs and in the Goldblatt 2K1C model [8, 71, 72].

HO-1 also potentially limits the availability of heme for CYP biosynthesis. Furthermore, the activity of both HO-1 and CYP is dependent on electron transfer from one and the same redox partner (CPR).

14.6.4
CYP Enzymes as Targets, Sources and Utilizers of Reactive Oxygen Species

14.6.4.1 Reactive Oxygen Species Affect CYP Activities
Reactive oxygen species are highly reactive, and may damage both the CYP enzyme itself and its membrane lipid environment. Moreover, O^{2-} is able to react with NO to peroxynitrite, which can directly attack tyrosine and cysteine residues of CYP and CPR protein [73–77].

On the other hand, high O^{2-} levels frequently contribute to NO deficiency due to the very rapid reaction of these two radicals to peroxynitrite and related compounds. As described above, this may lead to an unmasking of those CYP enzymes that are otherwise inhibited by NO (see Section 14.6.2).

14.6.4.2 Reactive Oxygen Species Production by CYP Enzymes
A matter of frequent concern is the fact that several CYP isoforms are unable to ensure a strict coupling of electron transfer, oxygen activation and substrate conversion. Thus, activated oxygen species may escape from the catalytic cycle (Figure 14.2B). This may have *in vivo* relevance as indicated by a recent study demonstrating that the enhanced oxidation of serum proteins as present in hypertensive patients can be ameliorated by treating them with a specific CYP2C9 inhibitor [78, 79]. However, other AA epoxygenases (CYP2C8 and CYP2J2) do not share the high degree of uncoupling with CYP2C9.

14.6.4.3 CYP Enzymes Can Use Reactive Oxygen Species and Hydroperoxides for Substrate Oxygenation
This phenomenon can be considered as a reversal of the ROS escape from reaction cycle. Experimentally, it is utilized to make a CYP protein active without any need of NADPH and CPR by adding a donor of activated oxygen such as coumene

hydroperoxide. Under *in vivo* conditions, fatty acid hydroperoxides (hydroperoxy-ETEs) as produced by LOX enzymes may be interesting candidates [80].

14.7
Biological Activities of EETs and 20-HETE

Considering EETs and 20-HETE as second messengers of various hormones and growth factors, it is not surprising that they have a myriad of different biological activities comparable to that of the extracellular signals triggering their formation. Currently, these biological effects of the CYP-dependent eicosanoids cannot be systematically evaluated and predicted since an important piece of information is missing – the identity of their receptors and/or other primary targets. Therefore, most of the effects discussed below are actions on intermediate signaling components or final effectors or even physiological net responses.

14.7.1
Regulation of Vascular Tone

EETs show EDHF-like properties. They are produced in ECs in response to vasodilator hormones and subsequently activate BK channels in the underlying VSMCs (Figure 14.5) [17]. Activation of BK channels leads to K^+ efflux and hyperpolarization of the membrane potential. As a result, Ca^{2+} influx via voltage-gated Ca^{2+} channels is inhibited, thus reducing Ca^{2+}-triggered Ca^{2+} release from the sarcoplasmic reticulum. Eventually, this pathway results in VSMC relaxation. The relevance of this pathway for EET-mediated vasodilation as well as the general significance of EETs in the whole vascular system is still under discussion [14–16].

20-HETE plays an opposite role and mediates vasoconstriction. There are apparently two mechanisms: (i) inhibition of BK channels and [81, 82] and (ii) activation of Rho kinase [83]. Inhibition of BK channels leads, via the same components described for EET action, to increased intracellular Ca^{2+} levels and vasoconstriction. 20-HETE-mediated Rho kinase activation leads to phosphorylation and inactivation of the myosin light chain (MLC) phosphatase. This results in an increased MLC phosphorylation and thus increased Ca^{2+} sensitivity of the contractile apparatus.

In renal microvessels, 20-HETE mediates the development of myogenic tone (i.e. it helps to keep the blood flow constant in response to changes in perfusion pressure). Moreover, 20-HETE mediates tubulo-glomerular feed back (i.e. the adjustment of blood and urine flow in the kidney) [81].

14.7.2
Regulation of Renal Tubular Function

EETs and 20-HETE are produced in various segments of the nephron [48, 81]. Their net effect is inhibition of salt reabsorption (Figure 14.6). In proximal tubules,

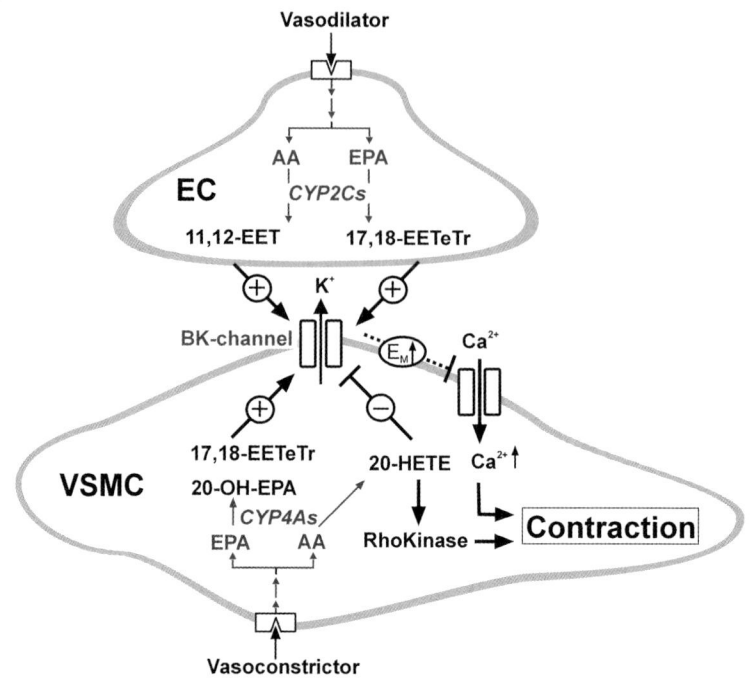

Figure 14.5 Role of CYP-dependent eicosanoids in the regulation of vascular tone. EETs are produced in ECs by CYP2C and CYP2J enzymes in response to vasodilator hormones. EETs activate calcium-dependent potassium (BK) channels in VSMCs. BK channel activation results in membrane hyperpolarization, reduced Ca^{2+}-influx via voltage gated Ca^{2+} channels and eventually in VSMC relaxation. 20-HETE is produced by CYP4A enzymes in VSMCs in response to vasoconstrictor hormones. 20-HETE triggers vasoconstriction via inhibition of BK channels and activation of Rho kinase. Dependent on nutrition, n-3 PUFAs such as EPA may become involved as alternative substrates of the CYP enzymes. This results in the production of potent BK channel activators such as 17,18-epoxyeicosatetraenoic acid (17,18-EETeTr) in both ECs and VSMCs.

20-HETE is an inhibitor of the Na^+-K^+-ATPase via protein kinase C-dependent phosphorylation. It mediates the effects of free AA released in response to stimuli such as dopamine and parathyroid hormone. In the thick ascending limb of the loop of Henle, the Na^+–K^+–$2Cl^-$ cotransporter is inhibited by 20-HETE. 20-HETE generation in the thick ascending limb of the loop of Henle is stimulated by various extracellular signals such as Ang II, BK and Ca^{2+}.

EETs regulate sodium reabsorption both in proximal tubules and in collecting ducts. In the latter, EETs are second messengers of adenosine, and perhaps other factors, and serve as inhibitors of ENaC which plays a predominant role in sodium reabsorption in this part of the nephron [49, 50].

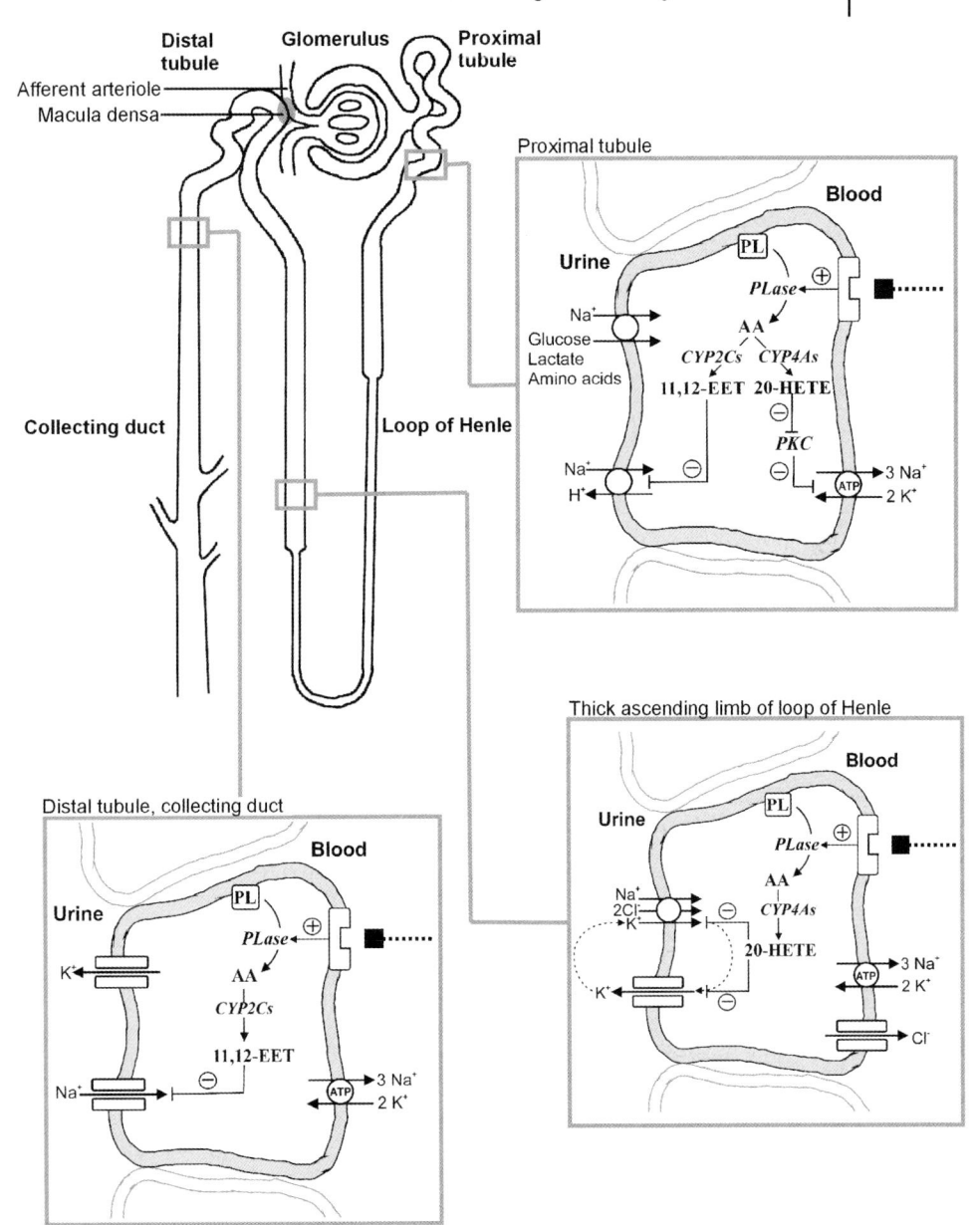

Figure 14.6 Role of CYP-dependent AA metabolites in the renal tubular system. 20-HETE and EETs are produced in various parts of the nephron and modulate the activity of ion channels involved in salt reabsorption. The net effect of these pathways is promotion of salt excretion. For further details, see text.

14.7.3
Cardiac Function

In the heart, EETs are involved in the regulation of L-type Ca^{2+}-, ATP-sensitive potassium (K_{ATP}) and Na^+ channels [84–87]. This has important implications for cardiomyocyte contractility, the functional recovery from I/R and the severity of myocardial infarction. In contrast, 20-HETE appears to play a detrimental role and aggravates the extent of I/R injury.

In isolated cardiomyocytes, 11,12-EET was shown to induce Ca^{2+} influx and contraction via a cAMP/protein kinase A-dependent mechanism [84]. In isolated hearts, 11,12-EET perfusion prior to global ischemia improved functional cardiac recovery in the subsequent reperfusion phase. Moreover, Zeldin *et al.* established a transgenic mouse model with heart-specific overexpression of the human cardiac EET synthase (CYP2J2). The transgenic mice showed improved functional recovery after I/R [88]. Cardiomyocytes isolated from these animals displayed enhanced L-type Ca^{2+} currents [89].

Activation of K_{ATP} channels is essential for improved functional recovery after I/R as shown by pharmacological inhibition experiments in CYP2J2 transgenic mice. Moreover, activation of p42/p44 MAPKs appears to contribute to the improved recovery in this model [88].

Recent results demonstrate that EETs have cardioprotective properties beyond functional recovery [90]. In canine hearts, perfusion with 11,12- and 14,15-EET significantly reduced the infarct size after I/R. Also this effect was blocked by K_{ATP} inhibition [91].

The beneficial effects of EETs are not shared by their sEH-mediated hydrolysis products (DHETs). This was shown *in vitro* for infarct size reduction in isolated hearts [91]. Moreover, recent studies also revealed improved cardiac functional recovery after I/R in sEH knockout compared to wild-type mice. This effect was inhibited by the EET antagonist 14,15-EEZE and blocked by K_{ATP} channel inhibitors [92].

20-HETE shows detrimental effects under the same conditions. Nithipatikom *et al.* demonstrated that I/R-induced myocardial infarction is significantly ameliorated in canine and rat hearts by pretreatments with 20-HETE synthase inhibitors or 20-HETE antagonists [93, 94]. In contrast, infarct size increased by a pretreatment with 20-HETE. The protective effect of 20-HETE synthase inhibition was abolished by pharmacological blockade of the sarcolemmal K_{ATP} channel. Noteworthy, the beneficial effect of preconditioning was also enhanced by the 20-HETE synthase inhibition [95].

Taken together, these results demonstrate a strong negative role of 20-HETE and an opposite important beneficial role of EETs in myocardial infarction. Essential components of EET-mediated cardioprotection are probably both sarcolemmal and mitochondrial K_{ATP} channels.

14.7.4
General Cell- and Organ-Protective Properties of EETs

In addition to the cardioprotective role of EETs described above similar effects are also exerted by these metabolites in other parts of the cardiovascular system. These include protection against ischemic brain injury, hypoxia-reoxygenation injury of EC and vascular inflammation (for reviews, see [24, 25, 52, 96, 97]). The underlying mechanisms that may synergistically contribute to these effects involve:

(1) Inhibition of the inflammatory response via inhibition of NF-κB activation and subsequent monocyte adhesion [18].
(2) Antithrombotic effects by BK channel activation in platelets and repression of P-selectin expression leading to reduced platelet adhesion to endothelial cells [98].
(3) Induction of tissue plasminogen activator mediating fibrinolysis [99].
(4) Antiapoptotic effects via phosphatidylinositol-3-kinase and Akt kinase [24, 100, 101].
(5) Antimigratory effects on VSMCs via enhanced cAMP levels and protein kinase A [102].

Again, 20-HETE appears to play an opposite role in several of these mechanisms. Recent studies show that it contributes to ischemic brain injury, oxidative stress and pathologic vascular remodeling.

14.7.5
Biological Activities of Eicosanoids Originating from CYP-Dependent n-3 PUFA Oxygenation

Currently, there are only a few studies available addressing this question. Epoxides of n-3 PUFAs were shown to display vasodilatory properties and to be potent activators of BK channels in VSMCs [60, 103, 104]. BK channel activation is highly regio- and stereoselective in some vascular beds such as cerebral arteries. In these vessels, we found that only 17(R),18(S)-epoxy-EPA was active and that its effect exceeded that of the most potent AA epoxide (11R,12S-EET) [60]. EPA can be oxygenated to 17(R),18(S)-epoxy-EPA by EET and 20-HETE synthases. Therefore, we propose that the generation of this metabolite may contribute to both enhanced vasodilator and reduced vasoconstrictor responses (Figure 14.5).

Another important finding was the demonstration of n-3 PUFA epoxides as highly potent activators for the opening of cardiac K_{ATP} channels [86] that play an essential role in endogenous cardioprotective mechanisms under I/R conditions. Taken together, it is tempting to speculate that several of the beneficial effects attributed to n-3 PUFAs are actually mediated by CYP-dependent conversion to ω-3 eicosanoids. Based on our current knowledge, this may apply, in particular, to the improvements in vascular function and the cardioprotective effects observed in response to ω-3 fatty acid-rich diets.

14.8
Secondary Product Formation and the Metabolic Fate of CYP-Dependent Eicosanoids

Once formed, EETs and 20-HETE may directly exert their function in signaling cascades by interacting with their primary targets (so far unknown receptors or other eicosanoid binding proteins). On the other hand, they become subject to various pathways (Figure 14.7), eventually leading to either:

(1) Selective retention by binding to fatty acid binding proteins (FABPs). There are different types of FABPs in the heart, liver, adipocytes and other tissues. FABPs protect EETs from further metabolism and secretion and their ligand specificity may be important for the selective accumulation of certain EET isomers in a given tissue [105].
(2) Production of secondary metabolites via CYP-dependent pathways. EETs are hydroxylated by CYP4A enzymes to hydroxy-EETs (HEETs). HEETs are also

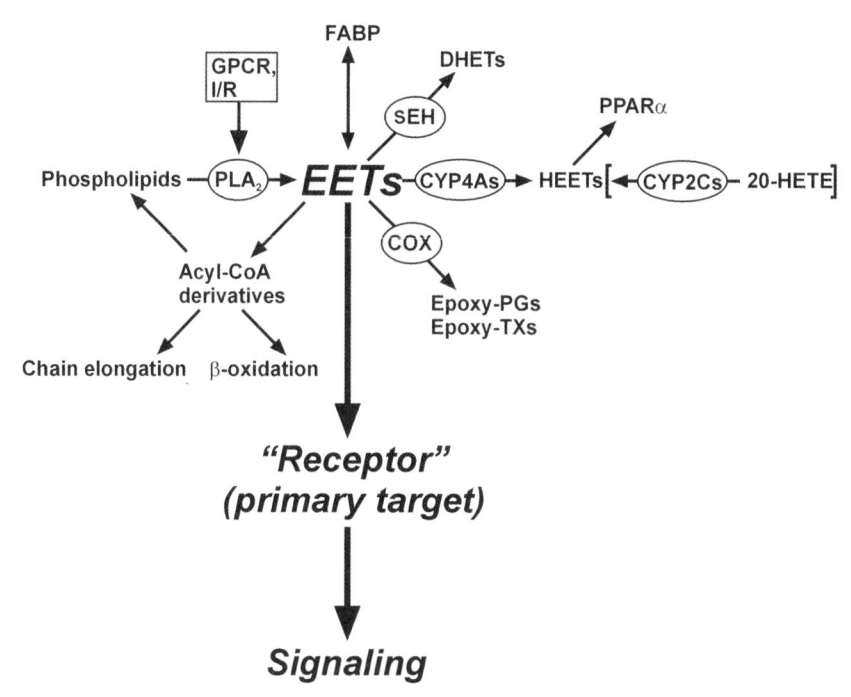

Figure 14.7 Metabolic fate of EETs. As described in detail in the text, EETs can recycle between a membrane bound and a free state, are specifically retained by FABPs, serve as precursors for secondary metabolites and are subject to degradation via β-oxidation. Both EETs and 20-HETE are precursors of HEETs, which are highly potent activators of PPARα.

efficiently produced by CYP2C-dependent epoxidation of 20-HETE. HEETs were shown to act as high-affinity ligands of peroxisome proliferator-activated receptor (PARP) ([106, 107].

(3) Production of prostaglandin and thromboxane analogs via COX-dependent pathways. Primary metabolites in question are 20-HETE, 5,6-EET and 17,18-epoxy-EPA. COX-dependent 20-HETE transformation plays an important role in renal physiology where the expression of COX-2 is under control of the renin–angiotensin–aldosterone system (RAAS). In several vascular beds, the vasoactivity of 5,6-EET consists of two components: a vasodilator component of the nonmetabolized 5,6-EET and a vasoconstrictor component of its COX-dependent metabolite [81, 108, 109].

(4) Incorporation into glycerophospholipids after formation of acyl-coenzyme A derivatives. This reaction directs EETs and 20-HETE into a membrane pool. The esterified eicosanoids may be then released by the activation of phospholipases as triggered by hormones, growth factors and I/R [51, 52].

(5) Chain elongation or partial and complete degradation by β-oxidation. The acyl-coenzyme A derivatives can become subject to peroxisomal and mitochondrial β-oxidation but also to chain elongation. Partial β-oxidation and chain elongation are a source of novel biologically active metabolites [51, 52].

14.9
CYP-Dependent AA Metabolism in Animal Models of Hypertension and End-Organ Damage

Genetic models of spontaneous and salt-sensitive hypertension (SHR and Dahl rats) were the starting point for an intensive research aimed at understanding the role of CYP-dependent AA metabolism in cardiovascular disease (compare Section 14.3). Even when including all the currently examined models of secondary hypertension (Table 14.2), there remain three major alterations in CYP-dependent AA metabolism that are associated with the disease state:

(1) Increased CYP4A expression and 20-HETE production affecting the function of renal microvessels and thus increasing renal vascular resistance. This may also apply to other organs, like brain and heart.
(2) Decreased CYP2C expression and EET production probably affecting both vascular resistance and tubular salt excretion. According to our recent knowledge the loss of general vascular and organ protective EET effects may synergistically aggravate the disease state and may contribute to hypertension-associated organ damage.
(3) Decreased CYP4A expression and 20-HETE production probably affecting primarily the renal tubular system and leading to impaired salt excretion.

Table 14.2 Animal models developing hypertension in association with alterations in CYP-dependent AA metabolism.

Animal model of hypertension	Alteration in CYP AA metabolism	Reference
SHR	increase in (vascular) 20-HETE level	Sacerdoti et al. [8]
DOCA-salt-treated		Oyekan et al. [110]
ANG II infusion		Muthalif et al. [111]
Cyp4a14 knockout mouse		Holla et al. [42]
Cyclosporin A-treated rats		Seki et al. [112]
Androgen-induced hypertension		Singh et al. [113]; Holla et al. [42]
Salt-sensitive Dahl rat	decreased EET level/inability to upregulate EET level	Makita et al. [10]
dTGR		Kaergel et al. [114], Muller et al. [107]
Ang II/salt-treated rats		Zhao et al. [115]
Rats fed high-fat diet		Wang et al. [31]
Young SD rats – salt-treated		Sankaralingam et al. [116]
Cyp4a10 knockout mouse		Nakagawa et al. [117]
Pregnant rat treated with EET synthase inhibitor		Huang et al. [118]
Salt-sensitive Dahl rat	decrease in (tubular) 20-HETE level	Roman et al. [11]
DOCA-salt mice		Honeck et al. [119]
HET0016/salt-treated		Hoagland et al. [120]
Ang II-infused mice		Vera et al. [121]

DOCA, desoxycorticosterone acetate.

Moreover, there are several combinations of these basic alterations in models showing both reduced EET and 20-HETE generation. Only some of the animal models will be discussed in detail below to demonstrate the general principles and the arising new treatment strategies for cardiovascular disease. Further examples are summarized in Table 14.2.

14.9.1
Prohypertensive and Proinflammatory Role of Vascular 20-HETE

14.9.1.1 Androgen-Induced Hypertension

Singh *et al.* demonstrated that androgen treatment elevates blood pressure in rat [113]. Androgens specifically induced CYP4A expression and 20-HETE production in renal arteries. Importantly, treatment of the animals with a selective CYP4A inhibitor (HET0016) reduced 20-HETE synthesis, diminished oxidative stress and ameliorated hypertension.

Stimulation of oxidative stress is a novel role of 20-HETE that goes beyond its vasoconstrictor function. Moreover, Malik *et al.* showed that 20-HETE stimulates intimal hyperplasia after endothelial denudation as produced by catheter procedures in animal models [122]. Thus, 20-HETE may also contribute to pathophysiological vascular remodeling under conditions of hypertension and associated end-organ damage.

There are also implications for human cardiovascular disease. Two papers indicated an association between enhanced urinary 20-HETE levels, oxidative stress and endothelial dysfunction in hypertensive patients [47, 123]. Moreover, androgens and male gender are risk factors for the development of hypertension and cardiac disease, whereas estrogens appear to provide protection in women [124].

14.9.1.2 Cyclosporin A-Induced Hypertension

Cyclosporin A is used as an immunosuppressive drug after heart and kidney transplantation. A major unwanted side-effect of this treatment is the development of hypertension and renal failure. This pathophysiology can be reproduced treating rats with cyclosporin A. In this animal model, hypertension and renal failure are associated with increased renal 20-HETE production [112, 125]. Oyekan *et al.* demonstrated that the development of the disease state can be attenuated by a simultaneous treatment with a 20-HETE synthase inhibitor (HET0016). Moreover, the same effect was exerted by L-arginine, the substrate of NO synthase. These findings reinforce the high relevance of CYP–NO interactions [125].

14.9.2
Antihypertensive Role of EETs in Salt-Sensitive Hypertension

Several animal models are characterized by EET deficiency (Table 14.2). The implications are best understood at the level of renal salt reabsorption. EETs inhibit salt reabsorption in different parts of the nephron. This is important in response to high-salt diets and any genetically determined or experimentally induced failure to upregulate EET production may be associated with salt-induced hypertension and renal damage. Dahl-salt sensitive rats provided the first example.

Ang II-infusion combined with a high-salt diet provokes the same effects: hypertension, renal damage and decreased CYP2C23-dependent EET generation in the rat kidney. Salt handling is largely determined by the RAAS system. Low salt intake activates RAAS to prevent salt deficiency in the body; however, high salt intake reduces the expression of RAAS components to stimulate salt excretion. Thus, Ang II infusion combined with high salt overrides an important physiological mechanism including the upregulation of CYP2C-dependent EET generation in the renal tubular system.

14.9.3
Antihypertensive Role of EETs in Pregnancy

14.9.3.1 Renal and Vascular Alterations During Pregnancy

Fitzgerald et al. reported increased urinary EET/DHET levels in healthy pregnant compared to nonpregnant women and a further increase in patients with pregnancy-induced hypertension [126]. The authors proposed that EETs contribute to the normal physiological adaptation during pregnancy but also to the pathophysiology of pregnancy-induced hypertension. Wang et al. demonstrated that normal pregnancy is associated with an upregulation of renal EET synthases in Sprague-Dawley rats (SD) rats [118, 127]. Pharmacological inhibition of EET generation resulted in a significant elevation of blood pressure in the pregnant animals and in a marked decrease in the body weight of fetal pups.

In human but also in animal models, vascular adaptation to normal pregnancy appears to involve a downregulation of the NO-dependent component of the vasodilator response and a compensatory increase of the EDHF component. This shift in the predominant mechanism of vasodilatation does nor occur in pre-eclamptic patients [128].

A recent study by Hagedorn et al. compared pregnancy-induced alterations in wild-type and pregnane X receptor (PXR) knockout mice [129]. Mesenteric arteries from pregnant wild-type mice showed increased vasodilator and decreased vasoconstrictor responses compared to nonpregnant controls. These adaptations were absent in PXR knockout mice. Treatment of wild-type but not of knockout mice with a PXR activator further enhanced vasodilator responses. This effect was attenuated by EET synthase inhibition. This study yields a first mechanistic understanding for the pregnancy-induced increase in EET synthesis. Several genes encoding EET synthases contain PXR-responsive promoter elements, which may become activated due to the high circulating levels of progesterone and its metabolites during pregnancy.

Granger et al. used a chronic reduction in uterine perfusion pressure (RUPP) model to mimic conditions of preeclampsia [130]. RUPP rats featured hypertension, increased renal vascular resistance and enhanced renal cortical 20-HETE production. Inhibition of 20-HETE synthesis by 1-aminobenzotriazole (a general CYP inhibitor) attenuated the development of hypertension and normalized vascular resistance. This study disclosed renal 20-HETE production as an important effector in the response of maternal circulation to restrictions in uteroplacental perfusion.

14.9.3.2 Role of EETs in Placenta, Decidua and Trophoblasts

These tissues and cells are characterized by a highly active LOX- and CYP-dependent AA metabolism. Several studies identified 5,6-EET as the major epoxygenase product in different human intrauterine tissues [131–133]. 5,6-EET formation was also shown in trophoblasts [134]. The predominant formation and/or retention of 5,6-EET in intrauterine tissues appears unique when compared to other parts of the cardiovascular system. It may be proposed that 5,6-EET specifically contributes to the regulation of placental and decidual vascular resistance and remodeling, trophoblast differentiation and invasion and also to steroidogenesis.

Taken together, CYP-dependent eicosanoids play an important role in the adaptation of maternal circulation and renal function to pregnancy. Moreover, they may contribute to the function of placenta, decidua and trophoblasts.

14.9.4
20-HETE Deficiency and Salt-Sensitive Hypertension

Renal tubular 20-HETE deficiency appears to represent an important contributor to salt-sensitive hypertension. Substantiating this hypothesis, recent work showed that normal SD rats can be rendered to a salt-sensitive phenotype simply by infusing a selective inhibitor of 20-HETE formation (HET0016) [120].

A further study of the Roman group allows a detailed insight into the relations between salt sensitivity, 20-HETE and oxidative stress in Dahl salt-sensitive rats [135]. The authors treated these animals with tempol – a chemical superoxide dismutase mimetic. This resulted in increased urinary 20-HETE levels and in an amelioration of hypertension. The antihypertensive effect of tempol was blunted by a simultaneous treatment with HET0016.

14.9.5
Role of EETs, HEETs and PPARα in Inflammatory Renal Damage

In several animal models, EET deficiency is combined with a reduced capacity to produce 20-HETE. Examples are Dahl-salt sensitive rats, double transgenic rats overproducing the human genes for angiotensinogen and renin (dTGR), and rats receiving a high-fat diet. A combined EET and 20-HETE deficiency may synergistically induce or aggravate problems with salt handling. Moreover, EET deficiency alone may have detrimental effects on the regulation of vascular tone and inflammatory response.

In addition, we have shown that a combined deficiency of CYP4A and CYP2C enzymes strongly reduces the capacity of HEET generation in the dTGR model [107]. HEETs are highly potent activators of PPARα [106]. PPARα inhibits the inflammatory response by suppressing NF-κB activation [136, 137]. We found that hypertension and end-organ damage is abolished by treating the dTGR animals with the PPARα activator fenofibrate [107]. Moreover, we observed that fenofibrate strongly induced the capacity of EET and HEET generation. Thus, it appears that fenofibrate treatment triggered the production of an endogenous PPARα ligand.

This positive-feedback loop may have significantly contributed to the marked success of fenofibrate treatment. A general beneficial effect of PPARα activation was also shown in other animal models including SHR, salt-sensitive Dahl rats, obese Zucker rats, as well as Ang II infusion and high fat-diet hypertensive rats [138–141].

Wang et al. used SD and fed them a high-fat diet [142]. This treatment induced hypertension and the blood pressure increase was completely prevented by fenofibrate. These authors demonstrated that the beneficial effect of fenofibrate was abolished by a simultaneous treatment of the animals with methylsulfonyl-propargyloxyphenylhexanamide, a specific EET synthase inhibitor. Moreover, they showed that the fenofibrate effect can be mimicked by a treatment with sEH inhibitors. Thus, both enhanced EET generation and stabilization of the EET (and HEET?) levels mediated the same beneficial effects.

14.10
Structure and Cardiovascular Functions of the Soluble Epoxide Hydrolase

14.10.1
Enzymatic Activities

The sEH protein consists of two domains [143, 144] (Figure 14.8). The N-terminal domain acts as a phosphatase that dephosphorylates test substrates (4-nitrophenyl

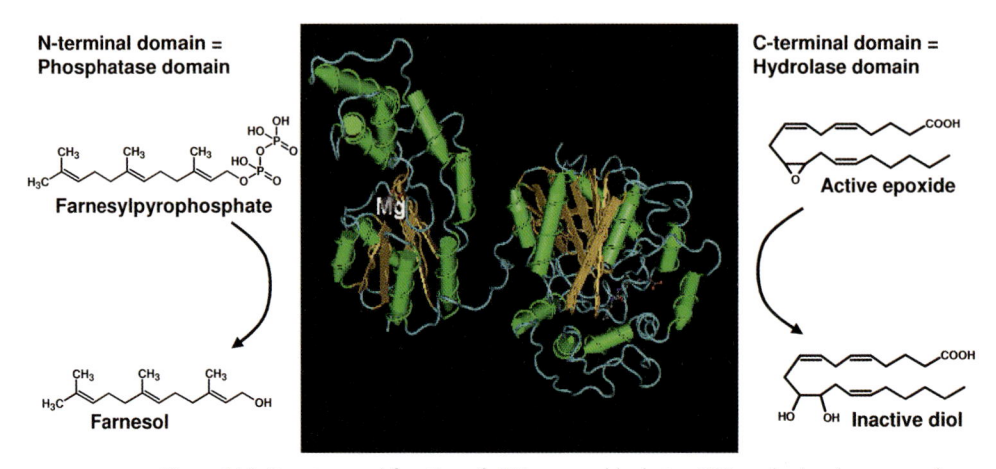

Figure 14.8 Structure and function of sEH. The sEH protein consists of two domains. The N-terminal domain displays phosphatase activity, and dephosphorylates isoprenoid pyrophosphates and related compounds. The C-terminal domain functions as a hydrolase, and hydrates EETs and related compounds. Crystal structures of human and mouse sEH are available from http://www.ncbi.nlm.nih.gov/structure. The example shown has the accession number 1ZD5.

phosphate) and different lipid phosphates [145]. The latter include isoprenoid pyrophosphates such as farnesyl pyrophosphate and geranylgeranyl pyrophosphate, and the corresponding monophosphates [146]. This offers the possibility that the N-terminal sEH domain interferes with prenylation and activation of signaling components, such as RAS and RhoA, and reduces the availability of precursors for cholesterol biosynthesis.

The C-terminal domain carries the name-giving hydrolase activity and converts EETs to the corresponding diols (DHETs). This reaction results (i) in a loss of several biological activities of the EETs [29], and (ii) in a shift between active and inactive EET enantiomers due to sEH stereoselectivity [147, 148]. sEH can be localized to both the cytosol (presumed site of EET hydrolysis) and peroxisomes (site of farnesyl pyrophosphate and cholesterol biosynthesis) [149]. Ang II induces sEH expression via AP-1 activation providing an important link to cardiovascular disease conditions [150].

14.10.2
sEH – A Novel Target for the Treatment of Cardiovascular Disease?

In animal models, pharmacological sEH inhibition ameliorated Ang II infusion hypertension, blocked the development of TAC-induced cardiac hypertrophy, reduced ischemic cerebral infarct size and exerted antinociceptive effects [28–30, 52, 151–153]. Direct evidence is partially missing that these beneficial effects of pharmacological interventions rely only on sEH inhibition. In the vasculature some sEH inhibitors themselves act as vasodilators [154]. Moreover, they do also act as PPARα activators [155]. However, also targeted disruption of the sEH gene exerted beneficial effects such as improved cardiac functional recovery after I/R in the knockout mice [92].

Substantiating an important role of sEH in the cardiovascular system, genetic variations in the human EPHX2 gene were found to be associated with coronary heart disease. Moreover, certain allelic variants appear to increase the susceptibility to and others to protect from ischemic stroke (see Section 14.11).

14.11
General Conclusions on Cause–Effect Relationships Associating Alterations in CYP-Dependent Eicosanoid Production and Cardiovascular Disease

14.11.1
Any Alteration in CYP-Dependent AA Metabolism May Contribute to Cardiovascular Disease

The animal models summarized above revealed a surprising fact: any alteration in CYP-dependent eicosanoid generation and metabolism can be associated with hypertension and end-organ damage. At the same time, however, the effect of each individual alteration can be explained based on our current understanding of the multiple and tissue-specific roles of EETs and 20-HETE, and by disease-related modulations of their production and action.

The apparent paradox has been resolved that 20-HETE can play both pro- and antihypertensive roles. This is due to the different functions of 20-HETE in the vascular and renal tubular system. 20-HETE mediates vasoconstriction, and promotes oxidative stress and pathophysiological vascular remodeling. On the other hand, it is required for the regulation of salt excretion.

EETs seem to possess only beneficial properties in the cardiovascular system. They serve as mediators of vasodilation, and protect against vascular, renal and cardiac damage. A matter of concern may be, however, their role in angiogenesis and cell proliferation considering the potential implications for tumor growth (compare Section 14.3).

14.11.2
EET and 20-HETE Availability – A Bottleneck of Various Signaling Pathways?

EETs and 20-HETE serve as second messengers in many signaling pathways stimulated by hormones and growth factors. Moreover, modulations in the EET and 20-HETE levels have profound pathophysiological effects. Thus, we have to assume that the availability of EETs and 20-HETE is a major determinant of the rate and capacity of whole signaling pathways. This is again surprising and not fully understood, since there are many other components that may play a similar role in the same pathways. On the other hand, this provides the basis for a future treatment of cardiovascular disease by targeting CYP-dependent eicosanoid production.

14.11.3
Is There a Primary Role of CYP-Dependent Eicosanoids and of CYP Gene Polymorphism in the Development of Cardiovascular Disease?

Some studies indeed indicate that such a role may exist. In favor of this idea, hypertension and end-organ damage are induced by genetic and environmental factors that directly interfere with the generation and maintenance of CYP-dependent eicosanoids. There are, however, only a few of "pure" examples. Salt-sensitive hypertension in SD rats upon 20-HETE synthase inhibition may be one and pregnancy-induced hypertension upon EET synthase inhibition a second (see Section 14.9). Moreover, also the fact may be counted that the beneficial effects of fenofibrate treatment were abolished upon EET synthase inhibition but mimicked by sEH inhibition in high-fat diet induced hypertension (see Section 14.9.5).

Moreover, certain allelic variants of the genes encoding 20-HETE (CYP4A11) and EET-generating CYP isoforms (CYP2J2 and CYP2C8) as well as the sEH (EPHX2) were shown to be associated with cardiac artery disease and stroke in human [34, 44, 156–164]. The associations of the currently examined polymorphisms with disease risk are relatively weak and population-dependent, but are significant. Moreover, a functional link was provided by showing that the allelic variants encode either less active CYP isoforms or are less efficiently expressed.

14.11.4
What Is the Cause for Alterations in the Production and Effects of CYP-Dependent Eicosanoids in Disease States?

In general, a role of EETs and 20-HETE becomes only obvious in disease states or other imbalances resulting from high salt loading, RAAS activation, inflammation, obesity, drug treatment and other conditions.

Thus, we have also to understand the cause for alterations in CYP-dependent eicosanoid production and action in cardiovascular disease. These questions concern the (i) mechanisms of how EET and 20-HETE levels are affected by the disease state, and (ii) the possible alterations in physiological processes rendering them EET and 20-HETE sensitive.

The first question is largely open in detail and has to be clarified for each individual case or model. Some general mechanisms of CYP enzyme inactivation include (i) damage of the CYP enzymes and their lipid environment under conditions of oxidative stress, (ii) inhibition by CO and NO, (iii) phosphorylation and subsequent degradation probably by the ubiquitin–proteasome system, and (iv) downregulation of CYP gene expression.

The eicosanoid levels depend on their generation (see Section 14.5), retention, membrane storage, and metabolic fate (see Section 14.8). All these conditions can be altered in disease states. A recent example was provided by Hammock *et al.* who showed that sEH expression is upregulated via AP-1-responsive elements in the sEH gene promoter [150]. This may be an important link explaining how low EET levels result from both reduced generation (CYP enzyme inactivation) and enhanced metabolism (sEH-mediated hydrolysis) under conditions of RAAS activation.

The second question about alterations in physiological responses remained largely unaddressed in many studies. In part, an answer may come from what we know about the function of EETs in the vasculature. There, one extracellular signal (BK or acetylcholine) is able to trigger at least three different pathways (NO, prostacyclin and EET formation) with the same net effect, namely vasodilation. If one or two of these pathways (NO and prostacyclin) fail, EETs could maintain vasoreactivity and then may indeed determine the efficiency of the whole process.

On the other hand, EETs and 20-HETE may become predominant factors in suddenly occurring pathophysiological states such as I/R (see Section 14.5). They may then trigger signaling pathways, which are preformed but largely inactive under normal conditions.

A better understanding of these complex relationships between alterations in CYP-dependent eicosanoid formation and cardiovascular disease will be an important basis to exploit the full therapeutic potential of this new branch of the AA cascade.

Finally, we would like to refer the reader to some excellent reviews on the role of EETs, 20-HETE and sEH in the cardiovascular system: [24, 25, 28, 29, 48, 52, 64, 81, 82, 96, 165].

More information on the structure and function of CYP enzymes can be found in: [38, 166–173].

References

1 Funk, C.D. (2001) Prostaglandins and leukotrienes: advances in eicosanoid biology. *Science*, **294**, 1871–5.
2 Cinti, D.L. and Feinstein, M.B. (1976) Platelet cytochrome P450: a possible role in arachidonate-induced aggregation. *Biochemical and Biophysical Research Communications*, **73**, 171–9.
3 Pessayre, D., Mazel, P., Descatoire, V., Rogier, E., Feldmann, G. and Benhamou, J.P. (1979) Inhibition of hepatic drug-metabolizing enzymes by arachidonic acid. *Xenobiotica*, **9**, 301–10.
4 Capdevila, J., Chacos, N., Werringloer, J., Prough, R.A. and Estabrook, R.W. (1981) Liver microsomal cytochrome P-450 and the oxidative metabolism of arachidonic acid. *Proceedings of the National Academy of Sciences of the United States of America*, **78**, 5362–6.
5 Capdevila, J., Parkhill, L., Chacos, N., Okita, R., Masters, B.S. and Estabrook, R.W. (1981) The oxidative metabolism of arachidonic acid by purified cytochromes P-450. *Biochemical and Biophysical Research Communications*, **101**, 1357–63.
6 Morrison, A.R. and Pascoe, N. (1981) Metabolism of arachidonate through NADPH-dependent oxygenase of renal cortex. *Proceedings of the National Academy of Sciences of the United States of America*, **78**, 7375–8.
7 Oliw, E.H. and Oates, J.A. (1981) Oxygenation of arachidonic acid by hepatic microsomes of the rabbit. Mechanism of biosynthesis of two vicinal dihydroxyeicosatrienoic acids. *Biochimica et Biophysica Acta*, **666**, 327–40.
8 Sacerdoti, D., Escalante, B., Abraham, N.G., McGiff, J.C., Levere, R.D. and Schwartzman, M.L. (1989) Treatment with tin prevents the development of hypertension in spontaneously hypertensive rats. *Science*, **243**, 388–90.
9 McGiff, J.C. (1991) Cytochrome P-450 metabolism of arachidonic acid. *Annual Review of Pharmacology and Toxicology*, **31**, 339–69.
10 Makita, K., Takahashi, K., Karara, A., Jacobson, H.R., Falck, J.R. and Capdevila, J.H. (1994) Experimental and/or genetically controlled alterations of the renal microsomal cytochrome P450 epoxygenase induce hypertension in rats fed a high salt diet. *Journal of Clinical Investigation*, **94**, 2414–20.
11 Roman, R.J., Alonso-Galicia, M. and Wilson, T.W. (1997) Renal P450 metabolites of arachidonic acid and the development of hypertension in Dahl salt-sensitive rats. *American Journal of Hypertension*, **10**, 63S–67S.
12 Campbell, W.B., Gebremedhin, D., Pratt, P.F. and Harder, D.R. (1996) Identification of epoxyeicosatrienoic acids as endothelium-derived hyperpolarizing factors. *Circulation Research*, **78**, 415–23.
13 Fisslthaler, B., Popp, R., Kiss, L., Potente, M., Harder, D.R., Fleming, I. and Busse, R. (1999) Cytochrome P450 2C is an EDHF synthase in coronary arteries. *Nature*, **401**, 493–7.
14 Fleming, I. and Busse, R. (2006) Endothelium-derived epoxyeicosatrienoic acids and vascular function. *Hypertension*, **47**, 629–33.
15 Feletou, M. and Vanhoutte, P.M. (2006) Endothelium-derived hyperpolarizing factor: where are we now? *Arteriosclerosis Thrombosis and Vascular Biology*, **26**, 1215–25.
16 Kohler, R. and Hoyer, J. (2007) The endothelium-derived hyperpolarizing factor: insights from genetic animal

models. *Kidney International*, **72**, 145–50.
17. Campbell, W.B. and Falck, J.R. (2007) Arachidonic acid metabolites as endothelium-derived hyperpolarizing factors. *Hypertension*, **49**, 590–6.
18. Node, K., Huo, Y., Ruan, X., Yang, B., Spiecker, M., Ley, K., Zeldin, D.C. and Liao, J.K. (1999) Anti-inflammatory properties of cytochrome P450 epoxygenase-derived eicosanoids. *Science*, **285**, 1276–9.
19. Munzenmaier, D.H. and Harder, D.R. (2000) Cerebral microvascular endothelial cell tube formation: role of astrocytic epoxyeicosatrienoic acid release. *American Journal of Physiology – Heart and Circulatory Physiology*, **278**, H1163–7.
20. Jiang, M., Mezentsev, A., Kemp, R., Byun, K., Falck, J.R., Miano, J.M., Nasjletti, A., Abraham, N.G. and Laniado-Schwartzman, M. (2004) Smooth muscle-specific expression of CYP4A1 induces endothelial sprouting in renal arterial microvessels. *Circulation Research*, **94**, 167–74.
21. Amaral, S.L., Maier, K.G., Schippers, D.N., Roman, R.J. and Greene, A.S. (2003) CYP4A metabolites of arachidonic acid and VEGF are mediators of skeletal muscle angiogenesis. *American Journal of Physiology – Heart and Circulatory Physiology*, **284**, H1528–35.
22. Pozzi, A., Macias-Perez, I., Abair, T., Wei, S., Su, Y., Zent, R., Falck, J.R. and Capdevila, J.H. (2005) Characterization of 5,6- and 8,9-epoxyeicosatrienoic acids (5,6- and 8,9-EET) as potent in vivo angiogenic lipids. *Journal of Biological Chemistry*, **280**, 27138–46.
23. Michaelis, U.R., Fisslthaler, B., Medhora, M., Harder, D., Fleming, I. and Busse, R. (2003) Cytochrome P450 2C9-derived epoxyeicosatrienoic acids induce angiogenesis via cross-talk with the epidermal growth factor receptor (EGFR). *FASEB Journal*, **17**, 770–2.
24. Medhora, M., Dhanasekaran, A., Gruenloh, S.K., Dunn, L.K., Gabrilovich, M., Falck, J.R., Harder, D.R., Jacobs, E.R. and Pratt, P.F. (2007) Emerging mechanisms for growth and protection of the vasculature by cytochrome P450-derived products of arachidonic acid and other eicosanoids. *Prostaglandins and Other Lipid Mediators*, **82**, 19–29.
25. Fleming, I. (2007) Epoxyeicosatrienoic acids, cell signaling and angiogenesis. *Prostaglandins and Other Lipid Mediators*, **82**, 60–7.
26. Jiang, J.G., Chen, C.L., Card, J.W., Yang, S., Chen, J.X., Fu, X.N., Ning, Y.G., Xiao, X., Zeldin, D.C. and Wang, D.W. (2005) Cytochrome P450 2J2 promotes the neoplastic phenotype of carcinoma cells and is up-regulated in human tumors. *Cancer Research*, **65**, 4707–15.
27. Jiang, J.G., Ning, Y.G., Chen, C., Ma, D., Liu, Z.J., Yang, S., Zhou, J., Xiao, X., Zhang, X.A., Edin, M.L., Card, J.W., Wang, J., Zeldin, D.C. and Wang, D.W. (2007) Cytochrome p450 epoxygenase promotes human cancer metastasis. *Cancer Research*, **67**, 6665–74.
28. Imig, J.D. (2006) Cardiovascular therapeutic aspects of soluble epoxide hydrolase inhibitors. *Cardiovascular Drug Reviews*, **24**, 169–88.
29. Inceoglu, B., Schmelzer, K.R., Morisseau, C., Jinks, S.L. and Hammock, B.D. (2007) Soluble epoxide hydrolase inhibition reveals novel biological functions of epoxyeicosatrienoic acids (EETs). *Prostaglandins and Other Lipid Mediators*, **82**, 42–9.
30. Xu, D., Li, N., He, Y., Timofeyev, V., Lu, L., Tsai, H.J., Kim, I.H., Tuteja, D., Mateo, R.K., Singapuri, A., Davis, B.B., Low, R., Hammock, B.D. and Chiamvimonvat, N. (2006) Prevention and reversal of cardiac hypertrophy by soluble epoxide hydrolase inhibitors. *Proceedings of the National Academy of Sciences of the United States of America*, **103**, 18733–8.
31. Wang, M.H., Smith, A., Zhou, Y., Chang, H.H., Lin, S., Zhao, X., Imig, J.D. and Dorrance, A.M. (2003) Downregulation of renal CYP-derived eicosanoid synthesis in rats with diet-induced hypertension. *Hypertension*, **42**, 594–9.
32. Barbosa-Sicard, E., Markovic, M., Honeck, H., Christ, B., Muller, D.N. and Schunck, W.H. (2005) Eicosapentaenoic acid metabolism by cytochrome P450

enzymes of the CYP2C subfamily. *Biochemical and Biophysical Research Communications*, **329**, 1275–81.

33 Theuer, J., Shagdarsuren, E., Muller, D.N., Kaergel, E., Honeck, H., Park, J.K., Fiebeler, A., Dechend, R., Haller, H., Luft, F.C. and Schunck, W.H. (2005) Inducible NOS inhibition, eicosapentaenoic acid supplementation, and angiotensin II-induced renal damage. *Kidney International*, **67**, 248–58.

34 Dai, D., Zeldin, D.C., Blaisdell, J.A., Chanas, B., Coulter, S.J., Ghanayem, B.I. and Goldstein, J.A. (2001) Polymorphisms in human CYP2C8 decrease metabolism of the anticancer drug paclitaxel and arachidonic acid. *Pharmacogenetics*, **11**, 597–607.

35 King, L.M., Gainer, J.V., David, G.L., Dai, D., Goldstein, J.A., Brown, N.J. and Zeldin, D.C. (2005) Single nucleotide polymorphisms in the CYP2J2 and CYP2C8 genes and the risk of hypertension. *Pharmacogenet Genomics*, **15**, 7–13.

36 Omura, T. and Sato, R. (1964) The carbon monoxide-binding pigment of liver microsomes. II. Solubilization, purification, and properties. *Journal of Biological Chemistry*, **239**, 2379–85.

37 Capdevila, J.H., Falck, J.R. and Estabrook, R.W. (1992) Cytochrome P450 and the arachidonate cascade. *FASEB Journal*, **6**, 731–6.

38 Capdevila, J.H. and Falck, J.R. (2002) Biochemical and molecular properties of the cytochrome P450 arachidonic acid monooxygenases. *Prostaglandins and Other Lipid Mediators*, **68–69**, 325–44.

39 Nelson, D.R. (1998) Cytochrome P450 nomenclature. *Methods in Molecular Biology Journals*, **107**, 15–24.

40 Nelson, D.R. (2006) Cytochrome P450 nomenclature, 2004. *Methods in Molecular Biology Journals*, **320**, 1–10.

41 Nelson, D.R., Zeldin, D.C., Hoffman, S.M., Maltais, L.J., Wain, H.M. and Nebert, D.W. (2004) Comparison of cytochrome P450 (CYP) genes from the mouse and human genomes, including nomenclature recommendations for genes, pseudogenes and alternative-splice variants. *Pharmacogenetics*, **14**, 1–18.

42 Holla, V.R., Adas, F., Imig, J.D., Zhao, X., Price, E. Jr, Olsen, N., Kovacs, W.J., Magnuson, M.A., Keeney, D.S., Breyer, M.D., Falck, J.R., Waterman, M.R. and Capdevila, J.H. (2001) Alterations in the regulation of androgen-sensitive Cyp 4a monooxygenases cause hypertension. *Proceedings of the National Academy of Sciences of the United States of America*, **98**, 5211–16.

43 Muller, D.N., Schmidt, C., Barbosa-Sicard, E., Wellner, M., Gross, V., Hercule, H., Markovic, M., Honeck, H., Luft, F.C. and Schunck, W.H. (2007) Mouse Cyp4a isoforms: enzymatic properties, gender- and strain-specific expression, and role in renal 20-hydroxyeicosatetraenoic acid formation. *Biochemical Journal*, **403**, 109–18.

44 Gainer, J.V., Bellamine, A., Dawson, E.P., Womble, K.E., Grant, S.W., Wang, Y., Cupples, L.A., Guo, C.Y., Demissie, S., O'Donnell, C.J., Brown, N.J., Waterman, M.R. and Capdevila, J.H. (2005) Functional variant of CYP4A11 20-hydroxyeicosatetraenoic acid synthase is associated with essential hypertension. *Circulation*, **111**, 63–9.

45 Powell, P.K., Wolf, I., Jin, R. and Lasker, J.M. (1998) Metabolism of arachidonic acid to 20-hydroxy-5,8,11, 14-eicosatetraenoic acid by P450 enzymes in human liver: involvement of CYP4F2 and CYP4A11. *Journal of Pharmacology and Experimental Therapeutics*, **285**, 1327–36.

46 Lasker, J.M., Chen, W.B., Wolf, I., Bloswick, B.P., Wilson, P.D. and Powell, P.K. (2000) Formation of 20-hydroxyeicosatetraenoic acid, a vasoactive and natriuretic eicosanoid, in human kidney. Role of Cyp4F2 and Cyp4A11. *Journal of Biological Chemistry*, **275**, 4118–26.

47 Ward, N.C., Rivera, J., Hodgson, J., Puddey, I.B., Beilin, L.J., Falck, J.R. and Croft, K.D. (2004) Urinary 20-hydroxyeicosatetraenoic acid is associated with endothelial dysfunction in humans. *Circulation*, **110**, 438–43.

48 Roman, R.J. (2002) P-450 metabolites of arachidonic acid in the control of

cardiovascular function. *Physiological Reviews*, **82**, 131–85.
49 Wei, Y., Sun, P., Wang, Z., Yang, B., Carroll, M.A. and Wang, W.H. (2006) Adenosine inhibits ENaC via cytochrome P-450 epoxygenase-dependent metabolites of arachidonic acid. *American Journal of Physiology–Renal Physiology*, **290**, F1163–8.
50 Sun, P., Lin, D.H., Wang, T., Babilonia, E., Wang, Z., Jin, Y., Kemp, R., Nasjletti, A. and Wang, W.H. (2006) Low Na intake suppresses expression of CYP2C23 and arachidonic acid-induced inhibition of ENaC. *American Journal of Physiology–Renal Physiology*, **291**, F1192-200.
51 Spector, A.A., Fang, X., Snyder, G.D. and Weintraub, N.L. (2004) Epoxyeicosatrienoic acids (EETs): metabolism and biochemical function. *Progress in Lipid Research*, **43**, 55–90.
52 Spector, A.A. and Norris, A.W. (2007) Action of epoxyeicosatrienoic acids on cellular function. *American Journal of Physiology–Cell Physiology*, **292**, C996–1012.
53 Simopoulos, A.P. (2006) Evolutionary aspects of diet, the omega-6/omega-3 ratio and genetic variation: nutritional implications for chronic diseases. *Biomedicine and Pharmacotherapy*, **60**, 502–7.
54 Kris-Etherton, P.M., Harris, W.S. and Appel, L.J. (2002) Fish consumption, fish oil, omega-3 fatty acids, and cardiovascular disease. *Circulation*, **106**, 2747–57.
55 Leaf, A., Kang, J.X., Xiao, Y.F. and Billman, G.E. (2003) Clinical prevention of sudden cardiac death by *n*-3 polyunsaturated fatty acids and mechanism of prevention of arrhythmias by *n*-3 fish oils. *Circulation*, **107**, 2646–52.
56 Lands, W.E. (2005) Dietary fat and health: the evidence and the politics of prevention: careful use of dietary fats can improve life and prevent disease. *Annals of the New York Academy of Sciences*, **1055**, 179–92.
57 Marchioli, R., Barzi, F., Bomba, E., Chieffo, C., Di Gregorio, D., Di Mascio, R., Franzosi, M.G., Geraci, E., Levantesi, G., Maggioni, A.P., Mantini, L., Marfisi, R.M., Mastrogiuseppe, G., Mininni, N., Nicolosi, G.L., Santini, M., Schweiger, C., Tavazzi, L., Tognoni, G., Tucci, C. and Valagussa, F. (2002) Early protection against sudden death by *n*-3 polyunsaturated fatty acids after myocardial infarction: time-course analysis of the results of the Gruppo Italiano per lo Studio della Sopravvivenza nell'Infarto Miocardico (GISSI)-Prevenzione. *Circulation*, **105**, 1897–903.
58 VanRollins, M. (1990) Synthesis and characterization of cytochrome P-450 epoxygenase metabolites of eicosapentaenoic acid. *Lipids*, **25**, 481–90.
59 Oliw, E.H. and Sprecher, H.W. (1991) Metabolism of polyunsaturated (*n*-3) fatty acids by monkey seminal vesicles: isolation and biosynthesis of omega-3 epoxides. *Biochimica et Biophysica Acta*, **1086**, 287–94.
60 Lauterbach, B., Barbosa-Sicard, E., Wang, M.H., Honeck, H., Kargel, E., Theuer, J., Schwartzman, M.L., Haller, H., Luft, F.C., Gollasch, M. and Schunck, W.H. (2002) Cytochrome P450-dependent eicosapentaenoic acid metabolites are novel BK channel activators. *Hypertension*, **39**, 609–13.
61 Stark, K., Wongsud, B., Burman, R. and Oliw, E.H. (2005) Oxygenation of polyunsaturated long chain fatty acids by recombinant CYP4F8 and CYP4F12 and catalytic importance of Tyr-125 and Gly-328 of CYP4F8. *Archives of Biochemistry and Biophysics*, **441**, 174–81.
62 Schwarz, D., Kisselev, P., Ericksen, S.S., Szklarz, G.D., Chernogolov, A., Honeck, H., Schunck, W.H. and Roots, I. (2004) Arachidonic and eicosapentaenoic acid metabolism by human CYP1A1: highly stereoselective formation of 17(*R*),18(*S*)-epoxyeicosatetraenoic acid. *Biochemical Pharmacology*, **67**, 1445–57.
63 Harmon, S.D., Fang, X., Kaduce, T.L., Hu, S., Raj Gopal, V., Falck, J.R. and Spector, A.A. (2006) Oxygenation of omega-3 fatty acids by human cytochrome P450 4F3B: effect on 20-hydroxyeicosatetraenoic acid production. *Prostaglandins, Leukotrienes, and Essential Fatty Acids*, **75**, 169–77.

64 Quilley, J., Fulton, D. and McGiff, J.C. (1997) Hyperpolarizing factors. *Biochemical Pharmacology*, **54**, 1059–70.

65 Huang, A., Sun, D., Wu, Z., Yan, C., Carroll, M.A., Jiang, H., Falck, J.R. and Kaley, G. (2004) Estrogen elicits cytochrome P450-mediated flow-induced dilation of arterioles in NO deficiency: role of PI3K-Akt phosphorylation in genomic regulation. *Circulation Research*, **94**, 245–52.

66 Bolad, I. and Delafontaine, P. (2005) Endothelial dysfunction: its role in hypertensive coronary disease. *Current Opinion in Cardiology*, **20**, 270–4.

67 Huang, A., Sun, D., Yan, C., Falck, J.R. and Kaley, G. (2005) Contribution of 20-HETE to augmented myogenic constriction in coronary arteries of endothelial NO synthase knockout mice. *Hypertension*, **46**, 607–13.

68 Alonso-Galicia, M., Sun, C.W., Falck, J.R., Harder, D.R. and Roman, R.J. (1998) Contribution of 20-HETE to the vasodilator actions of nitric oxide in renal arteries. *American Journal of Physiology*, **275**, F370–8.

69 Sun, C.W., Falck, J.R., Okamoto, H., Harder, D.R. and Roman, R.J. (2000) Role of cGMP versus 20-HETE in the vasodilator response to nitric oxide in rat cerebral arteries. *American Journal of Physiology – Heart and Circulatory Physiology*, **279**, H339–50.

70 Abraham, N.G. and Kappas, A. (2005) Heme oxygenase and the cardio-vascular-renal system. *Free Radical Biology and Medicine*, **39**, 1–25.

71 Botros, F.T., Laniado-Schwartzman, M. and Abraham, N.G. (2002) Regulation of cyclooxygenase- and cytochrome p450-derived eicosanoids by heme oxygenase in the rat kidney. *Hypertension*, **39**, 639–44.

72 Botros, F.T., Schwartzman, M.L., Stier, C.T. Jr, Goodman, A.I. and Abraham, N.G. (2005) Increase in heme oxygenase-1 levels ameliorates renovascular hypertension. *Kidney International*, **68**, 2745–55.

73 Fisslthaler, B., Michaelis, U.R., Randriamboavonjy, V., Busse, R. and Fleming, I. (2003) Cytochrome P450 epoxygenases and vascular tone: novel role for HMG-CoA reductase inhibitors in the regulation of CYP 2C expression. *Biochimica et Biophysica Acta*, **1619**, 332–9.

74 Oyekan, A. (2002) Nitric oxide inhibits renal cytochrome P450-dependent epoxygenases in the rat. *Clinical and Experimental Pharmacology and Physiology*, **29**, 990–5.

75 Wang, M.H., Wang, J., Chang, H.H., Zand, B.A., Jiang, M., Nasjletti, A. and Laniado-Schwartzman, M. (2003) Regulation of renal CYP4A expression and 20-HETE synthesis by nitric oxide in pregnant rats. *American Journal of Physiology – Renal Physiology*, **285**, F295–302.

76 Kuncewicz, T., Sheta, E.A., Goldknopf, I.L. and Kone, B.C. (2003) Proteomic analysis of S-nitrosylated proteins in mesangial cells. *Molecular and Cellular Proteomics*, **2**, 156–63.

77 Lin, H.L., Kent, U.M., Zhang, H., Waskell, L. and Hollenberg, P.F. (2003) Mutation of tyrosine 190 to alanine eliminates the inactivation of cytochrome P450 2B1 by peroxynitrite. *Chemical Research in Toxicology*, **16**, 129–36.

78 Fleming, I., Michaelis, U.R., Bredenkotter, D., Fisslthaler, B., Dehghani, F., Brandes, R.P. and Busse, R. (2001) Endothelium-derived hyperpolarizing factor synthase (cytochrome P450 2C9) is a functionally significant source of reactive oxygen species in coronary arteries. *Circulation Research*, **88**, 44–51.

79 Fichtlscherer, S., Dimmeler, S., Breuer, S., Busse, R., Zeiher, A.M. and Fleming, I. (2004) Inhibition of cytochrome P450 2C9 improves endothelium-dependent, nitric oxide-mediated vasodilatation in patients with coronary artery disease. *Circulation*, **109**, 178–83.

80 Pfister, S.L., Spitzbarth, N., Zeldin, D.C., Lafite, P., Mansuy, D. and Campbell, W.B. (2003) Rabbit aorta converts 15-HPETE to trihydroxyeicosatrienoic acids: potential role of cytochrome P450. *Archives of Biochemistry and Biophysics*, **420**, 142–52.

81 McGiff, J.C. and Quilley, J. (1999) 20-HETE and the kidney: resolution of old problems and new beginnings.

American Journal of Physiology, **277**, R607–23.
82 Imig, J.D. (2000) Eicosanoid regulation of the renal vasculature. *American Journal of Physiology – Renal Physiology*, **279**, F965–81.
83 Randriamboavonjy, V., Busse, R. and Fleming, I. (2003) 20-HETE-induced contraction of small coronary arteries depends on the activation of Rho-kinase. *Hypertension*, **41**, 801–6.
84 Xiao, Y.F., Huang, L. and Morgan, J.P. (1998) Cytochrome P450: a novel system modulating Ca^{2+} channels and contraction in mammalian heart cells. *Journal of Physiology*, **508** (Pt 3), 777–92.
85 Lee, H.C., Lu, T., Weintraub, N.L., VanRollins, M., Spector, A.A. and Shibata, E.F. (1999) Effects of epoxyeicosatrienoic acids on the cardiac sodium channels in isolated rat ventricular myocytes. *Journal of Physiology*, **519**, 153–68.
86 Lu, T., VanRollins, M. and Lee, H.C. (2002) Stereospecific activation of cardiac ATP-sensitive K(+) channels by epoxyeicosatrienoic acids: a structural determinant study. *Molecular Pharmacology*, **62**, 1076–83.
87 Xiao, Y.F. (2007) Cyclic AMP-dependent modulation of cardiac L-type Ca^{2+} and transient outward K^+ channel activities by epoxyeicosatrienoic acids. *Prostaglandins and Other Lipid Mediators*, **82**, 11–18.
88 Seubert, J., Yang, B., Bradbury, J.A., Graves, J., Degraff, L.M., Gabel, S., Gooch, R., Foley, J., Newman, J., Mao, L., Rockman, H.A., Hammock, B.D., Murphy, E. and Zeldin, D.C. (2004) Enhanced postischemic functional recovery in CYP2J2 transgenic hearts involves mitochondrial ATP-sensitive K^+ channels and p42/p44 MAPK pathway. *Circulation Research*, **95**, 506–14.
89 Xiao, Y.F., Ke, Q., Seubert, J.M., Bradbury, J.A., Graves, J., Degraff, L.M., Falck, J.R., Krausz, K., Gelboin, H.V., Morgan, J.P. and Zeldin, D.C. (2004) Enhancement of cardiac L-type Ca^{2+} currents in transgenic mice with cardiac-specific overexpression of CYP2J2. *Molecular Pharmacology*, **66**, 1607–16.
90 Seubert, J.M., Zeldin, D.C., Nithipatikom, K. and Gross, G.J. (2007) Role of epoxyeicosatrienoic acids in protecting the myocardium following ischemia/reperfusion injury. *Prostaglandins and Other Lipid Mediators*, **82**, 50–9.
91 Gross, G.J., Hsu, A., Falck, J.R. and Nithipatikom, K. (2007) Mechanisms by which epoxyeicosatrienoic acids (EETs) elicit cardioprotection in rat hearts. *Journal of Molecular and Cellular Cardiology*, **42**, 687–91.
92 Seubert, J.M., Sinal, C.J., Graves, J., DeGraff, L.M., Bradbury, J.A., Lee, C.R., Goralski, K., Carey, M.A., Luria, A., Newman, J.W., Hammock, B.D., Falck, J.R., Roberts, H., Rockman, H.A., Murphy, E. and Zeldin, D.C. (2006) Role of soluble epoxide hydrolase in postischemic recovery of heart contractile function. *Circulation Research*, **99**, 442–50.
93 Gross, E.R., Nithipatikom, K., Hsu, A.K., Peart, J.N., Falck, J.R., Campbell, W.B. and Gross, G.J. (2004) Cytochrome P450 omega-hydroxylase inhibition reduces infarct size during reperfusion via the sarcolemmal KATP channel. *Journal of Molecular and Cellular Cardiology*, **37**, 1245–9.
94 Nithipatikom, K., Gross, E.R., Endsley, M.P., Moore, J.M., Isbell, M.A., Falck, J.R., Campbell, W.B. and Gross, G.J. (2004) Inhibition of cytochrome P450omega-hydroxylase: a novel endogenous cardioprotective pathway. *Circulation Research*, **95**, e65–71.
95 Nithipatikom, K., Endsley, M.P., Moore, J.M., Isbell, M.A., Falck, J.R., Campbell, W.B. and Gross, G.J. (2006) Effects of selective inhibition of cytochrome P-450 omega-hydroxylases and ischemic preconditioning in myocardial protection. *American Journal of Physiology – Heart and Circulatory Physiology*, **290**, H500–5.
96 Larsen, B.T., Campbell, W.B. and Gutterman, D.D. (2007) Beyond vasodilatation: non-vasomotor roles of epoxyeicosatrienoic acids in the cardiovascular system. *Trends in Pharmacological Sciences*, **28**, 32–8.
97 Spiecker, M. and Liao, J.K. (2005) Vascular protective effects of cytochrome

p450 epoxygenase-derived eicosanoids. *Archives of Biochemistry and Biophysics*, **433**, 413–20.

98 Krotz, F., Riexinger, T., Buerkle, M.A., Nithipatikom, K., Gloe, T., Sohn, H.Y., Campbell, W.B. and Pohl, U. (2004) Membrane-potential-dependent inhibition of platelet adhesion to endothelial cells by epoxyeicosatrienoic acids. *Arteriosclerosis Thrombosis and Vascular Biology*, **24**, 595–600.

99 Node, K., Ruan, X.L., Dai, J., Yang, S.X., Graham, L., Zeldin, D.C. and Liao, J.K. (2001) Activation of Galpha s mediates induction of tissue-type plasminogen activator gene transcription by epoxyeicosatrienoic acids. *Journal of Biological Chemistry*, **276**, 15983–9.

100 Chen, J.K., Capdevila, J. and Harris, R.C. (2001) Cytochrome p450 epoxygenase metabolism of arachidonic acid inhibits apoptosis. *Molecular and Cellular Biology*, **21**, 6322–31.

101 Yang, S., Lin, L., Chen, J.X., Lee, C.R., Seubert, J.M., Wang, Y., Wang, H., Chao, Z.R., Tao, D.D., Gong, J.P., Lu, Z.Y., Wang, D.W. and Zeldin, D.C. (2007) Cytochrome P-450 epoxygenases protect endothelial cells from apoptosis induced by tumor necrosis factor-α via MAPK and PI3K/Akt signaling pathways. *American Journal of Physiology–Heart and Circulatory Physiology*, **293**, H142–51.

102 Sun, J., Sui, X., Bradbury, J.A., Zeldin, D.C., Conte, M.S. and Liao, J.K. (2002) Inhibition of vascular smooth muscle cell migration by cytochrome p450 epoxygenase-derived eicosanoids. *Circulation Research*, **90**, 1020–7.

103 Zhang, Y., Oltman, C.L., Lu, T., Lee, H.C., Dellsperger, K.C. and VanRollins, M. (2001) EET homologs potently dilate coronary microvessels and activate BK(Ca) channels. *American Journal of Physiology–Heart and Circulatory Physiology*, **280**, H2430–40.

104 Hercule, H.C., Salanova, B., Essin, K., Horst, H., Falck, J.R., Sausbier, M., Rut, P., Schunck, W.H., Luft, F.C. and Gollasch, M. (2007) The vasodilator 17,18-epoxyeicosatetraenoic acid targets the pore-forming BK α channel subunit. *Experimental Physiology*, **92**, 1067–76.

105 Widstrom, R.L., Norris, A.W., Van Der Veer, J. and Spector, A.A. (2003) Fatty acid-binding proteins inhibit hydration of epoxyeicosatrienoic acids by soluble epoxide hydrolase. *Biochemistry*, **42**, 11762–7.

106 Cowart, L.A., Wei, S., Hsu, M.H., Johnson, E.F., Krishna, M.U., Falck, J.R. and Capdevila, J.H. (2002) The CYP4A isoforms hydroxylate epoxyeicosatrienoic acids to form high affinity peroxisome proliferator-activated receptor ligands. *Journal of Biological Chemistry*, **277**, 35105–12.

107 Muller, D.N., Theuer, J., Shagdarsuren, E., Kaergel, E., Honeck, H., Park, J.K., Markovic, M., Barbosa-Sicard, E., Dechend, R., Wellner, M., Kirsch, T., Fiebeler, A., Rothe, M., Haller, H., Luft, F.C. and Schunck, W.H. (2004) A peroxisome proliferator-activated receptor-alpha activator induces renal CYP2C23 activity and protects from angiotensin II-induced renal injury. *American Journal of Pathology*, **164**, 521–32.

108 Ferreri, N.R., McGiff, J.C., Carroll, M.A. and Quilley, J. (2004) Renal COX-2, cytokines and 20-HETE: tubular and vascular mechanisms. *Current Pharmaceutical Design*, **10**, 613–26.

109 Moreland, K.T., Procknow, J.D., Sprague, R.S., Iverson, J.L., Lonigro, A.J. and Stephenson, A.H. (2007) Cyclooxygenase (COX)-1 and COX-2 participate in 5,6-epoxyeicosatrienoic acid-induced contraction of rabbit intralobar pulmonary arteries. *Journal of Pharmacology and Experimental Therapeutics*, **321**, 446–54.

110 Oyekan, A.O., McAward, K., Conetta, J., Rosenfeld, L. and McGiff, J.C. (1999) Endothelin-1 and CYP450 arachidonate metabolites interact to promote tissue injury in DOCA-salt hypertension. *American Journal of Physiology*, **276**, R766–75.

111 Muthalif, M.M., Karzoun, N.A., Gaber, L., Khandekar, Z., Benter, I.F., Saeed, A.E., Parmentier, J.H., Estes, A. and Malik, K.U. (2000) Angiotensin II-induced hypertension: contribution of Ras

GTPase/mitogen-activated protein kinase and cytochrome P450 metabolites. *Hypertension*, **36**, 604–9.

112 Seki, T., Ishimoto, T., Sakurai, T., Yasuda, Y., Taniguchi, K., Doi, M., Sato, M., Roman, R.J. and Miyata, N. (2005) Increased excretion of urinary 20-HETE in rats with cyclosporine-induced nephrotoxicity. *Journal of Pharmacological Sciences*, **97**, 132–7.

113 Singh, H., Cheng, J., Deng, H., Kemp, R., Ishizuka, T., Nasjletti, A. and Schwartzman, M.L. (2007) Vascular cytochrome P450 4A expression and 20-hydroxyeicosatetraenoic acid synthesis contribute to endothelial dysfunction in androgen-induced hypertension. *Hypertension*, **50**, 123–9.

114 Kaergel, E., Muller, D.N., Honeck, H., Theuer, J., Shagdarsuren, E., Mullally, A., Luft, F.C. and Schunck, W.H. (2002) P450-dependent arachidonic acid metabolism and angiotensin II-induced renal damage. *Hypertension*, **40**, 273–9.

115 Zhao, X., Pollock, D.M., Inscho, E.W., Zeldin, D.C. and Imig, J.D. (2003) Decreased renal cytochrome P450 2C enzymes and impaired vasodilation are associated with angiotensin salt-sensitive hypertension. *Hypertension*, **41**, 709–14.

116 Sankaralingam, S., Desai, K.M., Glaeser, H., Kim, R.B. and Wilson, T.W. (2006) Inability to upregulate cytochrome P450 4A and 2C causes salt sensitivity in young Sprague-Dawley rats. *American Journal of Hypertension*, **19**, 1174–80.

117 Nakagawa, K., Holla, V.R., Wei, Y., Wang, W.H., Gatica, A., Wei, S., Mei, S., Miller, C.M., Cha, D.R., Price, E. Jr, Zent, R., Pozzi, A., Breyer, M.D., Guan, Y., Falck, J.R., Waterman, M.R., and Capdevila, J.H. (2006) Salt-sensitive hypertension is associated with dysfunctional *Cyp4a10* gene and kidney epithelial sodium channel. *Journal of Clinical Investigation*, **116**, 1696–702.

118 Huang, H., Chang, H.H., Xu, Y., Reddy, D.S., Du, J., Zhou, Y., Dong, Z., Falck, J.R. and Wang, M.H. (2006) Epoxyeicosatrienoic Acid inhibition alters renal hemodynamics during pregnancy. *Experimental Biology and Medicine (Maywood)*, **231**, 1744–52.

119 Honeck, H., Gross, V., Erdmann, B., Kargel, E., Neunaber, R., Milia, A.F., Schneider, W., Luft, F.C. and Schunck, W.H. (2000) Cytochrome P450-dependent renal arachidonic acid metabolism in desoxycorticosterone acetate-salt hypertensive mice. *Hypertension*, **36**, 610–16.

120 Hoagland, K.M., Flasch, A.K. and Roman, R.J. (2003) Inhibitors of 20-HETE formation promote salt-sensitive hypertension in rats. *Hypertension*, **42**, 669–73.

121 Vera, T., Taylor, M., Bohman, Q., Flasch, A., Roman, R.J. and Stec, D.E. (2005) Fenofibrate prevents the development of angiotensin II-dependent hypertension in mice. *Hypertension*, **45**, 730–5.

122 Yaghini, F.A., Zhang, C., Parmentier, J.H., Estes, A.M., Jafari, N., Schaefer, S.A. and Malik, K.U. (2005) Contribution of arachidonic acid metabolites derived via cytochrome P4504A to angiotensin II-induced neointimal growth. *Hypertension*, **45**, 1182–7.

123 Ward, N.C., Puddey, I.B., Hodgson, J.M., Beilin, L.J. and Croft, K.D. (2005) Urinary 20-hydroxyeicosatetraenoic acid excretion is associated with oxidative stress in hypertensive subjects. *Free Radical Biology and Medicine*, **38**, 1032–6.

124 Regitz-Zagrosek, V. (2006) Therapeutic implications of the gender-specific aspects of cardiovascular disease. *Nature Reviews Drug Discovery*, **5**, 425–38.

125 Blanton, A., Nsaif, R., Hercule, H. and Oyekan, A. (2006) Nitric oxide/cytochrome P450 interactions in cyclosporin A-induced effects in the rat. *Journal of Hypertension*, **24**, 1865–72.

126 Catella, F., Lawson, J.A., Fitzgerald, D.J. and FitzGerald, G.A. (1990) Endogenous biosynthesis of arachidonic acid epoxides in humans: increased formation in pregnancy-induced hypertension. *Proceedings of the National Academy of Sciences of the United States of America*, **87**, 5893–7.

127 Zhou, Y., Chang, H.H., Du, J., Wang, C.Y., Dong, Z. and Wang, M.H. (2005) Renal epoxyeicosatrienoic acid synthesis during pregnancy. *American Journal of*

Physiology – Renal Physiology, **288**, F221–6.

128 Kenny, L.C., Baker, P.N., Kendall, D.A., Randall, M.D. and Dunn, W.R. (2002) Differential mechanisms of endothelium-dependent vasodilator responses in human myometrial small arteries in normal pregnancy and pre-eclampsia. *Clinical Science*, **103**, 67–73.

129 Hagedorn, K.A., Cooke, C.L., Falck, J.R., Mitchell, B.F. and Davidge, S.T. (2007) Regulation of vascular tone during pregnancy: a novel role for the pregnane X receptor. *Hypertension*, **49**, 328–33.

130 Llinas, M.T., Alexander, B.T., Capparelli, M.F., Carroll, M.A. and Granger, J.P. (2004) Cytochrome P-450 inhibition attenuates hypertension induced by reductions in uterine perfusion pressure in pregnant rats. *Hypertension*, **43**, 623–8.

131 Patel, L., Sullivan, M.H. and Elder, M.G. (1989) Production of epoxygenase metabolite by human reproductive tissues. *Prostaglandins*, **38**, 615–24.

132 Schafer, W.R., Zahradnik, H.P., Arbogast, E., Wetzka, B., Werner, K. and Breckwoldt, M. (1996) Arachidonate metabolism in human placenta, fetal membranes, decidua and myometrium: lipoxygenase and cytochrome P450 metabolites as main products in HPLC profiles. *Placenta*, **17**, 231–8.

133 Zhang, J.H., Pearson, T., Matharoo-Ball, B., Ortori, C.A., Warren, A.Y., Khan, R. and Barrett, D.A. (2007) Quantitative profiling of epoxyeicosatrienoic, hydroxyeicosatetraenoic, and dihydroxyeicosatrienoic acids in human intrauterine tissues using liquid chromatography/electrospray ionization tandem mass spectrometry. *Analytical Biochemistry*, **365**, 40–51.

134 Zosmer, A., Elder, M.G. and Sullivan, M.H. (2003) The regulation of arachidonic acid metabolism in human first trimester trophoblast by cyclic AMP. *Prostaglandins and Other Lipid Mediators*, **71**, 43–53.

135 Hoagland, K.M., Maier, K.G. and Roman, R.J. (2003) Contributions of 20-HETE to the antihypertensive effects of Tempol in Dahl salt-sensitive rats. *Hypertension*, **41**, 697–702.

136 Delerive, P., De Bosscher, K., Besnard, S., Vanden Berghe, W., Peters, J.M., Gonzalez, F.J., Fruchart, J.C., Tedgui, A., Haegeman, G. and Staels, B. (1999) Peroxisome proliferator-activated receptor alpha negatively regulates the vascular inflammatory gene response by negative cross-talk with transcription factors NF-kappaB and AP-1. *Journal of Biological Chemistry*, **274**, 32048–54.

137 Delerive, P., Gervois, P., Fruchart, J.C. and Staels, B. (2000) Induction of IkappaBalpha expression as a mechanism contributing to the anti-inflammatory activities of peroxisome proliferator-activated receptor-alpha activators. *Journal of Biological Chemistry*, **275**, 36703–7.

138 Roman, R.J., Ma, Y.H., Frohlich, B. and Markham, B. (1993) Clofibrate prevents the development of hypertension in Dahl salt-sensitive rats. *Hypertension*, **21**, 985–8.

139 Shatara, R.K., Quest, D.W. and Wilson, T.W. (2000) Fenofibrate lowers blood pressure in two genetic models of hypertension. *Canadian Journal of Physiology and Pharmacology*, **78**, 367–71.

140 Diep, Q.N., Amiri, F., Touyz, R.M., Cohn, J.S., Endemann, D., Neves, M.F. and Schiffrin, E.L. (2002) PPARalpha activator effects on Ang II-induced vascular oxidative stress and inflammation. *Hypertension*, **40**, 866–71.

141 Zhao, X., Quigley, J.E., Yuan, J., Wang, M.H., Zhou, Y. and Imig, J.D. (2006) PPAR-alpha activator fenofibrate increases renal CYP-derived eicosanoid synthesis and improves endothelial dilator function in obese Zucker rats. *American Journal of Physiology – Heart and Circulatory Physiology*, **290**, H2187-95.

142 Huang, H., Morisseau, C., Wang, J., Yang, T., Falck, J.R., Hammock, B.D. and Wang, M.H. (2007) Increasing or stabilizing renal epoxyeicosatrienoic acid production attenuates abnormal renal function and hypertension in obese rats. *American Journal of Physiology – Renal Physiology*, **293**, F342–9.

143 Newman, J.W., Morisseau, C., Harris, T.R. and Hammock, B.D. (2003) The

soluble epoxide hydrolase encoded by EPXH2 is a bifunctional enzyme with novel lipid phosphate phosphatase activity. *Proceedings of the National Academy of Sciences of the United States of America*, **100**, 1558–63.
144 Cronin, A., Mowbray, S., Durk, H., Homburg, S., Fleming, I., Fisslthaler, B., Oesch, F. and Arand, M. (2003) The N-terminal domain of mammalian soluble epoxide hydrolase is a phosphatase. *Proceedings of the National Academy of Sciences of the United States of America*, **100**, 1552–7.
145 Tran, K.L., Aronov, P.A., Tanaka, H., Newman, J.W., Hammock, B.D. and Morisseau, C. (2005) Lipid sulfates and sulfonates are allosteric competitive inhibitors of the N-terminal phosphatase activity of the mammalian soluble epoxide hydrolase. *Biochemistry*, **44**, 12179–87.
146 Enayetallah, A.E. and Grant, D.F. (2006) Effects of human soluble epoxide hydrolase polymorphisms on isoprenoid phosphate hydrolysis. *Biochemical and Biophysical Research Communications*, **341**, 254–60.
147 Zeldin, D.C., Kobayashi, J., Falck, J.R., Winder, B.S., Hammock, B.D., Snapper, J.R. and Capdevila, J.H. (1993) Regio- and enantiofacial selectivity of epoxyeicosatrienoic acid hydration by cytosolic epoxide hydrolase. *Journal of Biological Chemistry*, **268**, 6402–7.
148 Zeldin, D.C., Wei, S., Falck, J.R., Hammock, B.D., Snapper, J.R. and Capdevila, J.H. (1995) Metabolism of epoxyeicosatrienoic acids by cytosolic epoxide hydrolase: substrate structural determinants of asymmetric catalysis. *Archives of Biochemistry and Biophysics*, **316**, 443–51.
149 Enayetallah, A.E., French, R.A., Barber, M. and Grant, D.F. (2006) Cell-specific subcellular localization of soluble epoxide hydrolase in human tissues. *Journal of Histochemistry and Cytochemistry*, **54**, 329–35.
150 Ai, D., Fu, Y., Guo, D., Tanaka, H., Wang, N., Tang, C., Hammock, B.D., Shyy, J.Y. and Zhu, Y. (2007) Angiotensin II up-regulates soluble epoxide hydrolase in vascular endothelium in vitro and in vivo. *Proceedings of the National Academy of Sciences of the United States of America*, **104**, 9018–23.
151 Jung, O., Brandes, R.P., Kim, I.H., Schweda, F., Schmidt, R., Hammock, B.D., Busse, R. and Fleming, I. (2005) Soluble epoxide hydrolase is a main effector of angiotensin II-induced hypertension. *Hypertension*, **45**, 759–65.
152 Dorrance, A.M., Rupp, N., Pollock, D.M., Newman, J.W., Hammock, B.D. and Imig, J.D. (2005) An epoxide hydrolase inhibitor, 12-(3-adamantan-1-yl-ureido)-dodecanoic acid (AUDA), reduces ischemic cerebral infarct size in stroke-prone spontaneously hypertensive rats. *Journal of Cardiovascular Pharmacology*, **46**, 842–8.
153 Inceoglu, B., Jinks, S.L., Schmelzer, K.R., Waite, T., Kim, I.H. and Hammock, B.D. (2006) Inhibition of soluble epoxide hydrolase reduces LPS-induced thermal hyperalgesia and mechanical allodynia in a rat model of inflammatory pain. *Life Sciences*, **79**, 2311–19.
154 Olearczyk, J.J., Field, M.B., Kim, I.H., Morisseau, C., Hammock, B.D. and Imig, J.D. (2006) Substituted adamantyl-urea inhibitors of the soluble epoxide hydrolase dilate mesenteric resistance vessels. *Journal of Pharmacology and Experimental Therapeutics*, **318**, 1307–14.
155 Ng, V.Y., Morisseau, C., Falck, J.R., Hammock, B.D. and Kroetz, D.L. (2006) Inhibition of smooth muscle proliferation by urea-based alkanoic acids via peroxisome proliferator-activated receptor alpha-dependent repression of cyclin D1. *Arteriosclerosis Thrombosis and Vascular Biology*, **26**, 2462–8.
156 Koerner, I.P., Jacks, R., DeBarber, A.E., Koop, D., Mao, P., Grant, D.F. and Alkayed, N.J. (2007) Polymorphisms in the human soluble epoxide hydrolase gene EPHX2 linked to neuronal survival after ischemic injury. *Journal of Neuroscience*, **27**, 4642–9.
157 Lee, C.R., North, K.E., Bray, M.S., Couper, D.J., Heiss, G. and Zeldin, D.C. (2007) CYP2J2 and CYP2C8 polymorphisms and coronary heart disease risk: the Atherosclerosis Risk

in Communities (ARIC) study. *Pharmacogenet Genomics*, **17**, 349–58.

158 Liu, P.Y., Li, Y.H., Chao, T.H., Wu, H.L., Lin, L.J., Tsai, L.M. and Chen, J.H. (2006) Synergistic effect of cytochrome P450 epoxygenase CYP2J2*7 polymorphism with smoking on the onset of premature myocardial infarction. *Atherosclerosis*.

159 Spiecker, M. and Liao, J. (2006) Cytochrome P450 epoxygenase CYP2J2 and the risk of coronary artery disease. *Trends in Cardiovascular Medicine*, **16**, 204–8.

160 Lee, C.R., North, K.E., Bray, M.S., Fornage, M., Seubert, J.M., Newman, J.W., Hammock, B.D., Couper, D.J., Heiss, G. and Zeldin, D.C. (2006) Genetic variation in soluble epoxide hydrolase (EPHX2) and risk of coronary heart disease: the Atherosclerosis Risk in Communities (ARIC) study. *Human Molecular Genetics*, **15**, 1640–9.

161 Wei, Q., Doris, P.A., Pollizotto, M.V., Boerwinkle, E., Jacobs, D.R. Jr, Siscovick, D.S. and Fornage, M. (2007) Sequence variation in the soluble epoxide hydrolase gene and subclinical coronary atherosclerosis: interaction with cigarette smoking. *Atherosclerosis*, **190**, 26–34.

162 Spiecker, M., Darius, H., Hankeln, T., Soufi, M., Sattler, A.M., Schaefer, J.R., Node, K., Borgel, J., Mugge, A., Lindpaintner, K., Huesing, A., Maisch, B., Zeldin, D.C. and Liao, J.K. (2004) Risk of coronary artery disease associated with polymorphism of the cytochrome P450 epoxygenase CYP2J2. *Circulation*, **110**, 2132–6.

163 Przybyla-Zawislak, B.D., Srivastava, P.K., Vazquez-Matias, J., Mohrenweiser, H.W., Maxwell, J.E., Hammock, B.D., Bradbury, J.A., Enayetallah, A.E., Zeldin, D.C. and Grant, D.F. (2003) Polymorphisms in human soluble epoxide hydrolase. *Molecular Pharmacology*, **64**, 482–90.

164 Fornage, M., Lee, C.R., Doris, P.A., Bray, M.S., Heiss, G., Zeldin, D.C. and Boerwinkle, E. (2005) The soluble epoxide hydrolase gene harbors sequence variation associated with susceptibility to and protection from incident ischemic stroke. *Human Molecular Genetics*, **14**, 2829–37.

165 Kroetz, D.L. and Zeldin, D.C. (2002) Cytochrome P450 pathways of arachidonic acid metabolism. *Current Opinion in Lipidology*, **13**, 273–83.

166 Hasemann, C.A., Kurumbail, R.G., Boddupalli, S.S., Peterson, J.A. and Deisenhofer, J. (1995) Structure and function of cytochromes P450: a comparative analysis of three crystal structures. *Structure*, **3**, 41–62.

167 Estabrook, R.W. (2003) A passion for P450s (rememberances of the early history of research on cytochrome P450). *Drug Metabolism and Disposition*, **31**, 1461–73.

168 Lewis, D.F. (2003) P450 structures and oxidative metabolism of xenobiotics. *Pharmacogenomics*, **4**, 387–95.

169 Omura, T. (2005) Heme-thiolate proteins. *Biochemical and Biophysical Research Communications*, **338**, 404–9.

170 Coon, M.J. (2005) Cytochrome P450: nature's most versatile biological catalyst. *Annual Review of Pharmacology and Toxicology*, **45**, 1–25.

171 Johnson, E.F. and Stout, C.D. (2005) Structural diversity of human xenobiotic-metabolizing cytochrome P450 monooxygenases. *Biochemical and Biophysical Research Communications*, **338**, 331–6.

172 Otyepka, M., Skopalik, J., Anzenbacherova, E. and Anzenbacher, P. (2007) What common structural features and variations of mammalian P450s are known to date? *Biochimica et Biophysica Acta*, **1770**, 376–89.

173 Ruckpaul, K. and Rein, H. (1989–1994) *Frontiers in Biotransformation*, Vols 1–9, Taylor & Francis, London.

15
Nucleotides and the Purinergic System

Vera Jankowski and Joachim Jankowski

15.1
Introduction

Nucleotides and the underlying purinergic system have attracted increasing interest in cardiovascular physiology and pathophysiology [1]. The purinergic system comprises mononucleosides, mononucleoside polyphosphates and dinucleoside polyphosphates as agonists, as well as the respective purine receptors (also named "purinoceptors") showing very different functions [2]. The purinergic signaling system controlling vascular regulation displays a high degree of complexity. The complexity of the purinergic signaling system is given, on the one hand, by the large number of agonists, and, on the other hand, especially by the diversity of the purinoceptors including the subtypes of the P_1, P_{2X} and P_{2Y} receptors as well as formation of heteropolymeric P_{2X} ion channels, P_{2X} splicing variants, and the numerous soluble and ecto-nucleotidases not inactivating but transforming an active purinoceptor agonist into just another active purinoceptor agonist. Essential new knowledge on the role of the purinergic system has been gained in recent years, for example, by identifying new purinergic agonists and by cloning of a large number of purine receptor subtypes.

15.2
Mononucleoside Polyphosphates

The potent actions of purine nucleotides and nucleosides in the cardiovascular system vessels were first described in the year 1929 [3]. There is now good evidence that extracellular mononucleoside polyphosphates are involved in various vasoregulatory processes, and immunomodulatory and prothrombotic responses in the cardiovascular system [4, 5].

Extracellular ATP, acting at the macrophage purinoceptor, drives the rapid Ca^{2+}-dependent formation and release of phosphatidylserine-rich microvesicles that enhance the assembly of the prothrombinase complex and subsequent formation of thrombin. ATP and adenosine are very much involved in the mechanisms underlying local control of vessel tone [6] as well as cell migration, proliferation, differentiation and death during angiogenesis, atherosclerosis and restenosis following angioplasty [7, 8]. Furthermore, ATP plays a significant cotransmitter role in sympathetic nerves supplying hypertensive blood vessels. The purinergic component is increased in spontaneously hypertensive rats [2]. The increase in sympathetic nerve activity in hypertension is well established, and there is an associated hyperplasia and hypertrophy of arterial walls [9]. ATP, released as a cotransmitter from sympathetic nerves, constricts vascular smooth muscle, whereas ATP released from sensory-motor nerves dilates or constricts vascular smooth muscle. Furthermore, ATP released from endothelial cells (ECs) during changes in flow (shear stress) or hypoxia acts on endothelial receptors to release nitric oxide (NO), resulting in relaxation.

Moreover, ATP has a strong mitogenic effect in aortic vascular smooth muscle cells (VSMCs) [10]. ATP activated both 42- and 44-kDa mitogen-activated protein kinases (MAPKs), and tyrosine is phosphorylated and the VSMCs express protein kinase C (PKC)-α, -δ and -ζ. The mitogenic effect of ATP is triggered by activation of the G_q-protein-coupled purinoceptor that leads to the formation of inositol trisphosphate and activation of PKC. PKC and, in turn, Raf-1 and MAPK are then activated, leading eventually to DNA synthesis and cell proliferation.

ATP release from red blood cells is increased in pathological conditions such as subarachnoid hemorrhage, largely because there is widespread blood cell lysis [11]. This leads to transient constriction of arterioles and sustained constriction of large cerebral vessels. Adenosine produced by the breakdown of extracellular ATP causes vasodilatation via smooth muscle purinergic receptors. Receptors involved in these processes are also present on ECs, and appear to be associated with cell adhesion and permeability.

Adenosine is the most potent product of the ATP metabolism. Adenosine produces changes in cAMP and DNA synthesis in cultured arterial smooth muscle cells, which results in the regulation of cell proliferation [12]. There is now good evidence that adenosine does regulate smooth muscle cell proliferation, but its properties differ from those for ATP and ADP [13, 14].

Furthermore, mononucleosides and mononucleoside polyphosphates are released at sites of inflammation. The mononucleosides and mononucleoside polyphosphates are involved in the development of inflammation through a combination of actions: release of histamine from mast cells (provoking production of prostaglandins) and the production and release of cytokines from immune cells [15]. In contrast, mononucleosides exert anti-inflammatory actions [15, 16].

In addition to the roles of purines in inflammation, purines have a broad range of functions carried out through purinoceptors on immune cells, including killing intracellular pathogens by inducing apoptosis of host macrophages, chemoattraction and cell adhesion [17, 18]. Purinergic compounds may turn out to be useful for the treatment of neurogenic inflammation, rheumatoid arthritis and periodontitis [19].

15.3
Dinucleoside Polyphosphates

In recent years, dinucleoside polyphosphates have gained increasing interest as a further group of strong purinergic agonists. Dinucleoside polyphosphates are involved as extracellular and intracellular mediators in the regulation of vascular tone as well as mediators of vascular smooth muscle proliferation and mesangial cell proliferation [20, 21].

The group of dinucleoside polyphosphates (Xp_nX with X = adenosine, guanosine or uridine; n = number of phosphate groups (p)) consists of nucleotides (riboyslated nucleic acids) formed by adenosine and/or guanosine (or uridine), respectively, which are connected by a polyphosphate chain of two to seven phosphates through phosphoester bonds at the 5'-position of two ribose moieties (Figure 15.1, showing P1,P4-diadenosine tetraphosphate (Ap_4A) as an example).

Dinucleoside polyphosphates with an adenine and/or guanine or uridine, respectively, bases have previously been identified [22–27]. Diadenosine polyphosphates have a relatively long half-life and, after degradation, may serve as a potential source of extracellular ATP and other purines.

Ap_4A was the first diadenosine polyphosphate to be identified in mammalian tissue [28]. Some years later this compound was also identified in human platelets [29, 30]. In the following years, a multitude of different nucleotides has been isolated from human tissues [e.g. diadenosine polyphosphates (Ap_nAs, with $n = 2–7$)] were isolated from body fluids and cells (e.g. [23, 24, 31–33]). Dinucleoside polyphosphates can be released into the circulation from activated platelets [23, 29, 30, 34, 35] from chromaffin cells of the adrenal glands [31, 36–38] or from synaptic vesicles [39]. Ap_nAs serve as important neurotransmitter molecules in the nervous system [40] and, in addition, stimulate different responses in the cardiovascular system, controlling vascular tone and preventing platelet aggregation [41]. After dinucleoside polyphosphate release, local concentrations in the range of 10^{-5} mol/l or even higher can be assumed [42].

Recently, uridine (5')–adenosine (5') tetraphosphate (Up_4A) (Figure 15.2) has gained increasing interest [26]. Up_4A was isolated from the supernatant of stimulated human endothelium and was identified by mass spectrometry. Stimulation with ATP, UTP, acetylcholine, endothelin, A23187 and mechanical stress releases

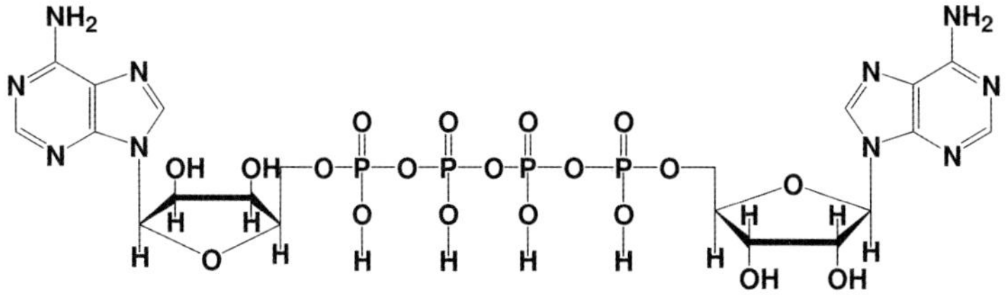

Figure 15.1 Molecular structure of Ap_4A.

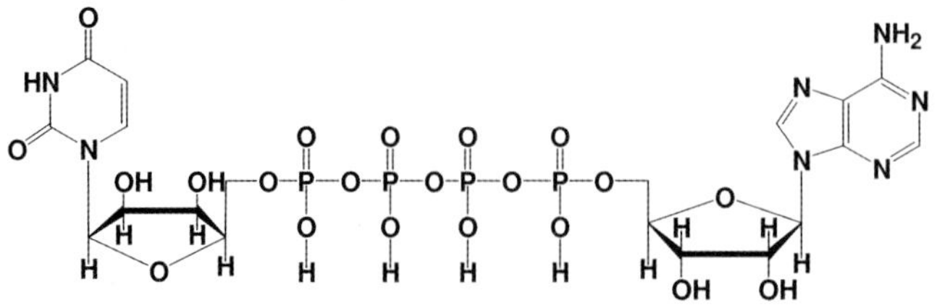

Figure 15.2 Molecular structure of Up$_4$A.

Up$_4$A from endothelium, suggesting that Up$_4$A contributes to vascular autoregulation. In healthy subjects Up$_4$A plasma concentrations are found which cause vasoconstriction. Up$_4$A most likely exerts vasoconstriction via purinoceptors. Up$_4$A is the first dinucleoside polyphosphate containing both purine and pyrimidine moieties isolated from living organisms. Its vasoactive effects, plasma concentrations and its release upon endothelial stimulation strongly suggest that Up$_4$A has a functional role in the cardiovascular system. Up$_4$A has a strong effect on human VSMC proliferation and the plasma Up$_4$A concentration is significantly increased in juvenile hypertensives compared to juvenile normotensives. The increased Up4A concentration and the proliferative effect result in an increased left ventricular mass and increased intima media wall thickness.

Table 15.1 gives a characteristic overview of some dinucleoside polyphosphates identified in human tissues and cells.

Diadenosine polyphosphates have a direct effect on the vascular tone [23, 32, 34, 48, 53–58]. In the vasculature of isolated perfused rat kidney, Ap$_5$A and Ap$_6$A were active at a concentration of 10^{-9} mol/l; in aortic rings, contractions were elicited at 10^{-8} mol/l. Intra-aortic injection in the rat caused a prolonged increase in blood pressure [34]. The vasoconstrictive action of Ap$_7$A on the vasculature of the isolated perfused rat kidney Ap$_7$A is slightly lower than that of Ap$_6$A. The threshold of the vasoconstrictive action of Ap$_7$A is in the range of 10^{-5} mol/l [23]. Vasoconstriction induced by the diadenosine phosphates is mediated by an increase in intracellular free calcium ions, $[Ca^{2+}]_i$ [59, 60].

Other dinucleoside polyphosphates have a vasodilatory effects (e.g. Ap$_2$A on the tone of isolated mesenteric arterial bed of rats [61]). When arteries with an intact endothelium are perfused with Ap$_3$A and Ap$_4$A, both induce vasodilatation, whereas Ap$_4$A causes vasoconstriction in arteries from which the endothelium has been removed [53]. Arterial infusion of Ap$_4$A produced a dose-dependent decrease of systemic blood pressure and coronary vascular resistance [62].

The vasodilatory effects of dinucleoside polyphosphates are mediated via endothelial A$_2$ receptors [61] and endothelial metabotropic P$_{2Y}$ receptors [54, 58, 63–65]. Stimulation of endothelial P$_{2Y1}$, P$_{2Y2}$ and P$_{2Y4}$ receptors causes endothelium-derived hyperpolarizing factors and NO-mediated dilatation [64–66]. Ohata et al. reported that in aortic strips *in situ* $[Ca^{2+}]_i$ waves within ECs are induced by ATP via P$_{2Y1}$ purinoceptor, but not P$_{2Y2}$ purinoceptor [67].

Table 15.1 Characteristic isolation, identification and characterization of dinucleoside polyphosphates in human tissues and receptor-mediated vascular effects.

Compound	Isolation from human tissue	Known vascular effects and receptor type involved (in parentheses)	References
Ap_2A	myocardium; platelets; placenta; adrenals	vasodilation (A_2) and vasoconstriction (A_1) in coronary arteries und renal vessels; proliferation of VSMCs (P_{2Y})	[24, 25, 32, 43–46]
Ap_2G	platelets	proliferation of VSMCs (P_{2Y})	[24]
Gp_2G	platelets	proliferation of VSMCs (P_{2Y})	[24]
Ap_3A	myocardium; platelets; placenta; adrenals; plasma	vasodilation (P_{2Y} and A_2) and vasoconstriction (A_1 and P_{2X}) in coronary arteries, renal vessels and mesenteric vessels; proliferation of VSMCs (P_{2Y})	[24, 25, 30, 33, 34, 43–47]
Ap_3G	platelets	vasoconstriction in renal vessels (P_{2X}); proliferation of VSMCs (P_{2Y})	[35]
Gp_3G	platelets	proliferation of VSMCs (P_{2Y})	[35]
Ap_4A	adrenals; plasma; platelets; brain	vasoconstriction in renal vessels (P_{2X} and A_1) and mesenteric vessels (P_{2X}); proliferation of VSMCs (P_{2Y})	[22, 24, 31, 33, 43, 44, 46, 48]
Ap_4G	platelets	vasoconstriction in renal vessels (P_{2X}); proliferation of VSMCs (P_{2Y})	[35]
Gp_4G	platelets	vasoconstriction of renal vessels (P_{2X}); proliferation of VSMCs (P_{2Y})	[35]
Up_4A	ECs	vasoconstriction of in renal vessels (P_{2X})	[26]
Ap_5A	adrenals; plasma; platelets; placenta; brain	proliferation in VSMCs (P_{2Y}); vasoconstriction in renal vessels (P_{2X}) and coronary arteries (P_{2X}); vasodilation in coronary arteries (P_{2Y})	[24, 31, 33, 34, 44–46, 49–52]

Table 15.1 Continued

Compound	Isolation from human tissue	Known vascular effects and receptor type involved (in parentheses)	References
Ap_5G	platelets	vasoconstriction in renal vessels (P_{2X}); proliferation of VSMCs (P_{2Y}); vasoconstriction in renal vessels (P_{2X}) and coronary arteries (P_{2X}); vasodilation in coronary arteries (P_{2Y})	[35, 50, 51]
Gp_5G	platelets	proliferation of VSMCs (P_{2Y})	[35, 50, 51]
Ap_6A	adrenals; plasma; platelets; red blood cells; placenta	proliferation of VSMCs (P_{2Y}); vasoconstriction in renal vessels (P_{2X}) and coronary arteries (P_{2X}); vasodilation in coronary arteries (P_{2Y})	[24, 27, 33, 34, 43–46, 48, 50, 51]
Ap_6G	platelets	vasoconstriction in renal vessels (P_{2X}) and coronary vessels (P_{2X}); vasodilation in coronary arteries (P_{2Y}); proliferation of VSMCs (P_{2Y})	[35, 50, 51]
Gp_6G	platelets	proliferation of VSMCs (P_{2Y})	[35, 50, 51]
Ap_7A	platelets	vasoconstriction in renal vessels (P_{2X})	[23]

Not only an endothelium-dependent but also direct dilating effects of Ap_4A and Ap_5A on VSMCs were demonstrated. Dilation of both dinucleoside polyphosphates was mediated by cAMP, which effects a decrease of $[Ca^{2+}]_i$ and possibly through activation of K^+ channels [68]. A direct, endothelium-independent vasodilation induced by purinoceptor agonists was observed also in newborn piglets isolated intrapulmonary arteries. The action was mediated by a P_{2Y} receptor on the VSMCs [69].

Dinucleoside polyphosphates do not only directly influence the vascular physiology, but also increase the proliferation rate of VSMCs. Growth-stimulating effects of nucleoside polyphosphates have been demonstrated in numerous types of vascular beds [70]. The ATP-induced proliferation of vascular smooth muscles is coupled to a G_q-protein and triggered phosphoinositide hydrolysis with subsequent activation of PKC, Raf-1, and MAPK [10, 71]. These results are in accordance with those of Tu et al., who observed that the stimulation of the P_{2Y2} receptor involves the activation of the Ras/Raf/MEK/MAPK pathway, which is modulated by $[Ca^{2+}]_i$, PKC and tyrosine kinase [72]. In coronary artery smooth muscle cells ATP-stimulated proliferation requires independent activation of both the extracellular signal-regulated kinase cascade and phosphatidylinositol 3-kinase signaling pathways [71]. P_{2Y2} receptor stimulation results in an increase of the expression of c-*fos* mRNA in cultured aortic smooth muscle cells [73].

Purinoceptor stimulation increases the expression of c-*fos* mRNA in cultured aortic smooth muscle cells [73] and stimulates proliferation of vascular tissue; Ap_4A is equipotent to ATP [74]. Ap_3A, Ap_4A, Ap_5A [43] as well Ap_2A, Ap_2G and Gp_2G [24] induce cell proliferation in VSMCs and furthermore stimulate c-*fos* proto-oncogene expression. The proliferative effect of the diguanosine polyphosphates Gp_nG (with $n = 3–6$) is significantly stronger than that of ATP in vascular tissues [35]. Ap_3A, Ap_4A, Ap_5A and Ap_6A also stimulate growth in rat glomerular mesangial cells in micromolar concentrations [21, 75]. Moreover, they potentiate the growth response to platelet-derived growth factor, but not to insulin-like growth factor-1 [21].

The dinucleoside polyphosphates are, furthermore, potent antagonists of ADP-induced platelet aggregation [23]. The interaction with ADP occurs at the P_{2T} receptor and appears to be a competitive inhibition, with Ap_4A having a K_i of approximately 0.7 mmol/l [76]. A comparison of the homologous series of Ap_nA compounds with phosphate chain lengths from two to six shows that Ap_5A is the most potent inhibitor of ADP-induced platelet aggregation, and that Ap_6A and Ap_4A are more potent than Ap_3A or Ap_2A. These dinucleoside polyphosphates inhibit the release of ADP from blood platelets, with a potency that decreases with decreasing chain length. Thus, dinucleoside polyphosphates in platelets may fulfill an antiaggregatory role. In human neutrophils, Ap_3A, Ap_4A, Ap_5A and Ap_6A all produce an increase in intracellular free calcium via a G-protein-coupled receptor [77].

Dinucleoside polyphosphates are stored in secretory granules from platelets, in adrenal chromaffin cells and in central nervous synaptosomes, reaching up to millimolar concentrations [29, 30, 37, 38]. The intracellular concentration of Ap_4A during normal growth [78] correlates directly with the proliferative state of the cell or tissue [28, 79]. The levels of Ap_4A are known to respond to cellular stresses, such as oxidation and heat shock, and they have been described as alarmones signaling the onset of cellular and metabolic stress, although their precise role has never been clear [79–81].

Dinucleoside polyphosphates are present in human plasma in physiologically relevant concentrations. The mean total plasma diadenosine polyphosphate concentrations (µmol/l; mean ± SEM) in cubital veins of normotensive subjects was 0.8 ± 0.59 for Ap_3A, 0.72 ± 0.72 for Ap_4A, 0.33 ± 0.24 for Ap_5A and 0.18 ± 0.18 for Ap_6A. It was recently demonstrated that in adrenal venous plasma significantly higher diadenosine polyphosphate concentrations are detectable than in plasma from the infrarenal and suprarenal vena cava. Adrenal medullae are obviously a source of plasma diadenosine polyphosphates in humans.

15.4
Purinoceptor System

The physiologic and pathophysiologic effects of mononucleosides, mononucleoside polyphosphates and dinucleoside polyphosphates are mediated via nucleotide-selective receptors; on the basis of pharmacological, functional and cloning data, major receptor subfamilies have been described [5, 82]. There are essentially two large families of purine receptors: those that are preferably activated by adenosine are summarized as P_1 receptors, and those that are activated by ATP, ADP, UTP, UDP and also by dinucleoside polyphosphates have been termed P_2 receptors. The group of P_1 receptors has been further subdivided into four subgroups A_1, A_{2A}, A_{2B} and A_3 according to molecular, biochemical and pharmacological criteria. The group of P_2 receptors is divided into P_{2X} and P_{2Y} receptors according to their molecular structure and their signal transduction pathways. At present there are no agonists or antagonists that discriminate adequately between families of P_{2X} and P_{2Y} receptors or between subtypes of receptors within each of these groups.

Purinoceptors are characterized by high plasticity and they are dynamically regulated during development. The purinoceptor system plays a general role as a sympathetic regulator of vasomotor tone [82, 83]. This is supported by the observation that in P_{2X1} receptor-deficient mice, where responses to P_{2X} receptor agonists are abolished, constriction of the vas deferens to sympathetic nerve stimulation is reduced by up to 60% [84]. The purinoceptor system controlling vascular homeostasis displays a high degree of complexity.

P_{2X} receptors are ligand-gated ion channels that are opened by purinergic messengers [85]. Thereby, rapid changes in the membrane permeability of monovalent and divalent cations are mediated [86–88]. Stimulation of ionotropic P_{2X} receptors induces an influx of Na^+ and Ca^{2+} ions into the cytosol. The increase of the concentration of these ions triggers a depolarization of the membrane potential, which opens potential operated Ca^{2+} channels [89]. The resulting Ca^{2+} influx increases $[Ca^{2+}]_i$, which effects the constriction of the VSMCs [60]. There are eight P_{2X} receptor cDNAs currently known, P_{2X1}–P_{2X8} [85, 90]. When expressed individually in heterologous systems, P_{2X1} and P_{2X3} subunits form channels activated by ATP or α,β-methylene ATP, whereas P_{2X2}, P_{2X4} and P_{2X5} form channels activated by ATP but not α,β-methylene ATP [90]. P_{2X1} and P_{2X3} receptor are characterized by strong and rapid desensitization, whereas P_{2X2}, P_{2X4} and P_{2X6} receptors desensitize relatively little [2]. Therefore, P_{2X1}-induced vasoconstrictions are transient, whereas P_{2X2}- or P_{2X4}-mediated vasoconstrictions are permanent as long as the agonist is present.

The ion-gating pore is formed by the aggregation of three P_{2X} monomers [91]. Coexpression of P_{2X} subtypes in heterologous expression systems can result in the formation of heteropolymeric P_{2X} trimers [92]. Therefore, it can be assumed that heteropolymeric channel formation is also possible *in vivo*. Heteropolymeric P_{2X} receptors clearly differ from their homomeric relatives [93, 94], and may therefore constitute an important mechanism for generating functional diversity of ATP-

and dinucleoside polyphosphate-mediated responses. The P_{2X1} forms large approximately elliptical clusters on the smooth muscle cells of mesenteric, renal and pulmonary arteries as well as in the aorta and in veins, which are restricted to the adventitial surface of the media. At the adventitial surface, the large clusters are immediately apposed to sympathetic varicosities. In the pulmonary artery large receptor clusters were found throughout the media of the vessel. Smaller spherical P_{2X1} cluster occur throughout the media of arteries of all sizes. The small P_{2X1} clusters are not associated with varicosities [95]. This observation may point to a paracrine role of purinergic agonists. P_{2X1}, P_{2X2} and P_{2X4} are coexpressed in smooth muscle cells of coronary vessels as well as in peripheral vessels, including the aorta, pulmonary artery, internal and external iliac arteries, renal artery, and femoral artery. The coexpression of P_{2X} receptor subtypes substantiates the possibility of a heteropolymeric assembly of the P_{2X} ion channels. In contrast, no mRNA transcripts of P_{2X1}, P_{2X2} and P_{2X4} were found in the superior mesenteric artery [92].

A further aspect contributing to the complexity of the purinergic system is the existence of P_{2X} splice variants. Hardy *et al.* detected a splice variant of the P_{2X1} receptor in the human bladder that is lacking part of the second transmembrane domain. It was suggested that isoforms may be potential sites for modifying or regulating putative purinergic activation [96]. This view is supported by the observation of Chen *et al.*, who noted that ATP-induced currents in cells expressing P_{2X2-1} and P_{2X2-2} variants were large and desensitized rapidly, whereas the current in those cells expressing the P_{2X2-3} variant was much smaller and desensitized slower [97]. Alternatively spliced P_{2X4} RNAs in human smooth muscle cells were identified [98].

In human vascular ECs in the umbilical vein, aorta, pulmonary artery and skin microvessels, the P_{2X4} receptor is the most prominent P_{2X} receptor. Vascular ECs are continuously exposed to variations in blood flow, which plays an important role in vessel growth or regression and in the local development of atherosclerosis. The shear stress that occurs during changes in blood flow leads to substantial release of ATP and UTP from ECs [99], and these purines might mediate alterations in the balance between proliferation and apoptosis. Atherosclerotic damage results in the disappearance of endothelium-dependent responses to ATP. Consequently, mRNA expression was much higher in these cells than was the expression of other subtypes, including P_{2X1}, P_{2X3}, P_{2X5} and $P_{2X}7$ [100]. P_{2X2} receptors are located on nerves and on ECs in rat blood vessels [95].

The calcium-permeable P_{2X1} receptor is considered the principal mediator of vasoconstriction [83], with P_{2X1} protein clusters on the adventitial surface of blood vessels immediately adjacent to sympathetic nerve varicosities [95]. However, P_{2X1} transcripts colocalize with mRNA for P_{2X2}, P_{2X4} and P_{2X5} in muscle cells of a number of blood vessels, and this points to the added presence of heteromeric P_{2X} receptors [56, 101–103]. For example, heteromeric $P_{2X1/5}$ receptors have been implicated in vasoconstriction of submucosal arterioles in the guinea pig [104].

P_{2Y} receptors are seven-transmembrane proteins [86–88]. Common mechanisms of signal transduction are shared by most seven-transmembrane receptors, including activation of phospholipase C (PLC) and/or regulation of adenyl cyclase activity [86–88]. P_{2Y} receptors do not act directly by inducing a cation influx, but via a downstream signaling cascade including G-proteins and inositol triphosphate among other factors [105].

Not only P_{2X} but also G-protein-coupled P_{2Y} receptors expressed in VSMCs were reported to mediate constrictions [55, 69, 106, 107]. The vasoconstriction inducing P_{2Y} receptor is probably coupled to PLCβ1 via $G\alpha_{q/11}$ and to PLC-β3 via $G\beta\gamma_3$ [108, 109]. In human coronary arteries extracellular nucleotides elicit constriction primarily by activation of P_{2Y2}, in addition to P_{2X} receptors, whereas a role for P_{2Y1} and P_{2Y6} receptors was excluded [64]. UTP- and ATP-induced vasoconstrictions in isolated intrapulmonary arteries in piglets are consistent with activation of the P_{2Y4} receptor [69].

UTP- and ATP-induced vasoconstriction in the intrapulmonary artery is consistent with activation of the P_{2Y4} receptor subtype [69], which is sensitive to Ap_4A. It has been proposed that dinucleoside polyphosphate vasoconstriction is also mediated by the adenosine A_1 receptor [48, 49, 54, 110, 111].

It is these principally different mechanisms that led to the terminology of "metabotropic" and "ionotropic" used for P_{2Y} and P_{2X} receptors, respectively. The purine receptor subtypes have been characterized according to their molecular structures, which show considerable differences, and according to their pharmacological characteristics, which also show large differences. Different purinoceptor subtypes may be expressed in the same cell type and heteromeric receptors are formed among different purinoceptor subtypes [105]. These purinoceptor heteromeres may show pharmacologic properties quite different from those of purinoceptor homomeres [93]. The ability to form heteromeres has been demonstrated both in P_{2X} receptors [112] and in P_{2Y} receptors [105], and by this property of receptor heteromerization the diversity of pharmacologic properties is manifold within the family of purinoceptor.

In particular, vascular endothelium has recently been shown to express both P_{2X} and P_{2Y} receptors. ECs are activated in a different way by various purine nucleotides such as UTP and ATP, suggesting that at least two different endothelial purinoceptors exist. Pirotton et al. [113] show that both P_{2Y1} and P_{2Y2} receptors are coexpressed in ECs from bovine aorta. This finding has been confirmed by other groups [114]. P_{2Y} receptors have likewise been identified in ECs from rat cerebral vessels [115, 116] and from mesenteric arteries [64, 66]. These findings have been paralleled by those in rat renal glomeruli [117]. Currently it has not been clarified whether still other purinoceptor subtypes are expressed in ECs. Undoubtedly, the endothelial P_{2Y1} receptor subtype mediates vasodilation [64, 114, 115]. In addition, also the P_{2Y2} and P_{2Y4} receptor subtypes may play a role in vasodilation [64]. Currently, little is known about the distribution of P_{2X} receptors in ECs. In ECs from rat cerebral arteries only the P_{2X2} receptor subtype has been demonstrated so far [118].

Dinucleoside polyphosphates have the capacity to potentiate signaling effects via P_2 receptors (primarily, via P_{2X1}, P_{2X3} and P_{2Y1} subtypes), although the existence of specific dinucleoside polyphosphate receptors have also been proposed [40, 41]. However, the specificity of dinucleoside polyphosphate receptors and their implication into multiple extracellular signals are still poorly understood, primarily due to the complexity of the purinergic signaling cascade. The effects of the dinucleoside polyphosphates are blocked by P_{2X} receptor antagonists such as pyridoralphosphate-6-azophenyl-2',4'-disulfonic acid or suramin, or by desensitization of P_{2X} receptors with α,β-methylene ATP subtype [76, 119, 120] or blockade by Ip_5I [121].

15.5
Metabolism of Nucleotides

Subsequent inactivation of the released nucleotides is thought to be mainly regulated by vascular endothelial [122] and lymphoid [123] membrane-bound nucleoside triphosphate diphosphohydrolase (NTPDase; also known as ecto-ATPDase, CD39) and ecto-5'-nucleotidase (CD73). Ectohydrolases are present on a broad variety of cell types, including aortic ECs [124, 125], chromaffin cells [126], and rat mesangial, bovine corneal epithelial, human Hep-G2 and peridontal cells [127]. A human diphosphorylated inositol phosphate phosphohydrolase shows a clear preference for Ap_5A and Ap_6A as substrates [128]. The human diphosphorylated inositol phosphate phosphohydrolase was shown to be a candidate for regulating signaling of diadenosine polyphosphates by hydrolysis of Ap_5A and Ap_6A in preference to other diadenosine polyphosphates [128]. The enzymatic breakdown of dinucleoside polyphosphates leads to the generation of mononucleotides and nucleotides that, in turn, are biologically active in vascular tissues.

In contrast to these traditional paradigms that focus on nucleotide-inactivating mechanisms, it has now become clear that nucleotide-phosphorylating enzymes adenylate kinase and nucleoside diphosphate kinase are also coexpressed on the cell surface and finely control the purinergic signaling cascade via two counterbalancing, ATP-inactivating and ATP-regenerating, pathways [129, 130]. The identification of a complex mixture of nucleotide pyrophosphatase/phosphodiesterase, NTPDase, adenylate kinase and other soluble purinergic enzymes freely circulating in the bloodstream adds another level of complexity to the understanding of the regulatory mechanisms of purine homeostasis within the vasculature [131–133]. The agonistic effects of nucleotides are obviously mediated by complex mechanisms: (1) by the dinucleoside polyphosphate or mononucleoside polyphosphate receptor pathway, (2) by a powerful inhibition of ecto-adenylate kinase activity, and (3) by generation of biologically active ATP and adenosine in the course of their degradation.

15.6
Therapeutic Aspects of the Purinergic System

There have been promising developments concerning purinergic antithrombotic drugs. Platelets are known to express P_{2Y1}, P_{2Y12} and P_{2X1} receptors [134]. Clinical trials like CURE [135] and CREDO [136] have provided clear evidence that the purinergic antithrombotic drugs clopidogrel and ticlopidine reduce the risks of recurrent strokes and heart attacks, especially when combined with aspirin [137, 138].

Moreover, diuridine tetraphosphate has been shown to have beneficial properties in the treatment of various diseases, such as chronic obstructive pulmonary disease. It has been demonstrated to facilitate the clearance of mucous secretions from the lungs of mammals (including humans) in need of treatment for various reasons, including cystic fibrosis, chronic bronchitis, asthma, bronchiectasis, postoperative mucous retention, pneumonia, primary ciliary dyskinesia [139], and the prevention and treatment of pneumonia in immobilized patients [140]. Further therapeutic uses include treatment of sinusitis [141–143], otitis media [144], dry eye [145], retinal detachment and nasolacrimal duct obstruction [146], the treatment of female infertility and irritation due to vaginal dryness via increased mucus secretions and hydration of the epithelial surface [147], and enhancing the performance of athletes. Last, but not least, recent findings showing increased dinucleoside polyphosphate concentration in hypertensive patients will provide new therapeutic approaches of hypertension.

Acknowledgments

Supported by a grant from the German Research Foundation (DFG, Ja-972 /11-1; J.J.) and by a Rahel Hirsch scholarship from the Charité (V.J.).

References

1 Gabriels, G., Rahn, K.H., Schlatter, E. and Steinmetz, M. (2002) Mesenteric and renal vascular effects of diadenosine polyphosphates (AP_nA). *Cardiovascular Reviews*, **56**, 22–32.
2 Ralevic, V. and Burnstock, G. (1998) Receptors for purines and pyrimidines. *Pharmacological Reviews*, **50**, 413–92.
3 Drury, A.N. and Szent-Györgyi, A. (1929) The physiological activity of adenine compounds with especial reference to their action upon the mammalian heart. *Journal of Physiology*, **68**, 213–37.
4 Moore, S.F. and MacKenzie, A.B. (2007) Murine macrophage P_{2X7} receptors support rapid prothrombotic responses. *Cellular Signalling*, **19**, 855–66.
5 Burnstock, G. (2002) Purinergic signaling and vascular cell proliferation and death. *Arteriosclerosis Thrombosis and Vascular Biology*, **22**, 364–73.
6 Burnstock, G. and Ralevic, V. (1994) New insights into the local regulation of blood flow by perivascular nerves and endothelium. *British Journal of Plastic Surgery*, **47**, 527–43.
7 Erlinge, D., Hou, M., Webb, T.E., Barnard, E.A. and Moller, S. (1998) Phenotype changes of the vascular smooth muscle cell regulate P_2 receptor expression as measured by quantitative RT-PCR. *Biochemical and Biophysical Research Communications*, **248**, 864–70.
8 Burnstock, G. (2001) Purinergic signalling in gut, in *Handbook of Experimental Pharmacology, Purinergic and Pyrimidinergic Signalling II – Cardiovascular, Respiratory, Immune, Metabolic and Gastrointestinal Tract Function* (eds M.P. Abbracchio and M.

9 Julius, S. and Nesbitt, S. (1996) Sympathetic overactivity in hypertension. A moving target. *American Journal of Hypertension*, **9**, 113S-20S.
10 Yu, S.M., Chen, S.F., Lau, Y.T., Yang, C.M. and Chen, J.C. (1996) Mechanism of extracellular ATP-induced proliferation of vascular smooth muscle cells. *Molecular Pharmacology*, **50**, 1000–9.
11 Sprague, R.S., Ellsworth, M.L. and Detrich, H.H. (2003) Nucleotide release and purinergic signaling in the vasculature driven by the red blood cell. *Current Topics in Membranes*, **54**, 243–68.
12 Jonzon, B., Nilsson, J. and Fredholm, B.B. (1985) Adenosine receptor-mediated changes in cyclic AMP production and DNA synthesis in cultured arterial smooth muscle cells. *Journal of Cell Physiology*, **124**, 451–6.
13 Burnstock, G. (2001) Purinergic signalling in development, in *Handbook of Experimental Pharmacology, Purinergic and Pyrimidinergic Signalling I– Molecular, Nervous and Urinogenitary System Function* (eds M.P. Abbracchio and M. Williams), Springer-Verlag, Berlin, pp. 89–127.
14 Di Virgilio, F. and Solini, A. (2002) P_2 receptors: new potential players in atherosclerosis. *British Journal of Pharmacology*, **135**, 831–42.
15 Di Virgilio, F., Falzoni, S., Mutini, C., Sanz, J.M. and Chiozzi, P. (1998) Purinergic P_{2X7} receptor: a pivotal role in inflammation and immunomodulation. *Drug Development Research*, **45**, 207–13.
16 Luttikhuizen, D.T., Harmsen, M.C., de Leij, L.F. and van Luyn, M.J. (2004) Expression of P_2 receptors at sites of chronic inflammation. *Cell and Tissue Research*, **317**, 289–98.
17 Burnstock, G. (2001) Overview of P_2 receptors: possible functions in immune cells. *Drug Development Research*, **53**, 53–9.
18 Di Virgilio, F., Chiozzi, P., Ferrari, D., Falzoni, S., Sanz, J.M., Morelli, A., Torboli, M., Bolognesi, G. and Baricordi, O.R. (2001) Nucleotide receptors: an emerging family of regulatory molecules in blood cells. *Blood*, **97**, 587–600.
19 Dubyak, G.R. and el-Moatassim, C. (1993) Signal transduction via P_2-purinergic receptors for extracellular ATP and other nucleotides. *American Journal of Physiology*, **265**, C577–606.
20 McLennan, A.G. (1992) *Ap_4A and Other Dinucleoside Polyphosphates*, CRC Press, Boca Raton, FL.
21 Heidenreich, S., Tepel, M., Schlüter, H., Harrach, B. and Zidek, W. (1995) Regulation of rat mesangial cell growth by diadenosine phosphates. *Journal of Clinical Investigation*, **95**, 2862–7.
22 Ogilvie, A., Luthje, J., Pohl, U. and Busse, R. (1989) Identification and partial characterization of an adenosine(5') tetraphospho(5')adenosine hydrolase on intact bovine aortic endothelial cells. *Biochemical Journal*, **259**, 97–103.
23 Jankowski, J., Tepel, M., van der Giet, M., Tente, I.M., Henning, L., Junker, R., Zidek, W. and Schlüter, H. (1999) Identification and characterization of P1,P7-diadenosine-5'-heptaphosphate from human platelets. *Journal of Biological Chemistry*, **274**, 23926–31.
24 Jankowski, J., Hagemann, J., Tepel, M., van der Giet, M., Stephan, N., Henning, L., Gouni-Berthold, I., Sachinidis, A., Zidek, W. and Schlüter, H. (2001) Dinucleotides as growth promoting extracellular mediators: presence of dinucleoside diphosphates Ap2A, Ap2G and Gp2G in releasable granlues of platelets. *Journal of Biological Chemistry*, **276**, 8904–9.
25 Luo, J., Jankowski, J., Knobloch, M., van der Giet, M., Gardanis, K., Russ, T., Vahlensieck, U., Neumann, J., Schmitz, W., Tepel, M., Deng, M.C., Zidek, W. and Schlüter, H. (1999) Identification and characterization of diadenosine 5',5'''-P1,P2-diphosphate and diadenosine 5',5'''-P1,P3-triphosphate in human myocardial tissue. *FASEB Journal*, **13**, 695–705.
26 Jankowski, V., Tölle, M., Vanholder, R., Schönfelder, G., van der Giet, M., Henning, L., Schlüter, H., Paul, M., Zidek, W. and Jankowski, J. (2005) Identification of uridine adenosine

tetraphosphate (Up$_4$A) as an endothelium-derived vasoconstrictive factor. *Nature Medicine*, **11**, 223–7.

27 Pintor, J., Rotllan, P., Torres, M. and Miras-Portugal, M.T. (1992) Characterization and quantification of diadenosine hexaphosphate in chromaffin cells: granular storage and secretagogue-induced release. *Analytical Biochemistry*, **200**, 296–300.

28 Rapaport, E. and Zamecnik, P.C. (1976) Presence of diadenosine 5′,5‴-P1,P4-tetraphosphate (Ap4A) in mamalian cells in levels varying widely with proliferative activity of the tissue: a possible positive "pleiotypic activator". *Proceedings of the National Academy of Sciences of the United States of America*, **73**, 3984–8.

29 Flodgaard, H. and Klenow, H. (1982) Abundant amounts of diadenosine 5′,5‴-P1,P4-tetraphosphate are present and releasable, but metabolically inactive, in human platelets. *Biochemical Journal*, **208**, 737–42.

30 Lüthje, J. and Ogilvie, A. (1983) The presence of diadenosine 5′,5‴-P1,P3-triphosphate (Ap3A) in human platelets. *Biochemical and Biophysical Research Communications*, **115**, 253–60.

31 Pintor, J., Diaz-Rey, M.A., Torres, M. and Miras-Portugal, M.T. (1992) Presence of diadenosine polyphosphates – Ap4A and Ap5A – in rat brain synaptic terminals. Ca^{2+} dependent release evoked by 4-aminopyridine and veratridine. *Neuroscience Letters*, **136**, 141–4.

32 Hoyle, C.H., Ziganshin, A.U., Pintor, J. and Burnstock, G. (1996) The activation of P$_1$- and P$_2$-purinoceptors in the guinea-pig left atrium by diadenosine polyphosphates. *British Journal of Pharmacology*, **118**, 1294–300.

33 Jankowski, J., Jankowski, V., Laufer, U., van der Giet, M., Henning, L., Tepel, M., Zidek, W. and Schlüter, H. (2003) Identification and quantification of diadenosine polyphosphate concentrations in human plasma. *Arteriosclerosis Thrombosis and Vascular Biology*, **23**, 1231–8.

34 Schlüter, H., Offers, E., Brüggemann, G., van der Giet, M., Tepel, M., Nordhoff, E., Karas, M., Spieker, C., Witzel, H. and Zidek, W. (1994) Diadenosine phosphates and the physiological control of blood pressure. *Nature*, **367**, 186–8.

35 Schlüter, H., Gross, I., Bachmann, J., Kaufmann, R., van der Giet, M., Tepel, M., Nofer, J.R., Assmann, G., Karas, M., Jankowski, J. and Zidek, W. (1998) Adenosine (5′) oligophospho-(5′) guanosines and guanosine(5′) oligo-phospho-(5′) guanosines in human platelets. *Journal of Clinical Investigation*, **101**, 682–8.

36 Castillo, C.J., Moro, M.A., Del Valle, M., Sillero, A., Garcia, A.G. and Sillero, M.A. (1992) Diadenosine tetraphosphate is co-released with ATP and catecholamines from bovine adrenal medulla. *Journal of Neurochemistry*, **59**, 723–32.

37 Pintor, J., Torres, M. and Miras-Portugal, M.T. (1991) Carbachol induced release of diadenosine polyphosphates – Ap4A and Ap5A – from perfused bovine adrenal medulla and isolated chromaffin cells. *Life Sciences*, **48**, 2317–24.

38 Rodriguez del Castillo, A., Torres, M., Delicado, E.G. and Miras-Portugal, M.T. (1988) Subcellular distribution studies of diadenosine polyphosphates – Ap4A and Ap5A – in bovine adrenal medulla: presence in chromaffin granules. *Journal of Neurochemistry*, **51**, 1696–703.

39 Zimmermann, H., Volknandt, W., Wittich, B. and Hausinger, A. (1993) Synaptic vesicle life cycle and synaptic turnover. *Journal of Physiology – Pairs*, **87**, 159–70.

40 Delicado, E.G., Miras-Portugal, M.T., Carrasquero, L.M., Leon, D., Perez-Sen, R. and Gualix, J. (2006) Dinucleoside polyphosphates and their interaction with other nucleotide signaling pathways. *Pflugers Archiv*, **452**, 563–72.

41 Flores, N.A., Stavrou, B.M. and Sheridan, D.J. (1999) The effects of diadenosine polyphosphates on the cardiovascular system. *Cardiovascular Research*, **42**, 15–26.

42 Ogilvie, A. (1992) Extracellular functions of ApnA, in *Ap$_4$A and Other Dinucleoside Polyphosphates* (ed. A.G. McLennan), CRC Press, Boca Raton, FL, pp. 229–73.

43 Jankowski, J., Hagemann, J., Yoon, M.S., van der Giet, M., Stephan, N., Zidek, W., Schlüter, H. and Tepel, M. (2001) Increased vascular growth in hemodialysis patients induced by platelet- derived diadenosine polyphosphates. *Kidney International*, **59**, 1134–41.

44 Jankowski, J., Jankowski, V., Seibt, B., Henning, L., Zidek, W. and Schlüter, H. (2003) Identification of dinucleoside polyphosphates in adrenal glands. *Biochemical and Biophysical Research Communications*, **304**, 365–70.

45 Jankowski, J., Yoon, M.S., Stephan, N., Zidek, W. and Schlüter, H. (2001) Vasoactive diadenosine polyphosphates in human placenta: possible candidates in the pathophysiology of pre-eclampsia? *Journal of Hypertension*, **19**, 567–73.

46 van der Giet, M., Khattab, M., Borgel, J., Schlüter, H. and Zidek, W. (1997) Differential effects of diadenosine phosphates on purinoceptors in the rat isolated perfused kidney. *British Journal of Pharmacology*, **120**, 1453–60.

47 Ogilvie, A. and Jakob, P. (1983) Diadenosine 5′,5′′′-P1,P3-triphosphate in eukaryotic cells: identification and quantitation. *Analytical Biochemistry*, **134**, 382–92.

48 van der Giet, M., Jankowski, J., Schlüter, H., Zidek, W. and Tepel, M. (1998) Mediation of the vasoactive properties of diadenosine tetraphosphate via various purinoceptors. *Journal of Hypertension*, **16**, 1939–43.

49 van der Giet, M., Cinkilic, O., Jankowski, J., Tepel, M., Zidek, W. and Schlüter, H. (1999) Evidence for two different P_{2X}-receptors mediating vasoconstriction of Ap_5A and Ap_6A in the isolated perfused rat kidney. *British Journal of Pharmacology*, **127**, 1463–9.

50 van der Giet, M., Westhoff, T., Cinkilic, O., Jankowski, J., Schlüter, H., Zidek, W. and Tepel, M. (2001) The critical role of adenosine and guanosine in the affinity of dinucleoside polyphosphates to P_{2X}-receptors in the isolated perfused rat kidney. *British Journal of Pharmacology*, **132**, 467–74.

51 van der Giet, M., Schmidt, S., Tölle, M., Jankowski, J., Schlüter, H., Zidek, W. and Tepel, M. (2002) Effects of dinucleoside polyphosphates on regulation of coronary vascular tone. *European Journal of Pharmacology*, **448**, 207–13.

52 Jovanovic, A., Jovanovic, S., Mays, D.C., Lipsky, J.J. and Terzic, A. (1998) Diadenosine 5′,5′′′-P1,P5-pentaphosphate harbors the properties of a signaling molecule in the heart. *FEBS Letters*, **423**, 314–18.

53 Busse, R., Ogilvie, A. and Pohl, U. (1988) Vasomotor activity of diadenosine triphosphate and diadenosine tetraphosphate in isolated arteries. *American Journal of Physiology*, **254**, H828–32.

54 Gabriels, G., Endlich, K., Rahn, K.H., Schlatter, E. and Steinhausen, M. (2000) In vivo effects of diadenosine polyphosphates on rat renal microcirculation. *Kidney International*, **57**, 2476–84.

55 Inscho, E.W., Cook, A.K., Mui, V. and Miller, J. (1998) Direct assessment of renal microvascular responses to P_2-purinoceptor agonists. *American Journal of Physiology*, **274**, F718–27.

56 Lewis, C.J. and Evans, R.J. (2000) Lack of run-down of smooth muscle P_{2X} receptor currents recorded with the amphotericin permeabilized patch technique, physiological and pharmacological characterization of the properties of mesenteric artery P_{2X} receptor ion channels. *British Journal of Pharmacology*, **131**, 1659–66.

57 Luo, J., Jankowski, J., Tepel, M., von der Giet, M., Zidek, W. and Schlüter, H. (1999) Identification of diadenosine hexaphosphate in human erythrocytes. *Hypertension*, **34**, 872–5.

58 Ralevic, V., Hoyle, C.H. and Burnstock, G. (1995) Pivotal role of phosphate chain length in vasoconstrictor versus vasodilator actions of adenine dinucleotides in rat mesenteric arteries. *Journal of Physiology*, **483**, 703–13.

59 Tepel, M., Lowe, S., Nofer, J.R., Assmann, G., Schlüter, H. and Zidek, W. (1996) Diadenosine polyphosphates regulate cytosolic calcium in human fibroblast cells by interaction with P_{2X}

purinoceptors coupled to phospholipase C. *Biochimica et Biophysica Acta*, **1312**, 145–50.

60 Tepel, M., Jankowski, J., Schlüter, H., Bachmann, J., van der Giet, M., Ruess, C., Terliesner, J. and Zidek, W. (1997) Diadenosine polyphosphates' action on calcium and vessel contraction. *American Journal of Hypertension*, **10**, 1404–10.

61 Ralevic, V. and Burnstock, G. (1996) Discrimination by PPADS between endothelial P_{2Y}- and P_{2U}-purinoceptors in the rat isolated mesenteric arterial bed. *British Journal of Pharmacology*, **118**, 428–34.

62 Nakae, I., Takahashi, M., Takaoka, A., Liu, Q., Matsumoto, T., Amano, M., Sekine, A., Nakajima, H. and Kinoshita, M. (1996) Coronary effects of diadenosine tetraphosphate resemble those of adenosine in anesthetized pigs: involvement of ATP-sensitive potassium channels. *Journal of Cardiovascular Pharmacology*, **28**, 124–33.

63 Hilderman, R.H. and Christensen, E.F. (1998) P1,P4-diadenosine 5' tetra-phosphate induces nitric oxide release from bovine aortic endothelial cells. *FEBS Letters*, **427**, 320–4.

64 Malmsjo, M., Hou, M., Harden, T.K., Pendergast, W., Pantev, E., Edvinsson, L. and Erlinge, D. (2000) Characterization of contractile P_2 receptors in human coronary arteries by use of the stable pyrimidines uridine 5'-O-thiodiphosphate and uridine 5'-O-3-thiotriphosphate. *Journal of Pharmacology and Experimental Therapeutics*, **293**, 755–60.

65 Rump, L.C., Oberhauser, V. and von Kugelgen, I. (1998) Purinoceptors mediate renal vasodilation by nitric oxide dependent and independent mechanisms. *Kidney International*, **54**, 473–81.

66 Malmsjo, M., Edvinsson, L. and Erlinge, D. (2000) P_{2X} receptors counteract the vasodilatory effects of endothelium derived hyperpolarising factor. *European Journal of Pharmacology*, **390**, 173–80.

67 Ohata, H., Ujike, Y. and Momose, K. (1997) Confocal imaging analysis of ATP-induced Ca^{2+} response in individual endothelial cells of the artery in situ. *American Journal of Physiology*, **272**, C1980–7.

68 Sumiyoshi, R., Nishimura, J., Kawasaki, J., Kobayashi, S., Takahashi, S. and Kanaide, H. (1997) Diadenosine polyphosphates directly relax porcine coronary arterial smooth muscle. *Journal of Pharmacology and Experimental Therapeutics*, **283**, 548–56.

69 McMillan, M.R., Burnstock, G. and Haworth, S.G. (1999) Vasoconstriction of intrapulmonary arteries to P_2-receptor nucleotides in normal and pulmonary hypertensive newborn piglets. *British Journal of Pharmacology*, **128**, 549–55.

70 Erlinge, D. (1998) Extracellular ATP a growth factor for vascular smooth muscle cells. *General Pharmacology*, **31**, 1–8.

71 Wilden, P.A., Agazie, Y.M., Kaufman, R. and Halenda, S.P. (1998) ATP-stimulated smooth muscle cell proliferation requires independent ERK and PI3K signaling pathways. *American Journal of Physiology*, **275**, H1209–15.

72 Tu, M.T., Luo, S.F., Wang, C.C., Chien, C.S., Chiu, C.T., Lin, C.C. and Yang, C.M. (2000) P_{2Y2} receptor-mediated proliferation of C_6 glioma cells via activation of Ras/Raf/MEK/MAPK pathway. *British Journal of Pharmacology*, **129**, 1481–9.

73 Malam-Souley, R., Seye, C., Gadeau, A.P., Loirand, G., Pillois, X., Campan, M., Pacaud, P. and Desgranges, C. (1996) Nucleotide receptor P_{2u} partially mediates ATP-induced cell cycle progression of aortic smooth muscle cells. *Journal of Cell Physiology*, **166**, 57–65.

74 Erlinge, D., You, J., Wahlestedt, C. and Edvinsson, L. (1995) Characterisation of an ATP receptor mediating mitogenesis in vascular smooth muscle cells. *European Journal of Pharmacology*, **289**, 135–49.

75 Schulze-Lohoff, E., Zanner, S., Ogilvie, A. and Sterzel, R.B. (1995) Vasoactive diadenosine polyphosphates promote growth of cultured renal mesangial cells. *Hypertension*, **26**, 899–904.

76 Kunapuli, S.P. and Daniel, J.L. (1998) P_2 receptor subtypes in the cardiovascular system. *Biochemical Journal*, **336**, 513–23.

77 Pintor, J., Gualix, J. and Miras-Portugal, M.T. (1997) Dinucleotide receptor modulation by protein kinases (protein kinases A and C) and protein phosphatases in rat brain synaptic terminals. *Journal of Neurochemistry*, **68**, 2552–7.

78 Garrison, P.N. and Barnes, L.D. (1992) Determination of dinucleoside polyphosphates, in *AP₄A and Other Dinucleoside Polyphosphates* (ed. A.G. McLennan), CRC Press, Boca Raton, pp. 29–61.

79 Remy, P. (1992) *Intracellular Functions of Ap_nN: Eukaryotes*, CRC Press, Boca Raton.

80 Brevet, A., Plateau, P., Best-Belpomme, M. and Blanquet, S. (1986) Variation of Ap4A and other dinucleoside polyphosphates in stressed *Drosophila* cells. *Journal of Biological Chemistry*, **260**, 15566–70.

81 Garrison, P.N., Mathis, S.A. and Barnes, L.D. (1986) *In vivo* levels of diadenosine tetraphosphate and adenosine tetraphospho-guanosine in *Physarum polycephalum* during the cell cycle and oxidative stress. *Molecular and Cellular Biology*, **6**, 1179–86.

82 Ralevic, V. (2000) P_2 receptors in the central and peripheral nervous systems modulating sympathetic vasomotor tone. *Journal of the Autonomic Nervous System*, **81**, 205–11.

83 Kennedy, C. (1996) ATP as a cotransmitter in perivascular sympathetic nerves. *Journal of Autonomic Pharmacology*, **16**, 337–40.

84 Mulryan, K., Gitterman, D.P., Lewis, C.J., Vial, C., Leckie, B.J., Cobb, A.L., Brown, J.E., Conley, E.C., Buell, G., Pritchard, C.A. and Evans, R.J. (2000) Reduced vas deferens contraction and male infertility in mice lacking P_{2X1} receptors. *Nature*, **403**, 86–9.

85 Bo, X., Schoepfer, R. and Burnstock, G. (2000) Molecular cloning and characterization of a novel ATP P_{2X} receptor subtype from embryonic chick skeletal muscle. *Journal of Biological Chemistry*, **275**, 14401–7.

86 Cattaneo, M. (2006) ADP receptors: inhibitory strategies for antiplatelet therapy. *Drug News and Perspectives*, **19**, 253–9.

87 Gachet, C. (2006) Regulation of platelet functions by P_2 receptors. *Annual Review of Pharmacology and Toxicology*, **46**, 277–300.

88 Gachet, C., Leon, C. and Hechler, B. (2006) The platelet P_2 receptors in arterial thrombosis. *Blood Cells, Molecules and Diseases*, **36**, 223–7.

89 Usune, S., Katsuragi, T. and Furukawa, T. (1996) Effects of PPADS and suramin on contractions and cytoplasmic Ca^{2+} changes evoked by AP4A, ATP and alpha, beta-methylene ATP in guinea-pig urinary bladder. *British Journal of Pharmacology*, **117**, 698–702.

90 North, R.A. and Surprenant, A. (2000) Pharmacology of cloned P_{2X} receptors. *Annual Review of Pharmacology and Toxicology*, **40**, 563–80.

91 Nicke, A., Baumert, H.G., Rettinger, J., Eichele, A., Lambrecht, G., Mutschler, E. and Schmalzing, G. (1998) P_{2X1} and P_{2X3} receptors form stable trimers: a novel structural motif of ligand-gated ion channels. *EMBO Journal*, **17**, 3016–28.

92 Nori, S., Fumagalli, L., Bo, X., Bogdanov, Y. and Burnstock, G. (1998) Coexpression of mRNAs for P_{2X1}, P_{2X2} and P_{2X4} receptors in rat vascular smooth muscle: an in situ hybridization and RT-PCR study. *Journal of Vascular Research*, **35**, 179–85.

93 Torres, G.E., Haines, W.R., Egan, T.M. and Voigt, M.M. (1998) Co-expression of P_{2X1} and P_{2X5} receptor subunits reveals a novel ATP-gated ion channel. *Molecular Pharmacology*, **54**, 989–93.

94 Bianchi, B.R., Lynch, K.J., Touma, E., Niforatos, W., Burgard, E.C., Alexander, K.M., Park, H.S., Yu, H., Metzger, R., Kowaluk, E., Jarvis, M.F. and van Biesen, T. (1999) Pharmacological characterization of recombinant human and rat P_{2X} receptor subtypes. *European Journal of Pharmacology*, **376**, 127–38.

95 Hansen, M.A., Dutton, J.L., Balcar, V.J., Barden, J.A. and Bennett, M.R. (1999) P_{2X} (purinergic) receptor distributions in rat blood vessels. *Journal of the Autonomic Nervous System*, **75**, 147–55.

96 Hardy, L.A., Harvey, I.J., Chambers, P. and Gillespie, J.I. (2000) A putative

alternatively spliced variant of the P_{2X1} purinoreceptor in human bladder. *Experimental Physiology*, **85**, 461–3.

97 Chen, C., Parker, M.S., Barnes, A.P., Deininger, P. and Bobbin, R.P. (2000) Functional expression of three P_{2X2} receptor splice variants from guinea pig cochlea. *Journal of Neurophysiology*, **83**, 1502–9.

98 Dhulipala, P.D., Wang, Y.X. and Kotlikoff, M.I. (1998) The human P_{2X4} receptor gene is alternatively spliced. *Gene*, **207**, 259–66.

99 Burnstock, G. (1999) Release of vasoactive substances from endothelial cells by shear stress and purinergic mechanosensory transduction. *Journal of Anatomy*, **194**, 335–42.

100 Yamamoto, K., Korenaga, R., Kamiya, A., Qi, Z., Sokabe, M. and Ando, J. (2000) P_{2X4} receptors mediate ATP-induced calcium influx in human vascular endothelial cells. *American Journal of Physiology-Heart and Circulatory*, **279**, H285–92.

101 Lewis, C.J. and Evans, R.J. (2001) P_{2X} receptor immunoreactivity in different arteries from the femoral, pulmonary, cerebral, coronary and renal circulations. *Journal of Vascular Research*, **38**, 332–40.

102 Pulvirenti, T.J., Yin, J.L., Chaufour, X., McLachlan, C., Hambly, B.D., Bennett, M.R. and Barden, J.A. (2000) P_{2X} (purinergic) receptor redistribution in rabbit aorta following injury to endothelial cells and cholesterol feeding. *Journal of Neurocytology*, **29**, 623–31.

103 Turner, C.M., Vonend, O., Chan, C., Burnstock, G. and Unwin, R.J. (2003) The pattern of distribution of selected ATP-sensitive P_2 receptor subtypes in normal rat kidney: an immuno-histological study. *Cells, Tissues, Organs*, **175**, 105–17.

104 Surprenant, A., Schneider, D.A., Wilson, H.L., Galligan, J.J. and North, R.A. (2000) Functional properties of heteromeric $P_{2X1/5}$ receptors expressed in HEK cells and excitatory junction potentials in guinea-pig submucosal arterioles. *Journal of the Autonomic Nervous System*, **81**, 249–63.

105 Barnard, E.A. and Simon, J. (2001) An elusive receptor is finally caught: P_{2Y12}, an important drug target in platelets. *Trends in Pharmacological Sciences*, **22**, 388–91.

106 Fukumitsu, A., Takano, Y., Iki, A., Honda, K., Saito, R., Katsuragi, T. and Kamiya, H. (1999) Endogenous ATP released by electrical field stimulation causes contraction via P_{2X}- and P_{2Y}-purinoceptors in the isolated tail artery of rats. *Japanese Journal of Pharmcology*, **81**, 375–80.

107 Hillaire-Buys, D., Dietz, S., Chapal, J., Petit, P. and Loubatieres-Mariani, M.M. (1999) Involvement of P_{2X} and P_{2U} receptors in the constrictor effect of ATP on the pancreatic vascular bed. *Journal de la Societe de Biologie*, **193**, 57–61.

108 Murthy, K.S. and Makhlouf, G.M. (1998) Differential regulation of phospholipase A2 (PLA2)-dependent Ca^{2+} signaling in smooth muscle by cAMP- and cGMP-dependent protein kinases. Inhibitory phosphorylation of PLA2 by cyclic nucleotide-dependent protein kinases. *Journal of Biological Chemistry*, **273**, 34519–26.

109 Murthy, K.S. and Makhlouf, G.M. (1998) Coexpression of ligand-gated P_{2X} and G protein-coupled P_{2Y} receptors in smooth muscle. Preferential activation of P_{2Y} receptors coupled to phospholipase C (PLC)-β1 via $G\alpha_{q/11}$ and to PLC-β3 via $G\beta\gamma_{i3}$. *Journal of Biological Chemistry*, **273**, 4695–704.

110 Vahlensieck, U., Boknik, P., Knapp, J., Linck, B., Muller, F.U., Neumann, J., Herzig, S., Schlüter, H., Zidek, W., Deng, M.C., Scheld, H.H. and Schmitz, W. (1996) Negative chronotropic and inotropic effects exerted by diadenosine hexaphosphate (AP6A) via A_1-adenosine receptors. *British Journal of Pharmacology*, **119**, 835–44.

111 van der Giet, M., Khattab, M., Börgel, J., Schlüter, H. and Zidek, W. (1997) Differential effects of diadenosine phosphates on purinoceptors in the rat isolated perfused kidney. *British Journal of Pharmacology*, **120**, 1453–60.

112 Le, K.T., Boue-Grabot, E., Archambault, V. and Seguela, P. (1999) Functional and biochemical evidence for heteromeric ATP-gated channels composed of P_{2X1}

and P$_{2X5}$ subunits. *Journal of Biological Chemistry*, **274**, 15415–9.

113 Pirotton, S., Communi, D., Motte, S., Janssens, R. and Boeynaems, J.M. (1996) Endothelial P$_2$-purinoceptors: subtypes and signal transduction. *Journal of autonomic pharmacology*, **16**, 353–6.

114 Bultmann, R., Hansmann, G., Tuluc, F. and Starke, K. (1997) Vasoconstrictor and vasodilator effects of guanine nucleotides in the rat aorta. *Naunyn-Schmiedeberg's Archives of Pharmacology*, **356**, 653–61.

115 Vigne, P., Breittmayer, J.P. and Frelin, C. (2000) Diadenosine polyphosphates as antagonists of the endogenous P$_{2Y1}$ receptor in rat brain capillary endothelial cells of the B7 and B10 clones. *British Journal of Pharmacology*, **129**, 1506–12.

116 Webb, T.E., Feolde, E., Vigne, P., Neary, J.T., Runberg, A., Frelin, C. and Barnard, E.A. (1996) The P$_{2Y}$ purinoceptor in rat brain microvascular endothelial cells couple to inhibition of adenylate cyclase. *British Journal of Pharmacology*, **119**, 1385–92.

117 Harada, H., Chan, C.M., Loesch, A., Unwin, R. and Burnstock, G. (2000) Induction of proliferation and apoptotic cell death via P$_{2Y}$ and P$_{2X}$ receptors, respectively, in rat glomerular mesangial cells. *Kidney International*, **57**, 949–58.

118 Loesch, A. and Burnstock, G. (2000) Ultrastructural localisation of ATP-gated P$_{2X2}$ receptor immunoreactivity in vascular endothelial cells in rat brain. *Endothelium*, **7**, 93–8.

119 Bo, X., Sexton, A., Xiang, Z., Nori, S.L. and Burnstock, G. (1998) Pharmacological and histochemical evidence for P$_{2X}$ receptors in human umbilical vessels. *European Journal of Pharmacology*, **353**, 59–65.

120 Wang, L., Karlsson, L., Moses, S., Hultgardh-Nilsson, A., Andersson, M., Borna, C., Gudbjartsson, T., Jern, S. and Erlinge, D. (2002) P$_2$ receptor expression profiles in human vascular smooth muscle and endothelial cells. *Journal of Cardiovascular Pharmacology*, **40**, 841–53.

121 Hoyle, C.H., Pintor, J., Gualix, J. and Miras-Portugal, M.T. (1997) Antagonism of P$_{2X}$ receptors in guinea-pig vas deferens by diinosine pentaphosphate. *European Journal of Pharmacology*, **333**, R1–2.

122 Marcus, A.J., Broekman, M.J., Drosopoulos, J.H., Islam, N., Pinsky, D.J., Sesti, C. and Levi, R. (2003) Metabolic control of excessive extracellular nucleotide accumulation by CD39/ecto-nucleotidase-1: implications for ischemic vascular diseases. *Journal of Pharmacology and Experimental Therapeutics*, **305**, 9–16.

123 Heptinstall, S., Johnson, A., Glenn, J.R. and White, A.E. (2005) Adenine nucleotide metabolism in human blood – important roles for leukocytes and erythrocytes. *Journal of Thrombosis and Haemostasis*, **3**, 2331–9.

124 Mateo, J., Rotllan, P., Marti, E., Gomez De Aranda, I., Solsona, C. and Miras-Portugal, M.T. (1997) Diadenosine polyphosphate hydrolase from presynaptic plasma membranes of *Torpedo* electric organ. *Biochemical Journal*, **323**, 677–84.

125 Mateo, J., Miras-Portugal, M.T. and Rotllan, P. (1997) Ecto-enzymatic hydrolysis of diadenosine polyphosphates by cultured adrenomedullary vascular endothelial cells. *American Journal of Physiology*, **273**, C918–27.

126 Gasmi, L., Cartwright, J.L. and McLennan, A.G. (1998) The hydrolytic activity of bovine adrenal medullary plasma membranes towards diadenosine polyphosphates is due to alkaline phosphodiesterase-I. *Biochimica et Biophysica Acta*, **1405**, 121–7.

127 von Drygalski, A. and Ogilvie, A. (2000) Ecto-diadenosine 5′,5‴-P1,P4-tetraphosphate (Ap4A)-hydrolase is expressed as an ectoenzyme in a variety of mammalian and human cells and adds new aspects to the turnover of Ap4A. *Biofactors*, **11**, 179–87.

128 Safrany, S.T., Ingram, S.W., Cartwright, J.L., Falck, J.R., McLennan, A.G., Barnes, L.D. and Shears, S.B. (1999) The diadenosine hexaphosphate hydrolases from *Schizosaccharomyces pombe* and *Saccharomyces cerevisiae* are homologues

of the human diphosphoinositol polyphosphate phosphohydrolase. Overlapping substrate specificities in a MutT-type protein. *Journal of Biological Chemistry*, **274**, 21735–40.
129 Yegutkin, G.G. and Burnstock, G. (2000) Inhibitory effects of some purinergic agents on ecto-ATPase activity and pattern of stepwise ATP hydrolysis in rat liver plasma membranes. *Biochimica et Biophysica Acta*, **1466**, 234–44.
130 Yegutkin, G., Bodin, P. and Burnstock, G. (2000) Effect of shear stress on the release of soluble ecto-enzymes ATPase and 5'-nucleotidase along with endogenous ATP from vascular endothelial cells. *British Journal of Pharmacology*, **129**, 921–6.
131 Birk, A.V., Bubman, D., Broekman, M.J., Robertson, H.D., Drosopoulos, J.H., Marcus, A.J. and Szeto, H.H. (2002) Role of a novel soluble nucleotide phospho-hydrolase from sheep plasma in inhibition of platelet reactivity: hemostasis, thrombosis, and vascular biology. *Journal of Laboratory and Clinical Medicine*, **139**, 116–24.
132 Yegutkin, G.G., Samburski, S.S. and Jalkanen, S. (2003) Soluble purine-converting enzymes circulate in human blood and regulate extracellular ATP level via counteracting pyrophosphatase and phosphotransfer reactions. *FASEB Journal*, **17**, 1328–30.
133 Yegutkin, G.G., Samburski, S.S., Mortensen, S.P., Jalkanen, S. and Gonzalez-Alonso, J. (2007) Intravascular ADP and soluble nucleotidases contribute to acute prothrombotic state during vigorous exercise in humans. *Journal of Physiology*, **579**, 553–64.
134 Hollopeter, G., Jantzen, H.M., Vincent, D., Li, G., England, L., Ramakrishnan, V., Yang, R.B., Nurden, P., Nurden, A., Julius, D. and Conley, P.B. (2001) Identification of the platelet ADP receptor targeted by antithrombotic drugs. *Nature*, **409**, 202–7.
135 Yusuf, S., Zhao, F., Mehta, S.R., Chrolavicius, S., Tognoni, G. and Fox, K.K. (2001) Effects of clopidogrel in addition to aspirin in patients with acute coronary syndromes without ST-segment elevation. *New England Journal of Medicine*, **345**, 494–502.
136 Beinart, S.C., Kolm, P., Veledar, E., Zhang, Z., Mahoney, E.M., Bouin, O., Gabriel, S., Jackson, J., Chen, R., Caro, J., Steinhubl, S., Topol, E. and Weintraub, W.S. (2005) Long-term cost effectiveness of early and sustained dual oral anti-platelet therapy with clopidogrel given for up to one year after percutaneous coronary intervention results: from the Clopidogrel for the Reduction of Events During Observation (CREDO) trial. *Journal of the American College of Cardiology*, **46**, 761–9.
137 Kam, P.C. and Nethery, C.M. (2003) The thienopyridine derivatives (platelet adenosine diphosphate receptor antagonists), pharmacology and clinical developments. *Anaesthesia*, **58**, 28–35.
138 Kunapuli, S.P., Ding, Z., Dorsam, R.T., Kim, S., Murugappan, S. and Quinton, T.M. (2003) ADP receptors–targets for developing antithrombotic agents. *Current Pharmaceutical Design*, **9**, 2303–16.
139 Stutts, I., Monroe, J., Boucher, J., Richard, C. and Eduardo, R. (1995) A.C. Dinucleotides useful for the treatment of cystic fibrosis and for hydrating mucus secretions, US Patent 5,635,160.
140 Jacobus, K.M. and Leighton, H.J. (1996) Method of preventing or treating pneumonia in immobilized patients with uridine triphosphates and related compounds, US Patent 5,763,447.
141 Jacobus, K., Rideout, J., Yerxa, B., Pendergast, W., Siddiqi, S. and Drutz, D. (1996) Method of treating sinusitis with uridine triphosphates and related compounds, US Patent 5,789,391.
142 Jacobus, K., Rideout, J., Yerxa, B., Pendergast, W., Siddiqi, S. and Drutz, D. (1998) Method for treating sinusitis with uridine triphosphates and related compounds, US Patent 5,981,506.
143 Jacobus, K., Rideout, J., Yerxa, B., Pendergast, W., Siddiqi, S. and Drutz, D. (1998) Method of treating sinusitis with uridine triphosphates and related compounds, US Patent 5,958,897.

144 Drutz, D., Rideout, J. and Jacobus, K. (1996) Method of treating otitis media with uridine triphosphates and related compounds, US Patent 6,423,694.

145 Yerxa, B., Jacobus, K., Pendergast, W. and Rideout, J. (1997) Method of treating dry eye disease with uridine triphosphates and related compounds, US Patent 5,900,407.

146 Yerxa, B. and Brown, E. (2003) Di(uridine 5′)-tetraphosphate and salts thereof, US Patent 20040014713.

147 Pendergast, W., Shaver, S., Drutz, D., Rideout, J. and Yerxa, B. (1998) Method of promoting cervical and vaginal secretions, US Patent 6,462,028.

16
Nitric Oxide

Valérie B. Schini-Kerth and Paul M. Vanhoutte

Endothelial cells (ECs) form a cobblestone monolayer covering the luminal surface of all types of blood vessels. They constitute a semipermeable barrier, which prevents the contact of blood cells and proteins with the subendothelial procoagulant and prothrombotic matrix, and regulates the exchange of oxygen, nutrients and macromolecules between circulating blood and the underlying blood vessel wall. In addition, they contribute to regulate the local concentrations of hormones including 5-hydroxytryptamine (5-HT; serotonin) and adrenaline, which are metabolized by monoamine oxidases and catechol-*O*-methyl transferases into inactive metabolites. ECs have also a crucial role in controlling vascular tone and structure. They express at their luminal surface angiotensin-converting enzyme, which catalyzes the formation of angiotensin (Ang) II (a proinflammatory and vasoconstrictor peptide that also promotes cell proliferation and hypertrophy) from its biologically poorly active precursor Ang I and the degradation of the vasodilator bradykinin (BK) into inactive peptides.

In 1980, Furchgott and Zawadzki, studying the reactivity of isolated arterial rings in conventional organ chambers, made the seminal observation that profoundly changed the understanding of the mechanisms regulating vascular tone and homeostasis [1]. They observed that acetylcholine caused full relaxation of contracted intact arterial rings in a concentration-dependent manner, whereas no such effect was observed in rings following mechanical removal of the EC layer. Using a "sandwich" preparation, they demonstrated that a labile factor termed initially endothelium-derived relaxing factor (EDRF) is released by ECs and diffuses to the underlying vascular smooth muscle to inhibit its tone. The combined effort of numerous research laboratories has led to the chemical identification of EDRF as nitric oxide (NO), which is generated by the conversion of L-arginine into L-citrulline by an enzyme termed endothelial NO synthase (eNOS, Figure 16.1). Apart from NO, ECs release several additional potent vasoactive factors including the vasodilators prostacyclin and endothelium-derived hyperpolarizing factor (EDHF), but also substances including the peptide endothelin-1 and several vasoconstrictor prostanoids (for reviews, see [2, 3], Figure 16.1). This chapter will focus on the mechanisms regulating the endothelial formation of NO and its direct

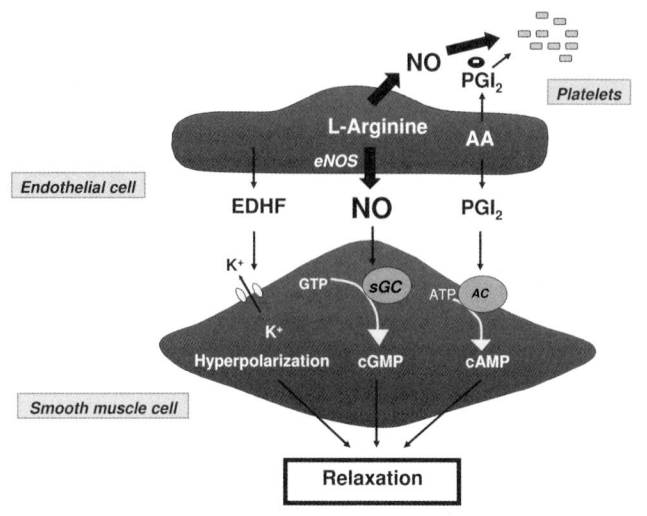

Figure 16.1 Schematic representation indicating that in healthy blood vessels ECs have a key role in the control of vascular homeostasis mostly by releasing potent vasodilator factors such as NO, EDHF and prostacyclin (PGI$_2$). NO is produced from L-arginine by the enzyme termed eNOS in ECs. NO is a potent vasodilator that also prevents platelet activation though activation of the cGMP pathway. EDHF hyperpolarizes the vascular smooth muscle by inducing K$^+$ efflux following opening of K$^+$ channels; this, in turn, leads to inhibition of voltage-gated Ca^{2+} channels, which promotes relaxation. Prostacyclin produced from arachidonic acid (AA) by prostacyclin synthase relaxes the vascular smooth muscle and inhibits platelet aggregation via the cAMP pathway in target cells. In addition, NO and PGI$_2$ act in synergy to prevent platelet activation. sGC, soluble guanylyl cyclase; AC, adenylyl cyclase.

biological effects that play an essential role in the maintenance of vascular homeostasis and the prevention of atherogenesis.

16.1
Regulation of the Endothelial Formation of NO

16.1.1
Hemodynamic Forces

The hemodynamic forces exerted by the streaming blood on the intimal layer of the blood vessel wall constitute a physiologically most important stimulus resulting in the continuous endothelial formation of NO (Figure 16.2). These forces include mostly wall shear stress resulting from the viscous drag of flowing blood on the luminal surface of ECs and wall cyclic stretch induced by the pulsatile changes of blood pressure. ECs respond within seconds to an increase in fluid

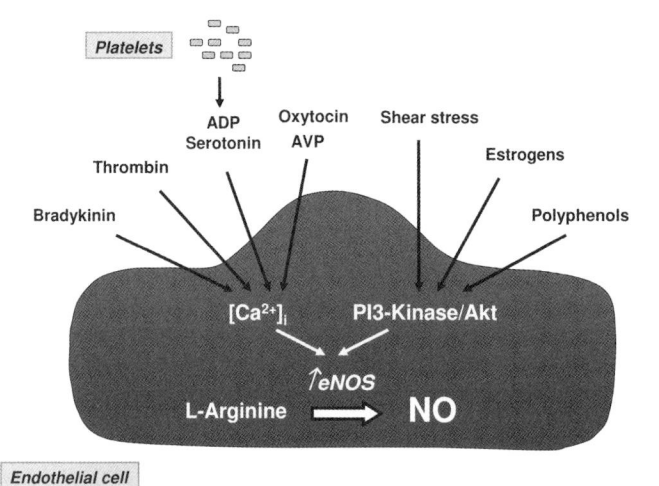

Figure 16.2 The endothelial formation of NO can be increased within seconds by a variety of stimuli including hemodynamic forces (i.e. shear stress), blood and platelet-derived substances (i.e. thrombin, ADP, serotonin), autocoids (i.e. BK), circulating hormones [i.e. arginine vasopressin (AVP), estrogens], and also substances contained in food and beverages such as polyphenols. NO formation is increased subsequently to the Ca^{2+}-calmodulin-dependent activation of eNOS and/or the PI3K/Akt-dependent phosphorylation of eNOS at Ser1177.

shear stress by a 2- to 3-fold enhancement of the normally basal production of NO and this effect persists as long as the stimulus is applied (for review, see [4]). The characterization of the signal transduction cascade leading to the response has indicated that the sustained formation of NO in response to fluid shear stress is a Ca^{2+}-independent event involving several kinases. Thus, activation of phosphatidylinositol 3-kinase (PI3K)/Akt, protein kinase A and AMP-activated protein kinase (AMPK), which all phosphorylate eNOS at Ser1177, leads to increased eNOS activity during shear stress-induced production of endothelial NO [5–8]. Although the molecular mechanisms that transduce mechanical force into eNOS activation are not well understood, a crucial role may be played by several signaling molecules including adaptor protein Gab1, tyrosine phosphatase SH2 and platelet EC adhesion molecule-1, as well as by a pathway involving the Src kinase-dependent transactivation of vascular endothelial growth factor (VEGF) receptor 2 [9–11]. Fluid shear stress, besides increasing eNOS activity, also can induce a Src kinase-dependent upregulation of the expression of eNOS via divergent pathways involving a short-term increase in eNOS transcription and a longer-term stabilization of eNOS mRNA [12]. Thus, mechanical forces have a central role in the regulation of the endothelial production of NO and, hence in vascular protection, by controlling locally the activity and the expression of eNOS.

16.1.2
Blood- and Platelet-Derived Factors

Events associated with blood coagulation have also a determinant role in the local regulation of the endothelial formation of NO (Figure 16.2). Indeed, activation of the coagulation cascade generates ultimately the serine protease thrombin, which catalyzes the conversion of fibrinogen to fibrin in order to stabilize developing blood clots. However, apart from its key role in hemostasis, thrombin can also exert direct effects on blood vessels. Indeed, thrombin induces endothelium-dependent relaxations in several types of canine arteries, whereas endothelium-dependent contractions were observed predominantly in veins [13, 14]. The kinetics of the endothelial formation of NO in response to thrombin are different from those in response to increases in fluid shear stress. Thrombin elicits within seconds a burst of formation of NO, about 10–20 the basal production, that peaks within 1 min, whereafter the production of NO declines to reach baseline levels within 5 min [15]. The stimulatory effect of thrombin is mediated by the N-terminal proteolytic activation of protease-activated receptor (PAR)-1, a G-protein-coupled receptor (GPCR), leading to an increase in $[Ca^{2+}]_i$, which, in turn, causes the Ca^{2+}-calmodulin-dependent activation of eNOS [16]. Thrombin might also enhance eNOS activity by phosphorylation of Ser1177 via an AMPK mediated pathway [17]. In addition to thrombin, an increased endothelial formation of NO is observed also in response to the serine protease factor Xa – an effect apparently independent of PAR-1, but mediated by the effector cell protease receptor-1 [18].

During hemostasis and thrombosis, besides products of the coagulation cascade, the activation of platelets also appears to be an important event regulating locally the endothelial formation of NO (Figure 16.2). Indeed, platelet activation elicits endothelium-dependent NO-mediated relaxations in several isolated arteries such as canine and porcine coronary arteries [19, 20]. The activation and aggregation of platelets leads to an increase in their $[Ca^{2+}]_i$ concentration resulting in release of several potent vasoactive factors. The characterization of the endothelium-dependent relaxations to platelets has indicated a major role for serotonin and adenine nucleotides (ADP and ATP) released from dense granules during platelet aggregation [19, 20]. The concentration of ADP and serotonin in the vicinity of a developing blood thrombus are likely to be elevated since dense granules contain about 650 mM ADP and 65 mM serotonin, respectively. Similarly to thrombin, serotonin and ADP act on GPCRs of the $5-HT_{1D}$ and P_{2Y} subtypes, respectively, on the EC membrane to induce a burst of NO, which vanishes after several minutes, most likely subsequent to the Ca^{2+}-calmodulin-dependent activation of eNOS [21, 22].

Thus, the development of a thrombus at the endothelial surface will generate and/or release several potent substances that, in turn, can act on ECs to cause an immediate and robust release of NO, thereby counteracting platelet adhesion and aggregation, and dilating the blood vessel to prevent vascular occlusion.

16.1.3
Local and Circulating Hormones, Growth Factors, and Neurotransmitters

Due to their strategic localization at the interface between blood and the vascular wall, ECs are continuously exposed to a variety of factors either circulating in the blood (hormones: catecholamines, neurohypophyseal hormones and sex hormones), formed locally (autacoids: BK and histamine), synthesized by the endothelial and the vascular smooth muscle cells themselves (growth factors), and possibly released by antonomic nerve endings (neurotransmitters). All these substances most likely act in concert at the level of the ECs to locally regulate both the acute and the long-term formation of NO (see Figure 16.2).

Among these factors, BK is a potent activator of ECs causing endothelium-dependent relaxations in many blood vessels, including coronary arteries and cerebral arteries. These responses to BK are mediated predominantly by the release of a burst of endothelial NO. The kinin acts on GPCRs of the B_2-kinin subtype on the endothelial surface to activate eNOS by both calcium-dependent and independent mechanisms leading to the phosphorylation of Ser1177 and the dephosphorylation of Thr495 of eNOS. Both these effects are associated with an increase in eNOS activity [23–26]. In addition, BK might also enhance NO formation by causing the transactivation of VEGF receptor KDR/Flk-1 [27].

The neurohypophyseal hormones vasopressin and oxytocin might also contribute to regulate NO formation in some blood vessels since both hormones cause marked endothelium-dependent NO-mediated relaxation in the canine basilar artery by activating V_1-vasopressinergic and oxytocinergic receptors of the ECs, respectively [28–30].

Estrogens are also physiologically important modulators of the endothelial formation of NO. Indeed, 17β-estradiol is a rapid activator of eNOS through a c-Src/PI3K/Akt-dependent pathway, mediated by estrogen receptor α via a nongenomic mechanism [31–33]. In addition, 17β-estradiol following conversion to its catechol metabolites possibly activates eNOS via an AMPK-mediated pathway [34]. Estrogens, in addition to acutely enhancing eNOS activity, also induce a long-term 2- to 3-fold increase in the formation of NO subsequent to an upregulation of mRNA expression of eNOS [35].

A modulatory role on the endothelial formation of NO has also been attributed to growth factors including VEGF, basic fibroblast growth factor, platelet-derived growth factor and insulin. Indeed, all these factors increase to some extent eNOS activity mostly via an initial calcium-dependent mechanism followed by changes in the phosphorylation level of eNOS [36–40]. In addition, an upregulation of eNOS expression has been observed in response to VEGF [41].

The activity of eNOS also may be regulated by substances released by nerve endings in the arterial wall. However, such a mechanism implies that the nerve-derived substances diffuse through the blood vessel wall to activate ECs. Potential neurotransmitters involved in such a regulatory mechanism include noradrenaline, substance P and vasoactive intestinal polypeptide, which all can induce endothelium-dependent NO-mediated relaxations of some blood vessels [42–44].

16.1.4
Polyphenols

Substances present in food and beverages also may be potent modulators of endothelial NO formation by their direct action on ECs resulting in increased eNOS activity. Thus, polyphenols derived from green tea, red wine and cocoa as well as authentic polyphenols such as epigallocatechin gallate, cause endothelium-dependent, NO-mediated relaxations of several blood vessels [45–48]. Moreover, ingestion of polyphenol-rich beverages is associated with an enhanced flow-mediated vasodilatation and acute elevations in the circulating levels of NO species in healthy humans [47, 49–51]. Like increases in shear stress, polyphenols induce within seconds a substantial formation of NO in ECs, an effect which persists for several hours [45, 46]. The stimulatory effect of polyphenols on eNOS activity is mediated both by the Src/PI3K/Akt pathway, which in turn induces the phosphorylation of eNOS at Ser1177, and the dephosphorylation of eNOS at Thr495 [46]. Although polyphenolic structures are well known to possess antioxidant properties, the polyphenol-induced activation of eNOS via the Src/PI3K/Akt pathway is initiated by a paradoxical intracellular formation of reactive oxygen species in ECs, possibly derived from the polyphenolic structure itself [46, 48]. Similarly to shear stress, polyphenols not only acutely increase eNOS activity but they can also up-regulate the protein level of eNOS by stimulating the expression of the enzyme and by enhancing the stability of its mRNA. This results in a 2- to 3-fold increased formation of NO for several days [52–53].

Altogether, these findings suggest that intake of food and beverages rich in polyphenols is likely to participate in both the acute and the long-term regulation of the endothelial formation of NO and, hence, in determining the level of vascular protection of the organism.

16.2
Vasoprotective Effects of NO

16.2.1
Regulation of Vascular Tone and Structure

NO is a short-lived reactive radical gas, which once produced by eNOS in ECs instantaneously diffuses toward the underlying vascular smooth muscle to effectively decrease vascular tone (Figure 16.3). The major target in the vascular smooth muscle is the heme-containing soluble guanylyl cyclase, the stimulation of which results in the increased formation of cGMP. cGMP, in turn, activates cGMP-dependent protein kinase I, which changes the phosphorylation level of several

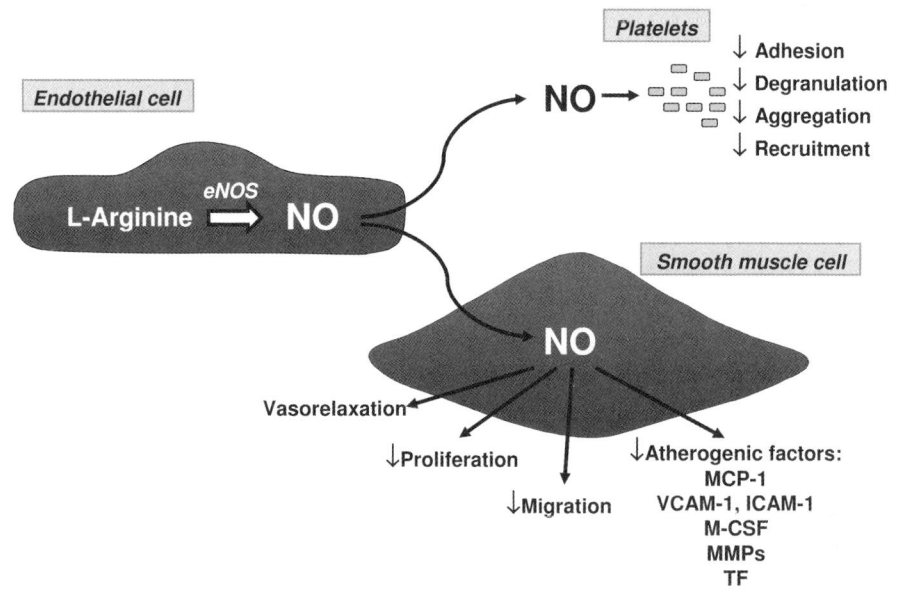

Figure 16.3 Endothelium-derived NO has a pivotal role in the maintenance of vascular homeostasis. NO helps to maintain blood fluidity by preventing procoagulant (i.e. tissue factor (TF) expression) and prothrombotic responses (i.e. platelet activation). NO also maintains the vascular smooth muscle cells in a quiescent state favoring dilatation of blood vessels and preventing proliferation, migration and expression of proatherogenic factors.

intracellular protein targets leading to a reduction of $[Ca^{2+}]_i$ with subsequent relaxation of the vascular smooth muscle (for review, see [54]). The reduction of the activator Ca^{2+} signal has been attributed to several mechanisms including activation of Ca^{2+}-ATPases to increase Ca^{2+} uptake or extrusion from the cytoplasm, inhibition of inositol 1,4,5-trisphosphate formation by inhibition of phospholipase C activation, inhibition of G-protein coupling to phospholipase C and inhibition of Ca^{2+} release from the endoplasmic reticulum. In addition, NO might also cause relaxation of vascular smooth muscle by inducing K^+ efflux following opening of K^+ channels, which in turn leads to hyperpolarization, inhibition of voltage-gated Ca^{2+} channels, and promotion of relaxation through a cGMP-dependent mechanism but possibly also through direct action of NO on K^+_{Ca} channels [55, 56]. Moreover, NO also can reduce vascular tone in resistance arteries without changes in $[Ca^{2+}]_i$ concentration. The latter relaxations result from the activation of myosin light chain phosphatase in a cGMP-dependent manner, leading ultimately to a reduced sensitivity of the contractile apparatus [57].

In addition to its direct relaxing effects, NO also favors vasodilatation by preventing the release and the action of endothelin-1 and endothelium-derived contracting prostanoids [3, 58, 59].

In addition to constantly adjusting the level of vascular tone, endothelium-derived NO might also prevent vascular remodeling, which contributes to the development of major cardiovascular diseases including hypertension (Figure 16.3). Indeed, NO is a potent inhibitor of vascular smooth muscle cell proliferation, in part by preventing the expression of cyclin D1 and A, and by enhancing the expression of p21, an inhibitor of cyclin-dependent kinase 2 [60–62]. As a consequence, cells will remain in the G_0/G_1 phase of the cell cycle. Moreover, NO also inhibits the migration of smooth muscle cells and modulates the synthesis of the extracellular matrix collagen type I and III, both responses contributing to vascular remodeling [63, 64].

16.2.2
Regulation of Coagulant and Thrombotic Responses

In addition to affecting the deeper layers of the blood vessel wall, NO will also diffuse towards the lumen where at the interface of ECs and blood it might contribute to regulate procoagulant and prothrombotic responses (Figure 16.3). The anticoagulant properties of the endothelial mediator are supported by the observation that enhanced NO production reduces endotoxin- and cytokine-induced tissue factor expression and activity in ECs [65]. Long-term inhibition of NO synthesis in the rat also increased arterial thrombogenicity as indicated by the upregulation of tissue factor expression and the resultant generation of thrombin [66]. NO also inhibits several key functions of platelets including adhesion, degranulation, aggregation and platelet recruitment by preventing the activator intracellular calcium signal [67–69]. The fact that NO is also effective in disaggregating small platelet aggregates provides a further mechanism to counteract the development of unwanted thrombus formation [67].

16.2.3
Regulation of Atherogenic Responses

Atherothrombosis, the leading cause of mortality worldwide, is characterized by an unpredictable atherosclerotic plaque disruption leading to an instantaneous development of a thrombus, which can ultimately cause the total occlusion of the blood vessel lumen, thereby inducing major cardiovascular events such as myocardial infarction, stroke and peripheral vascular disease. NO has potent antiatherosclerotic properties by affecting several key events in the formation, development and possibly also rupture of atherosclerotic plaques (Figure 16.3). Indeed, NO inhibits basal and lipopolysaccharide- and oxidized low-density lipoprotein-induced expression of monocyte chemotactic protein (MCP)-1, thought to be the major chemotactic factor mediating the recruitment of monocytes into the arterial wall, one of the earliest events in the genesis of atherosclerosis [70, 71]. NO might not only affect the recruitment but possibly also the adhesion of circulating leucocytes to the blood vessel wall by preventing the expression of EC adhesion molecules including vascular cell adhesion molecule (VCAM)-1 and intercellular adhesion molecule (ICAM)-1 [72, 73]. Moreover, since NO prevents the expression of oxidized low-density lipoprotein- and tumor necrosis factor-α-induced expression of macrophage-colony stimulating factor (M-CSF; that regulates growth and differentiation of macrophages), it may contribute to prevent the development of macrophage-derived foam cells in atherosclerotic lesions [74]. The ability of NO to prevent the expression of these proatherogenic factors is due, at least in part, to the inhibition of NF-κB activation subsequent to the induction and stabilization of the NF-κB inhibitor, IκBα [74]. In addition, NO might also play a key role in the prevention of plaque rupture by reducing the expression of matrix metalloproteinase (MMP)-2 and -9, which degrade most types of collagens thereby promoting plaque vulnerability [75, 76]. The protective effect of NO in atherosclerosis is further supported by the fact that chronic inhibition of NO formation with an inhibitor of NO synthase accelerates the development of atherosclerotic lesions in cholesterol-fed rabbits [77].

16.3
Conclusions

In healthy blood vessels, ECs continuously release small amounts of NO, which can be rapidly, within seconds, increased following activation of ECs by a variety of physical and chemical stimuli not only to continuously adjust vascular tone and, hence, blood flow to target organs, but also to maintain blood fluidity and to prevent proatherosclerotic responses.

Both experimental and clinical studies indicate that the protective function of endothelium-derived NO is altered, as evidenced by blunted endothelium-dependent, NO-mediated dilatations, by aging and in most cardiovascular diseases including hypertension, atherosclerosis, diabetes (for review, see [78]). Such reduced bioavailability of endothelium-derived NO generally is observed before the appearance of structural changes in the arterial wall, suggesting that the impaired formation of NO is likely to be an important initial event triggering the initiation and promoting the development of atherosclerosis. Thus, preserving the endothelial formation of NO throughout life is an attractive goal to prevent and retard alteration in vascular homeostasis promoting the development of cardiovascular diseases.

References

1 Furchgott, R.F. and Zawadzki, J.V. (1980) *Nature*, **299**, 373–6.
2 Vanhoutte, P.M. (2003) *Circulation Journal*, **67**, 572–5.
3 Vanhoutte, P.M. (2005) *British Journal of Pharmacology*, **144**, 449–58.
4 Busse, R. and Fleming, I. (2003) *Trends in Pharmacological Sciences*, **24**, 24–9.
5 Kuchan, M.J. and Frangos, J.A. (1994) *American Journal of Physiology*, **266**, C628–36.

6 Dimmeler, S., Fleming, I., Fisslthaler, B., Hermann, C., Busse, R. and Zeiher, A.M. (1999) *Nature*, **399**, 601–5.
7 Boo, Y.C., Sorescu, G., Boyd, N., Shiojima, I., Walsh, K., Du, J. and Jo, H. (2002) *Journal of Biological Chemistry*, **277**, 3388–96.
8 Zhang, Y., Lee, T.-S., Kolb, E.M., Sun, K., Lu, X., Sladek, M., Kassab, G.S., Garland, T. Jr and Shyy, J.Y.-J. (2006) *Arteriosclerosis Thrombosis and Vascular Biology*, **26**, 1281–7.
9 Dixit, M., Loot, A.E., Mohamed, A., Fisslthaler, B., Boulanger, C.M., Ceacareanu, B., Hassid, A., Busse, R. and Fleming, I. (2005) *Circulation Research*, **97**, 1236–44.
10 Fleming, I., Fisslthaler, B., Dixit, M. and Busse, R. (2005) *Journal of Cell Science*, **118**, 4103–11.
11 Jin, Z.G., Ueba, H., Tanimoto, T., Lungu, A.O., Frame, M.D. and Berk, B.C. (2003) *Circulation Research*, **93**, 354–63.
12 Davis, M.E., Cai, H., Drummond, G.R. and Harrison, D.G. (2001) *Circulation Research*, **89**, 1073–80.
13 De Mey, J. and Vanhoutte, P.M. (1982) *Circulation Research*, **51**, 439–47.
14 Ku, D.D. (1982) *Science*, **218**, 576–8.
15 Schini, V.B., Moncada, S. and Vanhoutte, P.M. (1990) *Nitric Oxide from L-Arginine: A Bioregulatory System*, Elsevier, Amsterdam.
16 Tesfamariam, B., Allen, G.T., Normandin, D. and Antonaccio, M.J. (1993) *American Journal of Physiology*, **265**, H1744–9.
17 Thors, B., Halldorsson, H. and Thorgeirsson, G. (2004) *FEBS Letters*, **573**, 175–80.
18 Papapetropoulos, A., Piccardoni, P., Cirino, G., Bucci, M., Sorrentino, R., Cicala, C., Johnson, K., Zachariou, V., Sessa, W.C. and Altieri, D.C. (1998) *Proceedings of the National Academy of Sciences of the United States of America*, **95**, 4738–42.
19 Houston, D.S., Shepherd, J.T. and Vanhoutte, P.M. (1985) *American Journal of Physiology*, **248**, H389–95.
20 Shimokawa, H., Aarhus, L.L. and Vanhoutte, P.M. (1987) *Circulation Research*, **61**, 256–70.
21 Boulanger, C.M., Schini-Kerth, V.B., Moncada, S. and Vanhoutte, P.M. (1990) *British Journal of Pharmacology*, **101**, 152–6.
22 McDuffie, J.E., Coaxum, S.D. and Maleque, M.A. (1999) *Proceedings of the Society for Experimental Biology and Medicine*, **221**, 386–90.
23 Lückhoff, A., Pohl, U., Mülsch, A. and Busse, R. (1988) *British Journal of Pharmacology*, **95**, 189–96.
24 Schini, V.B., Boulanger, C., Regoli, D. and Vanhoutte, P.M. (1990) *Journal of Pharmacology and Experimental Therapeutics*, **252**, 581–5.
25 Harris, M.B., Venema, H., Ju, V.J., Liang, H., Zou, R., Michell, B.J., Chen, Z.P., Kemp, B.E. and Venema, R.C. (2001) *Journal of Biological Chemistry*, **276**, 16587–91.
26 Fleming, I., Fisslthaler, B., Dimmeler, S., Kemp, B.E. and Busse, R. (2001) *Circulation Research*, **88**, E68–75.
27 Thuringer, D., Maulon, L. and Frelin, C. (2002) *Journal of Biological Chemistry*, **277**, 2028–32.
28 Katusic, Z.S., Shepherd, J.T. and Vanhoutte, P.M. (1984) *Circulation Research*, **55**, 575–9.
29 Katusic, Z.S., Shepherd, J.T. and Vanhoutte, P.M. (1986) *Journal of Pharmacology and Experimental Therapeutics*, **236**, 166–70.
30 Schini, V.B., Katusic, Z.S. and Vanhoutte, P.M. (1990) *Journal of Pharmacology and Experimental Therapeutics*, **255**, 994–1000.
31 Haynes, M.P., Sinha, D., Russell, K.S., Collinge, M., Fulton, D., Morales-Ruiz, M., Sessa, W.C. and Bender, J.R. (2000) *Circulation Research*, **87**, 677–82.
32 Hisamoto, K., Ohmichi, M., Kurachi, H., Hayakawa, J., Kanda, Y., Nishio, Y., Adachi, K., Tasaka, K., Miyoshi, E., Fujiwara, N., Taniguchi, N. and Murata, Y. (2001) *Journal of Biological Chemistry*, **276**, 3459–67.
33 Haynes, M.P., Li, L., Sinha, D., Russell, K.S., Hisamoto, K., Baron, R., Collinge, M., Sessa, W.C. and Bender, J.R. (2003) *Journal of Biological Chemistry*, **278**, 2118–23.
34 Schulz, E., Anter, E., Zou, M.H. and Keaney, J.F. (2005) *Circulation*, **111**, 3473–80.

35 MacRitchie, A.N., Jun, S.S., Chen, Z., German, Z., Yuhanna, I.S., Sherman, T.S. and Shaul, P.W. (1997) *Circulation Research*, **81**, 355–62.

36 Ku, D.D., Zaleski, J.K. and Brock, T.A. (1993) *American Journal of Physiology*, **265**, H586–92.

37 Cunningham, L.D., Brecher, P. and Cohen, R.A. (1992) *Journal of Clinical Investigation*, **89**, 878–82.

38 Tiefenbacher, C.P. and Chilian, W.M. (1997) *Cardiovascular Research*, **34**, 411–17.

39 Montagnani, M., Chen, H., Barr, V.A. and Quon, M.J. (2001) *Journal of Biological Chemistry*, **276**, 30392–8.

40 Brouet, A., Sonveaux, P., Dessy, C., Balligand, J.L. and Feron, O. (2001) *Journal of Biological Chemistry*, **276**, 32663–9.

41 Bouloumié, A., Schini-Kerth, V.B. and Busse, R. (1999) *Cardiovascular Research*, **41**, 773–80.

42 Miller, V.M. and Vanhoutte, P.M. (1985) *European Journal of Pharmacology*, **118**, 123–9.

43 Ignarro, L.J., Byrns, R.E., Buga, G.M. and Wood, K.S. (1987) *American Journal of Physiology*, **253**, H1074–82.

44 Berthiaume, N., Chaing, A., Regoli, D., Warner, T.D. and Juste-D'Orléans, P. (1995) *British Journal of Pharmacology*, **115**, 1319–25.

45 Lorenz, M., Wessler, S., Follmann, E., Michaelis, W., Düsterhöft, T., Baumann, G., Stangl, K. and Stangl, V. (2004) *Journal of Biological Chemistry*, **279**, 6190–5.

46 Ndiaye, M., Chataigneau, M., Lobysheva, I., Chataigneau, T. and Schini-Kerth, V.B. (2005) *FASEB Journal*, **19**, 455–7.

47 Schroeter, H., Heiss, C., Balzer, J., Kleinbongard, P., Keen, C.L., Hollenberg, N.K., Sies, H., Kwik-Uribe, C., Schmitz, H.H. and Kelm, M. (2006) *Proceedings of the National Academy of Sciences of the United States of America*, **103**, 1024–9.

48 Kim, J.A., Formoso, G., Li, Y., Potenza, M.A., Marasciulo, F.L., Montagnani, M. and Quon, M.J. (2007) *Journal of Biological Chemistry*, **282**, 13736–45.

49 Agewall, S., Wright, S., Doughty, R.N., Whalley, G.A., Duxbury, M. and Sharpe, N. (2000) *European Heart Journal*, **21**, 74–8.

50 Hashimoto, M., Kim, S., Eto, M., Iijima, K., Ako, J., Yoshizumi, M., Akishita, M., Kondo, K., Itakura, H., Hosoda, K., Toba, K. and Ouchi, Y. (2001) *American Journal of Cardiology*, **88**, 1457–60.

51 Engler, M.B., Engler, M.M., Chen, C.Y., Malloy, M.J., Browne, A., Chiu, E.Y., Kwak, H.-K., Milbury, P., Paul, S.M., Blumberg, J. and Mietus-Snyder, M.L. (2004) *Journal of the American College of Nutrition*, **23**, 197–204.

52 Leikert, J.F., Räthel, T.R., Wohlfart, P., Cheynier, V., Vollmar, A.M. and Dirsch, V.M. (2002) *Circulation*, **106**, 1614–17.

53 Wallerath, T., Poleo, D., Li, H. and Förstermann, U. (2003) *Journal of the American College of Cardiology*, **41**, 471–8.

54 Lincoln, T.M., Cornwell, T.L., Komalavilas, P., MacMillan-Crow, L.A. and Boerth, N. (1996) *Biochemistry of Smooth Muscle Contraction*, Academic Press, San Diego, CA.

55 Archer, S.L., Huang, J.M.C., Hampl, V., Nelson, D.P., Shultz, P.J. and Weir, E.K. (1994) *Proceedings of the National Academy of Sciences of the United States of America*, **91**, 7583–7.

56 Bolotina, V.M., Najibi, S., Palacino, J.J., Pagano, P.J. and Cohen, R.A. (1994) *Nature*, **368**, 850–3.

57 Bolz, S.-S., Vogel, L., Sollinger, D., Derwand, R., de Wit, C., Loirand, G. and Pohl, U. (2003) *Circulation*, **107**, 3081–7.

58 Vanhoutte, P.M. (2000) *Antonomic Nervous System*, **81**, 271–7.

59 Tang, E.H.C., Feletou, M., Huang, Y., Man, R.Y.K. and Vanhoutte, P.M. (2005) *American Journal of Physiology*, **289**, H2434–40.

60 Garg, U.C. and Hassid, A. (1989) *Journal of Clinical Investigation*, **83**, 1774–7.

61 Ishida, A., Sasaguri, T., Kosaka, C., Nojima, H. and Ogata, J. (1997) *Journal of Biological Chemistry*, **272**, 10050–7.

62 Kronemann, N., Nockher, W.A., Busse, R. and Schini-Kerth, V.B. (1999) *British Journal of Pharmacology*, **126**, 349–57.

63 Sakar, R., Meinberg, E.G., Stanley, J.C., Gordon, D. and Webb, R.C. (1996) *Circulation Research*, **78**, 225–30.

64 Myers, P.R. and Tanner, M.A. (1998) *Arteriosclerosis Thrombosis and Vascular Biology*, **18**, 717–22.
65 Yang, Y. and Loscalzo, J. (2000) *Circulation*, **101**, 2144–8.
66 Kubo-Inoue, M., Egashira, K., Usui, M., Takemoto, M., Ohtani, K., Katoh, M., Shimokawa, H. and Takeshita, A. (2002) *American Journal of Physiology*, **282**, H1478–84.
67 Mellion, T., Ignarro, L.J., Ohlstein, E.H., Pontecorvo, E.G., Hyman, A.L. and Kadowitz, P.J. (1981) *Blood*, **57**, 946–55.
68 Sneddon, J.M. and Vane, J.R. (1988) *Proceedings of the National Academy of Sciences of the United States of America*, **85**, 2800–4.
69 Freedman, J.E., Loscalzo, J., Barnard, M.R., Alpert, C., Keaney, J.F. and Michelson, A.D. (1997) *Journal of Clinical Investigation*, **100**, 350–6.
70 Zeiher, A.M., Fisslthaler, B., Schray-Utz, B. and Busse, R. (1995) *Circulation Research*, **76**, 980–6.
71 Tsao, P.S., Wang, B.-Y., Buitrago, R., Shyy, J.Y.-J. and Cooke, J.P. (1997) *Circulation*, **96**, 934–40.
72 Spiecker, M., Peng, H.B. and Liao, J.K. (1997) *Journal of Biological Chemistry*, **272**, 30969–74.
73 Lindemann, S., Sharafi, M., Spiecker, M., Buerke, M., Fisch, A., Grosser, T., Veit, K., Giere, C., Ibe, W., Meyer, J. and Darius, H. (2000) *Thrombosis Research*, **97**, 113–23.
74 Peng, H.B., Rajavashisth, T.B., Libby, P. and Liao, J.K. (1995) *Journal of Biological Chemistry*, **270**, 17050–5.
75 Knipp, B.S., Ailawadi, G., Ford, J.W., Peterson, D.A., Eagleton, M.J., Roelofs, K.J., Hannawa, K.K., Deogracias, M.P., Logsdon, C., Graziano, K.D., Simeone, D.M., Thompson, R.W., Henke, P.K., Stanley, J.C. and Upchurch, G.R. (2004) *Journal of Surgical Research*, **116**, 70–80.
76 Guriar, M.V., Sharma, R.V. and Bhalla, R.C. (1999) *Arteriosclerosis Thrombosis and Vascular Biology*, **19**, 2871–7.
77 Cayatte, A.J., Palacino, J.J., Horten, K. and Cohen, R.A. (1994) *Arteriosclerosis and Thrombosis*, **14**, 753–9.
78 Félétou, M. and Vanhoutte, P.M. (2006) *American Journal of Physiology*, **291**, H985–1002.

17
Acetylcholine

*Maria Cláudia Irigoyen, Catarina S. Porto, Pedro Paulo Soares,
Fernanda Consolin-Colombo, and Antônio Cláudio Nóbrega*

17.1
Muscarinic Acetylcholine Receptor: Subtypes and Intracellular Signaling

Acetylcholine (ACh) is, phylogenetically, an ancient signaling molecule [1] that has evolved to exert both neurotransmitter and autocrine functions in the embryo and adult [2]. The hormonal function of ACh differs from its roles in neurotransmission and includes regulation of cell functions, such as proliferation, differentiation and apoptosis [3]. The cellular actions of ACh are mediated by two structurally different families of membrane-bound proteins, the nicotinic and muscarinic ACh receptors (mAChRs) [4–7]. These receptors exhibit multiple subtypes, expressed pre- and postsynaptically, and are widely distributed throughout the central and peripheral nervous systems, being also expressed in several non-neuronal tissues [5].

mAChRs mediate distinct physiological functions according to location and receptor subtype [5–7]. mAChRs are class I heptahelical G-protein-coupled receptors, and comprise five distinct subtypes (M_1–M_5) that are encoded by intronless genes [8] and have been cloned from several species, including human, cow, pig, rat and mouse [4], exhibiting a high sequence homology across species. Indeed, the similarity in ligand-binding sites across the five subtypes explains why identification of subtype-selective ligands has been difficult [4]. In addition to the agonist-binding sites, mAChRs have allosteric sites where compounds can also modulate agonist activation [9]. The nature of these allosteric sites differs both from the agonist-binding site and between receptor subtypes, potentially allowing for the design of subtype-selective modulators [10].

The coupling of mAChRs to their cellular effector systems is mediated via heterotrimeric guanine nucleotide-binding proteins (also known as G-proteins) [11]. These cellular effectors depend upon which G_α subunit is activated. Furthermore, in some cases $G_{\beta\gamma}$ subunits play a role in cellular signaling of mAChRs. M_1, M_3 and M_5 mAChRs are preferentially coupled to $G_{q/11}$, mobilizing phosphoinositides to generate inositol 1,4,5-triphosphate and 1,2-diacylglycerol via activation of phosphoinositide-specific phospholipase Cβ, therefore increasing intracellular calcium. M_2 and M_4 mAChRs preferentially couple to $G_{i/o}$, inhibit elevated adenylate cyclase activity, as well as prolong potassium channel, nonselective cation and transient receptor potential channel opening [12]. M_2 mAChRs, via $G_{\beta\gamma}$, also activate phospholipase Cβ and modulate specific ionic conductances. mAChRs have also been shown to regulate several signaling intermediate pathways. Thus, both $G\alpha_{q/11}$- and $G\alpha_{i/o}$-coupled to mAChRs exert cytoskeleton effects by activating of small GTPase Rho and downstream effectors such as phosphoinositide-3 kinases, nonreceptor tyrosine kinases and mitogen-activated protein kinases [13]. The latter signaling pathway appears to play a major role in mAChR autocrine functions in terms of the control of cell growth and proliferation and apoptosis.

An extensive pharmacological literature, together with emerging data from studies in transgenic mice, indicates the function of all five mAChRs subtypes in neuronal and non-neuronal cholinergic systems [6, 7, 14].

M_1 mAChRs are highly expressed in cerebral cortex, hippocampus and striatum. Consistent with this distribution, M_1 mAChRs are implicated in learning and memory processes. Enhanced cholinergic receptor activation, either by the use of acetylcholinesterases or muscarinic agonists, ameliorates cognitive decline in preclinical and clinical studies.

M_2 mAChRs are widely expressed in both central and peripheral nervous systems. Selective M_2 receptor antagonist increases cholinergic overflow by reducing autoreceptor function in both brain and periphery. Thus, selective M_2 receptor antagonists or compounds possessing mixed M_2 antagonism/M_1 agonism are also therapeutic approaches to increase cholinergic function in Alzheimer's disease [15]. In the periphery, postjunctional M_2 mAChRs are expressed in the myocardium, although M_3 receptors also play a role in controlling cardiac contractility [14]. M_2 mAChRs are also postjunctionally expressed in smooth muscle from several tissues and regulate contractility in a synergic fashion to M_3 receptors [16]. For example, in the bladder, M_3 receptors induce direct smooth muscle contraction by a mechanism that relies on entry of extracellular calcium through L-type channels and activation of a Rho kinase, while M_2 receptors, which are predominant in number, not only appear to facilitate M_3-mediated contractions, but also produce bladder contractions indirectly by reversing cAMP-dependent β-adrenoceptor-mediated relaxation.

M_3 mAChR is widely distributed in the central nervous system, but at a markedly lower level than other muscarinic receptor subtypes [6, 17]. M_3 receptor-deficient mice are hypophagic and lean, suggesting a role in food intake regulation [18]. Classically, M_3 mAChRs mediate contraction of many smooth muscle types including the respiratory, gastrointestinal and genitourinary tracts [7, 19]. M_3 mAChR immunoreactivity is detected in peritubular smooth muscle of efferent ductules and along the different epididimal regions, with a strong reaction in the epididimal proximal and distal cauda, suggesting that this receptor subtype may play a role in smooth muscle contraction [20]. Reverse transcriptase–polymerase chain reaction has shown the presence of mRNA for five mAChRs (M_1–M_5) in rat seminal vesicles. M_3 mAChR is predominantly involved in the contraction of rat seminal vesicles [21].

M_4 mAChRs are distributed in the corpus striatum. The activation of spinal muscarinic receptors leads to potent antinociception [22]. The receptor subtype mediating this response is unclear.

Activation of M_5 mAChRs facilitates striatal dopamine release even though other muscarinic receptors, including M_4 receptor, are involved [23]. Brain microvasculature expresses mAChRs distributed in endothelial cells (M_2 and M_5 receptors) and vascular smooth muscle cells (M_1, M_2, M_3 and M_5), with M_5 receptors mediating cerebral vascular dilation [24].

This chapter focuses on the expression and role of mAChRs subtypes in the heart.

17.1.1
Muscarinic Receptors in the Heart

Parasympathetic innervation to the heart is provided by the vagal nerve. All heart regions are innervated by parasympathetic nerves although in most species, including the human heart, the supraventricular areas are more densely innervated than the ventricles [25]. The postganglionic release of ACh from parasympathetic nerve terminals activates postsynaptic mAChRs in the heart.

The details on the functional response elicited by mAChRs agonists in the heart depend on species, age, anatomic structure investigated and concentrations of used mAChR agonists [26]. Stimulation of mAChRs in the mammalian heart, specifically M_2 mAChR [4, 27], modulates pacemaker activity and atrioventricular conduction, and directly (in atrium) or indirectly (in ventricles) the force of contraction [26]. Indeed, bradycardia in response to carbachol in M_2 knockout mice is abolished [28], emphasizing the functional relevance of this subtype. Classically, stimulation of the vagal nerve or use of low concentrations of mAChR agonists produces characteristic negative chronotropic and inotropic effects on the heart. Paradoxically, under appropriate conditions, activation of cardiac mAChRs elicits stimulatory effects on the rate of beating and contractile force. These latter effects often require higher concentrations of agonist and occur after pretreatment with pertussis toxin to block M_2 mAChR cellular signaling. These contradictions are now known to be caused by different mAChR subtypes in the heart. The presence of M_1, M_3 and M_5 has been shown in human heart [29], and details of their functional roles are still emerging.

Functional M_1 mAChRs that increase heart rate have been reported in rodent cardiac tissue [30, 31] and human atrial myocytes [32, 33]. However, in mice lacking M_1 mAChRs, the basal heart function and cardiac stimulation due to catecholamine release from sympathetic neurons are unchanged compared with wild-type [34]. Other studies have also shown that non-M_2 receptors, in addition to M_2 mAChRs, play a role in the modulation of sympathetic neurotransmitter release in mouse atria [35].

Functional M_3 mAChRs have been identified in rodent and mammalian cardiac tissue [14, 33, 36], and in human atrial tissue [32]. M_3 mAChRs activate a delayed rectifying K^+ current I_{KM3} that participates in cardiac repolarization, negative chronotropic actions and antidysrhythmia, since it suppresses ischaemic dysrhythmias and prodysrhythmic actions (facilitating atrial fibrillation). M_3 mAChR also interacts with gap-junctional channel connexin 43 to maintain cell–cell communication and excitation propagation, and regulates intracellular phosphoinositides hydrolysis to improve cardiac contraction and hemodynamic function. Furthermore, M_3 mAChR activates antiapoptotic signaling molecules, enhances endogenous antioxidant capacity and decreases intracellular Ca^{2+} overload, contributing to heart protection against ischemic injuries [14]. On the other hand, studies using mice lacking either M_2 or M_3 mAChRs have indicated a predominant role for M_2 receptors in heart rate regulation, since no change was observed in basal heart rate of M_3 knockout mice [17, 28, 37].

The effects of pathological conditions on mAChRs subtypes and their role in modulating cardiac function have been reported. In chronic atrial fibrillation and experimental congestive heart failure, the number of M_3 mAChRs was increased, but M_2 receptors were decreased [33]. However, most studies have shown that the number of cardiac mAChRs in patients with chronic heart failure is not altered [38, 39]; only an *in vivo* positron emission tomography study has found a slight, but significant increase in cardiac mAChRs in patients with chronic heart failure [40]. M_2 and M_3 mAChRs are expressed in human coronary arteries [41], although the functional importance of these receptors is currently unclear. Studies using knockout mice lacking M_2, M_3 or M_5 receptors have suggested that M_3 mAChRs predominantly mediate ACh-induced dilation of mouse coronary arteries, endothelium dependently [18, 42].

In conclusion, available data indicate a prominent role of M_2 receptors in cardiac function. Further study is required to elucidate the role of other mAChR subtypes in the heart and how they may be altered in disease states.

17.2
Physiological Effects of ACh on the Heart

ACh action on the heart depends mainly on three factors: ACh release from vagal postsynaptic neurons, ACh degradation and its action on mAChRs on cardiac cell membranes. ACh released by parasympathetic neurons in the synaptic cleft binds to mAChR to produce its effects on the heart, and this depends on receptor affinity and intracellular signaling. ACh is rapidly inactivated in the synaptic cleft by the acetylcholinesterase enzyme and its action on circulation is not expected although nicotinic receptors are present in the endothelium [27].

ACh is the neurotransmitter in the ganglia of both limbs of the autonomic nervous system. It modulates postganglionic sympathetic and parasympathetic discharges to the periphery. Postganglionic neurons in the sympathetic branch release norepinephrine, while postganglionic parasympathetic neurons are cholinergic, that is, ACh mediates their action on the target organ. ACh is packaged into vesicles and, on depolarization of the terminal nerve, the influx of calcium through voltage-gated calcium channels promotes their fusion with the neuronal membrane and the transmitter release.

ACh released in the synaptic cleft binds to sinoatrial node cell M2 mAChRs, coupled to inhibitory heterotrimeric G-proteins [12]. This decreases the rate of spontaneous action potential generation by direct G-protein (G_k) gating of an inward-rectifying potassium current (I_{KACh}) as well as G-protein (G_i) inhibition of adenylyl cyclase to decrease cAMP-dependent stimulation of the hyperpolarization-activated current (I_f) and cAMP- and protein kinase A-dependent phosphorylation of the L-type calcium current (I_{CaL}). In fact, the exocytotic release of ACh and the muscarinic receptor regulation of pacemaker currents determine the parasympathetic control of heart rate, not only modulated by adrenergic tone (adrenergic-cholinergic cross talk) but also by endogenous production of nitric oxide (NO) [43].

Indeed, generation of NO by neuronal NO synthase (nNOS, also called NOS-1), the form of the synthase found in vagal nerve terminals, enhances ACh release [44]. Additional evidence regarding ACh release facilitation by NO was provided by experiments with mice knocked out for nNOS, in which bradycardic responses induced by vagal electrical stimulations were reduced [45]. Moreover, the increase in nNOS expression induced by exercise training was accompanied by an enhancement of bradycardic responses during vagal nerve stimulation in trained mice [46].

The intracellular mechanisms of sympathetic and parasympathetic modulation of sinus node function are associated with heart rate and heart rate variability (HRV). Changes in heart rate caused by sympathetic nervous system activity depend on cAMP intracellular concentration changes, which are linked to cAMP synthesis when β_1- and β_2-adrenoceptors couple to G_s protein to activate adenylyl cyclase. In contrast, M2 mAChRs in atrial myocytes can affect heart rate coupling to pertussis toxin-sensitive G-protein (G_i/G_o) inhibiting adenylyl cyclase and then, cAMP, counteracting sympathetic stimulation, and affecting inward rectifying potassium channels (I_{KACh}) through G-protein α- or $\beta\gamma$-subunits, resulting in hyperpolarization and reduced heart rate [47]. In transgenic mice with functionally reduced G-protein $\beta\gamma$-subunits, bradycardia induced by carbachol, baroreflex sensitivity and HRV indexes were reduced as compared to wild-type mice [48].

17.2.1
Reflex Control of Heart Rate by the Autonomic Nervous System

The neurohumoral regulation of the cardiovascular system includes parasympathetic (cholinergic) and sympathetic (adrenergic) influences on the heart, exerting tonic control on the sinoatrial pacemaker and, consequently, determining final cardiac activity.

Parasympathetic efferent activity to the heart depends on the central generation of vagal tonus that is modulated by many afferent inputs to the brainstem arising from baro-, chemo-, cardiopulmonary and nociceptors, and also from the central command [49]. Cholinergic receptors are also involved in central modulation of cardiovascular system, since the stimulation of central M2 mAChRs increases blood pressure and heart rate in conscious cats [50]. There are many autonomic markers used to evaluate parasympathetic and sympathetic interactions or dominance of one limb of the autonomic nervous system upon the other in the heart control. Resting heart rate, heart rate recovery from exercise, baroreflex sensitivity and HRV have been extensively employed in physiological and clinical studies showing the relevance of such measures [51].

From many cardiovascular reflexes, baroreflex is by far the most widely studied in terms of its clinical and physiological significance. Experimental data provide evidence of the rapid baroreflex buffering of blood pressure changes in order to maintain pressure levels in a narrow range, optimal for blood supply to tissues. In contrast, blood pressure lability, such as observed in sinoaortic denervated (SAD) rats, has been associated with target organ lesions in animal models and humans, indicating the homeostatic role of baroreflex modulation in autonomic activity to the heart and vessels. In SAD rats, disruption of baroreceptor afferences to the brainstem lead to reduced vagal outflow and increased sympathetic activity to the cardiovascular system, probably contributing to functional and morphological changes observed in this model. Indeed, in chronically SAD rats, there is mAChR upregulation that is accompanied by increased bradycardia produced by vagal electrical stimulation when compared to intact animals. These data show that increased mAChR density and affinity, along with greater functional response (bradycardia), might represent peripheral adjustments to central reduced vagal tonus in this model of blood pressure lability. Similar changes may be observed in pathological states associated with reduction of central parasympathetic activity as demonstrated in experimental diabetes [52].

There are several methods to access the baroreflex function in man and animal models [53] such as bolus or ramp infusions of vasoactive drugs, neck suction, low body negative or positive pressure, Valsalva maneuver and spontaneous baroreflex. Reflex bradycardia is highly dependent on vagal activity and reduced baroreflex sensitivity is a marker of reduced vagal activity present in many pathological states. After the demonstration that in patients with myocardial infarction higher baroreflex sensitivity was strongly associated with survival [54], the clinical relevance of baroreflex activity in the prognosis of different pathological states has been repeatedly highlighted.

Cardiovascular signals arise from a complex interplay of factors that modulate autonomic outflows to circulation presenting rhythmicity patterns of neural and cardiovascular variables. From many biological signals, heart rate is the most frequently investigated both in time and frequency domains. Spectral analysis of heart rate time series helps to describe healthy and pathological conditions associated with the cardiovascular system, allowing the assessment of autonomic balance to the heart, and identifying sympathetic and vagal contributions to heart rate modulation. Frequency domain analysis of heart rate is known to reveal sympathetic and parasympathetic interactions, and HRV has been used as a marker of autonomic activity that exhibits prognostic value after acute myocardial infarction [55]. After coronary artery bypass surgery, HRV indexes in time and frequency domains seem to be useful to identify the time course of vagal recovery after intervention, showing that autonomic balance moved from sympathetic dominance after surgery to control values under parasympathetic dominance after 90 days [56]. Low levels of vagal activity, as measured by HRV, both in time and frequency domains, are related to morbidity and mortality in a variety of pathologies such as myocardial ischemia, hypertension, heart failure and diabetes, and are also associated with risk factors to cardiovascular diseases such as obesity and cholesterol [51]. In fact, pharmacological blockade of mAChR by drugs such as atropine is known to produce tachycardia and reduce HVR. Interestingly, low doses of atropine reduce heart rate and increase HRV at the high frequency band that is recognized to be under vagal modulation. Although the mechanisms involved are not completely understood, these data suggest that atropine at low doses might modulate centrally generated vagal activity [57]. Functional effects of ACh might also be enhanced by controlling its degradation by acetylcholinesterase. Indeed, acetylcholinesterase inhibitors such as pyridostigmine bromide are known to prolong the cardiac effects of endogenous ACh released by vagal endings. In rats, oral administration of pyridostigmine for 7 days resulted in 40% decrease in acetylcholinesterase activity, reflecting increased vagal modulation of the heart as measured by the increased baroreflex sensitivity (Figure 17.1) and HRV [58].

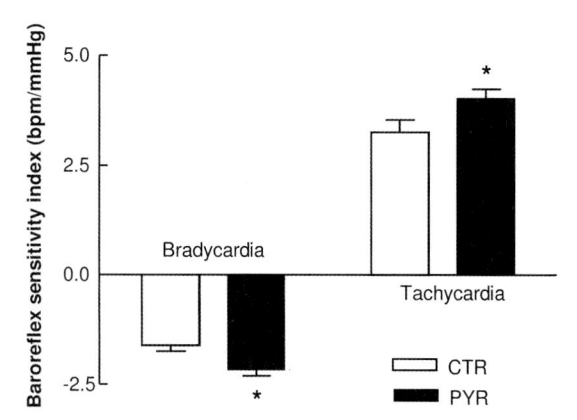

Figure 17.1 Baroreflex sensitivity indexes for control (CTR) or pyridostigmine treated rats (PYR) for reflex bradycardia and tachycardia after bolus infusions of phenylephrine and sodium nitroprusside, respectively. $*P < 0.05$ versus CTR. bpm, beats per minute.

17.3
Parasympathetic Dysfunction: Clinical Impact on the Cardiovascular System

Autonomic dysfunction, also known as dysautonomia, usually refers to the degenerative process involving autonomic ganglia and fibers as well as nuclei within the central nervous system that develops as a primary–pure autonomic failure and multiple system atrophy–or a secondary process to systemic diseases such as diabetes mellitus. Although no specific anatomical landmark has been identified, autonomic dysfunction is also part of the natural history of cardiovascular diseases, where different functional methods show increased sympathetic activity and reduced vagal modulation. Therefore, in patients with cardiovascular diseases, ACh is involved in the enhanced ganglionic transmission in the sympathetic chain and in the diminished release and action at the parasympathetic ganglion and postganglionic neuron–end-organ synaptic cleft.

It is well known that chronic activation of the sympathetic nervous system is a pathophysiological feature shared by many cardiovascular diseases such as arterial hypertension, heart failure and myocardial ischemia, contributing to cardiovascular remodeling, including left ventricular and vascular smooth muscle hypertrophy, as well as to the occurrence of life-threatening ventricular tachyarrhythmias and sudden death [59, 60]. This concept supported the clinical use of B-adrenoceptor antagonists to treat hypertension and myocardial ischemia and to blunt the progressive vicious cycle of neurohumoral activation seen in heart failure leading to reduced mortality. On the other hand, the clinical impact of the decreased cholinergic action on myocytes had been underestimated until the 1980s when the usage of indices of HRV showed that diminished parasympathetic modulation represents an independent and powerful risk factor for adverse cardiovascular events, including life-threatening ventricular arrhythmias and sudden death after myocardial infarction and heart failure [61, 62]. Currently, vagal dysfunction is an established risk factor for cardiovascular disease and mortality [51]. Accordingly, cholinergic stimulation protects against ventricular arrhythmias in experimental models employing either electrical stimulation of the vagus nerve [63] or injection of muscarinic agonists such as methacholine and oxotremorine [64]. In humans, vagal activation is known to interrupt ventricular tachycardia [65] and cholinomimetic drugs may have protective effects in patients with cardiovascular diseases [66, 67]. The development of a pharmacological treatment for parasympathetic dysfunction should consider the potential occurrence of undesired effects of direct stimulation of mAChRs on the heart as produced by methacholine (i.e. profound bradycardia, atrioventricular blockade and sinus arrest). In addition, excessive vagal activity may cause atrial fibrillation in susceptible patients [68]. Therefore, indirect cholinergic stimulation seems to be more appropriate for clinical treatment of the impaired parasympathetic modulation in humans.

Pyridostigmine bromide ($C_9H_{13}N_2O_2$) is a synthetic compound derived from carbamic acid with reversible anticholinesterase action and a dose-dependent cholinomimetic effect that is widely used in myasthenia gravis to counteract muscle weakness. As a quaternary amine, it can be orally administered, but is poorly absorbed (bioavailability approximately 8%) and does not cross the blood–brain barrier [69]. Considering that cholinesterase inhibitors increase the availability of ACh by inhibiting hydrolysis on autonomic ganglia, including the adrenergic branch, it is presumed that pyridostigmine enhances sympathetic ganglionic neurotransmission leading to increased vascular adrenergic tone. This is likely the mechanism responsible for pyridostigmine being able to prevent the fall in blood pressure in patients with orthostatic hypotension [70]. In addition to this effect, pyridostigmine also causes a vagal shift in cardiac sympathovagal balance, since both preganglionic and postganglionic neurons in the parasympathetic pathway are cholinergic and subjected to cholinesterase inhibition. In the heart, pyridostigmine acts on cardiac ganglion cells increasing the concentration of endogenous ACh, thus stimulating cardiac M2 mAChRs. This vagal neurocardiovascular effect of pyridostigmine causes predictable changes in markers of autonomic control of circulation. In healthy volunteers, single doses of pyridostigmine reduce resting heart rate [71] and promote homogenous ventricular repolarization, as shown by a reduction in corrected QT dispersion on the surface electrocardiogram [72], without showing any negative effect on left ventricular systolic or diastolic function [73]. In addition, the short-term administration of successive doses of pyridostigmine to healthy volunteers reduces mean heart rate and enhances 24-h HRV [74], mostly due to an increase in heart rate dynamic range of variation toward lower heart rates, consistent with a parasympathetic effect.

Mental stress and physical effort occur throughout our lives, both causing complex organic adjustments related to the well-known "fight or flight" response. The autonomic changes caused by mental and physical stress lead to increases in heart rate and blood pressure – a combination that causes augmented ventricular workload and oxygen demand, and may trigger myocardial ischemia and ventricular arrhythmias in susceptible patients [75]. Therefore, a pharmacological alternative acting on autonomic modulation of the cardiovascular system should consider its effect on the hemodynamic changes observed during these provocative situations. Accordingly, in a rat model of central activation of adrenergic drive, pyridostigmine blunted the increase in myocardial oxygen uptake demand [76]. In healthy volunteers, pyridostigmine inhibited the increase in rate–pressure product during mental stress [77] and blunted the chronotropic response to dynamic exercise without reducing functional capacity [78]. The administration of pyridostigmine to patients with cardiovascular diseases has also shown its potential as a therapeutic option. The short-term administration reduced ventricular arrhythmia in patients with heart failure, who exhibited a consistent 65% reduction of the density of isolated ventricular premature beats, despite the use of digoxin by all and B-adrenoceptor antagonists or amiodarone by some of the patients [79]. The double-blind, placebo-controlled, crossover design of the study made very unlikely the possibility that spontaneous variability of arrhythmia density could explain the results. The increase in vagal tone with pyridostigmine in patients with heart failure also improves heart rate recovery after maximal exercise [80] that is a well-known marker of worse prognosis [81]. It is known that vagal activation increases the ventricular fibrillation threshold, protects against reperfusion arrhythmias and may even terminate ventricular tachycardia [63, 65]. Therefore, cholinergic stimulation with pyridostigmine might induce prolongation of ventricular refractoriness by its direct effect or by antagonizing sympathetic effects on refractoriness [82]. In patients with coronary artery disease, cholinergic stimulation with pyridostigmine prevented the expected decrease in ejection fraction and the impairment in diastolic ventricular function (Figure 17.2) provoked by mental stress [83]. This protective effect does not seem to involve a lower afterload, since the chronotropic and pressor responses to mental stress were similar to the placebo condition. In addition, myocardial ischemia elicited by mental stress is known to occur at lower double-product as compared to exercise, suggesting that reductions in coronary blood supply, but not increased myocardial oxygen demand, is the major mechanism of mental stress-induced ischemia [84]. Reduced coronary dilation is probably caused by impaired endothelium-dependent, NO-mediated smooth muscle vasodilation via activation of ET_A receptors [85]. Since cholinergic stimulation is a well-known promoter of NO release from the endothelium [86], it is likely that indirect cholinergic stimulation with pyridostigmine limits coronary vasoconstriction (i.e. relative vasodilation) via enhanced NO release during mental stress, thus preventing myocardial ischemia and ventricular dysfunction. In addition, pyridostigmine may improve diastolic function by a direct action since NO has been shown to increase ventricular distensibility [87].

The vagal shift of autonomic balance caused by pyridostigmine lowers the heart rate response to progressive dynamic exercise, delaying the increase in myocardial oxygen demand and thus the onset of ischemia in patients with coronary artery disease [88]. Since the ischemic threshold was not altered, the protective effect of

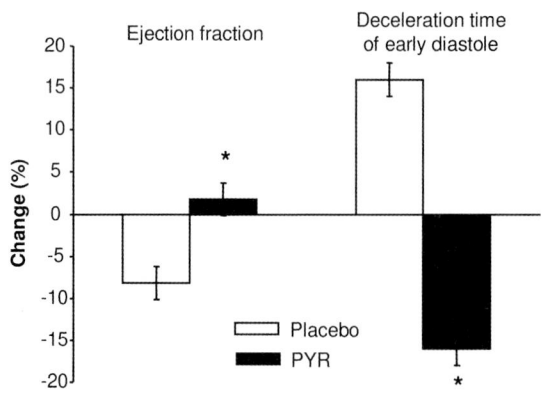

Figure 17.2 Cholinergic stimulation with pyridostigmine (PYR) prevents the impairment in left ventricular systolic and diastolic functions provoked by mental stress in patients with coronary artery disease. Mental stress decreased ejection fraction (left pair of bars) and increased deceleration time of early diastole (right pair of bars), whereas pyridostigmine preserved myocardial inotropism and improved diastolic function, as shown by decreased deceleration time of early diastole. *$P < 0.05$ versus placebo. (Modified from [91]).

pyridostigmine was not caused by changes in oxygen supply–demand balance, but it was rather an indirect consequence of negative chronotropic effect of cholinergic stimulation. Pyridostigmine is also effective in reducing the QTc interval on the electrocardiogram at rest [89] and in the recovery phase from maximal exercise in patients with coronary heart disease [68] that may denote a reduction in residual myocardial ischemia [90].

17.4
ACh and Vascular Function

The observation by Furchgott and Zawadzki [86] that ACh-induced vascular relaxation requires an intact endothelium has caused increasing interest in the role of ACh and the muscarinic receptors in this process. The contraction/relaxation effects of ACh (in vascular smooth muscle preparations) are now known to be associated with the activation of receptors present on the smooth muscle as well as on the endothelial cells [86].

The vasodilator response to ACh is assumed to be mediated, at least in part, by mAChRs-induced release of endothelium-derived relaxing factor, now known to be NO [91]. In the human forearm, a well-established *in vivo* model for peripheral resistance vasculature, ACh-induced vasodilation has also been shown to depend on NO release [92, 93]. In various conditions such as systemic arterial hypertension [94], menopause [95], heart failure [96], hypercholesterolemia [97], atherosclerosis [98] and insulin-dependent diabetes [99], the vascular responses to cholinergic agonists (ACh and methacoline) in this particular vascular bed have been shown to be impaired.

As previously stated, mAChRs form a heterogeneous population and mediate diverse effects on the vasculature [99]. mAChR subtypes involved in ACh-induced vasodilatation were extensively studied in arteries. In several studies using animal models, the ACh-induced relaxation in different segments of arterial bed was mediated by the M_3 receptor subtype localized on the endothelium [100–102]. Similar results were also described *in vivo* in the human forearm vasculature [103]. Nevertheless, pharmacologic and physiologic studies concluded that all three major subtypes of mAChRs (M_1, M_2 and M_3) are present in the vasculature, and elicit different effects on vessels (contraction or dilation) [104].

Few reports characterize the endothelial mAChRs in human veins. It has been suggested that muscarinic receptors are present on cultured human hand venous endothelial cells [105]. Vasodilation mediated by endothelial mAChRs was suggested in human venous vasculature in the forearm, in human isolated saphenous veins [106, 107] and in dorsal hand veins [108–110]. Furthermore, functional studies were performed using human isolated pulmonary veins. Few reports addressed the muscarinic receptor subtypes involved in the ACh-induced relaxation. A study performed on rabbit external jugular veins suggests that M_1 and/or M_3 receptors are involved in the ACh-induced relaxation [111], and the data obtained in human pulmonary veins suggests that ACh-induced endothelium-dependent relaxation was via activation of muscarinic M1 mAChRs [112]. The functional role of vascular mAChRs is still a matter of debate. Recently, the "non-neuronal cholinergic system" concept has brought important insights about the role of ACh in different tissues apart from the nervous system. Indeed, the major essential elements of cholinergic system have been demonstrated in various human tissues, including endothelial cells [113]. The possible functional consequences of the presence of endothelial ACh that may affect the local release of NO should be considered in local regulatory loop mechanisms of vascular tissue [113].

It is time to revise the biological role of Ach, expanding studies from neuronal cholinergic systems to local regulatory systems to understand how it contributes to cell and organ homeostasis.

References

1 Horiuchi, Y., Kimura, R., Kato, N., Fuji, T., Maseko, S., Endo, T., Kato, T. and Kawashima, K. (2003) Evolutional study on acetylcholine expression. *Life Sciences*, **72**, 1745–56.

2 Williams, B.P., Milligan, C.J., Street, M., Hornby, F.M., Deuchars, J. and Buckley, N.J. (2004) Transcription of the M_1 muscarinic receptor gene in neurons and neural progenitors of the embryonic rat forebrain. *Journal of Neurochemistry*, **88**, 70–7.

3 Wessler, I., Kilbinger, H., Bittinger, F., Unger, R. and Kilpatrick, C.J. (2003) The non-neuronal cholinergic system in humans: expression, function and patho-physiology. *Life Sciences*, **72**, 2055–61.

4 Hulme, E.C., Birdsall, N.J.M. and Buckley, N.J. (1990) Muscarinic receptor subtypes. *Annual Review of Pharmacology and Toxicology*, **30**, 633–73.

5 Caulfield, M.P. and Birdsall, N.J.M. (1998) International Union of Pharmacology. XV II. Classification of muscarinic acetylcholine receptors. *Pharmacological Reviews*, **50**, 279–90.

6 Wess, J. (2004) Muscarinic acetylcholine receptor knockout mice: novel phenotypes and clinical implications. *Annual Review of Pharmacology and Toxicology*, **44**, 423–50.

7 Eglen, R.M. (2006) Muscarinic receptor subtypes in neuronal and non-neuronal cholinergic function. *Autonomic and Autacoid Pharmacology*, **26**, 219–33.

8 Kubo, T., Fukuda, K., Mikami, A., Maeda, A., Takahashi, H., Mishina, M., Haga, T., Haga, K., Iciyama, A., Kangawa, K., Kojima, M., Matsuo, H., Hirose, T. and Numa, S. (1986) Cloning, sequencing and expression of complementary DNA encoding the muscarinic acetylcholine receptor. *Nature*, **323**, 411–16.

9 Christopoulos, A., Lanzafame, A. and Mitchelson, F. (1998) Allosteric interactions at muscarinic cholinoceptors. *Clinical and Experimental Pharmacology and Physiology*, **25**, 185–94.

10 Birdsall, N.J.M. and Lazareno, S. (2005) Allosterism at muscarinic receptors: ligands and mechanisms. *Mini Reviews in Medicinal Chemistry*, **5**, 523–43.

11 Lanzafame, A., Christopoulos, A. and Mitchelson, F. (2003) Cellular signalling mechanisms for muscarinic acetylcholine receptors. *Receptors and Channels*, **9**, 241–60.

12 Zholos, A.V., Zholos, A.A. and Bolton, T.B. (2004) G-protein-gated TRP-like cationic channel activated by muscarinic receptors: effect of potential on single-channel gating. *Journal of General Physiology*, **123**, 581–98.

13 Van Koppen, C.J. and Kaiser, B. (2003) Regulation of muscarinic acetylcholine receptor signalling. *Pharmacology and Therapeutics*, **98**, 197–220.

14 Wang, H., Lu, Y. and Wang, Z. (2007) Function of cardiac M_3 receptors. *Autonomic and Autacoid Pharmacology*, **27**, 1–11.

15 Longo, F.M. and Massa, S.M. (2004) Neuroprotective strategies in Alzheimer's disease. *NeuroRx*, **1**, 117–27.

16 Ehlert, F.J., Hsu, J.C., Leung, K., Lee, A.G., Shehnaz, D. and Griffin, M.T.J. (2005) Comparison of the antimuscarinic action of *p*-fluorohexahydrosiladifenidol in ileal and tracheal smooth muscle. *Pharmacology and Experimental Therapeutics*, **312**, 592–600.

17 Wess, J., Duttaroy, A., Zhang, W., Gomeza, J., Cui, Y., Miyakawa, T., Bymaster, F.P., McKinzie, L., Felder, C.C., Lamping, K.G., Faraci, F.M., Deng,C. and Yamada, M. (2003) M1–M5 muscarinic receptor knockout mice as novel tools to study the physiological roles of the muscarinic cholinergic system. *Receptor Channels*, **9**, 279–90.

18 Yamada, M., Miyakaw, T., Duttaroy, A., Yamanaka, A., Moriguchi, T., Makita, R., Mckinzie, D.L., Felder, C.C., Deng, C.X., Faraci, F.M. and Wess, J. (2001) Mice lacking the M_3 muscarinic acetylcholine receptor are hypophagic and lean. *Nature*, **410**, 207–12.

19 Eglen, R.M., Hegde, S.S. and Watson, N. (1996) Muscarinic receptors and smooth muscle function. *Pharmacological Reviews*, **48**, 531–65.

20 Siu, E.R., Yasuhara, F., Marostica, E., Avellar, M.C.W. and Porto, C.S. (2006) Expression and localization of muscarinic acetylcholine receptor subtypes in rat efferent ductules and epididymis. *Cell and Tissue Research*, 157–66.

21 Hamamura, M., Marostica, E., Avellar, M.C.W. and Porto, C.S. (2006) Muscarinic acetylcholine receptor subtypes in the rat seminal vesicle. *Molecular and Cellular Endocrinology*, 192–8.

22 Zhang, H.M., Li, D.P., Chen, S.R. and Pan, H.L. (2005) M_2, M_3, and M_4 receptor subtypes contribute to muscarinic potentiation of GABAergic inputs to spinal dorsal horn neurons. *Journal of Pharmacology and Experimental Therapeutics*, **313**, 697–704.

23 Eglen, R.M. and Nahorski, S.R. (2000) The muscarinic M_5 receptor: a silent or emerging subtype? *British Journal of Pharmacology*, **130**, 13–21.

24 Elhusseiny, A. and Hamel Cereb, E. (2000) Muscarinic–but not nicotinic–acetylcholine receptors mediate a nitric oxide-dependent dilation in brain cortical arterioles: a possible role for the M_5 receptor subtype. *Blood Flow and Metabolism*, **20**, 298–305.

25 Löffelholz, K. and Pappano, A.J. (1985) The parasympathetic neuroeffector junction of the heart. *Pharmacological Reviews*, **37**, 1–24.

26 Dhein, S., Van Koppen, C.J. and Brodde, O.E. (2001) Muscarinic receptors in the mammalian heart. *Pharmacological Reviews*, **44**, 161–82.

27 Caulfield, M.P. (1993) Muscarinic receptors–characterization, coupling and function. *Pharmacology and Therapeutics*, **58**, 319–79.

28 Stengel, P.W., Gomeza, J., Wess, J. and Cohen, M.L. (2000) M_2 and M_4 receptor knockout mice: muscarinic receptor function in cardiac and smooth muscle *in vitro*. *Journal of Pharmacology and Experimental Therapeutics*, **292**, 877–85.

29 Wang, H., Han, H., Zhang, L., Shi, H., Schiram, G., Nattel, S. and Wang, Z. (2001) Expression of multiple subtypes of muscarinic receptors and cellular distribution in the human heart. *Molecular Pharmacology*, **59**, 1029–36.

30 Islam, M.A., Nojima, H. and Kimura, I. (1998) Muscarinic M_1 receptor activation reduces maximum upstroke velocity of action potential in mouse right atria. *European Journal of Pharmacology*, **346**, 227–36.

31 Colecraft, H.M., Egamino, J.P., Sharma, V.K. and Sheu, S.S. (1998) Signaling mechanisms underlying muscarinic receptor-mediated increase in contraction rate in cultured heart cells. *Journal of Biological Chemistry*, **273**, 32158–66.

32 Dobrev, D., Knuschke, D., Richter, F., Wettwer, E., Christ, T., Knaut, M. and Ravens, U. (2003) Functional identification of M_1 and M_3 muscarinic acetylcholine receptors in human atrial myocytes: influence of chronic atrial fibrillation [abstract]. *Naunyn-Schmiedeberg's Archives of Pharmacology*, **367** (Suppl.1), R94.

33 Wang, Z., Shi, H. and Wang, H. (2004) Functional M_3 muscarinic acetylcholine receptors in mammalian hearts. *British Journal of Pharmacology*, **142**, 395–408.

34 Hardouin, S.N., Richmond, K.N., Zimmerman, A., Hamilton, S.E., Feigl, E.O. and Nathanson, N.M. (2002) Altered cardiovascular responses in mice lacking the M_1 muscarinic acetylcholine receptor. *Journal of Pharmacology and Experimental Therapeutics*, **301**, 129–37.

35 Trendelenburg, A.U., Gomeza, J., Klebroff, W., Zhou, H. and Wess, J. (2003) Heterogeneity of presynaptic muscarinic receptors mediating inhibition of sympathetic transmitter release: a study with M_2- and M_4-receptor-deficient mice. *British Journal of Pharmacology*, **138**, 469–80.

36 Pönicke, K., Heinroth-Hoffmann, I. and Brodde, O.E. (2003) Demonstration of functional M_3-muscarinic receptors in ventricular cardiomyocytes of adult rats. *British Journal of Pharmacology*, **138**, 156–60.

37 Stengel, P.W., Yamada, M., Wess, J. and Cohen, M.L. (2002) M_3-receptor knockout mice: muscarinic receptor function in atria, stomach fundus, urinary bladder, and trachea. *American Journal of Physiology – Regulatory, Integrative and Comparative*, **282**, R1443–9.

38 Böhm, M., Gierschik, P., Jakobs, K.H., Pieske, B., Schnabel, P., Ungerer, M. and Erdmann, E. (1990) Increase of G_{ia} in human hearts with dilated but not ischemic cardiomyopathy. *Circulation*, **82**, 1249–65.

39 Giessler, C., Dhein, S., Pönicke, K. and Brodde, O.E. (1999) Muscarinic receptors in the failing human heart. *European Journal of Pharmacology*, **375**, 197–202.

40 Le Guludec, D., Cohen-Solal, A., Delforge, J., Delahaye, N., Syrota, A. and Merlet, P. (1997) Increased myocardial muscarinic receptor density in idiopathic dilated cardiomyopathy. An *in vivo* PET study. *Circulation*, **96**, 3416–22.

41 Niihashi, M., Esumi, M., Kusumi, Y., Sato, Y. and Sakurai, I. (2000) Expression of muscarinic receptor genes in the human coronary artery. *Angiology*, **51**, 295–300.

42 Lamping, K.G., Wess, J., Cui, Y., Nuno, D.W. and Faraci, F.M. (2004) Muscarinic (M) receptors in coronary circulation: gene-targeted mice define the role of M_2 and M_3 receptors in response to acetylcholine. *Arteriosclerosis Thrombosis and Vascular Biology*, **24**, 1253–8.

43 Herring, N., Danson, E.J. and Paterson, D.J. (2002) Cholinergic control of heart rate by nitric oxide is site specific. *News in Physiological Sciences*, **17**, 202–6.

44 Herring, N. and Paterson, D.J. (2001) Nitric oxide–cGMP pathway facilitates acetylcholine release and bradycardia during vagal nerve stimulation in the guinea-pig *in vitro*. *Journal of Physiology*, **535**, 507–18.

45 Choate, J.K., Danson, E.J., Morris, J.F. and Paterson, D.J. (2001) Peripheral vagal control of heart rate is impaired in neuronal NOS knockout mice. *American Journal of Physiology – Heart and Circulatory Physiology*, **281**, H2310–17.

46 Danson, E.F. and Paterson, D.J. (2003) Enhanced neuronal nitric oxide synthase expression is central to cardiac vagal phenotype in exercise-trained mice. *Journal of Physiology*, **546**, 225–32.

47 Brodde, O.E. and Michel, M.C. (1999) Adrenergic and muscarinic receptors in the human heart. *Pharmacological Reviews*, **51**, 651–90.

48 Gehrmann, J., Meister, M., Maguire, C.T., Martins, D.C., Hammer, P.E., Neer, E.J., Berul, C.I. and Mende, U. (2002) Impaired parasympathetic heart rate control in mice with reduction of functional G protein βγ-subunits. *American Journal of Physiology – Heart and Circulatory Physiology*, **282**, H455–6.

49 Malliani, A. (1999) The pattern of sympathovagal balance explored in the

frequency domain. *News in Physiological Sciences*, **14**, 111–17.
50 Ally, A., Wilson, L.B., Nobrega, A.C. and Mitchell, J.H. (1995) Cardiovascular effects elicited by central administration of physostigmine via M_2 muscarinic receptors in conscious cats. *Brain Research*, **677**, 268–76.
51 Thayer, J.F. and Lane, R.D. (2007) The role of vagal function in the risk for cardiovascular disease and mortality. *Biological Psychology*, **74**, 224–42.
52 Dall'ago, P., Agord Schaan, B.D., da Silva, V.O., Werner, J., da Silva Soares, P.P., da Angelis, K. and Irigoyen, M.C. (2007) Parasympathetic dysfunction is associated with baroreflex and chemoreflex impairment in streptozotocin-induced diabetes in rats. *Autonomic Neuroscience: Basic and Clinical*, **131**, 28–35.
53 Farah, V.M., Moreira, E.D., Pires, M.D., Irigoyen, M.C. and Krieger, E.M. (1999) Comparison of three methods for the determination of baroreflex sensitivity in conscious rats. *Brazilian Journal of Medical and Biological Research*, **32**, 361–9.
54 La Rovere, M.T., Pinna, G.D., Hohnloser, S.H., Marcus, F.I., Mortra, A., Nohara, R., Bigger, J.T., Camm, A.J., Schwartz, P.J. and ATRAMI Investigators (2001) Autonomic tone and reflexes after myocardial infarction. Baroreflex sensitivity and heart rate variability in the identification of patients at risk for life-threatening arrhythmias: Implications for clinical trials. *Circulation*, **103**, 2072–7.
55 La Rovere, M.T., Pinna, G.D., Maestri, R., Mortara, R., Mortara, A., Capomolla, S., Febo, O., Ferrari, R., Franchini, M., Gnemmi, M., Opasich, C., Riccardi, P.G., Traversi, E. and Cobelli, F. (2003) Short-term heart rate variability strongly predicts sudden cardiac death in chronic heart failure patients. *Circulation*, **107**, 565–70.
56 Soares, P.P., Moreno, A.M., Cravo, S.L. and Nóbrega, A.C. (2005) Coronary artery bypass surgery and longitudinal evaluation of the autonomic cardiovascular function. *Critical Care*, **9**, R124–131.

57 Montano, N., Cogliati, C., Porta, A., Pagani, M., Malliani, A., Narkiewiez, K., Abboud, F.M., Birkett, C. and Somers, V.K. (1998) Central vagotonic effects of atropine modulate spectral oscillations of simpathetic nerve activity. *Circulation*, **98**, 1394–9.
58 Soares, P.P., da Nobrega, A.C., Ushima, M.R. and Irigoyen, M.C. (2004) Cholinergic stimulation with pyridostigmine increases heart rate variability and baroreflex sensitivity in rats. *Autonomic Neuroscience: Basic and Clinical*, **113**, 24–31.
59 La Rovere, M.T., Bigger, J.T., Marcus, F.I., Mortara, A. and Schwartz, P.J. (1998) Baroreflex sensitivity and heart rate variability in prediction of total cardiac mortality after myocardial infarction. *Lancet*, **351**, 478–84.
60 Schlaich, M.P., Kaye, D.M., Lambert, E., Sommerville, M., Socratous, F. and Esler, M.D. (2003) Relation between cardiac sympathetic activity and hypertensive left ventricular hypertrophy. *Circulation*, **108**, 560–5.
61 Kleiger, R.E., Miller, J.P., Bigger, J.T. and Moss, A.J. (1987) Decreased heart rate variability and its association with increased mortality after acute myocardial infarction. *American Journal of Cardiology*, **59**, 256–62.
62 Bonaduce, D., Petretta, M., Marciano, F., Vicario, M.L., Apicella, C., Rao, M.A., Nicolai, E. and Volpe, M. (1999) Independent and incremental prognostic value of heart rate variability in patients with chronic heart failure. *American Heart Journal*, **138**, 273–84.
63 Kent, K.M., Smith, E.R., Redwood, D.R. and Epstein, S.E. (1973) Electrical stability of acutely ischemic myocardium: influences of heart rate and vagal stimulation. *Circulation*, **47**, 291–8.
64 De Ferrari, G.M., Salvati, P., Grossoni, M., Ukmar, G., Vaga, L., Patrono, C. and Schwartz, P.J. (1993) Pharmacologic modulation of the autonomic nervous system in the prevention of sudden cardiac death: a study with propranolol, methacoline and oxotremorine in conscious dogs with a healed myocardial infarction. *Journal of the American College of Cardiology*, **22**, 283–90.

65 Waxman, M.B. and Wald, R.W. (1977) Termination of ventricular tachycardia by increase in cardiac vagal drive. *Circulation*, **56**, 385–91.

66 Sneddon, J.F., Bashir, Y. and Ward, D.E. (1993) Vagal stimulation after myocardial infarction: accentuating the positive. *Journal of the American College of Cardiology*, **22**, 1335–7.

67 Nobrega, A.C.L. and Castro, R.R.T. (2000) Parasympathetic dysfunction as a risk factor in myocardial infarction: what is the treatment? *American Heart Journal*, **140**, E23.

68 Castro, R.R., Mesquita, E.T. and Nobrega, A.C. (2006) Parasympathetic-mediated atrial fibrillation during tilt test associated with increased baroreflex sensitivity. *Europace*, **8**, 349–51.

69 Taylor, P. (2001) Anticholinesterase agents, in *Goodman and Gilman's Pharmacological Basis of Therapeutics* (eds J.G. Hardman, L.E. Limbird, P.B. Molinoff, R.W. Ruddon and A.F. Gilman), McGraw-Hill, New York, pp. 175–92.

70 Singer, W., Opfer-Gehrking, T.L., Nickander, K.K., Hines, S.M. and Low, P.A. (2006) Acetylcholinesterase inhibition in patients with orthostatic intolerance. *Journal of Clinical Neurophysiology*, **23**, 477–82.

71 Nobrega, A.C.L., Carvalho, A.C. and Bastos, B.G. (1996) Resting and reflex heart rate responses during cholinergic stimulation with pyridostigmine in humans. *Brazilian Journal of Medical and Biological Research*, **29**, 1461–5.

72 Castro, R.R., Serra, S.M. and Nobrega, A.C. (2000) Reduction of QTc interval dispersion. Potential mechanism of cardiac protection of pyridostigmine bromide. *Arquivos Brasileiros de Cardiologia*, **75**, 205–13.

73 Pontes, P.V., Bastos, B.G., Romeo Filho, L.J., Mesquita, E.T. and da Nobrega, A.C. (1999) Cholinergic stimulation with pyridostigmine, hemodynamic and echocardiographic analysis in healthy subjects. *Arquivos Brasileiros de Cardiologia*, **72**, 297–306.

74 Nobrega, A.C., dos Reis, A.F., Moraes, R.S., Bastos, B.G., Ferlin, E.L. and Ribeiro, J.P. (2001) Enhancement of heart rate variability by cholinergic stimulation with pyridostigmine in healthy subjects. *Clinical Autonomic Research*, **11**, 11–17.

75 Sheps, D.S., McMahon, R.P., Becker, L., Carney, R.M., Freedland, K.E., Cohen, J.D., Sheffield, D., Goldberg, A.D., Ketterer, M.W., Pepine, C.J., Raczynski, J.M., Light, K., Krantz, D.S., Stone, P.H., Knatterud, G.L. and Kaufmann, P.G. (2002) Mental stress-induced ischemia and all-cause mortality in patients with coronary artery disease: results from the Psychophysiological Investigations of Myocardial Ischemia study. *Circulation*, **105**, 1780–4.

76 Grabe-Guimarães, A., Alves, L.M., Tibirica, E. and Nobrega, A.C. (1999) Pyridostigmine blunts the increases in myocardial oxygen demand elicited by the stimulation of the central nervous system in anesthetized rats. *Clinical Autonomic Research*, **9**, 83–9.

77 Nobrega, A.C., Carvalho, A.C., Santos, K.B. and Soares, P.P. (1999) Cholinergic stimulation with pyridostigmine blunts the cardiac responses to mental stress. *Clinical Autonomic Research*, **9**, 1–6.

78 Serra, S.M., Costa, R.V., Bastos, B.G., Bousquet Santos, K., Ramalho, S.H. and da Nobrega, A.C. (2001) Cardiopulmonary exercise testing during cholinergic stimulation with single dose of pyridostigmine in healthy subjects. *Arquivos Brasileiros de Cardiologia*, **76**, 279–84.

79 Behling, A., Moraes, R.S., Rohde, L.E., Ferlin, E.L., Nobrega, A.C. and Ribeiro, J.P. (2003) Cholinergic stimulation with pyridostigmine reduces ventricular arrhythmia and enhances heart rate variability in heart failure. *American Heart Journal*, **146**, 494–500.

80 Androne, A.S., Hryniewicz, K., Goldsmith, R., Arwady, A. and Kats, S.D. (2003) Acetylcholinesterase inhibition with pyridostigmine improves heart rate recovery after maximal exercise in patients with chronic heart failure. *Heart*, **89**, 854–8.

81 Nishime, E.O., Cole, C.R., Blackstone, E.H., Pashkow, F.J. and Lauer, M.S. (2000) Heart rate recovery and treadmill exercise score as predictors of mortality

in patients referred for exercise ECG. *JAMA*, **284**, 1392–8.
82. Warner, M.R. and Zipes, D.P. (1994) Vagal control of myocardial refractoriness, in *Vagal Control of the Heart: Experimental Basis and Clinical Implications* (eds M.N. Levy and P.J. Schwartz), Futura Publishing, Armonk, NY, pp. 261–75.
83. Nobrega, A.C.L., Loures, D.L., Pontes, P.V., Sant'Anna, I.D. and Mesquita, E.T. (2008) Cholinergic stimulation with pyridostigmine prevents the impairment in ventricular function during mental stress in coronary artery disease patients. *International Journal of Cardiology*, **129**, 418–21.
84. Schöder, H., Silverman, D.H., Campisi, R., Sayre, J.W., Phelps, M.E., Schelbert, H.R. and Czernin, J. (2000) Regulation of myocardial blood flow response to mental stress in healthy individuals. *American Journal of Physiology – Heart and Circulatory Physiology*, **278**, 360–6.
85. Spieker, L.E., Hürlimann, D., Ruschitzka, F., Corti, R., Enseleit, F., Shaw, S., Hayoz, D., Deanfield, J.E., Lüscher, T.F. and Noll, G. (2002) Mental stress induces prolonged endothelial dysfunction via endothelin-A receptors. *Circulation*, **105**, 2817–20.
86. Furchgott, R.F. and Zawadzki, J.V. (1980) The obligatory role of the endothelial cells in the relaxation of arterial smooth muscle by acetylcholine. *Nature*, **288**, 373–6.
87. Paulus, W.J. and Shah, A.M. (1999) NO and cardiac diastolic function. *Cardiovascular Research*, **43**, 595–606.
88. Castro, R.R.T., Porphirio, G., Serra, S.M. and Nobrega, A.C.L. (2004) Cholinergic stimulation with pyridostigmine protects against exercise induced myocardial ischaemia. *Heart*, **90**, 1119–23.
89. Castro, R.R.T., Porphyrio, G., Serra, S.M. and Nobrega, A.C.L. (2002) Cholinergic stimulation with pyridostigmine reduces QTc interval in coronary artery disease. *Brazilian Journal of Medical and Biological Research*, **35**, 685–9.
90. Macieira-Coelho, E., Monteiro, F., da Conceição, J.M., Cunha, D., Cruz, J., Almeida, A. and da Sousa, T. (1983) Post-exercise changes of the Q–Tc interval in coronary heart disease. *Journal of Electrocardiology*, **16**, 345–9.
91. Palmer, R.M.L., Ferrige, A.G. and Moncada, S. (1987) Nitric oxide release accounts for the biological activity of endothelium-derived relaxing factor. *Nature*, **327**, 524–6.
92. Vallance, P., Collier, J. and Moncada, S. (1989) Effects of endothelium-derived nitric oxide on peripheral arteriolar tone in man. *Lancet*, **2**, 997–1000.
93. Bruning, T.A., Chang, P.C., Blauw, G.J., Vermeij, P. and VanZwieten, P.A. (1993) Serotonin-induced vasodilatation in the human forearm is mediated by The "nitric oxide-pathway": no evidence for involvement of the 5-HT$_3$-receptor. *Journal of Cardiovascular Pharmacology*, **22**, 44–51.
94. Panza, J.A., Quyyumi, A.A., Brush, J.E. Jr and Epstein, S.E. (1990) Abnormal endothelium-dependent vascular relaxation in patients with essential hypertension. *New England Journal of Medicine*, **323**, 22–7.
95. Lima, S.M., Aldrighi, J.M., Consolim-Colombo, F.M., Mansur Ade, P., Rubira, M.C., Krieger, E.M. and Ramires, J.A. (2005) Acute administration of 17beta-estradiol improves endothelium-dependent vasodilation in postmenopausal women. *Maturitas*, **50**, 266–74.
96. Katz, S.D., Schwarz, M., Yuen, J. and LeJemtel, T.H. (1993) Impaired acetylcholine-mediated vasodilation in patients with congestive heart failure. *Circulation*, **88**, 55–61.
97. Casino, P.R., Kilcoyne, C.M., Quyyumi, A.A., Hoeg, J.M. and Panza, J.A. (1993) The role of nitric oxide in endothelium-dependent vasodilation of hypercholesterolemic patients. *Circulation*, **88**, 2541–7.
98. Liao, J.K., Bettman, M.A., Sandor, T., Tucker, J.I., Coleman, S.M. and Creager, M.A. (1991) Differential impairment of vasodilator responsiveness of peripheral resistance and conduit vessels in humans

with atherosclerosis. *Circulation Research*, **68**, 1027–34.
99 Eglen, R.M. and Whiting, R.L. (1985) Determination of the muscarinic receptor subtype mediating vasodilatation. *British Journal of Pharmacology*, **84**, 3–5.
100 McCormack, D.G., Mak, J.C., Minette, P. and Barnes, P.J. (1988) Muscarinic receptor subtypes mediating vasodilation in the pulmonary artery. *European Journal of Pharmacology*, **158**, 293–7.
101 Hendriks, M.C.G., Pfaffendorf, M. and VanZwieten, P.A. (1992) Characterization of the muscarinic receptor subtype mediating vasodilation in the rat perfused mesenteric vascular bed preparation. *Journal of Autonomic Pharmacology*, **12**, 411–20.
102 Dauphin, F. and Hamel, E. (1990) Muscarinic receptor subtype mediating vasodilation in feline middle cerebral artery exhibits M_3 pharmacology. *European Journal of Pharmacology*, **178**, 203–13.
103 Bruning, T.A., Hendriks, M.G.C., Chang, P.C., Kuypers, E.A.P. and Van Zwieten, P.A. (1994) In vivo characterization of vasodilating muscarinic-receptor subtypes in humans. *Circulation Research*, **74**, 912–19.
104 Doods, H.N., Mathy, M.J., Davidesko, D., Van Charldorp, K.J., De Jonge, A. and VanZwieten, P.A. (1987) Selectivity of muscarinic antagonists in radioligand and *in vivo* experiments for the putative M_1, M_2 and M_3 receptors. *Journal of Pharmacology and Experimental Therapeutics*, **242**, 257–62.
105 Mahdy, Z., Otun, H.A., Dunlop, W. and Gillespie, I. (1998) The responsiveness of isolated human hand vein endothelial cells in normal pregnancy and in pre-eclampsia. *Journal of Physiology (London)*, **508**, 609–17.
106 Kemme, M.J., Bruning, T.A., Chang, P.C. and VanZwieten, P.A. (1995) Cholinergic receptor-mediated responses in the arteriolar and venous vascular beds of the human firearm. *Blood Press*, **4**, 293–9.
107 Hamilton, C.A., Berg, G., Mcintyre, M., Mcphaden, A.R., Reid, J.L. and Dominiczak, A.F. (1997) Effects of nitric oxide and superoxide on relaxation in human artery and vein. *Atherosclerosis*, **133**, 77–86.
108 Collier, J. and Vallance, P. (1990) Biphasic response to acetylcholine in human veins *in vivo*: the role of the endothelium. *Clinical Science (London)*, **78**, 101–4..
109 Plentz, R.D., Irigoyen, M.C., Muller, A.S., Casarini, D.E., Rubira, M.C., Moreno Junior, H., Mady, C., Ianni, B.M., Krieger, E.M. and Consolim-Colombo, F. (2006) Venous endothelial dysfunction in Chagas' disease patients without heart failure. *Arquivos Brasileiros De Cardiologia*, **86**, 466–71.
110 Giribela, C.R., Rubira, M.C., de Melo, N.R., Plentz, R.D., de Angelis, K., Moreno, H. and Consolim-Colombo, F.M. (2007) Effect of a low-dose oral contraceptive on venous endothelial function in healthy young women: preliminary results. *Clinics*, **62**, 151–8.
111 Martin, G.R., Bolofo, M.L. and Giles, H. (1992) Inhibition of endothelium-dependent vasorelaxation by arginine analogues: a pharmacological analysis of agonist and tissue dependence. *British Journal of Pharmacology*, **105**, 643–52.
112 Walch, L., Taisne, C., Gascard, J.P., Nashashibi, N., Brink, C. and Norel, X. (1997) Cholinesterase activity in human pulmonary arteries and veins. *British Journal of Pharmacology*, **121**, 986–90.
113 Kirkpatrick, C.J., Bittinger, F., Nozadze, K. and Wessler, I. (2003) Expression and function of the non-neuronal cholinergic system in endothelial cells. *Life Sciences*, **72**, 2111–16.

Index

a

acetylcholine (Ach)   310, 407ff
– degradation   410
– effect on heart   410, 411
– functions   407
– heart rate   411–413
– – vascular function   417, 418
Addison's disease   23
adenine nucleotides and platelet aggregation   398
adenisone   374
adenylyl cyclase   259
adhesion molecule expression   305, 306
adrenal corticotropic hormone (ACTH)   3, 6, 7, 9, 10, 203
– ACTH-R   10
adrenal gland   3, 5–7
– hypertrophy   7
– immunoreactive concentration   170
– zona fasciculate   5–7, 10
– zona glomerulosa   5–7
– zona reticularia   6
adrenaline   233ff
– adrenergic receptors   241–244
– biosynthesis and degradation   234–241
– release   233
adrenergic receptors   241–244
– α-adrenergic receptors, expression   19
– $α_1$-adrenoceptors   242
– $α_2$-adrenoceptors   243
– β-adrenoceptors   244
adrenergic system and cardiac disease states   233
adrenomedullin (AM)   169ff
– angiogenesis   179, 180
– cardiac contractility   172
– coronary vascular tone   173–175
– cytoprotection   177–179
– distribution and production   170
– expression   171
– heart failure   180–184
– myocardial ischemia   176–179
– receptors   171
– regulation of AM gene expression   170, 171
– structure   169, 170
– synthesis   175, 176
aging
– cognitive impairment   14
– dopamine   267
– renal function   267
– sex hormones   40, 41
albuminuria   152, 153
aldosterone   3–8, 73, 74, 134
– cardiovascular effects   16–21
– endothelial dysfunction   21
– increase   7, 8
– pathway   3, 4
– potassium depletion   9
– role in blood pressure homeostasis   6, 8
– synthase   5, 6
– synthesis   3–8
– vasodilation   21
aliskiren   8
allergy   308, 309
ALLHAT Trial   241
amino acid transporters   252–254
aminopeptidases   105
amnesia   85
amylin   169
andenylyl cyclase   259
androgens
– effect on lipoproteins   52, 53
– excess   6
– receptors   11
androstenedione   40
angiogenesis   112, 179, 180
– regulated by RAS   7, 8

Cardiovascular Hormone Systems. Edited by Michael Bader
Copyright © 2008 WILEY-VCH Verlag GmbH & Co. KGaA, Weinheim
ISBN: 978-3-527-31920-6

angiotensins 67ff
– Ang I 8, 67, 395
– Ang II 6–8, 19, 67, 69–72, 395
– – ARBS 71
– – blockade 76
– – cardiovascular system 72, 73
– – diabetes 73
– – endocrine system 73, 74
– – ovarian function 73
– – pregnancy 73
– – renal function 74–77
– Ang III 67, 85
– Ang IV 67, 85, 86
– Ang-(1–7) 68, 77–84
– – blood vessels and cardiac function 79–81
– – receptors 80, 82, 83
– – vasodilation 81–83
– Ang (1–9) 87
– Ang (1–12) 87
– And (2–10) 86
– Ang (3–7) 89
angiotensin converting enzyme (ACE) 8, 67–69, 105
– ACE1 and ACE2 78
– ACE inhibitors 8, 108–110, 112
– out–in inhibitors 69
angiotensinogen 8
antidiuretic hormone 9
ANZICS study 269
apelin 193, 194
– apelinergic system 199–202
– blood pressure 203
– distribution 196, 197, 203
– effect in brain 197–203
– fluid homeostasis 197–199, 202
– hypothalamic–adrenal–pituitary axis 202, 203
– precursor structure and processing 194
– receptor signaling 195–197
APPROVe study 320
arichidonic acid (AA) 315, 333
– cascade 335, 336
– metabolism 315, 335, 336
– release 341, 342
aromatic L-amino acid decarboxylase 234, 235
arosentan 155
arterial pressure
– elevation 70
– homeostasis 107, 108
arterial remodeling 72
ASCEND Trial 155

AT_1 receptors 8, 71–73, 86
– blockers 8
AT_2 receptors 71
AT_4 receptors 86
atherogenesis 321–323
atherosclerosis 20, 26, 27, 48, 51, 52, 299, 301, 307, 308
– EC dysfunction 72
– leukocyte accumulation 306
– prostaglandins 317
– protective role of estrogens 51
atherothrombosis 402
ATP 129, 130, 373, 374
– adenosine 374
– extracellular 375
– metabolism 374
– release 374
atrial fibrillation 414
atrial granules 125, 126
atrial natriuretic factor 125
atrial natriuretic peptide (ANP) 125, 126
– effect on cardiac hypertrophy and fibrosis 133
– effect on vascular relaxation and remodeling 133
autoimmune disease/response 308, 309
autonomic dysfunction 413
avonsentan 155

b

baroreflex 156, 412
blood flow
– BK 111
– coronary 111
– effect of aldosterone 20
– kinins 110, 111
– viscous drag 396
blood pressure
– control via MR 3
– effect of menstrual cycle 53
– effect of natriuretic peptides 125
– gender 53
– homeostasis 3, 6, 8
– kinins 109, 110
bosentan 146, 148, 150, 152, 154, 158, 159
BQ-123 151, 154, 155
BQ-788 151, 154, 155
bradycardia 412
– parasympathetic dysfunction 414
bradykinin 80, 82, 102, 108, 399
– cancer cell inhibition 82
– cardioprotective effect 111
– channel activation 347

– coronary blood flow   111
– effect of exertion   84
– hypotension   108
– renal function   83, 113, 114
– vasodilation   80, 81
brain, effect of serotonin   223, 224
brain natriuretic peptide   125, 126

c

C-type natriuretic peptide (CNP)   125, 126
calcitonin/calcitonin gene-related peptide (CGRP)   169
cAMP
– effect of AM   173, 175
– effect of apelin   195, 196
– effect of adenosine   374
– effect of dopamine   259
– effect of G-proteins   241
– effect of histamine   297
– effect on VSMCs   351
– synthesis and accumulation   259
carbon monoxide   346
cardiac disease, gender-based differences   47
cardiac fibrosis   17, 133, 150
cardiac function   350
cardiac hypertrophy   17, 49, 125, 133
– concentric and eccentric   50
– pathological   49
– reversible   49
– serotonin   223
cardiac remodeling   49
cardiomyocytes   47
cardiovascular growth, sex-differentiated   39, 40
cardiovascular remodeling   414
cardiovascular risk   10, 11
catecholamines
– cardiovascular function   233, 236
– lack of endogenous   236
– uptake   19
– VMAT transporters   217
cathechol-O-methyltransferase (COMT)   240, 254, 255
cathepsin K   105, 106
celecoxib   318–320
cholesterol   3, 4, 10
chronic obstructive pulmonary disease (COPD)   384
CLASS Trial   319
clopidogrel   384
coagulation   53
– cascade   301, 302, 398
Colorectal Adenoma Prevention Trial   320

congenital adrenal hyperplasia (CAH)
– androgen excess   6
– neonatal mortality   6
corticosteroids   73
– ligand access   12–15
– receptors   11–16
– regulation   6–8
– synthesis   3–8
corticosterone methyoxidase deficiency   6
corticotrophin-releasing hormone (CRH)   9
cortisol   3–6, 24
– deregulation   22
– pathway   3, 4
– release   10, 12
– synthesis   3–8
CREDO Trial   383
Cushing's syndrome   23, 24
CURE Trial   383
cyclooxegenase (COX)   315, 317
– COX-1   317
– COX-2   317, 318
– enzymes   334
– inhibitors   319
cysteine   337
cytochrome P450-dependent eicosanoids   333ff
– AA metabolism   336–341, 353, 354
– biological activity   351
– enzymes   333, 335–337
– epoxidam reaction   339
– family   340, 341
– formation   341–343
– hydroxylation reaction   339
– metabolism   352–354
– reactive oxygen species   346, 347
– role in pregnancy   357

d

darusentan   149, 152, 158, 159
decarboxylase   215
dehydration   201, 202
denucleoside polyphosphates   375–379
dexamethasone   14
diabetes   53
– diabetic nephropathy   152, 154, 155
– dopamine   266, 267
– EC dysfunction   72
– effect of Ang II   73, 74
– type 2   25
diclofenac   319
dihydroxytestosterone   40, 41
dionucleoside polyphosphates   375–379
diuresis   125, 134
diuridine tetraphosphate   384

DNA binding domain   41
dopamine   151ff, 215, 235
– aging   267
– availability   254, 255
– blood flow   261
– diabetes mellitus   266, 267
– distribution   257
– dopamine β hydroxylase (DBH)   236
– gastrointestinal tract   263
– heart failure   266, 268, 269
– hypertension   263–265
– peripheral effects   261–263
– receptors   255–257
– receptor function   264
– 'renal dose'   271, 272
– renal failure   265, 269–271
– renal function   261–263
– role in sodium transport   257, 258
– signaling   257–260
– synthesis   251, 252
– urinary flow   262
– use in cardiac surgery   271
– use in liver transplant surgery   272, 273
doxazosin   242
dysautonomia   414
dyslipidemia   25, 74

e

EARTH study   159
eiconoid formation   341
electrolyte homeostasis   3
ENABLE study   159
endothelial cells (ECs)   48
– dysfunction   72
endothelins   143ff
– blockade   149, 155
– cardiac function   350
– cardiovascular disease   145–148
– ET blockade   153
– ET-1   143, 144,
– – AM as antagonist   175
– – cardiovascular disease   145–148
– – growth factor   144, 145
– – heart failure   155, 156
– – hypertension   149–151
– – renal disease   152–154
– – vasoconstrictor   151, 152
– ET-2   143
– ET-3   143
– ET-4   143
– heart failure   155–159
– lung disease   148
– proinflammatory agent   149
– receptor antagonists (ERAs)   146
– renal disease   154, 155
endothelium
– dysfunction   20, 26, 27
– endothelium-derived contracting factor (EDCF)   298, 300, 301
– endothelium-derived relaxing factor (EDRF)   298–300, 395
– formation   396
– permeability   133
– remodeling   20
endothelial NO synthase (eNOS)   395, 399
– effect of thrombin   398
EPHESUS study   17, 20, 73
eplerenone   17, 20
epoxyeicosatrienoic acid (EET)
– bioactivity   347–351
– inflammatory response   335, 336
– blood pressure regulation   335
– metabolism and secretion   352
– organ protective properties   336, 351
– pregnancy   356, 357, 360
– renal disease   357, 358
– salt-sensitive hypertension   355, 356
– sources of   340, 341
– tissue repair   336
– tumor growth   336
– vasodilator response   345
ERα   41
– expression   45
estradiol   40, 52
– 17β estradriol, protective of vascular injury   47
– effect on coagulation   53
– effect on fibrinolysis   53
– effect on leukocyte adhesion   51
– effect on systemic and circulatory mediators of cardiovascular system   52
estrone   40, 41
estriol   40
estrogen   47, 399
– atherosclerosis   51, 52
– effect on fibroblasts   48
– hypertension   50
– production rate   40, 41
– receptors   39
– vasodilator   20

f

familial glucocorticoid deficiency (FGD)   10
fenoldopam   268, 270, 271, 273, 275

fetal development
– AADC   235, 236
– DBH   235, 236
– endothelins   144
– glucocorticoids   15, 16
– noradrenaline   235
fibrinolysis   53, 54
fibroblasts   48
fibrosis   17, 18
fluid
– homeostasis   202
– retention   74
– sheer stress   396–398
foam cells   307

g
gene expression regulation   170, 171
glomerular arteriolar resistance   74
glomerulosclerosis   153, 154
glucocorticoids
– cardiovascular effects   21–27
– excess   26
– glucocorticoid resistance syndrome (GRS)   25, 26
– receptors   8, 10, overexpression   23, 24, polymorphism   26
– resistance syndrome   26, 27
– synthesis   3–10
glucose sensitivity   26
G-proteins   258, 259
growth factors   399
growth receptors   126
guanylyl cyclases   128, 129

h
heart failure   49, 135
– AM   180–184
– dopamine   266, 268, 269
– EC dysfunction   72
– ERAs   157–160
– ET   155, 156
– NP receptors   135, 136
Heart Failure ET$_A$ Receptor Blockade Trial   158
heart muscaritic receptors   409
heart rate
– acetylcholine   410, 413
– parasympathetic control   411, 412
– parasympathetic dysfunction   414
heat shock protein   11, 42
heme
– degradation   346
– heme-binding   337
– oxygenase   346

20-HETE
– androgen–induced hypertension   355
– antagonists   350
– biological activity   347
– blood pressure   335
– deficiency   357
– effect of NO   345, 346
– myocardial infarction   350
– production   335, 345, 347
– proinflammatory role   355
– signaling pathways   360
– synthase inhibitors   350
high-density lipoprotein   52
histamine   295ff
– degradation   296
– distribution   305
– immune response   304
– platelet aggregation   304
– proinflammatory response   304
– receptors   297, 298
– synthesis   296
– tissue facto expression   301, 302
– vascular permeability   305
– vasomotion   298–301
histidine decarboxylase (HDC)   295
homocysteine   52
hormone replacement therapy   39, 48, 51, 52
hormone response elements   41, 42
– receptors   11
hormones, local and circulating   399
5-HT receptors   217–219
hydoxylase   235
11βhydroxysteroid dehydrogenase (11βHSD1)   2, 12–14, 25, 399
hyperkalemia   6, 21
hypertension
– androgen-induced   355
– Ang II-induced   149
– arterial   108
– Cushing's disease   24
– cyclosporine A-induced   355
– dopamine   263
– EC dysfunction   72
– essential   263
– KKS   109
– malignant   114
– pregnancy-induced   356
– prevalence   263
– primary   8, 151, 152
– role of estrogen   50
– salt-sensitive   16, 149, 263, 355, 357
– sodium retention   6

hypotension 6
– effect of BK 108
– post-exercise 109
– postural 16
hypothalamic–pituitary–adrenal axis 6, 19
– regulation 10
– signaling 11
hypoxia 148, 171, 374

i

icalibant 113
immunological response 295, 296, 304
indomethacin 318, 319
inflammation 304–307, 374
– kinin B_1 106
– vascular 20
– vascular remodeling 72
insulin
– effect of oral contraceptive 53
– resistance 74
irbesartan 150
ischemia 49
– release of kinins during 108

k

kaliuresis 8
kallidin 102
kallikreins 101
– excretion 112
– gene delivery 112
kallikrein-kinin system (KKS) 101
– and arterial pressure homeostasis 107, 108
kinins 101ff
– as anti-inflammatory 114
– blood pressure control 109, 110
– cardiac tissue 110
– cardiovascular role 107, 108
– coronary blood flow 110, 111
– degradation 108
– heart cell growth 111
– hypertension 109, 110
– inhibitors 108
– metabolism 111
– receptors 106, 107
– renal function 112–114
– renal protective 114
– synthesis 104
kininases 104–106
kininogens 103, 104

l

L-DOPA 235, 252, 253, 268
– amino acid transporters 252–254
– uptake and decarboxylation 251, 252

5-LO activating protein (FLAP) 315, 316
leptin 307
leukocyte
– accumulation 306
– adhesiveness 72, 305
– migration 305
leukotrienes
– athersclerosis 321–323
– cardiovascular action 320–324
ligand binding domain 41
ligand-dependent transcription factor 11
ligand-independent genomic pathway 42
lipoid congenital adrenal hyperplasia 3
lipoxygenase (LOX) enzymes 334
– inhibitors 334
low-density lipoprotein 52
lumiracoxib 319, 320
5-lypooxygenase(5-LO) 315–317, 321, 322

m

macrophage activation 307
melanocortin receptors 10
memory activation and retrieval 85
menopause
– cardiovascular events 39
– low- and high-density lipoprotein levels 52
mesangial cells 75
metabolic syndrome 14, 25, 74, 336
– effect of Ang II 73
metabolism 238–241
migraine 224
mineralocorticoids 3ff
– receptors 8, 11
– – antagonists 21
– – blockade 16–18, 73
– – deficiency 21
mito-agiogenic response 336
molecular-oxygen activation 337
monoamine oxidase 254
– inhibitors 216
– MAO-A 216, 240
– MAO-B 216
mononucleoside polyphosphates 373, 374
mononucleosites 374
myocardial infarction 49, 51, 108, 111
myocardial ischemia 176, 177
muscarinic acetylcholine receptor 407–410
– distribution 408, 419
– expression 408
– heart 409, 410

n

naproxen 319
natriuresis 75, 134
natriuretic peptides 125ff
– effect on blood pressure 132
– effect on cardiac hypertrophy and fibrosis 133
– effect on diuresis 134
– effect on endothelium permeability 133
– effect on natriuresis 134
– effect on vascular relaxation/remodeling 133, 134
– expression 126
– heart failure 135, 136
– – receptors 126–132
– – NPR-A 126–133
– – NPR-B 126, 130, 131
– – NPR-C 126, 131, 132
– renal function 134
– structure 126
neutral endopeptidase (NEP) 105
neutrophil adhesion 146
nifedipine 113
nimesulfide 318
nitric oxide (NO) 299, 345, 346, 395ff
– atherogenic response 402, 403
– coagulance response 402
– deficiency 76, 345
– effect of AM on 174
– eNOS activity 300
– formation 396–399
– regulation 396, 397
– release 48
– synthase blockade 110
– vascular tone 400–402
– vasoprotective effects 400–403
non-steroidal antiinflammatory drugs (NSAIDs) 319, 320
noradenaline 233ff
– adrenergic receptors 241–244
– biosynthesis 234
– metabolism 238–240
– transporter (NET) 238–240
nucleotides 273
– metabolism 383
– nucleotide-selective receptors 380

o

ω fatty acids 336, 343
– ω-3 fatty acid 343, 345
– ω-6 fatty acid 343
obesity 25, 74, 307
– abdominal 25

obstructive sleep apnea 152
oral contraception
– coagulation 54
– glucose/insulin resistance 53
organic cation transporters (OCTs) 240
ovarian function, effect of Ang II 73
oxytocin 399

p

peripheral adrenergic system 233
phenylethanolamine N-methyltransferase (PNMT) 237
pheochtomocytoma 169
phospholipase C 259
phytoestrogens 40, 48, 50
pituitary adenoma 24
plaque formation and rupture 51, 52, 301
plasma cortisol release 10
platelets 48
– activation and NO formation 398
– aggregation 146, 304, 398
– effect of serotonin 221
polycystic ovarian syndrome 39
polyphenols 400
potassium
– aldosterone 3–8
– depletion 6, 8
– wasting and GRA 6
pregnenolone 3, 4, 40
pressure overload 50
progesterone 40
– and fibroblasts 48
– receptors 11
prostacyclin 395
prostaglandins 315ff
– atherosclerosis 317–319
– cardiovascular risk 319–330
– cardiovascular system 317–320
– PGI_2 317
proteasomal degradation pathway 44, 45
protein kinase A 259
protein kinase C 259
protein transcription factor 12
proteinuria 152, 154
prothrombinase complex 373
pulmonary arterial hypertension (PAH) 147–149
pulmonary hypertension 156, 222
– hypoxia-induced 148
purines 374
– purinergic anti-thrombolic therapy 383, 384

– purinergic system 373ff
– purinoceptor system 380–383
pyridistigmine bromide 415–417

r
RALES study 17, 20, 73
Raynaud's phenomenon 221, 222
reactive oxygen species 18, 149, 264, 346, 347
renal disease
– ET-1 152–4; end-stage 152
– inflammatory 357, 358
renal failure
– dopamine 265, 269–271
– EC dysfunction 72
renal fibrosis 76, 114
renal function 134
– dopamine 261–263
renal glomerular filtration rate 75, 134, 261
renal glomerular hypertrophy 114
renal glomerular mesangial tone 75
renal medullary blood flow 75
renal transcapillary hydraulic pressure 75
renal tubular function 347, 348
renal/vascular alteration 356
renin-angiotensin system (RAS) 7, 52, 67, 68
– ACE 68, 69
– blockade 73, 74, 76
– enzymes 68, 69
– inhibitors 73
– local 67, 68, 71
renin/prorenin receptor 68
rofecoxib 318, 319

s
salt wasting 21
selective serotonin reuptake inhibitors (SSRIs) 216
sepsis 269, 274, 275
– septic shock 274
serotonin 211ff, 398
– blockers 216
– brain 223, 224
– cardiac hypertrophy 223
– cardiovascular actions 219–224
– enzymes 213, 216
– heart valve disease 223
– hypotension 221
– platelet aggregation 220
– production 211
– receptors 214, 217–219
– release 216
– synthesis 213–218
– transporters 216, 217
– vasoconstrictor 221, 222
sex steroid hormones 39ff
– cardiovascular cell types 47, 48
– receptors 39
– structure and function 41–46
– synthesis 40
sodium
– dopamine 257, 258
– excretion 70, 74, 76, 113
– reabsorption 8, 15, 75, 134, 347, 348
– retention 6, 75
– transport 3
soluble epoxide hydrolase 358, 359
spironolactane 17
steroid receptors 11
steroidogenic acute regulatory protein (StAR) 3
stress 415
– cortisol 3–6
– stress response 9, 21, 22
substance DK 101, 102
Syndrome of Apparent Mineralocorticoid Excess 14
syndrome X 25
systemic arterial hypertension 149–154

t
TARGET Trial 320
testosterone 40, 51
thrombin 374, 398
thrombosis 301, 398
ticlopidine 384
tissue factor expression 301, 302
triglyceride 52
typtophan hydrolases 213, 215
tumor necrosis factor 301
– inhibited by estradiol 48
tyrosine 234
tyrosite hydroxylase 235

u
ubiquitin-proteasome pathway 44

v
vagal dysfunction 414
valdecoxib 319
vascular endothelial growth factor 72
vascular injury
– role of 17β estradriol 47
– role of CNP 133
vascular integrity 72
vascular relaxation 133
vascular remodeling 72, 133

vascular resistance 8
vascular smooth muscle cells 19, 20
– AM secretion/synthesis 170
– effect of estradiol 47
– effect of estrogen 48
– effect of kinins 109
– NPR-B 133
vascular tone 22, 72, 151, 299, 347, 400–402
vasoconstriction 299
– serotonin 221, 222
vasoconstrictors, potentiation 19
vasodilation 299
– BK 80, 81
– effect of aldosterone 20
– endothelium-dependent 20
– role of estrogen 48
vasopressin 9, 193ff, 399
– fluid homeostasis 197–199
– hypothalamo–adrenal–pituitary axis 202, 203
– vasopressinergic system 197–199
ventricular arrhythmia 414
ventricular fibrillation 416
ventricular tachycardia 414
vesicular monoamine transporters (VMAT) 217
– VMAT1 217
– VMAT2 217, 241
VIGOR study 319
volume overload 50
von Bezold Jarisch reflex 223
von Willebrand disease 303
von Willebrand factor 54, 213

w

wall cyclic stretch 396
wall sheer stress 26, 396
Weibel-Palado bodies 303, 304